Developments in Structural Engineering

Volume One
Bridges and Space Structures

Books and Journals on Structural Engineering

The Behaviour and Design of Steel Structures
N.S. Trahair and M.A. Bradford

Expert Systems in Construction and Structural Engineering
Edited by H. Adeli

Fracture Mechanics of Concrete Structures: From Theory to Applications
RILEM Report
Edited by L. Elfgren

The Maintenance of Brick and Stone Masonry Structures
Edited by A.M. Sowden

Reinforced Concrete Chimneys and Towers
G.M. Pinfold

Reinforced Concrete Designers' Handbook
C.E. Reynolds and J.C. Steedman

Structural Analysis. A Unified Classical and Matrix Approach
A. Ghali and A.M. Neville

Vibration of Structures
J.W. Smith

Structural Engineering Review (Editor B.H.V. Topping)
A quarterly journal which provides a focal point for the publication
of innovative ideas in structural engineering and is primarily
concerned with analysis, design and construction of civil
engineering and building structures, emphasizing the
interdependence of architectural and engineering design.

Publisher's Note
This book has been compiled from camera ready copy provided by
the individual contributors. This method of production has allowed
us to supply finished copies to the delegates at the Conference.

Developments in Structural Engineering

Volume One
Bridges and Space Structures

Proceedings of the Forth Rail Bridge
Centenary Conference, held on 21–23 August
1990 at the Department of Civil Engineering,
Heriot–Watt University, Riccarton,
Edinburgh, Scotland, UK

Edited by
B.H.V. Topping

Reader in Structural Engineering,
Heriot–Watt University, Edinburgh

E. & F.N. SPON
An imprint of Chapman and Hall
LONDON · NEW YORK · TOKYO · MELBOURNE · MADRAS

UK   Chapman and Hall, 11 New Fetter Lane, London EC4P 4EE

USA   Van Nostrand Reinhold, 115 5th Avenue, New York NY10003

JAPAN   Chapman and Hall Japan, Thomson Publishing Japan,
          Hirakawacho Nemoto Building, 7F, 1-7-11 Hirakawa-cho,
          Chiyoda-ku, Tokyo 102

AUSTRALIA Chapman and Hall Australia, Thomas Nelson Australia, 480 La
          Trobe Street, PO Box 4725, Melbourne 3000

INDIA   Chapman and Hall India, R. Seshadri, 32 Second Main Road, CIT
          East, Madras 600 035

First edition 1990

© 1990 E. & F.N. Spon

Printed in Great Britain at the
University Press, Cambridge

ISBN 0 419 15240 7 (set)   0 442 31262 8 (set)
      0 419 15210 5

British Library Cataloguing in Publication Data
Available

Library of Congress Cataloguing-in-Publication Data
Available

Contents

VOLUME TWO
ANALYSIS, DESIGN AND ASSESSMENT OF STRUCTURES

Acknowledgements

I would like to thank Professor A.D. Edwards, Dean of the Faculty of Engineering, Heriot–Watt University who first suggested to me that I might consider organizing a conference to celebrate the Centenary of the Forth Rail Bridge.

I should also like to thank the invited speakers at the conference, many of whom assisted with the selection of papers and advised on the conference programme and in many other ways:

Dr M.R. Barnes, City University, London;

Dr P. Bulson, Advanced Mechanics and Engineering Limited, Guildford, Surrey;

Professor W.F. Chen, Purdue University, West Lafayette, United States of America;

Dr L.A. Clark, The University of Birmingham;

Dr J.F. Dickie, Department of Engineering, University of Manchester, Manchester;

Professor N.J. Gimsing, Technical University of Denmark, Lyngby, Denmark;

W.D.F. Grant, Scotrail, Edinburgh;

Professor H.B. Harrison, University of Sydney, Australia;

Professor E. Hinton, University College of Swansea, University of Wales, Swansea;

Dr D.W. Hobbs, British Cement Association, Wexham Springs, Slough;

D.J. Lee, G. Maunsell and Partners, London;

Professor A.E. Long, The Queen's University of Belfast;

D.G. McBeth, Kenchington Little plc, Edinburgh;

Professor D.A. Nethercot, University of Nottingham;

R.A. Paxton, Lothian Regional Highways Department, Edinburgh;

Professor R.M. Richard, The University of Arizona, United States of America;

Professor J.M. Rotter, University of Edinburgh;

Professor R.L. Sack, University of Oklahoma, United States of America;

Professor J. Schlaich, Universitat Stuttgart, West Germany;

Dr J.W. Smith, Department of Civil Engineering, University of Bristol.

Finally, I should like to thank Professor Paul W. Jowitt, Head of Department of Civil Engineering, Heriot–Watt University, for his collaboration, support and encouragement of this and other projects.

Barry Topping
Department of Civil Engineering
Heriot–Watt University
Riccarton
Edinburgh

Preface

Structural Engineering has been defined as 'the science and art of designing and making, with economy and elegance, buildings, bridges, frameworks, and other similar structures so that they can safely resist the forces to which they may be subjected'. The Forth Rail Bridge was opened on 4 March 1890 by the then Prince of Wales and has successfully resisted the forces to which it has been subjected for over a century. The Bridge has been described as 'the greatest monument to engineering skills of its era'. By contrast William Morris described it as 'the supremist specimen of all ugliness'. Regardless of your point of view, no one can dispute that the Bridge is an icon for engineers and that unlike the Eiffel Tower, with which it has frequently been compared, the structure has a function. It is of course a symbol to the Scottish nation and has been a source of great national pride. When the Bridge was opened it was the longest in the world and the construction had taken seven years with over 4000 men working on it.

Of the great engineers associated with its construction only the contractor Sir William Arrol was a Scot. The Bridge is regarded worldwide as a triumph of British and Scottish engineering. The Scottish people celebrated its centenary in a variety of ways. The object of this conference was to celebrate the centenary of the Forth Rail Bridge by reviewing the many aspects of civil engineering structures. Of course, as anticipated, many papers review aspects of bridge engineering. The full range of structural engineering topics considered by the papers is much wider and I would like to thank the authors of the papers included in these conference proceedings for writing their papers in celebration of the Forth Rail Bridge Centenary. They range from historical papers to those describing recent construction projects in various parts of the world. The research papers discuss recent progress on long standing problems such as alkali–silica reaction to the latest developments in structural engineering computation. It is particularly gratifying that papers have been written in celebration of the Bridge Centenary by authors not only from the United Kingdom but also from Europe, India, China, Japan, Australia, Canada and the United States of America.

It is particularly appropriate that this conference should have been held at Heriot–Watt University, since the Riccarton Campus is only a few miles away from the Bridge. Indeed the young civil engineers of tomorrow can view the Forth Bridges from the top floor of their building as they look northwards across the Forth Estuary to Fife.

Heriot–Watt University had its beginnings in Victorian Scotland in much the same ethos as characterizes the Forth Rail Bridge. The name Heriot–Watt commemorates two famous Scots. George Heriot was the jeweller and financier to King James VI of Scotland who, after the Union of the crowns in 1603, became King James I of England. Heriot bequeathed his fortune for educational purposes to the City of Edinburgh. James Watt was the great Scottish engineer and pioneer of steam power who died in 1819. Heriot–Watt University originated in the School of Arts of Edinburgh, Britain's first Mechanics' Institute, founded in 1821 two years after James Watt's death. The Watt Memorial subscription fund, established in 1824, was eventually applied towards the purchase of new premises for the School which, in 1852, was re-named the Watt Institution and School of Arts. Following the amalgamation of the endowments of the Watt Institution with those of George Heriot the name Heriot–Watt College was adopted in 1885. Even in those days the college was a unique educational institution and one of only three non-university institutions in the country permitted by Act of Parliament to appoint professors. The college was accorded university status in 1966 by Royal Charter.

Today Heriot–Watt University, Edinburgh's vibrant technological university on a 370 acre parkland campus at Riccarton, continues the spirit of the great Scottish engineers.

Barry Topping
Department of Civil Engineering
Heriot–Watt University
Riccarton
Edinburgh

PART ONE
THE HISTORY OF STRUCTURAL ENGINEERING

(with reference to the Forth Rail Bridge)

THE FORTH BRIDGES

D.G. McBETH
Kenchington Little plc, Consulting Engineers, Edinburgh,
Scotland, UK

The Forth Bridges are among the most famous bridges in the World.
Indeed on completion in 1890 the Rail Bridge was described as the
Eighth Wonder of the World. Its two clear spans were over 3½ times
as big as any similar bridge that had been built at that time. The
Road Bridge was completed in 1964 and was the largest bridge to be
built outside the USA. Both bridges were major milestones in the
development of Bridge Engineering. The paper describes the design and
construction of both bridges and highlights the developments in
structural engineering achieved in the construction of both these
structures.

Keywords: Bridges, Cantilever Bridges, Innovations in Design,
Suspension bridges, Steel Construction.

THE RAIL BRIDGE

The River Forth

Over the centuries the river has been criss-crossed with ferries, but
with the prevailing winds, currents and high seas experienced in the
Forth these could be often rather hazardous crossings. However, in
the 18th and 19th centuries the alternative was a long circuitous
coach trip via Stirling and so the ferries were an important and busy
means of communication.

Early Schemes For Crossing The Forth

Throughout the 19th century various schemes to construct a vehicluar
crossing were proposed. In 1805 the Scots Magazine published a report
on a scheme for twin tunnels and then in 1818 Anderson published his
paper on a proposed "Chain Bridge" over the river. However, the
economic climate wasn't right for an investment of the scale that was
necessary to construct a bridge over the Forth.

Railway Construction In The 19th Century

It was in the middle of the 19th century that the railway boom took
off as railways dramatically became the prime mode of transport in the
country. The railway fever resulted in people investing and
speculating in lines all over Britain and abroad. Among the more
successful of the many railway companies was the North British Company
who along with the Great Northern Railway Company, North Eastern

Railway Company and the Midland Railway Company formed the Forth Bridge Railway Company in order to bridge the river.

Bouch's Scheme
(See Fig No 1)
Thomas Bouch was appointed Engineer and commenced with a scheme for a multiple span bridge, similar to his Tay Bridge design, at Blackness Point. This got no further than a trial caisson in the river and he then proceeded with his scheme for a mighty double stayed suspension bridge at Queensferry. Work started on the bridge but with the collapse of Bouch's Tay Bridge in December 1879 the Company lost confidence in Bouch and the construction was halted. Following the Court of Inquiry into the Tay Bridge Disaster Bouch died a broken man. All that remains of the scheme is an isolated pier in the middle of the river adjacent to the central foundations of the present bridge.

Fowler and Baker Scheme
(See Fig No 2)
Two of the foremost engineers in the country, John Fowler and Benjamin Baker were eventually appointed to design another bridge for the crossing. They opted for the cantilever scheme known throughtout the world for its grandeur. The design was a major innovation in its day. Not only was its cantilever design concept unique but also in its use of steel, then a comparatively new building material, and by adopting tubular sections as the main compression members.

The basic layout of the bridge superstructure was dictated by the topography and geology of the river bed. The North Tower rests on the bed rock at North Queensferry. The Centre Tower utilises Inchgarvie and the South Tower is bedded in dense boulder clay at a depth of about 50 feet below high water.

Each of the towers are built on 4 piers. Those on the two side towers are at 120 feet by 260 feet centres. The towers taper to 33 feet centres at the top 330 feet above the foundations.

The cantilevers are each built out 680 feet from the towers and carry the central simply supported spans, each 350 feet giving a clearance of 150 feet above high water.

The basic dimensions of the bridge are as follows:-

Overall length	8296 feet
South approach	10 spans of 168 feet
North approach	5 spans of 168 feet
Length Portal to Portal	5329.5 feet
Length Tower to Tower	1912 feet
Cantilever lengths	680 feet
Simply supported spans	350 feet
Height of towers	330 feet
Rail level from high water mark	158 feet
Clear headway for shipping	150 feet

Based on researches carried out by Baker all the compression members are tubular and all tension members fabricated in lattice steelwork.

4

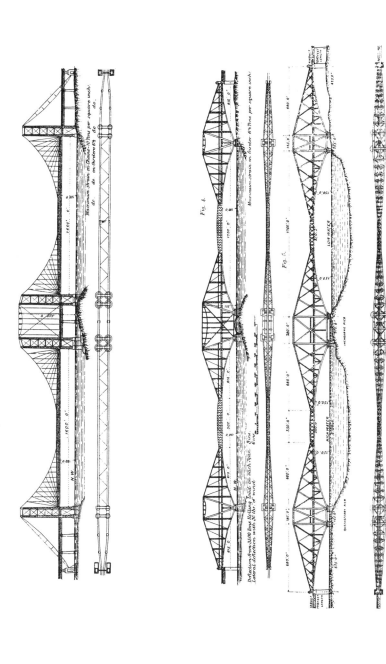

Figs 1 & 2 - Bouch's Scheme

The main columns are 12 feet diameters and the bottom booms of the cantilevers taper from 12 feet diameter at the skewback to 6 feet diameter at their extremity.

Construction
The contract was awarded in 1882 to Tancred Arrol and was carried out under the personal supervision of William Arrol.

Arrol had been the contractor for the abandoned Bouch scheme and had already established a working site at South Queensferry. The site, to the East of the line of the South approach viaduct, was greatly extended to accommodate the works necessary for the new structure. Furnaces were built, hydraulic benders and rivetters built. Rail sidings and stock piles laid out.

Amomg the more ingenious of the mechanical plant specially designed for the work were Arrol's tube drilling machines. By using these entire rows of holes could be drilled at one go.
(See Fig No 3)
An incline was constructed down to the river shore and sections of the bridge were winched down this before being loaded onto one of the eight barges or four steam launches used to transport materials to the working sites in the river.

Labour camps were established on both shores and also accommodation for workers was built in the old cast on Ichgarvie - (where 90 workmen lived during the sinking of the caissons).

Foundations
Fife and the north Inchgarvie piers are founded on rock. During construction it was found that the rock head was very irregular and before the cofferdams could be sealed divers had to cut out protruding edges and the bottom was plugged with cement bags and puddle clay.
(See Fig No 4)
The South Inchgarvie piers were constructed by first sinking pneumatic caissons to locate the rock head.

The Queensferry piers were constructed in boulder clay again using pneumatic caissons. These were 70 feet in diameter by 90 feet high constructed of an outer shell of rivetted plates over a cutting edge. Above this edge there was a buoyant collar used to float the caissons into position.

When in place the lower section was concreted leaving a 7 feet working space for excavation. Access to this was achieved by two tubes which were entered via air locks at the surface. The pressure under which the men were working when the caissons were at their deepest was 3 atmospheres ($42lb/ft2$)

When floating the Queensferry, North West caisson into place it was incorrectly anchored, tilted on the sea bed and then flooded. The accident took ten months from December 1884 to rectify and caused a serious delay to the main construction programme.
(See Fig No 5)
Piers were filled with concrete, up to the topmost 36 feet which consisted of a core built from Arbroath stone and faced in Aberdeen and Cornish Granite. To the top fo each pier was fitted a base plate held down by 48 no 21/2" diameter bolts each 25 feet long.

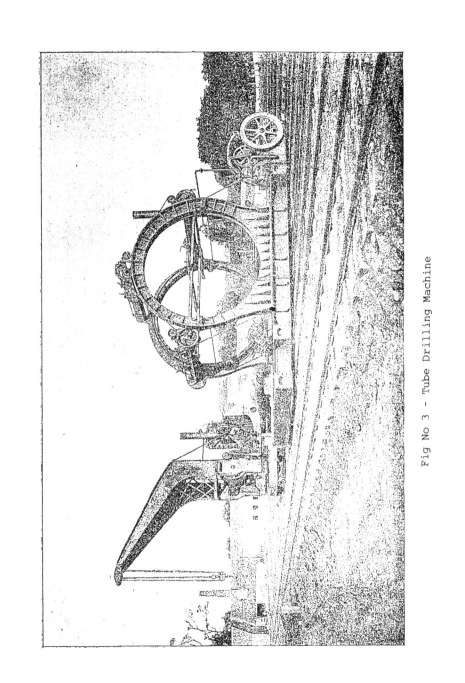

Fig No 3 - Tube Drilling Machine

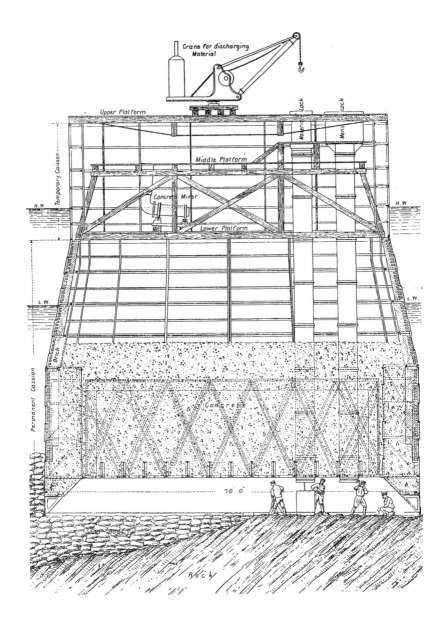

Fig No 4 - Inchgarvie Caisson

8

Fig No 5

Superstructure

The Forth Rail Bridge was the first major structure in the world to be built in steel. It was manufactured using the Siemens Open Hearth process and supplied by three foundries. Blochairn and Newton (Glasgow); Dalzells Works (Motherwell) and Landore Foundry (Swansea). In total there was 54,160 tons used in the superstructure plus 4,200 tons of rivets (6,500,000).

The initial assembly of the Tubular Sections was carried out on land in the fabrication yards. The bent plates were clamped into place in the tube and predrilled with Arrol's tube drilling machines. Sections were then numbered, dismantled and shipped out to the river to be erected.

From the skewbacks on the piers the main towers were erected with all three being brought up at the same time. All scaffolding and temporary works were kept to a minimum. Erection gantries were constructed so as to climb up the towers as these were erected. These gantries supported small 'Goliath' cranes and 'Scotch' derricks. Access to the gantry was by means of a hoist.

(See Fig No 6)

Plates were lifted by crane and held in position temporarily with bolts. A tube rivetting cage was assembled round the tubes and teams of rivetters worked in the comparative safety of these enclosures rivetting the tubes. These are approximately 100 rivets per foot on the bridge and to speed erection Arrol invented a hydraulic rivetting machine.

(See Fig No 7)

On completion of the towers, work started on the building out of the cantilevers. All six cantilevers were built together and great care was taken to ensure that cantilevers on either side of each tower were equally balanced. Again temporary scaffolding was kept to a minimum with each section being built out from the section previously erected. Specially designed cranes and gantries crawled out along the members. As with the towers the same method was adopted, of holding the sections in place temporarily and rivetting them in place by the team of men following up. The tube rivetting machines were altered because of the angle of the members. These were winched out along the booms to allow men to work in comparative safety.

To transport materials out to the working positions on the cantilevers, tramways were constructed along the outside of the bottom booms. Plates and structural sections could then be winched out from the jetties of each pier.

(See Fig No 8)

The central simply supported spans were also built out from the cantilevers by being connected temporarily at their point of support. The final closure was achieved by taking templates of the members on either side of the gap and forming plates to match. However, due to delays it was discovered, on one span, that when connected, the temperature had dropped and the bolt holes didn't match up. As this was in November and there was therefore little chance of getting a warm sun to help expand the bridge again, it was decided to stack timber on the cantiliver, cover the steelwork in naptha and set fire to it! This did the trick and the last plates were bolted together.

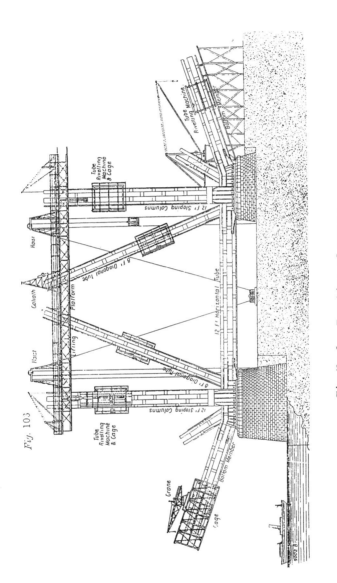

Fig. 103

Fig No 6 - Erection of Towers

Fig No 7 - Fife Pier Cantilever

Fig No 8 - Inchgarvie: South Cantilever

The guys temporarily supporting the simply supported spans could then be cut away and the bridge function as designed. In cutting away these guys some of the temporary bolting sheared dramatically with an explosion that caused panic among those still working on the bridge. However, an inspection revealed that there was no damage to the structure and all the rockers and sliding joints were functioning satisfactorily.

The official opening took place on March 20th, 1890 when the Prince of Wales ceremoniously drove the last rivet in the centre of the North span.

Labour

Over the eight years during which the bridge was being built there had been up to 4,600 men working on it. Some specialists in compressed air working came from the continent to construct the foundations, others were full time navvies who moved from job to job"on the tramp". The majority came from the Edinburgh area and were transported to the site either by train to Dalmeny or from Leith by steamer which departed at 4.00 am every morning returning at 7.00 pm in the evenings.

A welfare and sick fund was created by the contractor for his employees, Each contributed 8d a week towards this. There were many minor accidents and hospital cases. The total deaths toll on the contract was 57.

Each of the cantilevers was provided with a boat and boatman as a rescue service. It is recorded that these boatmen saved a total of eight lives, 8000 caps and "numerous other articles"

Painting

The painting of the Forth Bridge has become a legend. It is true when work finishes at one end then it has to recommence at the other. In all there are 145 acres of surface area to paint in the most awkward of locations both outside and inside the members. It takes from 4 to 5 years to paint the entire structure.

The specification was for two undercoats of Red Lead followed by two of Red Oxide. The same Red Oxide paint is used to day in this continuous operation.

Costs

Bridge and railway connections	£2,549,200
Abortive work on Bouch's scheme	£ 250,000
Parliamentary expenses and fees	£ 378,000
	————————
	£3,117,206
	————————
Cost of painting	£ 50,000
Saving on resale of mechanical plant and scrap	£ 120,000

The bridge is still used today for the same function for which it was designed and built. Trains have become heavier but still are somewhat lighter than the test load of 1800 tons that was trundled over the completed structure in January 1890.

An inspection, a few years ago, of the bridge gave it a clean bill of health and confirmed that it will continue to serve its function for many years to come as probably the most famous rail bridge in the world.

THE FORTH ROAD BRIDGE

History

In January 1926 Messrs Mott Hay and Anderson, Consulting Engineers, were appointed by the Ministry of Transport to survey and report on a possible road crossing of the Forth at Queensferry. Borehole investigations were carried out and in 1930 and they reported on 4 possible sites.   This concluded that the most economical location for a bridge was approximately 3/4 mile upstream from the Rail Bridge. This scheme made use of the Mackintosh Rock, a submerged rock on the north side of the estuary, which was considered suitable for founding a pier of a large suspension bridge.   For the main pier on the south side, borings indicated that good rock overlain by boulder clay would be met at a depth not greater than 110 ft below high water.   This would entail a bridge with a main central span of 3,000 ft and side spans of 1,350 ft costing £3.5m, the biggest suspension bridge in the world.   The national economic crisis in 1934 put paid to plans to construct the bridge at that time.

In parallel with these investigations, schemes were being considered for a bridge at Kincardine on Forth some 27 miles upstream from the Rail Bridge.   A bridge in this location was estimated to cost £311,000 and approval was given to proceed.   The Kincardine Bridge, designed by Sir Alexander Gibb, was completed in 1936 and was the largest swing bridge in the world with 20ft wide carriageway and two 6ft wide footpaths.

Shortly after the War, work started on proposals for a long span bridge over the Severn.   Messrs Mott Hay & Anderson and Freeman Fox & Partners were appointed joint consultants.   Research progressed into suspension bridge design and in connection with this a wind tunnel was constructed at Thurleigh near Bedford.   When design work started on the Forth Bridge much of the theoretical work and testing carried out for the Severn Bridge was involved in developing the design.

In 1947 the Forth Road Bridge Joint Board was set up and in 1948 they appointed Mott Hay & Anderson as Consulting Engineers for the bridge.   Site investigations, surveys and preliminary designs progressed throughout the 1950s and in 1958 authority was given by the Secretary of State to proceed with the scheme.   Messrs Mott Hay & Anderson invited Freeman Fox & Partners (their associates on the Severn Bridge project) to assist in the detailed design work.   the Consultant Architects were Sir Giles Scott, Son & Partners.]

Design

As has been previously stated, much of the early design work on the Forth Bridge benefitted from the early research carried out for the Severn Bridge. Various configurations of bridge deck were examined in the wind tunnel at the National Physical Laboratory and this resulted in an arrangement with the decks on the top of the main stiffening trusses.   The trusses are 27 ft 6 in in depth, giving a span/depth ratio of 120:1.   This arrangement gave a large improvement in the torsional stiffness of the suspended structure and a significant improvement of the aerodynamic behaviour.

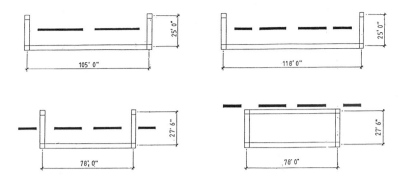

Fig No 9 - Evolution and Stiffening Truss and Deck Configuration

The panel lengths were 60 ft with 30 ft cross girders supporting the
deck.   This gave a good truss configuration and members of
reasonable size for transport and handling. The horizontal air gaps
were extremely useful when it came to the design of the structural
connections. The main cables and hangers are positioned between the
road deck and the cycle tracks.   In fact the towers too are neatly
located between these tracks. The deck to the main bridge span
consists of a fully welded battledeck in 60 ft panels, with expansion
joints located just beyond every second cross truss. The side spans
have a composite concrete deck, again in 60 ft panels.

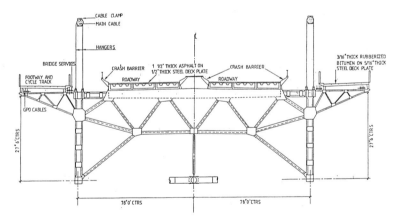

Fig No 10 - Full Cross Section of Bridge

    The ratio of cable sag to length of span had to be determined at an
early stage in the design.   Again the research carried out in
connection with the Severn Bridge was helpful.   The optimum sag ratio
for overall economy was estimated from a detailed analysis of the

17

whole structure for sag ratios varying from 1/8 to 1/14 taking into
account variations in weight of cable, tower height, stiffening
girders, etc.   The result of this indicated that the maximum economy
lay between 10½ and 11.   A ratio of 1/11 was adopted.   This has been
used for the Forth Bridge design.   It was decided that a continuous
wire spun cable would be adopted for the Forth Bridge rather than the
preformed strand system adopted for smaller suspension bridges.   The
cost of the specialised spinning plant, imported from the USA was
eventually shared with the Severn Bridge Contract.

The towers were designed as a system of welded box sections with
diagonal stiffening bracing.   The details were slightly modified in
discussion with the Contractor and an arrangement of 3 boxes with 4
stiffened connecting plates was adopted.   The steel towers were
designed in welded high tensile steel. The towers were founded on bed
rock in the river.   At the north advantage was taken of the
Mackintosh rock, a whinstone outcrop that was exposed at low tide.
The south tower was to be constructed on two caissons sunk under
compressed air through the mud onto the rock head.   At the
Contractor's suggestion these were amended to circular caissons that
were designed to be sunk without the use of compressed air.

The anchorages on the Forth Bridge took advantage of the natural
rock configuration to the north and south of the river and tunnels
(184 ft and 260 ft) driven at 30° angle into the rock.   Pre-stressing
cables were grouted in and the tunnel filled up with concrete to
resist the 13,800 tonne thrust.

(See Fig No 11)

Contractual Arrangements

The Forth Bridge was the first major suspension bridge to be built in
the United Kingdom.   The contract for the main bridge was negotiated
with a consortium of the 3 major bridge building firms in the country
- Sir Wm Arrol & Co Ltd, The Cleveland Bridge and Engineering Co Ltd
and Dorman Long (Bridge and Engineering) Co Ltd.   This was to become
the ACD Bridge Company.   They in turn engaged John A Roeblings Sons
Corporation of America as advisors on cable spinning.

The contracts were as follows:
    Preliminary contract to clear Port Edgar
                  - Farran, Edinburgh   £25,346
    Contract No 1 - Substructure and tunnel anchorages
                  - John Howard & Co Ltd   £2,201,726
    Contract No 2 - Main Bridge
                  - ACD Bridge Co Ltd   £8,694,410
    Contract No 3 - Approach roads, Dolphinton to Bridge
                  - A M Carmichael Ltd   £794,198
    Contract No 4 - Approach viaducts
                  - Reed & Mallik   £918,000
    Contract No 5 - North approach roads
                  - Whatlings Ltd   £2,220,014
    Contract No 6 - Approach roads, Cramond Bridge to
              Dolphinton
                  - A M Carmichael Ltd   £554,829

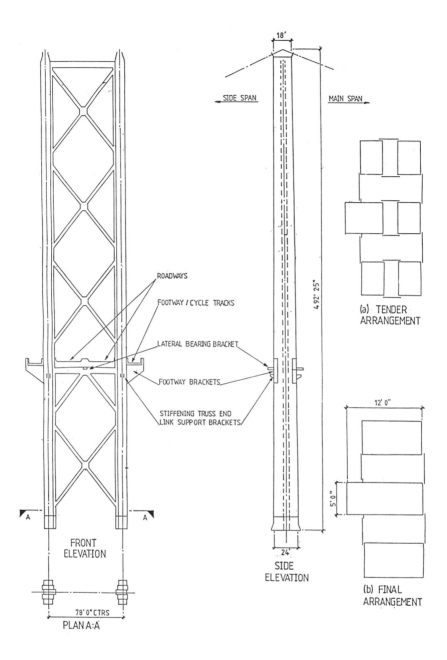

ROADWAYS

FOOTWAY / CYCLE TRACKS

LATERAL BEARING BRACKET

FOOTWAY BRACKETS

STIFFENING TRUSS END
LINK SUPPORT BRACKETS

SIDE SPAN

MAIN SPAN

18'

492' 25"

24'

FRONT
ELEVATION

SIDE
ELEVATION

78' 0" CTRS

PLAN A:A

A   A

(a) TENDER
ARRANGEMENT

(b) FINAL
ARRANGEMENT

12' 0"

5' 0"

Fig No 11 - Comparison of Tender and Final Tower Leg Cell
Arrangement and General Arrangement of a Main Tower

19

```
Contract No 7 - South approach roads, surfacing
         - Limmer & Trinidad Lake Asphalt          £81,088
Contract No 8 - North approach roads
         - Amalgamated Asphalte Company            £131,668
Contract No 9 - Toll registration equipment
         - Communication Systems Ltd              £43,869
Contract No 10 - Road Signs
         - Franco Traffic Signs Ltd              £25,660
Contract No 11 - Administration Building
         - Holland, Hannan & Cubitts             £160,687
Contract No 12 - Accommodation works - Rosyth
         - Hugh C Gibson Heirs                   £69,895
```

The main Bridge contract was negotiated on a "target price" basis.
In association with this a "fixed fee" for profit was negotiated.
Large plant was supplied by the contractor and paid for on a
depreciation rate rather than a plant hire rate.   Special equipment
such as the cable spinning equipment remained the property of the
Client.   A large part of this was subsequently transferred to the
Severn Bridge.

The works on the preliminary contract commenced during the summer
of 1958.   Contract No 1 for the foundations to the main bridge
commenced in November 1959.   The project was officially opened by the
Queen on 4 September 1966.

Construction

Foundations and Anchorages
Each anchorage is designed to resist a pull of 28,000 tons.   The
anchorage tunnels on both sides of the river are driven into the rock
at an inclination of 30°.   Each tunnel was approximately 25 ft in
diameter and 252 ft long.   115 tensioned anchorage strands were
grouted into ducts and the entire tunnels backfilled with concrete.
19 crossheads were provided within the anchorage chamber to receive
the main cables.

The original design of the **south pier** was for 2 rectangular
concrete caissons to be floated out into position and sunk.   The
contractor was confident that a cofferdam could be sealed in the
boulder clay and the caissons could then be built in the dry and sunk
in free air.   This scheme was accepted and work started in May 1959.

The cofferdam consisted of a figure of eight walling unit which was
floated out and fixed into position.   Seven 12 in x 12 in u/c piles
were driven inside the unit to secure the frame and following this the
cofferdam formed using 80 ft Larssen 4B piles and Apply Froddingham JA
piles.   Within this cofferdam a concrete ring beam seal was formed at
the toe and after it had been   de-watered the caissons were
constructed.   Each caisson was formed of a heavy steel cutting edge
surmounted by steel frame sections covered with a skin plating.   The
caisson walls were fitted with reinforced concrete to a height of 28
ft.

The caissons sunk under their own weight.   The boulder clay was
excavated from within the caisson by a mechanical digger.   The clay

20

proved to be very stiff and the rate of sinking was on average 1 ft per day. Each caisson was provided with an air deck, and air locks, but it proved not to be necessary to use these and both caissons were sunk in free air. Rock head was located at 95 ft and excavation continued to 99 ft to ensure sound rock. (Test bores were sunk for another 10 ft to confirm the soundness of the rock.)

On completion of the caissons they were backfilled with concrete and a capping slab formed in Class B reinforced concrete between -30 OD and -10 OD. The upper part of the pier (-10 to + 30 ft) consists of a 40 ft x 157 ft 6ins concrete block with cut water ends. This forms the main base to the tower. Each tower base is held down by 56 bolts, 15 ft long by 2 3/4ins diameter. The concrete surrounding the holding down bolts is pre-stressed to the capping slab.

The **north pier** was founded on the Mackintosh Rock. This was at a maximum depth of 40 ft below high water. The scheme called for a sheet piled cofferdam within which to construct the pier. The sealing of such a cofferdam was to present a problem. The Contractor proposed that the sheet piles to the cut water ends should be intrusion grouted and side walls of tremic concrete, pre-stressed to the rock formation would seal the cofferdam. This scheme was adopted, work started on the north pier foundations in December 1959.

Again, the capping slab and holding down bolt arrangements were similar to the south pier. The external surfaces of both piers were covered with Corennie granite exposed aggregate slabs.

Main Towers

The main towers consist of 3 box sections that are plated together to form a 5 cell structure. All the towers were fabricated at Arrols works in Glasgow, where sections of the complete towers and bracing could be laid out. The boxes were welded up in the shops, bearing surfaces ground smooth, connecting plates drilled and close tolerance bolts fitted. Three sections were built up on the shop floor. Two of these sections were then dismantled and the next sections built onto the one left. The dismantled section was taken to the storage yard at Drem for metallisation.

The towers were erected off a climbing structure. This was in the form of a steel box collar made up of welded box sections 9 ft deep. This collar was jack up the legs on hydraulic jacks. It embraced both legs and provided accommodation for men and stores. There was also a derrick crane on the frame to lift the box sections and other materials.

As the towers grew towards their full height they were subject to quite extensive swaying in high winds. This caused extensive delays to the work as not only the derrick cranes had to be guyed but the steel erectors suffered from sickness. They were stabilised by having the towers guyed to counterweights, which were on sloping planes. This proved to be very effective and these 'dampers' were kept in place until the main cables were spun and the towers, therefore, stabilised by the finished structure.

Cable Spinning

(See Figs Nos 12 - 18)

On all the major suspension bridges built in the United States, a system of continuous cable spinning had been adopted. This had been invented by John A Roebling and first used in 1855 on the Grand Trunk Bridge. Over the years the systems had been improved and much of the spinning process mechanised. It was therefore decided to consult the Roebling Bridge Division for advice. They were responsible for the design of the spinning plant and organising and carrying out of the work on site.

The cable spinning is done by carrying loops of wire over the bridge on a grooved spinning wheel, which runs on an overhead tramway. When the loops reach the other side of the river they are taken off the wheel and placed round a strand shoe that is fixed to the anchorage. The wheel then returns across the river empty to collect the next strand. Two wheels operate in opposite directions working back and forward across the river. Each wire is adjusted for level at the centre and at major reference points on the centre and side spans.

The cables are made up of galvanised hard drawn wires 0.196 in in diameter. There are 37 strands, with approximately 314 wires in each strand giving a total of 11,618 strands in total.

On completion of the cable spinning, the cable band suspension supports were bolted into place along the cables. These consisted of two semi-circular drums which were bolted together with high tensile steel bolts. The suspenders were made up of two lengths of steel wire ropes which were looped over grooves in the cable bands. They terminated in twin sockets at the lower end where they were bolted to the steel decks. No means of adjustment is provided in these suspenders, they had, therefore, to be made exactly to the right length.

For protection the main cables were wrapped with galvanised steel wires between the cable bands. This was done using 0.147 in diameter galvanised wires under tension. This was the last operation on the cables and as soon as it was finished the two top strands at the tops of the parapets on the catwalks were transferred to posts erected on the cable bands, so as to make permanent inspection walkways along the tops of the cables.

Steel Superstructure

The steel superstructure was prefabricated at the workshops of the companies in ACD Ltd. It was delivered to a site on the disused airfield at Drem in East Lothian. Here sections could be pre-assembled and checked for any fabrication discrepancies. Steelwork could be stored in the old hangers. A grit blast and paint treatment plant was set up in one of these hangers.

Steelwork was taken by road to the site where sections were taken out to the two towers along the jetties. The sections were then lifted into place at deck level. The initial sections were built out from the towers, carefully balancing the loads between the main spans and side spans. Once 18 panels had been erected work stopped on the main span and the two side spans were completed. Suspension bridges

of this nature at this stage in the construction have the two sections
of trusses in the main span curving upwards. The lower chords of the
stiffening trusses remain open and cannot be closed until a later
stage. The reason for this on the Forth Bridge is that the main
cables had to stretch about 18 ft in their length of 7,000 ft between
anchorages as the deck was erected and the full dead load came on
them.

There were approximately 80 steel erectors working on the bridge
with 4 fronts being constructed at the same time. Work commenced
from first light and proceeded to midnight during this period of the
works. The fixed final connection was made at a ceremony attended by
the Chairman of the Joint Board on 20 December 1963. The final
connection gaps of 8.125 in and 6.625 in were within .25 in of the
calculated figures.

On completion of the steel truss sections the decks were erected,
with a steel battledeck on the main span and reinforced concrete
decking to the side span. The hot rolled asphalt surfacing was
applied direct to these. On the main span the 1½ in wearing course
was laid by hand to the top surface of the battledeck decking, which
had been grit blasted. On the side span the asphalt was laid by
machine.

Approach Roads
At the same time as the main bridge was under construction work was
progressing on the approach roads. In all there were 8 miles of
approach road constructed. This included 7 principal junctions and 5
miles of connecting and slip roads. There are 24 bridges, 2 subways,
5 No 12 ft culverts, an administration building, toll booths and a car
park. The earthworks included the removal of 1 million of cubic
yards of oil shale at Dalmeny, the excavation of a 90 ft deep rock
cutting at St Margaret's Head, embankments, cuttings, etc.

All work was complete by the autumn of 1964 and the official
opening of the bridge was performed by the Queen on 4 September that
year.

Maintenance
Since then the bridge has continued to give good service and the
traffic values have increased. There are now some 14 million
vehicles crossing it each year. Like the Rail Bridge, there is a
continuous maintenance programme and checking carried out on the
bridge. This has been shown to be very cost effective and the only
major works/strengthening is progressing at present. The bracing
members of the main towers are being strengthened to meet the latest
loading requirements of the Department of Transport.

Conclusion
Both bridges are major milestones in the development of Bridge
Engineering. On the Forth we have them positioned in a dramatic and
photogenic location such that each has become a symbol of engineering
achievement in Scotland. Both bridges have recently had major
anniversaries with the Road Bridge 25 years old in September 1989 and
the Rail Bridge 100 years old in March 1990. It is right that we

should be celebrating these, acknowledging the achievements of the Engineers and Contractors who built them and looking forward to the developments in structural engineering into the next Century. I trust that we too are as inspired in the development of our designs as our forefathers.

Fig No 12 - Cable Spinning

Fig No 13 - Cable Spinning

Fig No 14 - Cable Spinning

Fig No 15 - Cable Spinning

Fig No 16 - Cable Spinning

29

Fig No 17 - Cable Spinning

Fig No 18 - Cable Spinning

Fig No 19 - Suspension Supports

Fig No 20 - Building out of Main Deck Trusses

Fig No 21 - Steelwork Erection

Fig No 22 - Bridge under Construction

FORTH CROSSING CHALLENGES AT QUEENSFERRY BEFORE THE RAIL BRIDGE

R.A. PAXTON
Vice Chairman Institution of Civil Engineers' Panel for Historical
Engineering Works
Senior Principal Engineer Lothian Regional Council, Highways Dept
Senior Research Fellow Heriot–Watt University

Abstract
In this paper engineering challenges and improvements to crossing
the Forth at or near Queensferry from c.208 to 1873 are identified
and assessed. Attention is given to the practice of various civil
engineers in the work contexts of harbours and ferries, tunnels, and
road and railway schemes. More particular consideration ranges from
the improvement of ferry landings by Smeaton, Rennie, Stevenson and
Telford, to impracticable proposals for tunnels and bridges, and
concludes with the railway triple challenge of Sir Thomas Bouch.
The subject is necessarily set in a context of the evolution of
structural practice and the whole constitutes a history of the
Queensferry crossing from a civil engineering standpoint.

1 Harbour and Ferry Improvements

1.1 Introduction

In 1760, although the Queensferry 'Passage Ferry' was the most
frequented in Scotland, the bad condition of the loading and landing
places, especially at low water, was "not only highly disagreeable
and inexpeditious, but even dangerous".(1) As the communications
improvements associated with the Industrial Revolution began to
gather pace nationally it became essential to improve the ferry. In
1772 a petition was sent to the Forfeited Estates Commissioners from
Fife J.P.s and the ferry owners requesting financial aid towards a
£980 package of improvements.(2) The name of the engineer, if any,
who prepared the plans has not been found. The Commissioners
consulted John Smeaton (1724-92), the 'father of civil engineering',
who was already making an important contribution to the Scottish
infrastructure. In addition to engineering the Forth & Clyde Canal,
he had already introduced major improvements to the machinery at
Carron Ironworks and built large bridges at Perth and Coldstream.(3)
He had also reported on numerous harbours.

1.2 Smeaton's Report on the Queensferry Landings 1772.(4)

Smeaton considered the principal defect of the ferry to be in its
landing places, which being "in a great measure furnished by nature
... require a little assistance from art". He drew particular
attention to the lack of low water landings by which "travellers are
often detained when the wind is fair and afterwards further
detained by the winds coming foul". Then as now the prevailing wind
was from the west and there were strong cross currents.

Smeaton recommended having a spread of landings on each shore to enable boats to cross more frequently without tacking, thus saving time. More particularly his recommendations included improving a 96yd length of the Grey Landing (contiguous to Queensferry Harbour) down to low water, to face both east and west. At the West Hall (Hawes) Pier he proposed part facing, part building on and part levelling the rock for 142yd down to a point from 5-6ft above the sand. On the north shore he advocated the extension of Craig End Pier (the town pier) by 53yd and that the East Ness Landing access should be improved by providing a smooth road across the rough rocks. This work was to be done by blasting or by bolting timber to the rock, to take the wheels of carriages in the manner of a rail-road.(4) It would appear that Smeaton's advice, or much of it, was heeded by the applicants and grant-aiding authorities as by July 1777, the Royal Burghs of Scotland had contributed £300(1) and the Forfeited Estates Commissioners £600; the latter on the basis of the ferry forming part of a military road and being the most frequented sea passage in Scotland.(2) In 1775, the Trustees for the Improvement of Fisheries and Manufacturers also contributed to the repair of Newhalls Pier and a landing east of North Ferry.(6)

1.3 Baird's Report on the Improvement of Queensferry Harbour 1817
In the latter part of the 18th century Queensferry Harbour consisted

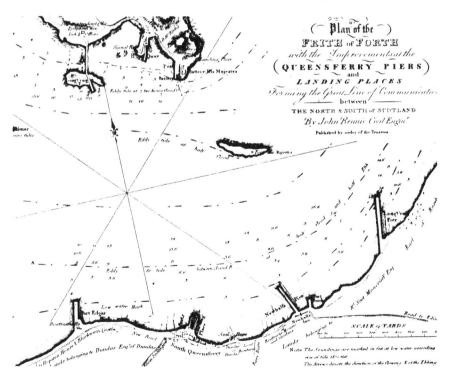

1 Queensferry landing improvements - Rennie 1809-17

of a pair of not quite parallel piers curving inwards at their
sea-ward ends to form an entrance from the north, with a ferry
landing place on the outside of the east pier.(fig 1) The harbour
was improved to a design of 1817 by Hugh Baird (1770-1827), engineer
to the Union Canal. He advised turning the west pier at a right
angle and running it eastwards to a new entrance in the north-east
corner of the harbour.(fig 10) This work, which involved rebuilding
the head of the east pier adjoining the ferry landing, was carried
out and thus the harbour was brought more or less to its present
form.(7)

1.4 Major Improvements to the Ferry 1808-17
The ferry improvements completed c.1777, which presumably resulted
in two good landings on each shore, sufficed for over two decades,
but with increased trade, commerce and travel, a better crossing
facility became necessary. In May 1809 an Act(8) was passed vesting
the ferry in new Trustees as part of the improvement of the Great
North Road from Edinburgh to Perth and beyond, and major development
ensued. With a capital of £18,500, after paying off the former
owners, the Trustees set to work improving the whole establishment
to the plans of the eminent engineer John Rennie (1761-1821). His
recent work in the locality had already included Musselburgh Bridge,
Bell Rock Lighthouse, Leith, Berwick and St Andrews harbours,
recommending improvements at Newhaven, Charlestown, Burntisland and
Perth harbours, and proposing the Berwick to Kelso railway and a
multi-span cast iron arch bridge over the Forth at Alloa.(9)
      Rennie developed Smeaton's principle of establishing a spread of
landings on each shore to facilitate boats crossing the river
diagonally with assistance from wind and tide without having to
tack. By 1812 Newhalls Pier had been enlarged to a length of about
240 yards and rebuilt with a central breakwater flanked by paved
roads at a total cost of £8,696. A new pier about 200yds long had
been built at Port Edgar for £4,763.(6) A small pier had been built
at Portnuick for use by cattle, involving blasting out rock to
provide sufficient water depth. On the north shore a landing place
and approach road had been constructed at the west side of the
Battery, also a new house for the ferry superintendent and a signal
house with accommodation for a boat's crew below. Other new
buildings included a boatman's house at Port Edgar, six boatmen's
houses at Newhalls and leading lights at the piers.(fig 1) The
improvements on the south side fulfilled the intended purpose of
encouraging the keeping of some boats there overnight. Previously
the general custom had been to berth boats at night only on the
north shore. These works were executed with the solidity and
excellence that characterised Rennie's practice and most of them
still exist. (fig 2) Unfortunately their cost considerably exceeded
the initial capital, nearly £34,000 having been spent, with two
piers still not constructed. To give an idea of the scale of use of
the improvements engineered by Rennie, in the year ending 15 May
1811, 33,220 persons, 5,769 carriages and carts, 44,365 horses,
cattle and sheep, and 5,520 barrels, crossed by the ferry.(10) (In
1989 about 30m persons crossed by road and 3m by rail!)
      A new Act(11) was obtained in July 1814 authorising expenditure

2 Longcraig Pier 1990 - Rennie, constructed 1816-17

of a further £20,000 to construct Longcraig Pier on the south shore
and Longcraig Island Pier on the north shore. The site of Longcraig
Pier was advertised to be determined on 13 May 1816(12) and by
October 1817 the work to Rennie's plan was almost completed.(13)
The completion date of 1812 given by Graham(7) is,
uncharacteristically, incorrect. Longcraig Island Pier was never
built.

Another engineer, Robert Stevenson (1772-1850), constructor of
the Bell Rock Lighthouse was called in by the ferry superintendent
in 1817 to advise on lighting arrangements. He recommended
repositioning the signal house reflector at the pier head at 12-15ft
above high water level. The reflector would probably have been of
the parabolic type of 21-24in dia. and the light source an Argand
oil lamp producing several thousand candle-power.

Just when costly near-perfection had been achieved at this ne
plus ultra of sailing establishments, the enterprise encountered
major competition from steam-boats which, not being so dependent on
wind and tide, were quicker in operation. They first started
operation on the Fife & Midlothian or 'Broad Ferry' between Newhaven
and Dysart in September 1819. By the autumn of 1820, the Fife and
Midlothian Ferry was operating three steam-boats from Newhaven and
the effect of this resulted in the Queensferry Passage losing about
two-thirds of its coach passenger traffic(14). Difficult tidal
conditions and the design of and spread of the piers were not
conducive to the general introduction of steam-boats on the
Queensferry Passage. Its Trustees, after considering various types
of paddle steamer, probably including Stevenson's novel 'Dalswinton'

internal-paddle steam-boat which he advocated for use on this
ferry,(15) commissioned a paddle steam-boat to the design of their
superintendent. The vessel, named the 'Queen Margaret' entered
service in October 1821, towing large and small sailing boats in its
wake. On the south side at low-water, only Longcraig Pier had
sufficient water depth to accommodate the boat and because of the
incompatibility of its external side paddles with the pier profile,
wheeled traffic could not be handled. In 1821 a fleet of new
sailing boats was introduced but the whole operation failed to meet
the increasing steam-boat challenge from the 'Broad Ferry' and in
1828 the Trustees consulted Britain's leading civil engineer Thomas
Telford (1757-1834) to see what could be done to improve matters.

1.5 Telford's Reports on the Forth Ferries 1828
Telford reported that the probable future revenue of the ferry was
incompatible with changing the whole mode of operation from a
sailing to a steam-boat system. He advised adopting only
improvements which could be accomplished at a justifiable expense,
adding, "that such are become indispensibily necessary the rapid
improvement of conveyance on all sides is sufficient evidence."(16)
Telford's recommendations included an extension of the Signal House
(Craig End) Pier into deeper water. This measure was intended to
provide a safer wharfage on its eastern side, to protect the
extremity of the Battery Pier, and to supply additional
accommodation. He commented, that to have extended this pier before
the introduction of steam-boats would have obstructed the necessary
tacks for sailing boats making passage to the south. On the south
side Telford considered it impracticable to obtain a greater
low-water depth at Newhall Pier without unwarrantable expense. For
low-water use he recommended Longcraig Pier where the water depth
was already sufficient, but because this pier was exposed to the
prevalent westerly winds and the force of the ebbing tide current,
he advised provision of a rubble stone breakwater alongside it at a
short distance to the west. Telford left the question of the detail
and estimates for these improvements to his Edinburgh civil
engineering associate James Jardine (1776-1858).
    From a comparison of Rennie's plan and the 1856 O.S. map Signal
House pier appears to have been extended. Longcraig breakwater was
not built. In 1828 Telford was also consulted about the Fife &
Midlothian Lower Ferry. He considered its revenue prospects very
good and supportive of nearly £61,000 of improvements including a
new pier at Burntisland and a new landing at Newhaven 400 yards out
from the existing pier head so as to achieve a 10ft low water depth
for steam-boats.(17) This work was not carried out but subsequently
major development occurred at Granton and Burntisland harbours.

1.6 Development of Engineering Practice 1770-1830
In structural engineering terms the works referred to so far would
have required little in the way of strength calculations, mainly
consisting of foundations, gravity masonry walls and timber piles
and beams in foundations, and timber and cast iron as struts and
tension members. Practices adopted were based on experience or
experiment. Piers generally consisted of a pair of masonry walls

with uncoursed stone hearting between them.(figs 2 & 3))
    From c.1800, cast iron beams, columns, plates and other castings

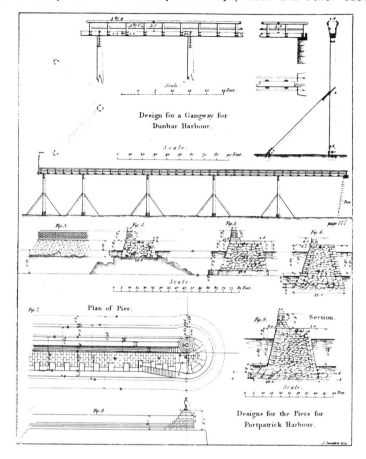

3 Typical harbour construction - Smeaton 1770-72

were available. Wrought iron was obtainable up to about 3in dia.
cross-section in long lengths and as narrow plates. From c.1800
portable steam-engines were used increasingly for powering pumps,
dredgers and other equipment. By 1830, the use of artificial
cement, mass concrete in foundations and more effectively preserved
timber was developing. The use of steel and reinforced concrete in
structures did not begin until the latter part of the century.
'Strength of materials' education for engineers from textbooks, as
distinct from 'word of mouth' and experience, was in its infancy
and gathered momentum from c.1817, developing rapidly in the 1820's
mainly on a practical and empirical basis. From 1822, Tredgold's
textbook on cast iron(18) with its empirically derived safe-load
tables was useful to engineers in designing beams of up to 30ft span
and columns up to 24ft high. The foundation of the Institution

of Civil Engineers as a forum for the exchange of knowledge in 1818
represented a landmark in the development of engineering education.
The reliable theoretical approach to engineering design now
practised universally had not evolved to any extent by 1830.

2 Tunnel Projects

2.1 Under-sea Tunnels c.1580-1805
Tunnels under the Forth existed at least four centuries ago. In
1618, John Taylor 'The Water Poet' wrote of Sir George Bruce of
Carnock's 'Moat' coal-pit at Culross with its sea cofferdam
entrance:

"I...went in by sea, and out by land", this being possible
because "at low water, the sea being ebd away, and a great part
of the sand bare; upon this same sand (being mixed with rockes
and cragges) did the master of this great worke build a round
circular frame of stone, very thicke, strong, and joined together

4 Horse-whim and machinery to mine shaft chain - Agricola 1556

with glutinous or bituminous matter so high withall that the sea at the greates flood...can neither dissolve the stones...or yet overflow the height of it. Within this round frame...hee did set workmen to digge with mattockes, pickaxes...They did dig forty feet downe right into...that which they expected, which was sea-cole...they following the veine of the mine did dig forward still: So that in the space of eight and twenty or nine and twenty yeeres, they have digged more then an English mile under the sea...the mine is most artificially cut like an arch or vault...that a man may walk upright in most places...The sea at certaines places doth leake...into the mine...is all conveyed to one well neere the land; where...a device like a horse-mill, that with three horses and a great chaine of iron going downeward many fathoms, with thirty-sixe buckets fastened to the chaine, of which eighteene goe down still to be filled, and eighteene ascend up...which doe emptie themselves (without any mans labour) into a trough that conveyes the water into the sea againe..."(19) (fig 4)

The works described are of outstanding significance in Scotland's industrial history and provide an insight into the entrepreneurial enterprise of Sir George Bruce, gentleman coal-owner who can be considered a civil engineer in all but name. (Smeaton is believed to have been the first to call himself 'civil engineer' nearly two centuries later.) When leasing the mine at Culross in 1575 Bruce's "great knowledge and skill in machinery" was acknowledged and he was considered the best person to re-open the then abandoned mine.(20) He adopted the best continental 'state of the art' practice of Georg Agricola and others.(21) By 1595 Bruce had constructed a storage reservoir on Culross Muir to guarantee water supply to a colliery water-mill at or near the horse-gin site. He also erected a windmill and a tide-mill as alternative power sources.(22) The workings are believed to have extended some two miles under the sea before the mine was flooded over the cofferdam in a storm in 1625.(23)

It has been written that a proposal for a tunnel under the Forth at Inchgarvie was mooted about 1790,(24) but it was not taken seriously, possibly because of the impracticability of mining through whinstone. Fifteen years later a proposal for a tunnel 1½ miles to the west did receive wide consideration.(25) The engineering case for it was supported by successful under-sea tunnelling precedents at the Culross, Bo'ness and Whitehaven mines, and operational canal tunnels at Harecastle and Sapperton. By 1805, the Bo'ness workings had extended about a mile under the Forth at depths from 20-80 fathoms. The Valleyfield under-sea workings of Sir Robert Preston at Culross were so dry that they could be drained "by a boy with a bucket".(27) At Whitehaven the workings were at a depth of from 80 to 150 fathoms under the sea with access via white-walled tunnels on a 1 in 6 gradient.

2.2 Forth Tunnel Proposals 1805-7
In November 1805 a William Vazie, possibly a mining engineer, sought the opinion of a leading Edinburgh mining engineer John Grieve as to

whether a tunnel under the Forth from Rosyth Castle to the opposite
shore was practicable. Grieve thought that it was, as the rock was

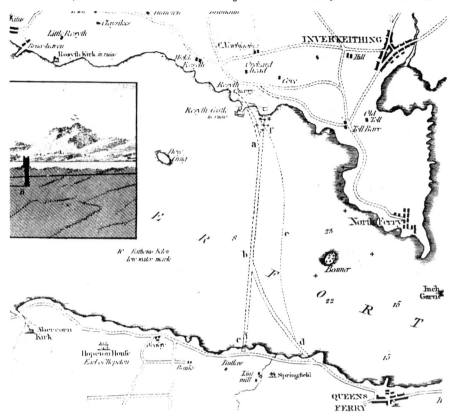

5 Forth Tunnel plan 1806. Inset : Moated shaft - looking west

likely to be passable freestone, but called for this to be confirmed
by borings all along the tunnel line. On the basis of a maximum
water depth of 11 fathoms (66ft) from a chart, Grieve suggested a
maximum depth for the tunnel sole of 30 fathoms (180ft). He
proposed twin 15ft wide arched tunnels with a central drain level
beneath. The tunnels were to have had 500yd entry sections parallel
to each shore with gradients of 1 in 25 so as to achieve 50ft of
cover before turning under the sea. From these turnings the main
tunnels would have descended for 1800 yards from each side at a
gradient of 1 in 45 meeting mid-way at the maximum depth. For
drainage Grieve proposed constructing two moated engine pits over
200ft deep at each low water mark. At the bottom of the pits
steam-engines and pumps were to have been installed. He estimated
the cost of the tunnel at £160,000-£170,000 with a four year
construction period.
     In summer 1806 Vazie and his associate Taylor reported in similar
vein after a site visit with Grieve. Some alterations were

suggested to meet objections from the Earl of Hopetoun. The proposed tunnel entrance on the south was moved westwards to within a few hundred yards of Queensferry.(fig 5) To obviate possible smoke nuisance from the steam-engine and to reduce activity near Hopetoun House it was proposed that any buildings associated with the project, including the permanent steam pumping installation, were to be located on the north shore. A busy little town was envisaged at Rosyth "with the Castle in its bosom". Alternative cross-sections were given both with separate carriageways for 'comers' and 'goers'. More thought had been given to passing under the deep part of the river.

> "If the boring should in any manner of way leave the
> investigation incomplete...it may become necessary to
> advance...with caution...by putting down pits at low water
> mark...to the necessary depth and cutting a communication by a
> level between them...Such a level will at all events be necessary
> as a drain...for drawing the water from the tunnel...Will require
> to have placed...the engines necessary for the great work...no
> new or additional expense...an expenditure would be incurred,
> including engines, from 12 to 15000 £..."(25)

The proposal was also supported by the civil engineer Robert Bald (c.1778-1861), who considered it highly prudent to make soundings and borings as a preparatory step. The Scots Magazine was "happy to see that this undertaking is in a great state of forwardness and that a number of noblemen and gentlemen of the first respectability have organised themselves into a regular body for the purpose of carrying it into effect".(26) In March 1807 a Dr Millar and Vazie republished an enlarged illustrated edition of the various reports with an economic case.(27) The tunnel was not started, probably more for economic reasons than doubts about its engineering practicability.

2.3 Assessment
It is fortunate for the promoters that the project did not proceed, as the ground under the deep part of the river would have proved very different from that which they imagined. The mining experts of the day expected the freestone to extend from shore to shore, a concept which was proved as late as 1964 several miles west when the Kinneil and Valleyfield mines were joined, but at a depth of about 1800ft.(28) At the depth of 180ft proposed for the Queensferry Tunnel, the miners would have encountered a deep channel in the bed of the river filled with sand and silt. H.M.Cadell of Grange, the Scottish geologist drew attention to this subject in 1913(23) and provided a dramatic sketch of his impression of the pre-glacial Forth valley, complete with mammoth and Forth Bridge.(fig 6) Although Cadell's concept of deeply buried pre-glacial river channels is no longer considered tenable,(29) there is no doubt that a channel containing a considerable depth of sand and silt does exist, whatever its origin, and his sketch serves to illustrate the difficulty the tunnellers would have had to contend with. The question now is whether the tunnel could have been constructed in

such material in 1807. A review of contemporary experience
indicates the answer.

6 View of pre-glacial Forth Valley at Queensferry - Cadell 1913
   NOTE: The approximate tunnel line has been added by the author

From 1796-98 an engineer Ralph Dodd proposed a tunnel under the
Tyne between North and South Shields.(30) Although this tunnel did
not proceed, it was the precursor of his ambitious scheme for a 16ft
dia. road tunnel under the Thames from Gravesend to Tilbury which
did start.(31) Difficulties with groundwater in the preparatory
operation of sinking a shaft for this tunnel in sandy material
proved so great that the entire capital for the project was consumed
without even achieving the shaft and in 1803 the project was
abandoned. Undaunted by this set-back, a Cornish mining engineer,
Robert Vazie (it is not known whether he was related to the William
Vazie previously referred to) commenced work on a tunnel under the
Thames at Limehouse in 1805. Difficulties experienced in sinking a
13ft diameter shaft through gravel and quicksand again proved so
great that operations were suspended. Rennie and another leading
engineer William Chapman were consulted but could not agree on a
course of action. Work eventually recommenced under the direction
of Richard Trevithick, notable Cornish mining engineer (and 'father
of the locomotive'), on a 5ft pilot driftway ultimately intended to
form a drain under the tunnel. A 30hp steam engine was used to pump
out water. For a time good progress was made until, when nearing
the far side of the river, sand and water frequently burst into the
driftway and in 1808 work stopped. In March 1809 a premium was
offered to any person furnishing a plan enabling the tunnel to be
completed. At least 53 plans were received and examined by the
eminent engineers Dr Charles Hutton and William Jessop who, after
due consideration, concluded that "an underground tunnel which would
be useful to the public and beneficial to the adventurers is
impracticable". The problem had confounded the experts. Many
thousands of pounds had been irretrievably lost and not a single
brick of the tunnel had been laid.(32-34)
   There can be no doubt that the proposed Forth Tunnel involving a
substantial length of construction in river-bed silt and sand was
beyond the technology of its time. A considerably deeper tunnel
with the same gradients and passing under the soft material would

have been ruled out on cost grounds. It would however be an option
to consider for a new crossing of the Forth today.

3   Road Bridge Schemes

3.1   Possible Roman Campaign Boat Bridge

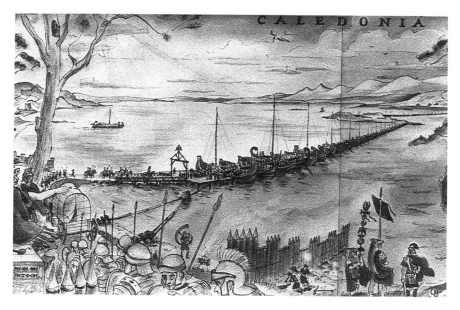

7   Possible Roman campaign boat bridge, Queensferry c.208. Drawn by
    D Cameron with advice on details from Dr G. Maxwell & author

A bridge across the Forth at Queensferry was probably considered by
the Romans, possibly c.208 during the campaigns of Emperor Severus
and his son Caracalla. One romanist has recently suggested that a
1¼ mile long boat bridge, divided near its middle by Inchgarvie, was
constructed under the guidance of Caracalla about where the Forth
Railway Bridge now stands.(35) (fig 7) In the absence of firm
evidence the case for such a bridge is conjectural, but the Romans
did have the technology, men and access to materials to have built
one. There are various precedents of boat bridges elsewhere, some
being depicted on Trajan's column. Several tens of thousands of
Roman soldiers are believed to have campaigned north of the Forth
and a bridge would have formed a useful link northwards from the
Severan base at Cramond three miles to the east. It is difficult to
imagine a boat bridge surviving winter storms; possibly assembling
it was a seasonal operation. The provision, positioning and
securing of some 500 boats would have been a major task. Would the
Romans have given such a project the necessary priority over a
ferrying operation?

3.2 Developments 1740-1817

A bridge may have been suggested as early as 1740(24) or 1758(36) but no details have come to hand. As the materials then available for construction were essentially timber and stone with limitations in use on bridge spans of about 100ft and a maximum foundation depth of about 10-15ft under shallow water, a bridge in deep water would have been impracticable. In 1772 Smeaton thought that it would be worth spending up to £100,000 (perhaps equivalent to £50m today) to bridge the Forth at Queensferry, but considered a bridge unachievable.(4) The considerably increased production of good quality wrought iron that followed implementation of Henry Cort's (1740-1800) inventions in iron manufacture after 1783 gave engineers scope for constructing bridges with tension members. Before Cort's improvements a tilt hammer working by water-power produced one ton of bars of doubtful quality in 12 hours. His rolling mill, absorbing approximately the same power, produced 15 tons of uniformly high quality iron in the same time. At a final stage, the iron was passed through grooved rollers producing uniform sections of various dimensions.(37) (fig 8) (38) The wrought iron link-bar

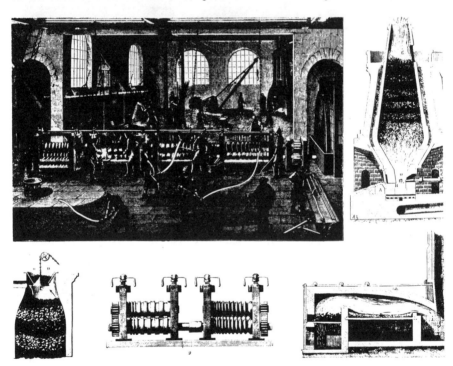

8   Iron making c.1850 - Rolling mill, blast and puddling furnaces

suspension bridge was adopted in North America from 1800.(39) Telford designed a bridge centering supported by inclined iron stays for crossing Menai Strait in 1811. By the summer of 1817 Scotland led Europe in having four iron tension footbridges erected. Their spans ranged from 110ft-261ft.(40) From 1814-17 Telford and Capt Samuel Brown (1774-1852) were taking the first steps in developing the long-span suspension bridge for carriage traffic based on experiments in connection with the Runcorn Bridge project.(41) At the end of 1817 the first practical 'strength of materials' textbook having any bearing at the subject was published and that was mainly about timber.(42)

3.3 Anderson's 'Chain Bridge' Designs, January 1818.(43)
It was against this primitive technological background that an Edinburgh land surveyor and civil engineer and former pupil of Jardine, James Anderson (c.1790-1856) proposed a wrought iron bridge on either the rod-stay or catenarian bar-cable principle.(fig 9) He envisaged spans of 2000ft, with estimated costs for alternative

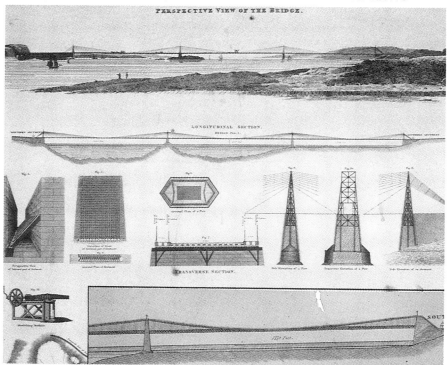

9 Proposed 'chain' bridge at Queensferry - Anderson 1818 - Note
   stay design and ironwork stretching machine. Inset and
   cross-section - Catenary cable design

heights of 90ft and 110ft above the river of from £144,000-£205,000. The site was to have been within about 300yds of the present rail bridge.(fig 10) The headroom for shipping was to have been 90ft or 110ft and the deck 33ft wide with a 25ft carriageway. In the rod-stay design the pairs of rods terminated at the outside of the deck at 100ft (or 50ft) intervals and at the other end fanned out laterally across the tower top to counteract "the effects of wind and any undulating or vibratory motion". The stays were to have had cross-sectional areas proportional to the strain induced. The pair of stays from the tower tops to mid-span would have a declination of

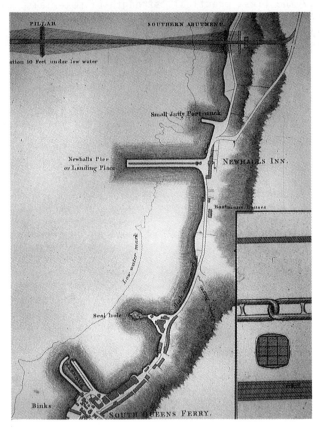

10 Proposed 'chain' bridge at Queensferry – Anderson 1818 – Plan at south side. Inset chain and cable details

100ft in 1000ft or just less than 6°.

For the catenarian cable or alternative design a curvature depth of one-thirtieth of the chord line (66ft 4in) was proposed. Twelve 3in nominal diameter cables were envisaged, each consisting of nine 5/8in square bars and 4 facing segments, the whole bound round with wire.(fig 10) For this proposal the iron stays of the first design were retained to inhibit deck undulation. In both designs masonry piers were proposed with cast iron tower frames above the roadway. The timber deck was to have rested on 20 (or for the stay design 40) principal bar members or 'basis chains' 1 x 1½in deep extending nearly 6000ft between abutments and tensioned to a sag of 20ft in 2000ft. The abutments and towers were to have been constructed

first, over which were to have been stretched a temporary catenarian
footway along which the stays which were to meet at mid-span would
have been conducted. The middle bearer with two 'basis chains' was
then to have been hoisted up from boats and the stay ends connected.

The cables and bars were to have been stretched into position
using a machine capable of exerting 65tons from 1cwt applied to the
handle, and terminate at a cast iron anchor beam on each side of the
bridge. These anchors were to have been positioned 150ft behind
each abutment face and 100ft below the roadway, stability being
provided by a superincumbent mass of masonry of these dimensions
40ft wide and weighing nearly 23,000tons. Anderson based his
proposed ironwork on simple experimental results,(fig 11) both his

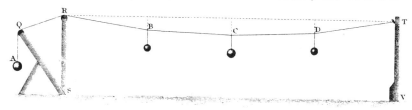

London. Published Aug.st 1s 1817. by J.Taylor. High Holborn.

11 Iron strength testing arrangement by Anderson (after Telford)(42)

own and Telford's, and assumed a design proportionality factor of
15-20tons in^2 or half of its breaking strain. He proposed using
local stone and 'excellent quality' lime from the Elgin Lime Works.
Anderson particularly emphasised the need for further experiments on
a larger scale before deciding a preference for either of the
designs, and reserved the right to modify and improve them.(43) He
sent a copy of his designs to Telford(44) who almost certainly
regarded them as over-ambitious.

3.4 Assessment
At the time of publication of his designs, Anderson was probably
approaching 30 years of age with more experience of land surveying
than civil engineering. His designs as illustrated were undoubtedly
over-ambitious for the technology of his time and justify
Westhofen's comment that the proposed structure was "so light indeed
that on a dull day it would hardly have been visible and after a
heavy gale probably no longer to be seen on a clear day either".(45)
Basically the cross-sectional areas of the iron cables and bars were
much too small for the elevations adopted which, with tower heights
of 67ft and 100ft above the roadway, were too flat. Unacceptably
high levels of stress would have been induced in the ironwork.
Anderson seems to have been unaware that as wrought iron was
stretched, it deformed permanently beyond a stress of between 9.5
and 11.5tons in^2.(46) The stress in the cables of his catenarian
design would have exceeded these figures under their self weight
alone. His design stress was three to four times greater than the 5

tons in^2 safe design stress which gained general acceptance later.
The provision against deck oscillation was also almost certainly
inadequate but at least he had made some allowance. Even at that
time Telford regarded 1000ft as a maximum span for suspension
bridges, modifying this to 800ft after 1825 when he had experienced
the deck oscillation phenomenon at Menai and 600ft in the case of
Clifton Gorge. It was not until 1931, with the completion of George
Washington Bridge, that a 2000ft span was reached and surpassed.

The nearest Anderson seems to have come to suspension bridge
construction was the successful renewal in 1830 of the timber
sea-ward abutment of Trinity Chain Pier erected by Capt Samuel Brown
for steam-boat use in 1821.(47,48)(fig 12) This was a difficult and

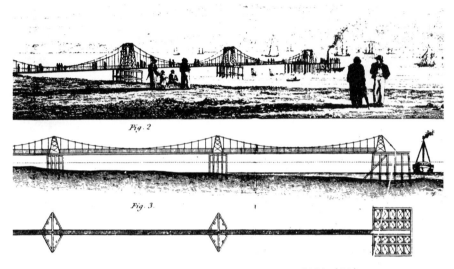

12 Trinity Pier, Newhaven - Sea-ward abutment 1821 (48)

hazardous operation involving the replacement of many sea-worm
ravaged piles whilst at the same time preserving the tension
supporting the structure.

It is doubtful whether Anderson would have promoted his designs
at all if he had not been encouraged by Telford's Runcorn Bridge
project with its 1000ft central span.(49,50) Unfortunately for him,
Telford's development of the long-span suspension bridge had not yet
matured and been translated into the elegant and long-lasting Menai
Bridge, a process which took a further five years to evolve at the
frontiers of technology. In consequence, Anderson adopted and even
compounded undesirable features from the 1814 Runcorn Bridge design
which Telford later abandoned e.g. the cable form, catenarian cables
of too flat curvature under as well as over the road-way and a
design stress that was too high.(51)

In conclusion, Anderson deserves some credit for correctly
foreseeing rock-founded cable-stayed or suspension bridges as the
means of achieving large spans. The proposal seems to have helped
his practice to flourish. From 1836-46 Anderson was an F.R.S.E.

13 Dee Bridge, Chester 1847 - Combined iron girder 100ft span (52)

14 Britannia Bridge, Menai Strait - Construction of tube 1848 (53)

4.1 Introduction

From 1830-50 most iron bridges on railways were of the cast iron
arch or beam types or combinations of cast and wrought iron, the
latter contributing additional tensile support, with spans rarely
exceeding 100ft.(fig 13) A number of failures involving cast iron
beams had occurred and from the mid-century wrought iron replaced
cast iron in general use for beams . The wrought iron plate girder,
precursor of the steel 'I' beam, developed c.1846. A railway
suspension bridge had been erected at Stockton in 1830. It was
under-designed and proved hopelessly inadequate, two waggons causing
a deflection of 18in, and after being propped for a time it was
replaced by a cast iron bridge in 1842.(41) This experience
discouraged engineers from adopting suspension bridges for railways.
The Britannia Tubular Bridge with its 460ft spans over the Menai
Strait constructed from 1846-50 under the superintendence of
Stephenson and Fairbairn represented a major step forward in the
evolution of the wrought iron girder bridge.(53)(fig 14) Crossing
the Forth and Tay was a bigger challenge and an interim solution was
adopted by Sir Thomas Bouch (1822-80). By 1850 he had designed and
successfully installed the world's first floating railway between
Granton and Burntisland.

4.2 The Granton-Burntisland 'Floating Railway' 1850

The ferry vessel was a specially designed end-loading paddle-steamer
called 'Leviathan' built by Robert Napier & Co. The 389-ton vessel
had a speed of 5 knots and commenced operation in February 1850. It
could carry up to 34 goods waggons and the average time for a single

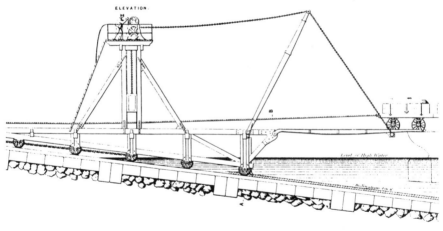

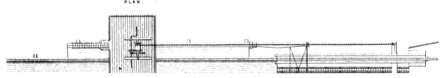

15 Forth Floating Railway - Bouch 1850 - Granton slip-way

trip, including loading and unloading an average of 21 waggons, was 56 minutes.(54) The waggon transference arrangement on each shore consisted of a slip-way travelling platform with horizontal top, at the end of which were four movable wrought iron girders that were lowered onto the end of the ferry boat when the platform was in position. The platform was moved up and down the slip-way by means of a 30hp. stationary steam engine which was also used for hauling the trains. The movable girders were operated manually from two powerful crab-winches above the platform.(fig 15)(55) In the early 1860s Bouch proposed a similar system at Queensferry to accommodate passenger trains, but he allowed his preference for a bridge to override this concept which, by comparison, he considered inefficient. The Granton to Burntisland ferry continued to operate until the Forth Bridge was opened in 1890.

Bouch credited Thomas Grainger (1794-1852), his predecessor as Engineer to the Edinburgh, Perth and Dundee Railway, with the original idea of floating trains across the Forth. Grainger proposed to use hydraulic cranes to transfer trains between shore and the ferry vessel. Bouch thought that this operation would be too slow. Another engineer J. F. Bateman (1810-89) claimed that he had originated the floating train concept with a proposal for Queensferry in 1845 when he was Engineer to the Edinburgh & Perth Railway. He had proposed installing stationary steam engines at the top of 1 in 12 ramps on each shore, trains being hauled over tail-pieces between the vessel and ramp.(55)

4.3 The Proposed Forth Bridge at Charlestown 1862-66

Bouch, now Engineer of the North British and Edinburgh & Glasgow Railway, first considered the Queensferry site for a bridge across the Forth. He ruled out a suspension bridge there as being inappropriate for railway traffic and rejected a girder bridge on account of the impracticability of founding piers in a depth of up to 240ft of water and because of the impediment to navigation.(56) The predisposition against using suspension bridges for railway traffic was not accepted by all engineers. In 1864 a 'Mr Thorntan of Edinburgh', probably Robert Thornton, prepared plans for a suspension bridge with three 2000ft spans at or near the site of the present railway bridge.(57) In 1862, a Charles Dowling published a proposal for a bridge with two continuous wrought iron tubes 5810ft long in seven 800ft spans at about the same site. Although he considered the tubes just self supporting at this span, he proposed adding suspension chains or cables, including some diagonals, to inhibit lateral movement.(58) Neither of these proposals was adopted.

In 1862 the Westminster consultant engineers G.R. Stephenson (1819-1905) and J.F. Tone(59) produced an outline report for consideration by the North British Railway directors on the means by which it might be possible to 'pass' the River Forth. Stephenson, nephew of the famous Robert Stephenson, had already had the experience of constructing a major iron bridge over the Nile and had assisted his uncle with the multi-span box girder bridge over the St Lawrence at Montreal. Stephenson and Tone strongly advised against the railway ferry concept which they considered inefficient. They

also advised against the construction of a bridge across the Forth
at Queensferry, considering a suspension bridge with minimum spans
of 1300ft to be impracticable for railway traffic, and cited the
speed limit of 3mph on the American engineer J.A. Roebling's 800ft
span Niagara Bridge (1855-92). Stephenson and Tone recommended
construction of an iron girder bridge across the Forth between
Blackness Castle and Charlestown at an estimated cost of £500,000
and envisaged a completion time of three years.(59)
    Bouch seems to have accepted or to have come to the same
conclusions as Stephenson and Tone and in 1863-64 was working on
designs for a single-track girder bridge across the Forth near
Charlestown. One design in 1864 was for a 3979yd viaduct rising to
100ft in height for two 290ft navigation spans. From the south the
spans were: 19 x 40ft; 42 x 40ft; 24 x 207ft; 2 x 290ft; 4 x 207ft;
44 x 40ft; and 4 x 40ft.(60) Another design had spans ranging from
100ft to two 600ft spans over the navigation channel.(56) The
proposed main spans were larger than those of Britannia Bridge and
Brunel's Saltash Bridge (1859) with its 455ft spans. The Company
prevailed on Bouch to reduce the large spans and the design for the
'Bridge of Forth' in the 1865 Bill was a 3837yd viaduct with 62
wrought iron close-lattice girder spans, rising to 125ft clearance
for four 500ft navigation spans. From the south end the spans were:
14 x 100ft; 6 x 150ft; 6 x 175ft; 15 x 200ft; 4 x 500ft; 2 x 200ft;
4 x 173ft; 4 x 150ft and 7 x 100ft.(61)(fig 16) The 500ft span
girders, each 64ft deep and weighing 1170 tons, were to have been
fabricated on land, floated to site on pontoons and elevated into
position by means of hydraulic jacks. The bridge was estimated to
cost £476,000 excluding the railway and contingencies. If it had

16 Proposed Forth Bridge, Charlestown - Bouch 1865 (61)

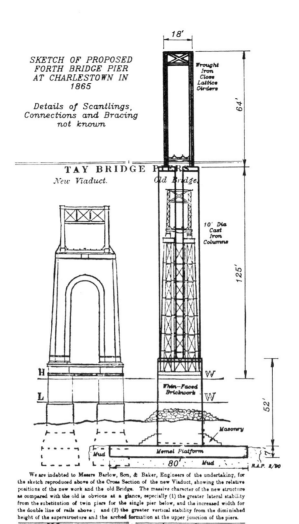

SKETCH OF PROPOSED
FORTH BRIDGE PIER
AT CHARLESTOWN IN
1865

*Details of Scantlings,
Connections and Bracing
not known*

18'

*Wrought
Iron
Close
Lattice
Girders*

64'

TAY BRIDGE PIERS

New Viaduct. Old Bridge.

10' Dia
Cast
Iron
Columns

125'

H   W

L   W

Thin-Faced
Brickwork

52'

Masonry

Mud

Memel Platform

Mud   Mud

80'

R.A.P. 2/90

We are indebted to Messrs Barlow, Son, & Baker, Engineers of the undertaking, for the sketch reproduced above of the Cross Section of the new Viaduct, showing the relative positions of the new work and the old Bridge. The massive character of the new structure as compared with the old is obvious at a glance, especially (1) the greater lateral stability from the substitution of twin piers for the single pier below, and the increased width for the double line of rails above ; and (2) the greater vertical stability from the diminished height of the superstructure and the arched formation at the upper junction of the piers.

17 Proposed Forth Bridge, Charlestown - Bouch 1865. Drawn on 1881 section showing old and present Tay Bridges for comparison(68)

been built it would have been the longest and largest railway bridge in the world.

The promoters were concerned about the difficulty of achieving adequate foundations for the great girder piers in soft ground. Of the many borings made on Bouch's behalf by Jessie Wylie (whose subsequent borings for the Tay Bridge indicated a non-existent rock shelf almost right across river and involved Bouch in considerable design changes and delay)(62) many easily penetrated through soft silt for more than 120ft. One bore even went to 231ft without reaching the bottom.(23) From several hundred borings there was not a single bit of stone.(56) In 1864 Bouch conducted experiments on site to determine the bearing capacity of the ground in the river bed using two 6ft dia. cylinders 48ft high, one with an open end and the other with a closed end. After loading with 60tons of pig iron one cylinder became top heavy and toppled over. On 7 November 1864 another cylinder was successfully loaded with 80tons and was expected to take 120tons or "if possible 5tons ft^2" later that day.(64) This outcome encouraged Bouch to proceed.

In 1865 prior to the Act being obtained the project was thoroughly examined by parliamentary referees. Bouch explained in evidence that he proposed to determine whether satisfactory foundations could be obtained for the piers of the large girders by building and load-testing an experimental pier in situ. He proposed reducing the pressure on the mud to less than $\frac{3}{4}$ton ft^2 by use of a platform of green beech 114ft x 80ft x 9ft thick, which being slightly denser than sea-water would sink without load. It was to be towed to site supported by floats and sunk into position. On top of the platform the masonry and brickwork were to be built up within a wrought iron cylinder to 12ft above high water level as the platform sank into position in the mud in 40ft of water. Twelve 8ft dia. tubes on the platform around the edge of the masonry and also the interior cavity of the piers were to be loaded with 10,000 tons of pig iron, equivalent to 2$\frac{1}{2}$ times the weight of the structure plus a standing train. The piers above the brickwork were to consist of a pair of 10ft dia. columns 1in thick.(56,63)(fig 17). Bouch envisaged the girder spans as continuous but had not designed them on this basis, considering this an additional safety factor.

On 14 June 1866, Bouch's trial platform was launched from Burntisland and towed into position off Charlestown. It was smaller than previously envisaged, now being 80ft x 60ft x 7ft thick and constructed of memel (pine).(63) Six weeks later, when the preparations for submersion of the platform were rapidly approaching completion, the Company abandoned the project for financial reasons. It is understood that they expected to lose northern 'through traffic' revenue following an amalgamation between their Caledonian Railway rival and the Scottish North Eastern Railway, which took place on 10 August 1866.(65-66) The workmen were paid off and the raft was towed back to Burntisland. The experiment had cost the North British Railway Company £34,390.(67)

4.4 Assessment of the Charlestown Bridge Proposal of 1865-6

The abandonment of the bridge was almost certainly fortunate, not only on account of the questionable nature of its structural continuity and foundations in 'mud' (60) but also because of its probable instability in strong wind. In cross-section, the bridge with its 64ft tall girders 125ft above the river would have been too narrow for its height.(fig 17). Although the proposal did survive searching parliamentary scrutiny, the wind problem was not properly appreciated at that time. Bouch envisaged a wind load of 180tons on a 500ft span, weighing 1170tons, based on a pressure of 30lbs ft^2, from which it seems reasonable to assume a lattice girder elevation consisting of 58% holes and 42% iron on the basis that the wind pressure was applied to the nett area of iron. Bouch would probably also have adopted a factor of safety against overturning.

If the rules drawn up by the Board of Trade Committee after the Tay Bridge disaster (69) had applied to this proposal, ie 56lbs ft ^{2} on the windward and 28lbs ft^2 on the leeward side of the girder, the wind load would have been 1200tons, based on the gross area of the girder, discounting the holes. The Committee specified a factor of safety of 2 against overturning where gravity provided the restoring force and a factor of safety of 4 was was to be applied to holding down connections resisting overturning. Modern practice would give give a wind load of 580tonnes, more than three times the figure that Bouch estimated, and a minimum factor of safety to be applied to this against overturning of 1.4. This figure is indicative of the considerable over-reaction by the Committee to the wind question following the Tay Bridge disaster. Both figures are however, considerably in excess of the 130tons assumed by Bouch. With hindsight, the design is also questionable from the standpoint of post construction settlement. The girder design was probably influenced by Runcorn Bridge (1863)(63)(fig 18).

18 Runcorn Bridge today - constructed 1863-68

Bouch's design cannot necessarily be assumed to have suffered a similar fate to that which befell the Tay Bridge thirteen years later. Its piers were more robust than the latter. The design might have been abandoned or modified as a result of the experiment. Without details of the scantlings, connection arrangements and bracings it is not possible to comment with certainty as to how the different elements of the design would have fared with time. By comparison with the present Tay Bridge, with its iron caissons sunk at least 20ft into the sandy silt, widely straddled piers, and designed to the post-disaster wind pressure code, there is no doubt which design is to be preferred.

4.5 The Forth Railway Suspension Bridge Project 1871-80

The pressure for bridging the Forth and Tay did not subside for

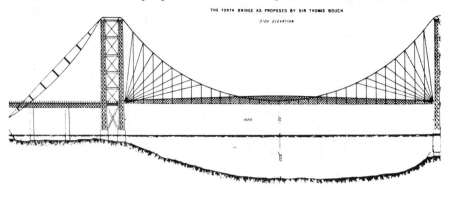

THE FORTH BRIDGE AS PROPOSED BY SIR THOMAS BOUCH

SIDE ELEVATION

19 Forth Suspension Bridge proposal - Bouch 1873

long. Under new management, the North British Railway Company took over the ferry at Queensferry in 1867. In 1868 the railway from Ratho Junction to Queensferry was completed thus establishing a rail-ferry link with Fife. This link was further developed in 1877-78 with the construction at Queensferry of a 900ft timber jetty and a 1,300ft whinstone breakwater.

In 1871 Bouch, it is said perhaps influenced by Anderson's earlier scheme(67), prepared several designs (45) and proposed a double-span steel suspension bridge with heavily stiffened deck and 1,600ft spans more or less on the line of the present bridge.(fig 19) After having been carefully examined and favourably reported on by the eminent engineers W.H. Barlow (1812-1902) and W. Pole (1814-1900),(70) the bridge received its authorising Act in August 1873. Work was slow to start and it was not until 30 September 1878 that the foundation stone of a brick pier was laid at Inchgarvie. Towards the close of 1879 William Arrol (1839-1913) was hard at work on preparations for the steelwork when the Tay Bridge fell. By the following summer Bouch's design had been abandoned, and all that survives on site is the base and a score or so courses of brickwork of Inchgarvie pier, now supporting a beacon.

Bouch was probably influenced to change his mind and adopt a suspension bridge by the success of Roebling's Cincinatti-Covington road suspension bridge. This bridge of 1075ft span was completed in 1866 and is still in use.(71) Bouch's bridge might have lasted too, but few engineers would doubt the superiority of its successor.

References

1. Extracts from the Records of the Convention of the Royal Burghs of Scotland 1759-79, Edinburgh, 1918.
2. Ms.Minutes of the Commissioners of Forfeited Estates 1745. 1768-82. Scottish Record Office E721/11.
3. Skempton, A.W. John Smeaton F.R.S., London, 1981.
4. Smeaton, J. Report on Queensferry Shipping Places, 15 August 1772, in Reports of the late John Smeaton, 2nd ed., London, 1837.
5. Smeaton, J. Reports of the late John Smeaton, 1837. II, pl 23.
6. Mason, J. The Story of the Water Passage at Queensferry, Dumbarton, 1962.
7. Graham, A. Archaeological Notes on some Harbours in Eastern Scotland, in Proc. Soc. Antiq. Scot. (1968-69) 101.
8. 49.Geo.III.c.83.
9. Boucher, C.G.T. John Rennie 1761-1821, Manchester, 1963.
10. Statement respecting the Queen's-Ferry Passage and the Great North Road, London, 1811.
11. 54.Geo.III.c.138.
12. Edinburgh Advertiser 13 April 1816.
13. Ms.Ltr. John Paterson to John Rennie, 29 October 1817. Nat.Lib. Scot. Ms.19795.
14. Brodie, I. Queensferry Passage, West Lothian History and Amenity Society, 1976.
15. (Stevenson, R.) Origin of Steam-boats and Description of Stevenson's Dalswinton Steam-boat, in Annals of Philosophy (April 1819) XIII.
16. Telford, T. Report on the Queensferry Passage, in Report of the Committee appointed by the Managing Trustees of the Queens-ferry Passage on the fifth April 1828, (Edinburgh?).
17. Telford, T. Report respecting the Lower Ferry between the Counties of Midlothian and Fife, (Edinburgh, 1828).
18. Tredgold, T. A Practical Essay on the Strength of Cast Iron, London, 1822.
19. Taylor, J. The Pennyles Pilgrimage. London, 1618. (reprinted in The Old Book Collector's Miscellany. ed Ch. Hindley, London, 1872).
20. Shaw, J. Water Power in Scotland 1550-1870, Edinburgh, 1984.
21. Agricola, G. De Re Metallica, Basileae, 1556.
22. Communicated to the author by The Earl of Elgin and Kincardine.
23. Cadell, H.M. The Story of the Forth, Edinburgh, 1913. 81-103.
24. History of the Forth Bridge, Edinburgh, Banks & Co., 1911.
25. Grieve, J., Taylor, J. and Vazie, W. Reports of Surveys made for ascertaining the practicability of making a Land Tunnel under the River Forth, (Edinburgh,) 1806.
26. Scots Magazine August 1806.
27. Miller, J. & Vazie, W. Observations on the Advantages and Practicability of Making Tunnels under Navigable Rivers, particularly applicable to the Proposed Tunnel under the Forth, Edinburgh, 1807.
28. The Kinneil Valleyfield Link Up, in Colliery Engineering, December 1964.

29. Francis, E.H. & Armstrong, M. The Geology of the Stirling District, I.G.S., London, H.M.S.O. 1970. 263
30. James, J.G. Ralph Dodd, the very Ingenious Schemer, in Trans. Newcomen Soc. (1976). 47.
31. Dodd, R. Reports...of the proposed Dry Tunnel...from Gravesend in Kent to Tilbury in Essex. London, 1798.
32. Law, H. A Memoir of the Thames Tunnel 1824, in Quarterly Papers on Engineering, ed by J. Weale, London, 1844-49.
33. Beamish, R. Memoir of the Life of Sir Marc Isambard Brunel, London, 1862.
34. Clements, P. Marc Isambard Brunel, London, 1970.
35. Reed, N. The Scottish Campaigns of Septimius Severus, in Proc. Soc. Antiq. Scot. (1975-76) 107, 96-102.
36. Weir, M. Ferries in Scotland, Edinburgh, 1988.
37. Pannell, J.P.M. An Illustrated History of Civil Engineering, London, 1964.
38. The National Cyclopaedia (undated) 1847-59?
39. Paxton, R.A. Menai Bridge (1818-26) and its influence on Suspension Bridge Development, in Trans. Newcomen Soc. 1977-78. 49.
40. Stevenson, R. Description of Bridges of Suspension, in Edin. Phil. J. (Apr.-Oct. 1821). V.
41 Paxton, R.A. The Influence of Thomas Telford...on the use of Improved Constructional Materials in Civil Engineering Practice. 1975. Thesis at Heriot-Watt Univ. and I.C.E.
42. Barlow, P. An Essay on the Strength and Stress of Timber. Also an appendix on the Strength of Iron, London, 1817.
43. Anderson, J. Plan and Sections of a Bridge of Chains proposed to be thrown over the Firth of Forth at Queensferry, Edinburgh, 1818. Illustrations from Hopetoun House copy.
44. The Institution of Civil Engineers' copy.
45. Westhofen, W. The Forth Bridge. London, 1890.
46. Kirkaldy, W.G. Illustrations of David Kirkaldy's System of Mechanical Testing, London, 1891.
47. P.P. Report from the Select Committee...Harbours of Leith and Newhaven with the Minutes of Evidence. H. of C. 6 July 1835.
48. Brown, Capt. S. Description of the Trinity Pier, in Edin. Phil. J. (1822). VI.
49. Telford, T. Report of Select Committee...Proposed Bridge at Runcorn, 13 March 1817, Warrington, 1817.
50. Telford, T. idem. Supplementary Report, 22 July 1817.
51. Paxton, R.A. Menai Bridge 1818-26, in Thomas Telford: Engineer, London, 1980.
52. Illustrated London News 12 June 1847.
53. Clark, E. The Britannia and Conway Tubular Bridges, London, 1850.
54. Scott, Bruce W. The Railways of Fife, Perth, 1980.
55. Hall, W. On the Floating Railways across the Forth and Tay Ferries in connection with the Edinburgh, Perth and Dundee Railway, in Min. Proc. I.C.E. (1861-2). XX.
56. P.P. House of Commons. Referees on Private Bills...North British and Edinburgh and Glasgow (Bridge of Forth) Railway Bill 8 May 1865. Minutes of Evidence. S.R.O. BR/PYB/(S)/1/36

57. Engineer 29 April 1864 17 258.
58. Dowling, C.H. Iron Work. Practical Formulae...with the description of a suggested Railway Bridge across the Queensferry, London, 1862. Forms Div. II of Formulae, Rules, and Examples for Candidates for the Military, Naval, and Civil Service Examinations, London, Weale, 1862.
59. Stephenson, G.R. & Tone J.F. Firth of Forth Bridge. Report..., London, 1826.
60. Ms. Forth Bridge Railway Plan Shewing Bores on Site of Proposed Viaduct 1864. Elgin Estates Archives.
61. Ms.idem. Cartoon Forth and Tay Bridges. 1865. Elgin Estates Archives.
62. Prebble, J. The High Girders, London, 1956.
63. Scotsman 14 June 1866.
64. Thomas, J. A Regional History of the Railways of Great Britain, Volume VI, Scotland The Lowlands and Borders, Newton Abbot, 1971.
65. Scotsman 6 August 1866.
66. Bradshaw's Railway Manual, London, 1874.
67. Douglas, H. Crossing the Forth, London, 1964.
68. Dundee Advertiser 18 October 1881.
69. Barlow, C. The New Tay Bridge, London, 1889.
70. Barlow, W.H. & Pole, W. Report on the Forth Bridge, designed by Thomas Bouch. Edinburgh, 1880.
71. Roebling, J.A. Report of..., in Annual Report of the... Covington & Cincinnati Bridge Company, Trenton, 1867. Bridge illustrated on front cover of A.S.C.E. calendar for 1990.

Additional Sources

Skempton, A.W. British Civil Engineering 1640-1840 : A Bibliography, London, 1987.
Marshall, J. A Biographical Dictionary of Railway Engineers, Newton Abbot, 1978.
Dictionary of National Biography.

Acknowledgements

The author acknowledges with thanks assistance received from:

Members of staff of the I.C.E. Library, the National Library of Scotland, the Royal Society, the Scottish Record Office; the Earl of Elgin and Kincardine; Hopetoun House Preservation Trust; Mr P. Cadell; Mr D. Cameron; Mr C. Johnston; Mr W.T. Johnston; Mr A. Loughlin; Dr G. Maxwell; Mr J.S. Shipway; Prof. A.W. Skempton; Mr C.J. Smith; Mrs M. Young; and Elton Engineering Books.

AMERICAN LONG SPAN METAL TRUSS EXPERIENCE PRIOR TO THE FIRTH OF FORTH

F.E. GRIGGS, Jr
Dept Civil Engineering, Merrimack College, North Andover,
Massachusetts, USA

Abstract
This article traces the development of long span metal truss building in the United States
in the time period 1840 to 1890. Squire Whipple, the Father of Iron Bridges, designed
and built 146' span double intersection (cancelled) railroad bridges as early as 1852 and
projected that it would be possible to span distances of 400' to 500' with trusses. His
book entitled "A Work on Bridge Building," published in 1847, described, for the first
time, methods for designing truss bridges. The Whipple, Pratt, Warren, Bollman, Fink
and Howe trusses will be covered as they made 500' simple span truss bridges possible
across the Mississippi, Missouri and Ohio rivers by engineers such as J.H. Linville and
George S. Morison. The High Bridge across the Kentucky River, designed by C.
Shaler Smith, the first major cantilever bridge built in the United States (1877) is
described in depth. The paper also describes the Niagara River cantilever (1883) by
Charles Schneider and the Hudson River cantilever at Poughkeepsie, N.Y. (1889) by the
Union Bridge Company.
Keywords: Bridges, Truss, Simple Span, Cantilever, 19th Century, Iron, Steel.

1 INTRODUCTION

The opening of the Firth of Forth Bridge was the culmination of a century which saw
bridge building move from an art using primarily stone and wood to a science using iron
and later steel. This paper will describe the evolution of long span metal truss building
in the United States prior to the Forth Bridge.

2 Simple Span Trusses

American bridge builders at the beginning of the century used wood as the primary
structural material as it was available, cheap, and easily worked with by the craftsmen
and tools available. Iron was used in some wooden bridges as tension members such as
in the Howe and Pratt trusses and as bolts, plates, washers etc. These early millwright-
engineer-architects, using intuition and feel for the material, were able to span distances
of up to 370'. The 340' span at Philadelphia by Louis Wernwag, appropriately called
the "Colossus", survived for over 30 years before it was destroyed by fire. The first
bridge to use iron in any significant amount was the cast iron arch bridge built on the
National Road at Brownsville, Pennsylvania in 1839. This bridge, with a span of 100',
was patterned after the early iron arch bridges built in England.

Men such as Palmer, Burr, Wernwag, Long, Howe, Pratt and Town made great
progress in bridging the rivers and streams of the United States with wooden bridges.
It, however, remained for Squire Whipple, Fig.1, to develop the method of
scientifically analyzing and designing a truss bridge and to instruct his fellow engineers

on the proper use of cast and wrought iron. After graduating from Union College in 1830 Whipple went to work as a rodman on the Baltimore and Ohio Railroad. This project, one of the earliest efforts undertaken to build a railroad over the Appalachian Mountains, tested the ingenuity and engineering skills of America's early engineers. Whipple in his capacity as one of the technical staff would have come in contact with Col. S. H. Long who had been assigned to the railroad by the United States Government. Long had designed and built a wooden truss in a pattern which he was later to patent. This bridge, the Jackson Bridge, was probably the first truss bridge built in the United States using some mathematical analysis. Later while working on the enlargement of the Erie Canal, Whipple along with everyone else connected with the enterprise, saw that the wooden bridges which were built in the early 1820's were falling into the canal and that they would have to be replaced with stronger bridges made

Fig. 1. Squire Whipple.

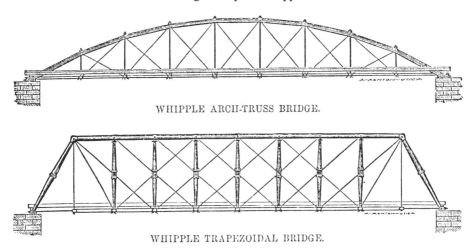

WHIPPLE ARCH-TRUSS BRIDGE.

WHIPPLE TRAPEZOIDAL BRIDGE.

Fig. 2. The Whipple Trusses

out of materials which would last longer. In 1841 he designed and built his first bowstring truss, Fig. 2. Whipple said:

> The design and intent were to construct an iron truss to be used in connection with a wooden floor system, the truss to have sufficient stability to stand of itself, without any dependence upon the wood, so that the latter could be renewed from time to time as might be required, without disturbance or danger to the iron work .

This patent was issued number 2064 and was the first successful iron truss bridge built in the United States. The parts were sized to carry the loads placed upon them and the materials used were the best available at the time, e.g. cast iron in compression, wrought iron in tension and replaceable wood as decking. Many such bridges were built across the Erie Canal in the 1840's-50's.

Whipple's greatest contribution to the development of truss bridge building came in 1847 when he published his **"A Work on Bridge Building."** This book, which for the first time anywhere described his method of analyzing a truss, also gave future bridge engineers a guide on methods of selecting materials and construction techniques. He also developed his plan for the trapezoidal, double cancelled truss, Fig. 2, at this time. He said:

> Prior to 1846, or thereabouts, I had regarded the arch-formed truss as probably, if not self-evidently, the most economical that could be adopted; and at about that time I undertook some investigations and computations with the expectation of being able to demonstrate such to be the fact, but on the contrary the result convinced me that the trapezoidal form, with parallel chords and diagonal members, either with or without verticals, was theoretically more economical than the arch , and that the trapezoid was more economical without than with vertical members-there being shown a less amount of action (sum of maximum strains into lengths of respective long members) under a given load.

Up until this time with spans being short the height of the truss was approximately equal to the panel length and the diagonals were of equal length and made 45 degree angles with the horizontal. As spans of necessity increased the height of the truss also had to increase. This increase in height resulted in very long panels if the diagonals were to remain at the most efficient inclination, 45 degrees plus or minus. To maintain the diagonal inclination he simply had the diagonals reach over two or more panels. He once again used cast iron for his top chord and verticals and wrought iron for his lower chord and diagonals and wood for his decking. He designed and built two of these bridges in Utica and West Troy, New York in 1852-54 for railroad use. Both of these bridges lasted well over 40 years and became a standard truss pattern up until the late 1880's. These bridges had spans of 146' which made them the longest iron railroad trusses in the world at that time. Subsequent to Whipple, Wendell Bollman and Albert Fink designed and built iron railroad bridges on the Baltimore and Ohio Railroad in the 1850's and 60's but these patterns were short-lived and not adopted by others.

In 1878 the American Society of Civil Engineers prepared an exhibit on American Civil Engineering for the Paris Exposition. Included as plate XI, Fig. 3, was a summary of the history of truss building in the United States. By this time period the only truss on the list still being used to any degree was the Whipple truss. Whipple's trusses were later to be modified by Murphy and Linville who changed the lower chord from forged wrought iron links to eye bars and later made all the metal parts out of wrought iron and in the 1880's out of steel. In the 1880's The pattern was used exclusively by George Morison when he built seven (7) railroad bridges, using first wrought iron and later steel, with spans of up to 518' across the Missouri River.

PLATE XL
TRANS. AM. SOC. CIV. ENGR'
VOL. VII. N° CLXXIV
PARIS COMMITTEE

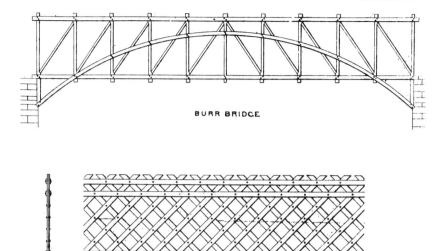

BURR BRIDGE

TOWN LATTICE

HOWE TRUSS

PRATT TRUSS

BOLLMAN TRUSS

FINK TRUSS

WHIPPLE TRUSS

POST TRUSS

Fig. 3. American Trusses

The record for span length reached out from Whipple's 146 feet as follows:

Table 1. Record Span Length

LOCATION	SPAN	DATE	TYPE	
Steubenville, Ohio	320'	1864	Linville Truss	
Louisville, Ky.	390'	1870	Truss by Al Fink	
Louisville, Ky.	420'	1872	Linville Truss	
Cincinnati Southern RR	515'	1876	Linville Truss	Fig. 4
Chesapeake & Ohio RR	545'	1888	Bowstring Truss	Fig. 5

Fig. 4. Cincinnati Southern Railway Bridge Over The Ohio.

Fig. 5. Cincinnati Bridge over The Ohio.

Whipple was to compare his bridges to other, what he considered to be inferior, trusses in both later editions of his book and in articles submitted to the American Railroad Journal, Appleton's and even to The Engineer and Architects Journal. Thus even though the names of Pratt, Warren, Howe and others may be better known to 20th century engineers, it was Whipple who showed the profession in his writings and practice how to design and build bridges and how to do it with precision and efficiency. He has been, and rightfully so, called **"the father of iron truss bridges"**.

The simple span truss, even when made of steel, reaches a maximum span length as the height of the truss required becomes unwieldy, even if double or triple intersections are made. The maximum span length prior to the Forth Bridge was, as shown in the table above, the 545' modified bowstring bridge built over the Ohio River at Cincinnati by L. F. G. Bouscaren in 1888, Fig 5. It had a height of over 85' at the center and and 60' at the end posts. All of these trusses were pin connected and were constructed in remarkably short periods of time.

3 CANTILEVER BRIDGES

Cantilever bridges had been used throughout time in the Far East using wood in a corbel fashion. Many engineers, including Baker and Fowler in England and Prof. W. P. Trowbridge in the United States had known about the points of contraflexure in continuous beams and had prepared designs to build cantilevered bridges. The first cantilever, however, was by Heinrich Gerber who built, on falsework, a 124' span over the Main River at Hassfurt, Germany in 1867. The first railroad bridge built to this design was at Posen, Prussia in 1876 (Fig. 6).

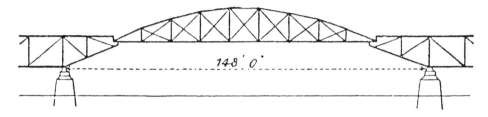

Fig. 6. Posen Railway Bridge.

In 1876, C. Shaler Smith, Chief Engineer of the Baltimore Bridge Co.,was asked to bid on a proposed bridge across the Kentucky River. The river gorge was over 1125' across and up to 276' deep. The river also was prone to flash flooding with a record of a 40' rise in water level in a 24 hour period These two facts made the inexpensive construction of a conventional truss built on falsework impracticable. John A. Roebling had been faced with these same conditions 22 years earlier and had begun the construction of a railroad suspension bridge with a suspended span length of 1236' at this site. He had built his masonry towers and anchorages and had his wire delivered when the Railroad Company ran out of funds to complete the project in 1858. After visiting the site Smith decided to use the cantilever method, with no falsework to construct his bridge. He started each span by suspending it from Roebling's towers until about half way out (196'-10") to the first permanent pier where he erected a temporary wooden support. He did not suspend the ironwork from the top of Roebling's tower with cables as had been done by others such as James Eads at St. Louis and Eiffel at Garabit. Instead he tied, using eye bars, the end of the top chord back horizontally and through the space between the bases of Roebling's towers. He also inserted powerful screw jacks which pushed against the rock formation and the end of the lower chord to adjust the elevation of the ends of the cantilevered truss. When he reached the temporary tower, which was built off the cantilever, he inserted screw jacks

Fig. 7. High Bridge.

at each of the four posts and jacked the truss up, thus relieving some of the tension in the top chord and compression in the bottom chord. From that point he cantilevered the truss out to the location of the permanent pier which had been built from the ground up to save time (Fig. 7). The next phase had him build the truss as a cantilever out from the pier to the center of the bridge. This procedure was completed first on one side of the bridge and then on the other side until they met in the middle. The mismatch in the middle of the bridge ranged between 2" and 5". These gaps were closed by moving the permanent iron towers which were on rollers at the foundation level and jacking while waiting for the right temperature.

The top chord was a continuous riveted member made up of plates and angles except for pins at the planned points of suspension of the shore spans. The bottom chord and verticals were also built up members, the diagonals being eye bars (Fig. 8). After completing the structure he cut the temporary rivets which held together a tendon joint which had been placed in the lower chords at the point of contraflexure about 75' landward from each permanent pier. This was done to convert the continuous truss, which would see stresses resulting from the changes in pier height due to temperature changes, to a three span bridge with the middle span being 525' long and the two side spans being 300' long. The spans of this bridge as originally built were 375', 375' and 375' or well within simple truss capability if falsework could have been placed. Perhaps the most noteworthy achievement of this bridge was that the superstructure was started on October 16,1876 and was completed on February 20, 1877 an elapsed time of four (4) months and four (4) days. The parts which were fabricated by The Edgemoor Iron Works went together,as was reported by the Railroad Gazette, "like a Springfield Rifle."

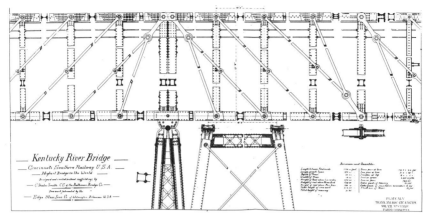

Fig. 8. High Bridge.

The cantilever method in the United States from that time forward was based in part upon the experience gained on this bridge. It is interesting to note that Smith used a Whipple Double Intersection Truss pattern for his bridge. He went on to design and build many cantilevered bridges such as the Lachine Bridge shown in Fig. 9, over the St. Lawrence River below Montreal in 1886. This time, however, he built the truss as a cantilever, again without falsework, and then riveted the parts together thus making it a continuous truss, or just the opposite sequence to his High Bridge. Span lengths on this bridge were 408' and the trusses were again double intersection. The curved upper and lower chords were designed to smooth the transition from a deck to a through truss which many observers thought was unsightly. The Railroad Gazette of May 28, 1886 stated, "Whether giving to the through spans the form of a bastard arch which is not an arch is any real remedy may plausibly be disputed, but we are inclined to think that the design shown will be generally considered a more pleasing, or rather less ugly, solution of the problem than the ordinary form, and so, on this ground alone, worthy of use in such locations."

Fig. 9. Lachine Rapids Bridge.

Charles C. Schneider was the next American to build a major cantilevered bridge, this one at Niagara Falls, New York. Schneider was educated at the Royal School of Technology in Chemnitz, Saxony. He came to the United States in 1876, the same year that Smith started his "High Bridge", and worked under Octave Chanute and with George Morison on his Missouri River bridges. This bridge was erected by building the anchor spans on falsework and the two piers from the falsework platform (Fig. 10). He then built each span outward from the piers to the middle of the Niagara River over 210' above the rapids. The bridge as completed had anchor spans of 195.2', cantilever spans of 175' and a suspended span of 120'. Schneider stated that he did not like to use the double intersection pattern but he did so in this case to give his long verticals some lateral support. This bridge was built just up the river from Roebling's suspension bridge which he had completed in 1855 with a span of 820' for railroad and carriage traffic (Fig. 11).

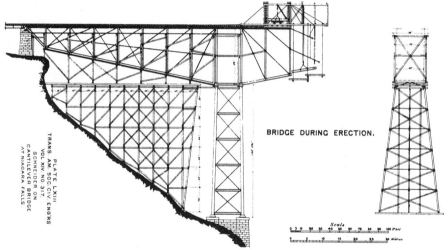

Fig. 10. Falsework and Cantilever Niagara Bridge.

The time of construction for the superstructure of Schneider's bridge was from April 15, 1883 to December 20, 1883 or 8 months 5 days.

Fig. 11. Niagara Bridge - Roebling's 1855 Suspension Bridge in Background.

Some would say that it was Schneider that introduced the modern cantilever bridge to the United States as he used the concept of anchor spans, cantilever spans and a suspended span. Smith's High Bridge, as noted, was originally built as a continuous truss and then later modified to act like a cantilever.

The next bridge which took cantilever bridge construction to a new level in the United States was the Poughkeepsie Bridge across the Hudson River in New York State. The Poughkeepsie Eagle in its October 1889 souvenir edition commemorating the opening of the bridge reported " Among the great bridges of the world the one just erected at Poughkeepsie, and now thrown open for use, is not surpassed by any other completed bridge in magnitude, boldness of design or beauty of shape and situation. The great bridge over the Firth of Forth, in Scotland, will when finished, be larger, but with this exception the Poughkeepsie bridge is without equal in the world". Maybe they overstated their case a little but there is no doubt that the bridge was a major railroad structure and was one which utilized the cantilever in new, if not necessary, ways. The river crossing portion of the bridge consisted of two spans of 548', one of 546' and two of 525' for a total length of 2692'. In addition there were two shore anchor spans of 201' (Fig. 12).

Fig. 12. Poughkeepsie Railroad Bridge.

The Hudson River was and still is a main transportation artery and the original charter issued to the bridge company stated that there could be no piers in the river which would impede shipping. That restriction was impossible to meet, even with a suspension bridge, as the longest railroad suspension bridge was Roebling's with a span of 820' at Niagara. In spite of this charter restriction Horatio Allen as Chief Engineer in 1871 proposed a suspension bridge with a 1200' main span and two side spans of about 700'. Allen was a prominent civil engineer of the time and was the man that John Jervis sent to England in the 1820's to study Stephenson's Stockton and Darlington Railway and to buy two locomotives for use on the Delaware and Hudson Canal system. He was also the first man to operate a steam locomotive, one of Stephenson's, on rails in the United States. The company, however, knew that this bridge would not be possible under the existing charter and maybe not even possible from a technical standpoint. Captain James Eads who was in the process of building his famous steel arch railroad bridge with three 500'+ spans over the Mississippi at St. Louis, Missouri suggested that an attempt be made to revise the charter permitting four piers in the river so that a truss bridge with spans of approximately 500' could be built.

After a long legislative battle the charter was amended in May of 1872 permitting the four river piers with a minimum navigable width of 500' between them. The new Chief Engineer was to be J. H. Linville who, as noted previously, was one of the preeminent iron truss builders in the world. After a slow start the company, in the financial panic of 1873, lost funding and work stopped. After another false start with the American Bridge Company the Union Bridge Company, which had been formed as a sort of "union" by some of the leading engineers of the day, took over the design and construction of the bridge. Men such as Thomas C. Clark, Charles Kellog, Charles MacDonald, George S. Field, and Charles Maurice were members of this "Union" of engineers. MacDonald was effectively partner in charge of the design of the bridge and determined that the structure would not be a series of simple trusses all built on falsework in the river. He built instead two simple spans on falsework in the river, two shore arms on falsework, and three river spans built on the cantilever method as shown in Fig. 13. Serious foundation work was begun in December of 1886 and

Fig. 13. Poughkeepsie Bridge Falsework and Cantilevers.

completed in May of 1887. The falseworks in themselves were major structures resting on 130' long wood piles and rising 130' above the water level to support the steel trusses. Pomeroy P. Dickinson became Chief Engineer during the construction of the superstructure. One of these trusses, with a 525' span, was the longest truss built of steel in the world at the time. It was started on June 19, 1888 and finished on July 13, 1888 or in a total of 24 calendar days. The cantilever arms, both from the shore spans and off the connecting spans were 160' and were built in an identical fashion to those at Niagara described above. That should come as no surprise as MacDonald was one of the consultants on that project. The suspended spans were 212' and were cantilevered out from the cantilever spans.

It may be of interest to many of the readers of this paper from England to know that all of the steel used in the construction of this bridge was made in their country. It turns out that all the American steel mills were operating at capacity and couldn't provide the amounts required. All of the English mills were also at capacity. Thanks, however, to the efforts of Lord Randolph Churchill who had attacked the construction bureau of the British Navy resulting in a slow down of steel ship construction, the English mills had steel to sell to the United States. The steel was sold to George S. Field, brother of Cyrus Field who was well known in England due to his work on the Atlantic Cable. The steel was fabricated in the yards of the Union Bridge Company in Athens, Pennsylvania and Buffalo, New York. The top and bottom chords and verticals were all built up of riveted plate stock. The tension members were eyebars and in accordance with American practice at the time, all connections were pinned.

The approach spans would be considered major structures in their own right if they didn't have to compete with the main span. On the East shore of the Hudson there were 24 spans with lengths of up to 175' and on steel piers up to 200' in height. On the western shore there were two trusses of 145', seven of 60' and one of 53'. The total length of bridge was therefore 6767'. This is in comparison to the 8295' total length of the mighty Forth bridge.

4 Conclusion

In conclusion, American bridge building in the first 90 years of the 19th century relied on the truss, first in wood and later in cast and wrought iron and by the 1880's on steel. The preferred erection method was to use a system which pinned the members together and massed the material in a few members. These assumptions and techniques made it possible for American engineers using the methods developed by Squire Whipple to accurately determine loads in members and to design them accordingly. In England the scarcity of wood had precluded them from building wood trusses such as developed in the United States. The iron arch, the inverted arch and wrought iron tubular structures were the primary bridge types. Trusses when built were of the lattice type with a disperse pattern of diagonals and after some debate Warren trusses were also built. There was an almost complete reliance on riveting with few pinned structures being designed or built. The difference in English and American bridge building was summed up quite well by Benjamin Baker, M. Inst. C. E., in his discussion of the Niagara Cantilever when he said:

> The execution of the works of the Niagara Bridge in so short a time was a feat of which the engineers and contractors of any country in the world might well be proud. Probably in this as in almost every other case, the experience gained would suggest certain modifications in the details of construction. As it stands, however, the bridge is a very interesting example of the adaptability of essentially American details of construction to a novel type of bridge. At the Forth Bridge, owing to the magnitude of the parts, the details in many respects present a greater analogy to the building of an Atlantic steamship than to that of an American Bridge.

It has been said that the the English and Americans are one people separated only by a common language. Using this analogy 19th century bridge engineers in America and England were one people separated only by their pins and rivets.

5 References

Cooper, Theodore (July 1889) American Railroad Bridges, in **Transactions ASCE**, Vol. XXI, pp. 574 - .

Morison, G.S., North , Edward P., Bogart, John (Nov. 1878) American Engineering, in **Transactions ASCE**, Vol. VII, pp. 321 - .

O'Rourke, John F. (June 1888) The Construction of the Poughkeepsee Bridge, in **Transactions ASCE**, Vol. XVIII, pp. 199 - 215, plus plates.

Schneider, Charles C. (Nov. 1885) The Cantilever Bridge at Niagara Falls, in **Transactions ASCE**, Vol XIV, pp. 499 - 539, plus plates.

Whipple, Squire (1847) **A work on Bridge Building**. H. H. Curtis, Utica, New York.

_____, The Kentucky River Bridge (Sept. 21, 1877) **Railroad Gazette**, 428-429.

_____, (Aug. 8, 1890) The Queen City of the West, **Engineering**, 149-150.

PART TWO

RECENT STRUCTURAL ENGINEERING PROJECTS IN SCOTLAND

RAILWAY ENGINEERING IN SCOTLAND TODAY

W.D.F. GRANT
Area Civil Engineer (Edinburgh) ScotRail, Edinburgh, Scotland, UK

Abstract
This paper describes some of the Mechanisation of Track Maintenance
and Relaying methods being used in Scotland (and indeed throughout
much of British Rail) to enable stringent Business Goals to be met.
It also describes Paved Track for Tunnels, and the work done on
Bridges prior to Electrification.
Keywords: Mechanisation, Track, Business Goals, Paved Track,
Electrification.

1 Introduction

The Civil Engineering Department as one of the "Functional"
departments is by the nature of our work the biggest spender in
normal maintenance and renewal.
 We are responsible for the maintenance, repair and renewal of
track, (commonly known to us as the "Permanent Way") and structures,
to the standards required to support the Business goals, and in a
safe and cost effective manner.   The last part of this statement
i.e. the "cost effectiveness" is the task to which we have had to
turn particular attention to over the years and especially over the
last 2 to 3 years.
 I, as Area Civil Engineer, Edinburgh, look after approximately
one third of ScotRail infrastructure, but the overall size of the
task in Scotland is represented roughly by the following

 2660 Track miles of Permanent Way
 9400 Bridges including Forth and Tay Railway Bridges
  540 Stations
   27 Miles of Tunnel

 The overall budget for our Department for 1990/91 is about
£70,000,000.   A great deal of this sum, something over one third,
is spent on labour costs.   There are currently some 3,100 staff
employed by the CE Dept in Scotland.   In the sixties this staff
numbered about 10,000.   Mechanisation is the obvious key to
reduction in manpower and great strides have taken place,
particularly on the Permanent Way front, to this end.
 Although I will give mention to some of the interesting Works
projects going on, I will be concentrating mostly on the Permanent
Way developments of recent years, and what might be foreseen for the
future.

2 Permanent Way

2.1 Tamping and Lining

Permanent Way, despite its name, is far from permanent - all of
its components:- track ballast, sleepers, fastenings and rail
are subject to axle loads up to 25T and trains travelling at
speeds up to 125mph.   Apart from wearing out, which of course
all of these components will do in time, fairly constant
attention is needed to pack and line the track so that trains
and their passengers can ride safely and comfortably.   The
packing of ballast is termed tamping, and the packing or tamping
process has come through many phases in 150 years or so of
railways.   "Beater" packing was the first, using flattened out
pickheads to force the stone under the sleepers.   Shovel
packing was the main method in many places where the track was
jacked up to the required height and then shovels used to ram
the ballast under the sleepers.

   Measured shovel packing was commonly in use when I was a
youngster in the railways in the 50's.   Here, a more scientific
approach was used.

   The track was levelled using sighting boards and again jacked
up to the required level.   The requisite number of cans of
granite chips were then inserted under the sleepers by shovel.
This gave a longer lasting and quite accurate method of packing.

   When mechanisation came in on BR in the early 60's the first
"Matisa" tampers used a small gang of men to lift the track
ahead on jacks, and vibrating tines pushed the ballast back
under the sleepers.   This was required as concrete sleepers
were then coming in to regular use, and they were really too
heavy for effective packing by traditional manpower methods.
Kango generators and guns (electrically powered beater picks in
essence) were used as well, particularly on switch and crossing
work, and are still used today.

   As mechanised tampers progressed they became more
sophisticated. First using a trolley which ran ahead of the
machine connected with taut wires to the sensors of the machine
which jacked the track up to the correct height before tamping
the sleepers. Later, electronic and laser methods were used and
the most sophisticated machine currently in use is the Plasser
09-32.   This machine can pack at the rate of 1100m/hr and is
computer controlled.   A feature of it is that the tamper itself
does not stop at each sleeper or pair of sleepers to tamp.   The
machine moves continuously and the tamping bank moves separately
on a frame.

   All of the tamping machines used over the last twenty years
or so have also had the ability of lining the track.   This is
usually done on a separate operation from tamping.   Again
computer controlled methods are available to give the very
accurate alignment required for high speed running.

ATA - Automatic Track Alignment

Principally for InterCity routes a number of Plassermatic

tampers have been fitted with equipment which can align the track automatically. The method used is as follows :-

a) Pre-measurement Survey
A pre-measurement survey of the required site is carried out by the machine (at up to 10mph). The versines of the curved track are taken and the tops and bottoms of transitions noted (as marked up on the track). Any restricted clearances to structures can also be noted.
b) Design of new alignment
The design of the new alignment is then carried out by the ATA computer. There are two modes for this:
i) Under the control of the machine supervisor
In this mode the slues are limited to 30mm and the positions of the transitions cannot be moved (ie the requirements of BR Handbook 11 are automatically met).
ii) Under the control of authorised technical staff
In this mode the alignment scheme is produced under the control of technical staff. There is no slue limit and the transition details can be amended. The technical staff are responsible for ensuring that the curving rules are obeyed.
c) Execution of new alignment
The new alignment is then executed by the tamper running back through the site. The ATA computer passes information to the slueing equipment to enable the tamper to achieve the slues required. Any adjustments required to make allowance for the corrections necessary on transitions are automatically applied.
d) Post Measurement Survey
In order to assist the supervisor/technical staff in checking that the design alignment has been achieved, a post measurement run can be carried out to resurvey the site. The new versines are compared with the design values and any locations where the requirements of Handbook 11 are not met are flagged up, but the decision as to what action is required is left with the person in charge.

Regretfully, despite the excellent job these tampers make of providing a good top and line, they have faults. These are twofold -

a) Each time the tines pack stone under the sleeper they crush part of the individual stones. It is calculated that some 9T of stone per mile is converted to material too small to be of use for tamping.
b) The packing produces a "pyramid" of tightly packed stone under each seating. After some 3 weeks to 12 weeks have passed it is considered that the sleeper has gone back to its original level.

Thus because of a) above and natural degradation of the stone by the heavy loading of trains, the ballast becomes contaminated by dirt, stone and cementacious dust. This results in water retention and a vicious circle of further ballast degradation. it is necessary to ballast clean the track every 7/12 years to replace dirty ballast with clean stone.

Because of both a) and b) we have been experimenting with a mechanised form of the former measured shovel packing procedure.

Hand held Stoneblowers have now been in use for about 3 years with some excellent results.   The track is jacked up to the required height using "air-bag" jacks which are obstructionless to trains.   A portable compressor supplies air to inject the required amount of chips under the rail seating area.

I consider that track so treated will last up to 3 times as long as with conventional tamping.   However the length of track to be packed cannot possibly all be tamped in this manner and the hand held stoneblowers tend to be used in trouble areas where conventional machine tamping does not last long - usually in areas awaiting ballast cleaning, or in pockets of poor formation.

BR Research Department have been developing, along with Messrs Plasser and Theurer of Austria, a Stoneblowing machine which, if successful, will revolutionise tamping and give longer lasting results.   My own view is that present conventional tampers will still have a place for many years to come, but the prospect of Stoneblowers is exciting.

2.2 Relaying
One Permanent Way problem which has long been recognised is "Track Memory".   Despite ballast-cleaning or even deep excavation of old ballast by traxcavator the locations of previous joints seem to be remembered when new ballast and new track is installed.   Once the track has been in and run over for some months a slight deflection can be found where old joints once existed.   There are many locations where this phenomena exists and these lead to increased maintenance costs and ultimately earlier ballast cleaning or relaying than should be necessary.

In addition we tend to build in a memory by using "plant" or "service" rails when relaying track by the panel method (this is when the old long welded rail is replaced by 60' serviceable rails prior to relaying so that the Track Relaying machines can remove panels and replace with new ones of the same length - the LWR is then either replaced or renewed).   Traffic running over these service rails whose joints were often mismatched was enough to induce a memory to even newly ballast cleaned track.

Our modern and more effective method is called "Sleeper beam relaying" and is carried out in the following stages.

1) The track is ballast cleaned.
2) The long welded rail is removed from its housing and slued to the cess and six foot side to form a 10ft gauge track after being aligned and packed to reasonable limits.
3) Twin gantries run on the spread rails having been offloaded from wagons using their own turntable.   These gantries

Fig.1. Placing of Sleepers by Sleeper Beam

carry a specially designed beam carrying chains and hooks.
They lift out the old sleepers which are hooked to the beam
and then placed in empty wagons.
4) One of two machines specially designed for the purpose is
off-loaded, again using the gantry and placed on the
ballasted formation.   The machine either Bruff or Z-cat
profiles the ballast, and vibrates it to leave a prepared
surface on which to lay the new sleepers.
The machines can use either "V" or "beacon" laser to
control the profile.
5) The beam then picks up sufficient new sleepers, usually 26 or
28, but sometimes 30, from the sleeper carrying wagons and
offloads them in two stages, first dropping off every second
sleeper, which gives exactly the correct spacing, then the
remainder equivalent in total to the number required for a
60ft track length. (Fig.1)
6) Using a pair of rail threaders, the long rails are then
manoeuvered into place on the concrete sleepers.   The rail
is then clipped up.
7) Packing by tamper then finalises the track height and the
track is fine lined, and can usually be reopened to traffic
at a minimum of 50mph, and often at higher speeds.

The increasing desire of passengers to travel at weekends
means pressure on us to give relayed and reballasted track back
at much higher speeds than we used to only a year or so ago.
Then it was usually at 40mph for passenger traffic.   Sleeper

beam and good compaction gave us useful increases, but for
Intercity trains slowing down to even 50mph from 125mph costs time
and money. (It is calculated that a high speed train braking to
20mph from 125mph and accelerating back up costs £25 in wear on
brake shoes and diesel fuel costs. This in addition to lost time
costs, calculated at about £15/minute).

The dynamic track stabiliser has been developed to compact the
ballast in a few passes equivalent to the passage of 100,000 tonnes
of traffic. This machine, (and I have one to cope with relaying
on E.Coast and W.Coast main lines) can produce a vibration on the
ballast ranging from 6 cycles/sec to 45 cycles/sec. One danger is
to arch underbridges and we have strict rules to avoid damaging
these.

We have now given back sections of renewed track on Sunday
afternoon at line speed (up to 125mph) or at least to 80mph on many
occasions, using this modern technology coupled to the sleeper beam
relaying method.

PAL/PUM

This equipment, produced by Geismar of France, is now being used
fairly extensively for certain types of work. There is a great
advantage when switch and crossing work (S&C), or track in a
platform, requires formation renewal, to be able to lift out the
whole S&C unit or, say, 800 or more metres of concrete sleepered
track in one go.

PAL and/or PUM allow this to be done. Each self contained unit
consists basically of a portal frame, the 'legs' of the portal
being vertical hydraulic jacks. The horizontal member is fitted
with a hydraulically operated screw thread which can slue two
clamps fitted to grip the rail head.

PAL is capable of lifting 7 tonnes/unit and the larger PUM   9
tonnes/unit.

To move track to the side, the required number of units are
installed over the track to give the clamps their maximum available
travel. The track is then lifted to the required height (a lift
of 1.2m can be obtained) and slued to the full extent of travel.
The track is then set down and the beams repositioned before
repeating the process. Thus the S&C unit or plain line can be
'walked' to the required distance before excavation is carried
out. The process is reversed to put the track back in place.

For longitudinal movement, (say to remove a length of continuous
welded track from a platform) two other pieces of equipment are
required. These are lightweight, but heavy duty trollies with
small diameter wheels, and a pair of channel section rails.
Without going into great detail, the track to be recovered is
raised by PAL, the channel rails pulled under it and resting on the
formation. The trollies are pushed along the channel rails under
the track and the track is lowered on to the trollies. The whole
section of track is then towed out using special ramps to place it
on top of the adjoining track. The process is reversed to replace
the track when drotting has been completed.

Each unit can be built up by experienced men in 30 minutes, and is operated from a hydraulic power pack with central control.

The benefits of this type of equipment are that track can be moved without the need for overhead electrified lines to be isolated.   Long lengths of track can be removed and replaced without cutting the rail into short lengths and new S&C units can be built up and the rails welded before installation, avoiding the need to break up the units for handling with conventional craneage.   These advantages have been used to reduce the number of Sundays required to carry out work, thus reducing costs and improving quality.

The equipment has successfuly been used for installing new S&C at Dunbar, reballasting S&C on the West Coast Main Line, and reballasting plain line throughout the Edinburgh Area.   The PAL and PUM have become so popular throughout BR that Geismar are doubling the number of units available in the UK and reducing hire costs, reflecting the increased useage.

One of the most versatile tools we have to assist in maintenance work is the TRAMM.   This is basically a powered flat wagon carrying a $1^1/_2$T capacity crane which can carry and load or offload 60ft rails, concrete sleepers and many of the other components and tools we need. They can travel at 40 mph and therefore run in between passenger trains.   In an emergency such as a broken weld or insulated joint the necessary material can be got quickly to the site.

The TRAMM is fitted with ballast ploughs which means that after dropping ballast from a hopper train it can be spread ready for tamping.   Various other options such as a small grab bucket and snowplough can be fitted.

Numerous ingenious pieces of small plant too are available, mostly from private railway suppliers, which help to make our task easier.   In my young days, the first petrol driven rail saws took 10/15 minutes to cut a rail (The old hand saw which I have seen used took an hour).   We now get through a rail in 7 minutes with a much more accurate cut using disc cutters.

3 Works Matters

3.1 Paved Track

In certain circumstances it may be economic to depart from conventional ballasted track in tunnels.   When an electrification scheme is proposed, the additional clearance required for the overhead equipment may well require more tunnel headroom.   In some tunnels this may safely be obtained by lowering the track, but in others, particularly those without inverts and with sidewalls whose footing is shallow, some other method must be found.

Several different methods have been tried e.g. Ladder track named from the shape of the precast units each 9.1m long which

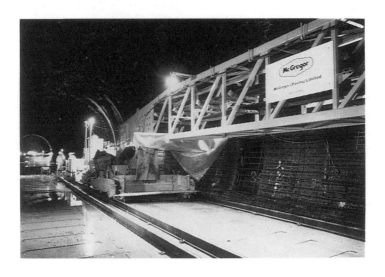

Fig.2. Paving Machine

have been laid on an asphalt or concrete base (Queen St tunnel approaching Glasgow is laid with this type).

The most common however is slip-form paving. Haymarket South tunnel in Edinburgh has recently been laid with this type of track, prior to electrification between Edinburgh and Carstairs. A potted description is as follows -

The existing track has to be removed, first the continuous welded rail is dragged out in sections, then the concrete sleepers are picked up and removed by use of service rails. These can of course be re-used elsewhere. Any remedial work and strengthening to the footings or tunnel walls is carried out first. The earthmoving equipment then moves in and the spoil is moved out by lorry. Any rock which is too high for the future base is then removed by Montabert or pneumatic drill. Once the correct level is obtained, a start is made to the sub base. This is composed either of Type 1 or lean mix concrete. The base slab is then laid to a +/- 10mm tolerance.

The next step is to pull back the long welded rails, which are set to a gauge of 2.8m and carefully lined and levelled. On these rails the paving machine hired from (MacGregors) is set up. This machine lays the top or rail slab to a fine tolerance.(Fig.2). The slab is reinforced with high yield weldable steel.

Holes are then drilled for the Pandrol rail fastenings which are set in the slab with an epoxy resin. When set the rails are then lifted back and placed on continuous rubber padding before being fixed down with Pandrol clips.

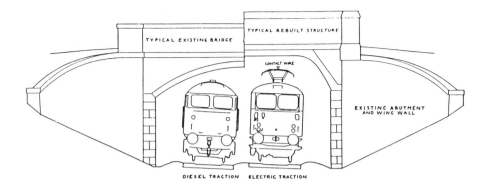

Fig.3. Electrification
Existing & Rebuilt Structures.

Water from the tunnel is normally dealt with by means of
channel drains in the six foot and cess.
Transition beams are normally installed at each end.   These
allow, on a sloping stone chip filled concrete tray, for the
transition from hard (concrete slab) to soft (normally)
ballasted track.

3.2 Electrification Work
The biggest work load likely for the Civil Engineering Dept
before any line can be successfully electrified is the
reconstruction of bridge superstructures.   The structure gauge
used in railway bridge construction in the mid to late 1800's is
generally too low to allow safe installation of the 25KVA system
which is used on BR for all high speed line electrification.
Although in some cases lowering the track may be the most
economic course, for the majority of masonry arch bridges
crossing over the railway the clearance has been increased by
reconstructing the bridge. (Fig.3).   The technique for this is
well established and involves construction of an adjacent
temporary crossing, usually during one overnight possession;
diversion of utilities and subequent demolition of the masonry
arch during a second overnight period.   Remaining abutments are
then cleaned off, an in situ springer beam is cast and precast
arch beams lifted into place.   Waterproofing, backfilling,
surfacing and the construction of reinforced spandrels and
parapets then follow with the minimum of interference with rail
traffic below.
With steel girder overbridges however it is normal, so long
as road levels and gradients can be accommodated, to raise the
existing bridge by jacking and raising the abutments by using
pre-cast concrete blocks.

Fig.4. Bathgate Station

When practicable the simplest method of increasing clearances in tunnels involves the singling and slueing of the railway track. This technique has been used through the twin bore Calton Tunnels in Edinburgh, where rationalisation of track requires only one line to pass through each bore in place of the two for which each was originally constructed.

In the last 3 years we have reconstructed 35 overbridges between Berwick and Edinburgh, and a further 16 will have been dealt with on the Edinburgh/Carstairs line by the end of this financial year.

Certain other work has to be carried out, even on bridges which have sufficient headroom not to require reconstruction. Parapets have to be raised and steeple copes fitted to avoid any chance of the public coming in contact with the overhead equipment.

3.3 New Stations
In co-operation with Regional and District Councils, ScotRail has had considerable success in building new stations in areas where large housing estates and schemes have changed peoples' pattern of travel. Some of the designs have been done by the Local Authorities and others by BR.

Many of these stations have been highly successful especially in the East of Scotland - South Gyle, Musselburgh and Livingston, Uphall and Bathgate (Fig.4) on the re-opened Bathgate Line.

Most of the stations are of simple construction, concrete
block platform walls with concrete copes and asphalt surface,
and simple waiting shelters, many of which require vandal proof
glazing. Wherever possible the stations have been built beside
overbridges thus doing away with the need for footbridges and
saving considerable expense. Ramped accesses for the use of the
disabled are provided.

3.4 Flood Risk - Assessment of Structures
Following recent bridge failures as a result of floods, BRB
sought the advice of Hydraulic Research Ltd to assess the
hydraulic aspects of bridges over water.
    Regional and Area engineers were remitted to appoint staff
dedicated to identify river bridges susceptible to
flooding/scour. Assessments are being made of possible causes
of failure based on bridge geometry and river and catchment
characteristics using BR Handbook 47 'Assessment of the Risk of
Scour'.
    Since the initial Consultants' reports identified bridges
liable to risk, site investigation contracts are proceeding to
find the foundation depths and conditions of the bridges and
their ground bearing capacity.
    A system is now in operation should a flood warning be
received from Police/River Purifcation Boards/Met Office, the
Public or BR staff to monitor all bridges considered "at
risk". These bridges are now all equipped with water gauges
and will be manned at all appropriate times during flood
conditions with staff directly linked by radio-telephone to the
appropriate Area Office and to the Railway Control Office.
    Evaluation is also underway on sonar, echo sounding,
ultrasonics, radar and telemetry equipment to establish river
bed depths, foundation depths and river behaviour to help
minimise failure through flooding.

4 What for the Future?

The shorter the possession time we require and the better the
handback speed, the less our passengers are inconvenienced.
This leads directly to greater revenue.
    Our current rate of relaying and ballast cleaning is really
very slow - no more at best than 200m/hour for ballast cleaning
and less than that probably 100m/hour for relaying even when
everything works perfectly - and so a kilometre of track takes a
minimum of 8 hours and more likely 12 hours necessitating
weekend Saturday/Sunday possessions.
If we can use short possessions on weeknights, say 5 hours, and
give the track back near line speed, we can be very much more
cost effective. Some Continental and many North American
railways have the equipment to do this and BR is currently
looking at investment in such machinery -

a) High Output Ballast Cleaners
These are machines capable of working at rates of up to
700m/hour, and BR's specification would demand a minimum rate of
400m/hour.   Currently available machines do not fit in to the
restrictive BR loading gauge and considerable modification to
current design is required.   It is planned to have a prototype
in operation in March 1991.
b) Relaying Trains
The origin of this technology is now approximately 20 years old
and there are many Relaying Trains in use on different Railways
around the world.   The equipment is a collection of relatively
simple tasks combined together into a continuously moving
production line.   Again a redesign to fit our load gauge is
necessary.
    The sequence of operations for a Relaying Train is generally
similar - regardless of the manufacturer or type -

    move the rails outside the sleeper ends
    remove the old sleepers
    grade the ballast bed
    lay the new sleepers
    place the new rails, or the original rails if they are
    to be re-used, within the fastening shoulders.

The old and new sleepers are transported to and from the
relaying machine in layers by gantry cranes running on the
sleeper wagons.
    Peak work rates of over 700 metres per hour and set-up times
of around 35 minutes have been observed and hence with these
production rates it is feasible to work such equipment for only
a few hours at a time because of the demand on materials supply.
    The machines are characteristically large and expensive and
can therefore be considered only where large and long term
workloads exist.   Similarly, the facility for daily
possessions is also required to achieve proper utilisation.
    The full package envisaged on BR includes -

    4 High Output Ballast Cleaners
    Single line Spoil Handling System
    2 Relaying Trains
    1 Mobile Flashbutt Welder

    The investment outlay is between £23m and £27m but should
lead to annual cost reduction of £15m to £25m.   This equipment,
carefully programmed to move around the whole of BR InterCity
lines, should be capable of carrying out the majority of the
annual programme of renewal work on these lines.
    There are exciting times ahead for the Civil Engineering
Department in British Rail.

THE ST ENOCH CENTRE, GLASGOW

D.S. BLACKWOOD
Ove Arup and Partners Scotland, South Queensferry, West Lothian, Scotland, UK

Abstract
The paper describes the structural engineering aspects of the design and construction of the St Enoch Centre, one of the largest and most exciting city centre commercial projects ever undertaken in Scotland. A feature of the project is the major steel roof structure which supports a glazed envelope of some 30,000m², believed to be the largest of its kind in Europe. The architectural concept which led to this unique design is also described as are the internal reinforced concrete structures, the substructure and foundations.
Keywords: Steelwork, Piling, Fast Track.

1 Introduction

The St Enoch Centre in Glasgow is a major city centre commercial development. It is one of the largest and most unusual ever undertaken in Scotland but was completed on time to a very short programme.

The development presented many structural engineering challenges. These included the construction of the foundations in variable ground conditions and the design and construction of the roof steelwork, an important visual element in the completed building.

Located in the heart of the city, next to the main shopping area, the site was that of a former railway terminus and hotel closed down in the late 1960s. Their demolition created a large gap site of some 6ha and a major urban regeneration task, which was not tackled until the Scottish Development Agency took the initiative, in the mid-1970s, to promote the investigation of possible developments for the site.

Over the following years many schemes were investigated, but finally the concept of a 75,000m² commercial development incorporating multi-storey parking for 750 cars emerged, and this was transferred in 1984 to the private sector in the form of the Church Commissioners for England and Sears Holdings plc, who became developers for the scheme.

2 Concept

In addition to providing economic commercial space, a key factor in the architect's concept prepared for the SDA was that the new development should make a strong architectural statement which would result

91

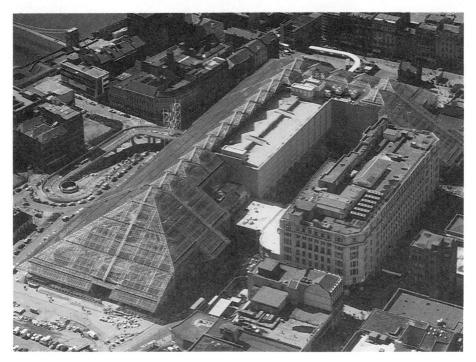

Fig.1. Aerial view of completed building.

in a civic building worthy of its important location.   Another prime
factor was the desire to ensure a light and airy interior.   This led
to the basic concept for the St Enoch Centre as a large, simple,
glazed enclosure virtually independent from the internal structures in
which the final layout of shop units could remain to be planned in
detail late in the construction period, and with maximum flexibility
to adapt to future changes without altering the external appearance.
The glass envelope itself was conceived as a simple wind and
weathertight barrier which would also provide an opportunity for solar
responsive passive energy systems to ensure a pleasant internal
environment without the need for air-conditioning in the malls.
     The final element in the concept was the desire to reflect
Glasgow's vigorous engineering tradition.   To this end, the steel
roof structure was to be strongly expressed to make a major contribution to the visual excitement of the interior, while externally a
cable-stayed bridge would mark the high level entrance to the car
park.

3 General description

The glazed envelope of some 30,000m², believed to be the largest of
its kind in Europe, encloses a volume of 250,000m³ above the lower,
main mall level.   Below this the substructure incorporates access for
the largest articulated lorries to unloading bays with goods holding

areas, plant spaces and shop servicing corridors, all at basement
level.   The ground floor, or podium, forms a roof over the basement
and acts as a base for the structures above.

The glazed enclosure is supported by structural steelwork which
provides a headroom of approximately 9m at the periphery of the
building and rises to 25m at the core structures.   These include the
seven-storey car park, a five-storey, multi-purpose building and a
three-storey structure which is linked to an existing adjacent department store.

Rising from the lower mall are the internal structures.   These
comprise four major stores, 75 unit shops, a 450-seat food court, and
leisure areas which operate on the two main mall levels.   The main
store has a third-level sales floor.   Elsewhere, at high level, are
the centre management suite and a restaurant.   An unusual feature is
the provision of an open ice rink at first floor level.

Cars enter the car park over the heads of the shoppers via an
elevated ramp which rises from ground level to bridge a main road and
passes through the glazed enclosure in twin tubes.   These are clad in
mirror glass which acts as a smoke separator.

The provision of permanent access systems to maintain the glass
envelope and service the high interior was an important consideration.
External access is achieved from ladder-type gantries at the 30° pitch
of the roof, traversing the length of the enclosure on rails supported
from the steelwork purlins.   Internal maintenance is combined with
the need to service lighting, clean internal glazing and to tend the
plants on the wall of the car park, which rises 20m above the upper
mall for a length of 86m.   The main system of access is from horizontal gantries which span between main trusses.   These traverse the
slope to a lateral transfer beam at high level.

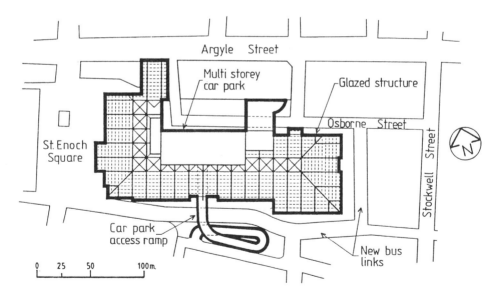

Fig.2. Plan of building on site

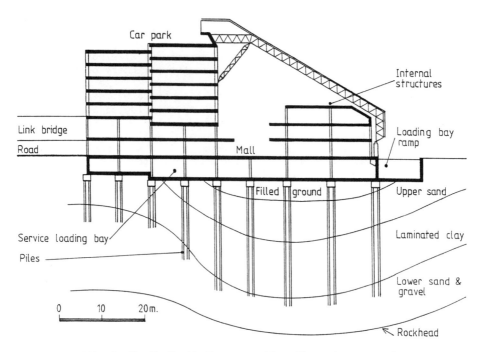

Fig.3. North-South Cross section through car park

The clear south-facing glazing ensures a light interior, but computer-controlled motorized solar shades are installed to prevent direct sunlight falling on populated areas of the shopping malls.

4 Site and substructure

Extensive site investigations comprising 27 boreholes, 38 cone penetrometer tests, 4 continuous 'Delft' soil samples and 8 trial pits revealed highly variable subsoil conditions. The upper layer, within 2-3m of the ground level, generally consists of fill, overlying upper sands 0-9m thick over laminated clay which occurs in thicknesses ranging from zero to 20m. Beneath the clay, lower sands and gravels occur above rockhead which is at a depth of approximately 35m. Shallow pad footings were considered but eventually discounted because of the variable thickness of the clays giving potential for large differential settlements resulting from the wide variety of loads imposed by the structure.

Discussions took place with piling contractors to establish which systems would be most suitable and economical for the site. Large bored piles to rock were eliminated on the grounds of cost and the uncertainty of being able to achieve a good end-bearing condition. Following further detailed design work, three piling firms were asked to tender for piles to carry the building loads to the sands and gravels either in friction or end-bearing. An important factor

influencing the choice of piles was the proximity of existing buildings, sewers and tunnels to certain parts of the development. It was
finally decided to construct two types of pile: driven cast-in-place
(Franki-type) up to a maximum length of 20m, with load being transferred to the sands and gravels in end-bearing; and continuous flight
auger piles up to 30m long with friction being the main load transfer
medium.   The cfa piles had the advantage of virtually vibrationless
installation and so were used close to the existing structures,
tunnels and sewers.

The in situ reinforced concrete basement is designed to resist
water to a head of approximately 2m, with the self-weight of the
basement slabs and the suspended podium above providing the necessary
balancing forces.   The structure is designed as watertight concrete
but, in the event of seepage, and as a protection against penetration
of dampness through the concrete structure, drained cavity construction is generally employed.   On top of the basement slab a 'no fines'
concrete layer beneath a damp-proof membrane and screed ensures that
no moisture reaches the 'habitable' environment.   One area of basement was protected by conventional tanking.

The podium slab supports the main mall and the shop units.   Cost
exercises indicated that in situ reinforced concrete was the most
economical solution and troughed construction with ribs at 1.8m
centres was adopted.   The heavy partition and live loads dictated the
adoption of a 125mm thick topping concrete and the subcontractor
elected to use polystyrene formers, usable two or three times.   Tree
pits, escalator and lift bases, fountains and water features at this
level introduce a high degree of complexity into the otherwise simple
structure.

5 Superstructure

The car park is of the split level type on 13 half-decks and was
planned to coincide with the 7.2m grid of shops below resulting in a
parking bay of 2.4m.   Discussions with Strathclyde Regional Council,
who operate the car park, led to outboard columns and a clear span of
16m.   The penalty for seeking the economy of inboard columns would
have been a bay size of 2.5m, which would have disrupted the vertical
planning.   In situ reinforced concrete, troughed slab construction
spanning 7.2m onto portal frames over the 16m span, has resulted in a
good open structure with 'slow up' and 'fast down' ramps at each end.

In the design of the structure the roof steelwork was by far the
most demanding element.   The combinations of load cases, including
temperature changes, are extensive, and the calculations were processed using plane frame analysis and three-dimensional analysis for
each combination.   To confirm wind loading criteria a 1:400 solid
block model of the building and its environs was constructed and
tested in a wind tunnel (Ref (i)).   Analysis of the readings from the
180 pressure tappings confirmed that interpretation of the British
Standard for wind loading was appropriate for the St Enoch glazed
enclosure.

The steelwork relies for lateral stability upon the car park and
the east and west core structures at the centre of the building.   1m

diameter reinforced concrete columns at the periphery provide vertical
support.   The triangulated main trusses, 1.8m deep, span 36m onto
'treehead' members which spring from the core structure and from
isolated columns at 14.4m centres.   The 'treehead' configuration
results from the decision to offset the support points from the
trusses.   The 18m high mall columns comprise 750mm dia steel tubes
(with wall thicknesses of 20 and 25mm).   The required 1½ hours fire
rating was obtained by wrapping the tubes with metal lathing and
coating with 'Pyrocrete'.   Otherwise the roof steelwork is not fire
protected.   Triangulated purlins 900mm deep span 14.4m between main
trusses and restraint to lateral and rotational forces is afforded to
main trusses by longitudinal lattice members.

    Pin joints are developed wherever possible to limit the extent of
site welding.   At external columns high shears and vertical forces
are carried through a steel 'top hat' column cap which is bolted and
grouted in place over the concrete.   Loads are transferred to the
column cap through a node 'can' which distributes forces to 75mm dia
pins, allowing rotations which arise from roof deflections.   Pin
joints are also used at connections between mall columns and tree head
members and at the intersection of tree heads with main trusses.

    Movement joints through the roof to control thermal stresses have
been provided in three positions to coincide with a main truss.   The
250m long roof structure thus comprises four structurally independent
sections.   By the use of twin trusses at each of these positions

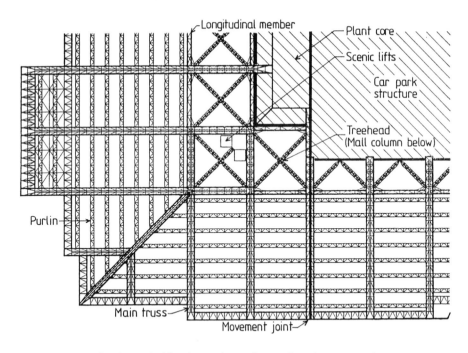

Fig.4. Detail plan of roof steelwork at south
west corner showing main elements

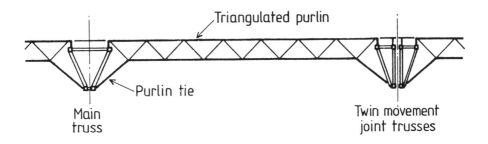

Cross section along roof

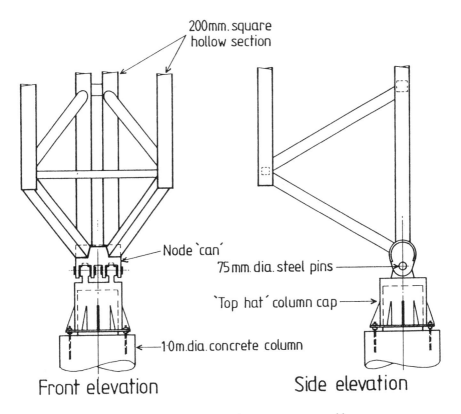

Main truss to column connection

Fig.5. Steelwork details

visual uniformity is maintained whilst allowing freedom of movement at the joint. Roof glazing is split across each joint with flexible flashings to accommodate up to 50mm lateral temperature movement.

A feature of the design of the enclosure steelwork has been the close co-ordination required with the glazing. The steel trusses were detailed on nodal geometry and accessory steelwork formed the link with the glazing geometry. The main benefit of separating these elements was to allow the main roof structure to be erected and depropped, thus enabling the internal structures to be built in parallel with the accessory steelwork and glazing overhead.

The decision to build these structures within the erected envelope was an important factor influencing the adoption of in situ reinforced

Fig.6. South elevation showing car park access ramp
and cable stayed bridge

concrete, ribbed slab construction (similar to the podium slab), since this minimized the access and handling problems.

A building of this scale and novelty has inevitably led to the involvement of many specialist consultants and the structural design subjected to close scrutiny by others. The clients' decision to obtain decennial insurance necessitated an external consultant checking the design and detail of the building fabric, while the local authorities have shown a very keen interest and Glasgow City Department of Building Control examined every aspect of the structure in detail.

6 External works

The external works include the long ramp of insitu reinforced concrete which gives access to the first level of the multi-storey car park at some 14m above ground level. Where the ramp enters the glazed enclosure the slab is supported by a cable stayed bridge. The towers are of tubular steel construction and form a striking visual gateway to the development for those arriving by car.

7 Enabling works

To ensure an effective start to the main works an advance works package was introduced. These works included new bus links, following closure of an existing road which intersected the site.

Along the line of the road, twin sewers had to be diverted and were replaced with a single 1.6m diameter concrete pipe. In the unusual circumstances of a major sewer being allowed beneath a new building, particularly with very limited cover available, it was necessary to provide large access chambers at either end of the diversion so that obstructions entering the sewer could be readily removed by the statutory authority. No contact could be permitted between the building structure and the sewer and the basement slab was carried from pile caps on either side of the new pipe with soft packing to prevent load transfer to the new sewer.

8 Construction

To achieve the completion of a project of this size and complexity, within a short timescale of three years it was necessary to adopt fast track construction methods. It was therefore decided to implement the project on a management contract basis. To gain the greatest benefit from this arrangement the management contractor was appointed early in 1985 at the design stage and became part of the design team advising on buildability and programming.

An important time-saving feature of the construction approach was to overlap the erection of the glazed envelope and the internal reinforced concrete structures. To this end the erection of the steel superstructure and glazing was programmed ahead of the internal structures which were constructed, to a slower pace, under cover.

Fig.7. Interior of completed building looking west
along mall

This called for careful sequencing of the work and created some access problems but the advantages in time and early weather protection were considerable.

Approval to start on site was given in September 1985 and two months later the advance works packages commenced. The start of the main contract followed in May 1986. Initially some difficulties were experienced when two preliminary cfa pile tests failed, but after investigation revised construction methods were introduced and a further seven trial piles were successfully load tested. This has been fully reported elsewhere (Ref ii). Subsequently the installation of the 1,500 piles was completed in five months. Prior to and during the piling operations inspections were made of the underground tunnel in St Enoch Square and the British Rail Tunnel in Argyle Street and vibrograph monitoring of the latter was carried out. Construction of the basement and podium slabs was straightforward but, to meet the demands of the programme, an average rate of $1,000m^3$ of concrete per week had to be achieved.

Fabrication of the 2,500 tonnes of steelwork was carried out in Manchester and Cambuslang, near Glasgow. Because of the complexity of many of the connections a great deal of skill and accuracy was required both in fabrication and erection. The temporary works required to support the steelwork during erection were major structures in their own right. Tall steel towers erected off the podium slabs (propped as necessary to take the additional loads) supported the roof truss sections until they were welded together to form complete trusses.

Steelwork erection commenced on programme in March 1987 and was completed in December 1987. Painting was further subcontracted and shop-applied. Because of the amount of additional work, constructed under subcontract packages, which followed the erected steelwork and took support from it, extensive corrosion protection touch-up and decorative finishing paintwork were necessary on site.

The subcontractor for the internal structures, constructed within the enclosure, elected to use track-mounted cranage, gaining access for materials through temporary openings in the glazing. At this stage of the design and, against the constantly-changing background of client needs, early decisions made as to the extent and configuration of piling were put to the test. By a combination of some minor replanning on occasion and re-routing loads, the clients' revised needs were met. There is flexibility in certain areas of the development for future extension and this will probably be exploited in the near future.

The fast track method of construction put all disciplines under severe pressure but by close co-operation the programme was achieved. In order to do so construction outputs of £3.5 million per month were reached and the centre was opened to the public in May 1989.

9 References

(i)   Everett, T.W. and Lawson, T.V. (1984) A Wind Tunnel Study of the
      Cladding Loads on the Proposed Development of St Enochs
      Square, Glasgow.   Dept of Aeronautical Engineering, Univer
      sity of Bristol.

(ii) Couldery, P.A.J. and Fleming, W.G.K. (1987) Continuous flight
      auger piling at St Enoch Square, Glasgow.   **Ground Engineer
      ing.**   September 1987, 17-28.

10 Credits

Clients: **Church Commissioners for England** and **Sears Property Glasgow
Ltd**
Development conceived and initiated by the **Scottish Development Agency**
Project manager: **Chesterton**
Architects: **Reiach and Hall + GMW Partnership**
Structural and civil engineers: **Ove Arup & Partners Scotland**
Services engineer: **Blyth & Blyth (M & E)**
Quantity surveyor: **Muirheads**
Management contractor: **Sir Robert McAlpine Management Contractors Ltd**

Principal structural subcontractors
Steelwork: **Redpath Engineering Ltd**
Reinforced concrete work: **Balfour Beatty**
Piling: **Cementation Piling & Foundations Ltd**

The Author wishes to record his thanks to the clients for their
permission to publish this paper and to his colleagues, Jack Carcas,
Cliff Kidd and Malcolm Somerville for their assistance in the preparation of the paper.

NEW VISITOR FACILITIES AT EDINBURGH CASTLE

T.H. DOUGLAS
James Williamson and Partners, Glasgow, Scotland, UK

Abstract
As part of a series of improvements to the facilities for visitors
to Edinburgh Castle, Phase 1 of the project includes a series of
unusual structures along the 200m long route of a single lane
vehicle and services tunnel. In addition, and within the environs
of the Castle, a new Toilet Block has been constructed near the
Guardhouse at the drawbridge and a new Gift Shop has been erected
in old buildings part of which have been used during the annual
Tattoo.

The paper outlines the various structures and concentrates on three
aspects. First the interesting but challenging need to work with
very sensitive historic buildings. Secondly, the special problems
which arise in the design and construction of the basement of the
4 storey Gift Shop. Finally, all the work was carried out in close
consultation with the Architect and detailing of a high order was
required.
Keywords: Edinburgh Castle; vehicle tunnel, unusual structures,
new Toilet Block and Gift Shop, historic buildings, proximity of
blasting.

1 Introduction

Edinburgh Castle occupies a commanding position in the centre of
the City and has provided a principally defensive role throughout
history. However, in recent times its military role has been
overtaken and it is now Scotland's premier tourist attraction
welcoming over one million visitors in 1989. The narrow defensive
entry ways, which are a legacy of historic times, conflict with the
increasing number of visitors who are on foot and the continued
need to service the Castle facilities by vehicle. In addition
there is a potential problem, as numbers increase, in providing
everyday access for fire engines and ambulances. (Ref. 2 and 3).

   During studies initiated by Historic Buildings and Monuments,
Scotland (HBM), to improve visitors facilities the need to cater

for vehicle access was considered an essential initial stage in the general development programme.

Phase 1 of the work includes the upgrading of various structures and the driving of a short tunnel through the Castle rock to provide the vehicle and services route into the inner facilities. The works comprise, also, new public toilets and a new Gift Shop.

To establish the route it has been necessary to reconstruct and widen the late-Victorian 3 span masonry arched bridge over the Dry Ditch, build two cut and cover structures under the pedestrian ways and to extend the cross section as a basement structure under the former shop at Mills Mount.

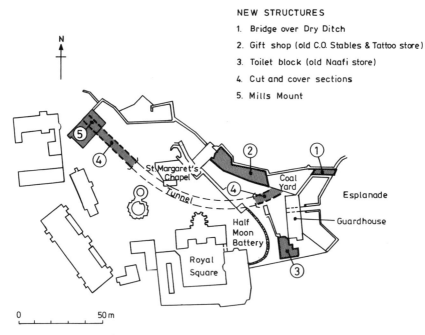

NEW STRUCTURES
1. Bridge over Dry Ditch
2. Gift shop (old C.O. Stables & Tattoo store)
3. Toilet block (old Naafi store)
4. Cut and cover sections
5. Mills Mount

Figure 1- Plan of Works

The overall development programme is expected to be carried out in 4 phases over a period of about 8 years. The Consultants for Phase 1 were appointed in the late Spring of 1988 and work on Site was initiated later that year. The main contract was originally planned to be undertaken during 1989 and work was largely completed by the end of the year. The cost of the works covered by this paper amount to approximately £2.5M.

2   Site Investigations

2.1 Geology
The rock of the Castle is formed from a basalt 'plug' intruded
sedimentary sandstones, shales and marls of Carboniferous series -
the basalt is predominately exposed with only isolated pockets of
superficial deposits but the sedimentary rocks are overlain by
glacial deposits and made ground.

2.2 Investigations
Preliminary investigations of the site were undertaken in 1987 and,
in view of the general exposure of the basalt and extensive records
held by HBM on the old buildings and their foundation, site work
was confined to 3 trial pits and 6 boreholes. These were directed
generally towards areas likely to have steeply sloping underlying
rock surfaces alongside the tunnel route. Only one borehole was
sunk below rockhead by rotary coring methods. During the final
design stage and following a shift in the proposed tunnel route, a
supplementary borehole investigation was carried out.

2.3 Interpretation of the Results
Data prepared and compiled in advance of the Contract included site
mapping, historical records, as well as the test pit, boring and
material property results. In addition, two special series of data
were made available. First, a series of archaeological digs were
organised by HBM in areas likely to be affected by the proposed
development. Geotechnical information was compiled also from these
zones. Secondly, due to requirements for blasting to remove rock
in open cut and, especially, during tunnelling operations in close
proximity to existing buildings, the results of blasting trials
were included. These had been carried out to determine the
limiting factors which it would be necessary to impose during the
Contract relating to noise and ground borne vibrations.

3   Tunnel Route

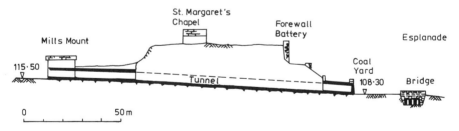

Figure 2 - Longitudinal Section along Tunnel Route

3.1 Bridge over Dry Ditch

The old bridge had been constructed as a 3-span masonry arch 3.5m
wide and 13m span and the structure had suffered some deterioration
and cracking, principally due to settlement of the glacial deposits
and made ground which rest on a steeply north sloping sedimentary
rockhead. At the feasibility study stage it had been thought that
the bridge width would be maintained and that it might be possible
to strengthen the structure with some limited foundation works
coupled with grouting and stitching the masonry. During the design
development stage it was ascertained that somewhat larger fire
engines than allocated earlier were coming into service with the
Lothian Fire Brigade and that these vehicles could not be expected
to use the narrow crossing. Furthermore the foundations of the old
bridge were found to be inadequate for the new and increased
loadings. It was decided, therefore, to reconstruct and widen the
entire bridge, using the existing stonework re-built in elevation
(see Fig. 3).

Figure 3 - Old Dry Ditch Bridge

A first proposal was prepared for new foundations to be formed
within the Ditch and the abutments to be cast against the old
walls. This solution was not only visually intrusive but the
foundations would have been expensive to found on the steeply
sloping sidelong rockhead. Instead minipies were drilled through
the masonry and overlying ground and grouted directly into the
rockhead below. The minipiles were debanded from the masonry by
greased sleeves. This enabled reinforced concrete bank seats to be
detailed to support a standard inverted T-beam Bridge Deck with the
whole structure designed to carry HA loading to BS5400.

106

3.2 Cut and Cover Sections

At each end of the tunnel, reinforced concrete box sections generally 4.75m wide x 4.5m high and 600mm thick have been formed to carry the tunnel route under the pedestrian route which was the only original route into the Castle (see Figs. 1 and 2).

The structure is designed similarly to emerge from the west portal of the tunnel and to cross over to the new basement construction at Mills Mount. To simplify shuttering the sections were designed as straight as possible with horizontal and vertical curvature taken out in the tunnel itself. 300 x 300mm fillets were formed to provide additional stiffening of the roof span and 25mm deep vertical ribs were inserted at 2.5m centres to cover shutter panel joints in otherwise plain concrete surfaces.

3.3 Mills Mount

It had been intended to continue the cut and cover box section through and under the existing shop, as a basement construction. The only difference being that the roof slab could be reduced to 250mm deep since it should only have to support floor loadings.

The historic and archeological operations, referred to in Section 6, unearthed an important storekeepers room with its fireplace, which was considered valuable to preserve. A redesign of the section was undertaken which incorporated a thinner base and walls - to ensure minimum excavations - and a large opening has been designed in the north east wall to surround the old fireplace. (See Fig. 4).

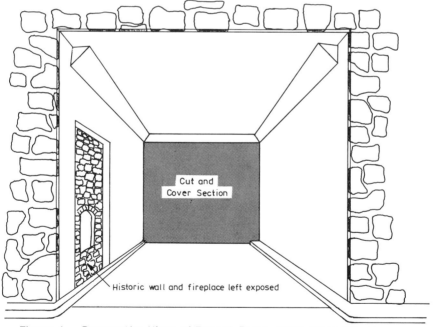

Cut and Cover Section

Historic wall and fireplace left exposed

Figure 4 - Perspective View of Tunnel Route under Mills Mount

4 Toilet Block

The area of the Castle to the left of the Guardhouse (see Fig. 1) had been used as an old NAFFI stove. New public toilets are an essential facility for the increasing number of visitors and a full range of uses was planned. Some use could be made of the existing structures but a new extension, incorporating old masonry gate pillars, was designed in the same style as that adopted for the new Gift Shop. This comprised twin circular steel hollow section columns supporting timber roof trusses providing a central ridge of glazing surrounded by a flat roof (see Fig. 5).

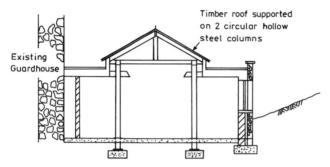

Figure 5 – Typical Section through Toilet Block

The basic design requirement was fairly straight forward but two special problems arose through the use of this general arrangement. First was the need to provide longitudinal stiffness and this was effected by the adoption of a pattern of cross tie rods bolted to the heads of the columns. The second problem arose due to using the shallow structural depth of the roof timbers to house all the main services. Considerable care, ingenuity and special co-ordination by the design team, the Contractor and his services operatives were required to achieve the service runs and cross-overs needed to supply the building.

5 Gift Shop

5.1 Conceptual Design
Limited use has been made in recent years of the area now comprising the Gift Shop, although the eastern portion had been designated for storage of equipment for the Tattoo while the western part had been used previously as C.O.'s stables.

Since it was intended to restrict delivery vehicle access to the Coal Yard (see Fig. 1) and two floors of shopping have been provided, the design development led to a four storey configuration

- the lower two storeys housing plant and storage space as well as providing access to the lift well situated against the south wall. (Figures 6 and 7). This was all built within what was originally a single storey building.

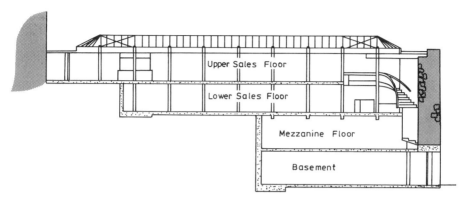

Figure 6 - Longitudinal Section Through Shop

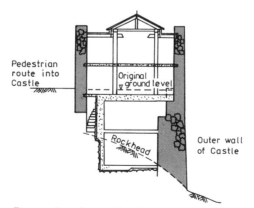

Figure 7 - Cross Section through Shop

5.2 Structural Form

An overriding constraint set by the Client was to restrict the overall height of the development to avoid altering the view of the Castle as seen from Princes Street. This constraint led to the adoption of reinforced concrete continuous flat slabs for the two shopping floors and the use of short spans in each direction. To restrict the span of the timber roof (similar to the form adopted in the Toilet Block – see Section 4 above) pairs of circular steel hollow section columns have been erected at 3m centres both ways. In the main shopping areas these are timber clad.

The design called for interconnection between the steel and concrete and this was achieved by adopting circular connecting/baseplates flush with the soffit of the upper floor (see Figs. 9 and 10).

Figure 9 – Gift Shop under Construction – Looking East

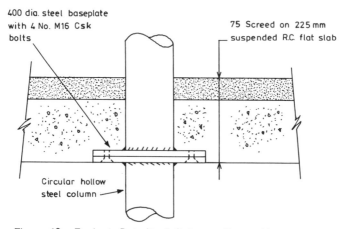

400 dia. steel baseplate with 4 No. M16 Csk bolts

75 Screed on 225 mm suspended R.C. flat slab

Circular hollow steel column

Figure 10 – Typical Detail of Column Connection

5.3 Basement Construction

From the site investigation data it had been expected that basement excavation could be effected using limited blasting of the bedrock and rock support measures such as bolting, dowelling and meshing of the exposed faces. In the event two problems exacerbated the situation. First the walls of the Castle were discovered to be relatively poorly founded and a considerable amount of underpinning was required before it was safe and practical to achieve basement formation level. Secondly, and more concerning, the form of the rock itself with markers of thin clay filled joints, gave rise to questions of temporary stabilty of the south (inner) wall and the main Castle Walls during excavation.

A number of options were investigated and eventually the Contractor proposed to form the outer wall of the Basement with mini piles. The Mezzanine floor would then form a strong brace between the walls and allow the basement excavation to proceed below while the remainder of the structure above could be built up from the floor. (See Figures 11 and 12).

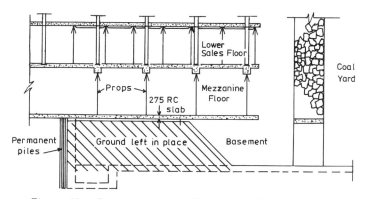

Figure 11 - Section showing Temporary Support

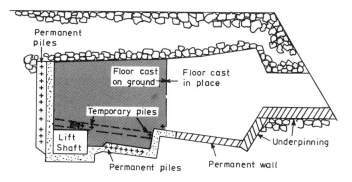

Figure 12 - Plan

111

Some concern was felt at the time and cost of the temporary works operation so that the final solution which was adopted restricted the piling to the deeper (western half) of the basement. The eastern half was excavated using the permanent access opening in the east wall and the basement walls and floor proceeded normally.

The western half of the Mezzanine floor was cast directly on the ground with the mini-piles in place. Construction proceeded immediately with the walls in the Mezzanine area and the lower sales reinforced concrete beam and slab floor was constructed with temporary propping off the basement floor. Erection of the structure of the two shopping floors proceeded concurrently with the careful removal of the basement excavations.

The complete exercise was carried out as a co-operative effort between the Engineer and the Contractor so as to incorporate as much as practical of the temporary works in the permanent walls. In the event, the roof of the shop was virtually complete before the basement had been fully excavated so that there was a considerable savings in overall programme time from that which would have arisen if the basement had been constructed first.

5.4 Special Features

It was considered essential by the Client to avoid cutting into, or damaging in any way, historic elements of the walls. Features which had to be protected included an old fireplace, the masonry cills of window openings, an historic stair and the access to the east walk. In order to avoid changes in the flat soffit of the slab, it was designed to span across window openings, and so forth.

Another special feature was the preferred use of curved floor openings and stairways. A spiral stair was used at the east end between the upper and lower sales floors, while curved stairs were used near the main door and in other areas readily seen by the public. Here again flat slabs were designed to avoid the introduction of beams or other changes in the soffit of the slabs.

6   History and Archaeology

As noted in Section 2.3, a series of archeological 'digs' were organised by HBM during the pre-contract investigation and preparatory work stage and these were directed specifically to clear the areas intended to be excavated under the Contract. Notwithstanding this well conceived and organised activity, a number of unexpected and interesting finds were encountered during the Contract.

In the area of the Coal Yard (see Figure 1), not only was a north-western extension to the dry ditch unearthed but also a civil-war cemetry was discovered immediately inside the ditch.

Through commendable and total day to day co-operation between the Contractor, the Resident Engineer the Castle Authorities and the Archaeologists, delays to the contract were minimised and these finds had virtually no effect on the Contract programme.

More extensive new discoveries were encountered underneath and just outside Mills Mount (see Figure 1). In the turning area at the end of the tunnel route, walls forming part of the 17th Century Western Defences were exposed. Under Mills Mount, a storekeepers room from Charles II's time was unearthed with a fireplace in one wall. The wall and fireplace are left exposed within the reinforced concrete box structure noted in Section 3 above.

A wealth of archaeological finds including substantial remains and artefacts from the Roman and Iron period 2000 years ago were uncovered in the fill material under the floor of Mills Mount and the work was carried out in close co-operation here again.

In spite of the co-operation between all parties some inevitable delays were experienced which has affected the Contract programme and cost. It is considered by HBM that this a small price to pay for the invaluable discoveries and for the greater knowledge of our history and our heritage.

7 Acknowledgements

The Client for the project is the Historic Buildings and Monuments, Scotland, and James Williamson & Partners are the lead consultants for Phase 1 of the development programme.

The Hurd Rolland Partnership are Project Architects for the Toilet Block and Gift Shop.

Hulley & Kirkwood are the Mechanical and Electrical engineering consultants for this phase of the work.

Lilley Construction Limited are the main Contractors for Phase 1 and the site investigation work was undertaken by Wimpey Laboratories Ltd and Ritchies Ltd.

The author would like to thank Historic Buildings and Monuments, Scotland, for permission to publish the paper and wishes to acknowledge the assistance and support received from all his colleagues and associates on the Project.

8 References

Douglas, T. H. and Keeble, S. (1990) Design and Construction of a New Service Tunnel at Edinburgh Castle. Proceedings Tunnel Construction '90 Conference, London.

Edinburgh Castle : Visitors Reception Feasibility Study Report (February, 1986)

Edinburgh Castle : Proposed Service and Utilities Tunnel Report (October 1987)

PART THREE
BRIDGE ENGINEERING

PROGRESS IN BRIDGE ENGINEERING

D.J. LEE
G. Maunsell and Partners, London, UK

Abstract
Progress in bridge engineering depends on the provision of a rational
and integrated procedure for dealing with empirical rules, working
stress methods, plastic theory, and techniques of probability,
statistics and even human psychology.
   Progress in bridge design has to take account of specified
loadings, changes in materials technology, structural forms,
workmanship quality and limit state design methods.
Keywords: Bridges, Codes, Limit States, Loading.

1 Introduction

Limit state design of bridges can be interpreted in a wide sense.
Design is an activity relating to finding an appropriate solution to
a given set of circumstances. Thus the use of relevant materials and
how they are handled in a construction project cannot be divorced
from a particular codification procedure. Conversely, a design
formula must be looked at in relation to the materials to which the
formula is applied and the Quality Assurance procedures, if any,
which are being used.
   In addition many bridge designers in the UK are having to devote
their time to inspection, repair and renovation of existing
structures as well as apply Codes of Practice to new projects. The
assessment of the strength and durability of existing bridges is an
important activity in the UK. Some of this work relates to ancient
or bridges more than a hundred years old, but much of the trouble
stems from work which is only decades old.

2 British Highway Bridge Loading

The historical development of design live load in Britain is
interesting. Before 1931, the normal practice was to design for the
effects of a standard loading train, although equivalent uniformly
distributed loads (constant for any span) had been suggested prior to
this date. The 1931 MoT Equivalent Loading Curve was the first
attempt to derive an equivalent uniformly

distributed loading which took account of the reduced probability of the maximum possible intensity of live load occuring on longer loaded lengths. However, no statistical analysis was carried out, the curve simply being judged between a determined value at the lower end and an assumed value at the upper end. An allowance for dynamic effects was included, varying from 50% at 23m to 15% at 122m and zero at 762m. These were arbitrary assumptions.

The 1954 loading was motivated by a recognition that the numbers, weights and nature of heavy vehicles had changed considerably since 1931. Also the normal lane width provided on bridges had increased from 10 feet to 12 feet. The basic intensity of live load determined for short spans (up to 23m) was therefore greater than the 1931 value, but this was accompanied by a reduction in the dynamic allowance from 50% to 25%, resulting in the same design load. For longer loaded lengths, the 1954 equivalent loading was derived from the affects of a train of vehicles consisting of 5 No. 22t vehicles each occupying 40 ft, followed and preceded by 4 No. 10t vehicles each occupying 35 ft, with further 5t vehicles occupying 35 ft, filling up the rest of the loaded length. The vehicle spacing is characteristic of freely moving traffic.

The 1954 equivalant loading curve, or one similar to it was used for the design of highway bridges in the UK until 1982. Since this loading considers the effect of only 5 vehicles over 10T and none over 22T in any lane for loaded lengths which nowadays may be in excess of 2000 metres, there was clearly a pressing need to reconsider the design loading under modern conditions such as increased weight and number of vehicles, motorway operation, contra-flow and urban conditions. The influence of contemporary bridge forms, widths and spans should also be taken into account.

The Interim Revised Loading Specification was introduced by the Department of Transport in 1982 to apply to the design of highway bridges in the UK with spans greater than 40m. For the first time in the UK, the design loading for long spans was intended to embody observations of actual vehicular traffic conditions, being based on the preliminary findings from such a study. One of these findings was that for all loaded lengths, the maximum traffic loading occured with stationary traffic. This criterion has also been observed in the U.S.A.

Studies of vehicular traffic on long span bridges resulted in the derivation of a new design loading curve for normal live load, based on a probabilistic analysis of the observed traffic behaviour using a mathmatical model. This was combined with a deterministic study of live load on short span bridges to produce a proposed loading for all loaded lengths.

The reduction of average live load intensity with carriageway width is just as significant as span or loaded length. The curves of maximum intensity of live load are applied in any one traffic lane. The 1954 design loading recognised that this was unlikely to occur simultaneously on a wide bridge, and therefore specified that the full intensity should be applied to any two lanes with 1/3 of that intensity applied to any remaining lanes. In the 1984 proposals not only has the maximum intensity of loading increased, but the factor applied to the remaining lanes was increased, generally to 0.6. However reduction in intensity on the second lane was permitted for longer loaded lengths on less than six lanes.

In his paper describing the derivation of the 1954 loading, Henderson recognised that the long span loading was rather speculative, and said it "should be looked on mainly as a guide". He then stated that quite large differences in live load are not of great significance, quoting as an example that for spans of 300 feet (91m) "the dead to live load ratio will be approximately 3 to 1."

This factor has become eroded over the years, particularly for steel bridges. As live load has increased, simultaneous developments in design have lead to lighter bridge decks.

3 Comparison of Highway Loadings

Contrary to what is frequently assumed, British HA loading is neither unduly heavy nor light for bending moments and shears. The lightest moments and shears for a single lane loading arise from such standards as AASHTO or the Austrian. The heaviest originate from Italy, West Germany and Holland. Most national standards have special vehicle loading to cater for heavy or abnormal loading - some of which are unduly complex.

It is difficult to compare one code with another because the variable action values have to be related to the resistance calculations. In simplest terms a heavy loading specification might be combatted by sophisticated strength calculations (Italian and German). Alternatively simple conservative analysis with lighter loading leads to a similar result. The essential factor therefore is the nature of the checks for serviceability limit states.

In general specifications are too variable and unnecessarily complicated. The range of characteristic values suggested in the F.I.P. Recommendations (Ref 1) envelop most national loadings.

4 CEB-FIP Recommendations

The CEB-FIP Model Code for Concrete Structures (MC 78) 1978 with a new edition due in 1990, represents over 30 years of work. The limit state design philosophy, partially at least, has been adopted by several national Codes. MC 78 is nevertheless too detailed for every day use and the FIP Commission on Practical Design prepared a document which placed MC 78 in a form suitable for general international use particularly for comparison between national codes (Ref 1).

So that there was a basis of comparison a simple range of characteristic values of variable actions was included. The FIP Practical Design Commission is also preparing for publication several bridge examples in accordance with the design recommendations as actually constructed . In this way the relationship of various national codes and loadings with the FIP proposals will be compared.

5 Designers' Choice

The division between Codes of Practice as guides to good practice and national standards which have the implications of mandatory specifications has unfortunately tended to become blurred in post war years. It can best perhaps be summarised as the difference between what one should do, in a code of practice, and what one shall do, in a mandatory specification. Although in the UK compliance with design codes is not strictly mandatory to meet statutory requirements, it is customary to use them: they are "deemed to satisfy" the requirements. This means that it is usually easier and quicker to obtain the necessary regulatory approvals. If the designer and his client has time it is usually possible to obtain approval for design to an unusual - or no - code. For buildings this flexibility is likely to disappear with the advent of European standards. With bridges it has not existed for many years.

6 Development of Concrete Codes

It is interesting to look at the evolution of the British reinforced concrete codes. The first, produced by the Department of Scientific and Industrial Research, was published in 1934. This was succeeded in 1948 by CP114, still a slim document. Both of these were true permissible stress codes.

1957 saw the emergence of a revised CP114. For most designers of reinforced concrete in UK it was the first sight of load factor design. It is instructive to see what the Code said about the method:

"(It) does not involve a knowledge of or use of the modular ratio and does not assume a linear relationship between the stress and strain in the concrete. It assumes instead that, as failure is approached, the compressive stresses will adjust themselves to give a total compression greater than that deduced from the elastic theory, the extent of this adjustment having been determined from test to destruction........."

"The resistance moments of beam and slab sections may be calculated to have a load factor generally of 1.8; in the calculations of the ultimate strength, however, the cube strength of the concrete should be taken as only 2-thirds of the actual cube strength........ It is necessary also to ensure that the stresses at working loads are not such as to cause excessive cracking........"

This goes some way towards introducing limit states. In the preface to the second edition of the 1957 Code warning was given that a completely "new look" version was to appear as the result of work being done following the publication of the recommendations of the European Concrete Committee (CEB). It was this work which gave rise to the so-called unified Code, CP110:1970, but it still retained the option of the elastic method.

This Code was now in three ring-binder volumes of A4. Compared with the slim documents which had been in use previously the new one was frightening - but not if we compare like with like. Two of the volumes were charts to make the designer's life easier. Volume 1 contained the meat of the Code - and more than this, embraced two other Codes, CP115 and CP116, respectively dealing with prestressed and precast concrete. From the first there were objections (though not normally directed to the parts dealing with prestressed and precast concrete). Although it had been intended that the building regulations would withdraw recognition to CP114, both Codes had "deemed to satisfy" status. Work was started in 1979 in the updating of CP110 and now available as BS8110.

Meanwhile work was proceeding in CEB: engineers in Europe, and in most of the other continents of the world, collaborated to produce invaluable bulletins containing the results of wide-ranging research, development and design investigations. This work was to become of crucial importance in the Eurocode Strategy.

7 Significance of Limit State Approach

In the past 40 years the approach to structural design has changed. This change has affected codes. Economic pressure to make the greatest use of the inherent strengths of materials and the development of plastic theory have led to an improved understanding of ultimate load-carrying capacity. New or improved materials and better statistical control of quality have allowed the attainment of higher stresses under working conditions; as a result of this it has become more important to check that the serviceability limit states (such as excessive deformation, cracking or vibration) have an acceptable probability of not being reached in service. To use an expression from statistics, the "confidence level" has risen.

The changes in method can increase the work of the designer by dint of more complex formuli for resistence, the need to consider modes of failure previously precluded by the use of lower stresses and restrictions on proportions, and by the need for explicit consideration of serviceability limits. In other respects, such as in the calculation of the design load effects in a continuous beam, plastic theory has simplified analysis.

There are numerous instances in which the elastic methods with allowable working stresses provide safe lower bounds. Within codes such as BS 8110 such methods are commonly incorporated for use within certain defined limitations. Some codes such as that for timber structures, apply to materials with little capacity for stress re-distribution and for these the elastic approach is retained - although it must be said that there are those who believe that limit states are applicable to all forms of construction. The arguments continue over whether it is logical to use limit states in foundation design.

The partial safety factor format enables the designer to adopt margins which allow rationally for the degrees of uncertainty in

the assumptions of loadings and permit readily the calculation of
combinations of load effects, for whatever limit state. The formats
adopted, although simpler than those necessary to produce uniform
liability in service, lead to a far better consistency in safety than
is achievable with permissible stress methods. In particular they
safeguard against failure and overload of structures where stresses
are not linearly related to loads such as the Ferrybridge cooling
towers. Those who favour permissible stress methods claim that they
are intrinsically simpler to use for those structures which might be
regarded as conventional and repetitive in nature. They might well
apply to most of the structure in a normal reinforced concrete
building.

The use of the new design methods has been shown to offer a
significant overall saving in the cost of material of construction.
The new codes include formuli which facilitate the use of computers
and design aids to alleviate the apparent complication of the new
procedures. Much of the complexity of the codes has resulted from
the demand for their more comprehensive application to diverse
structural types, methods of construction and materials. However,
preparation of simpler rules for everyday structures from bulky codes
is perfectly possible.

The changes in design approach outlined above are being adopted
world-wide. Their current adoption in the drafting of Eurocodes
which are intended eventually to supersede the UK National Codes is
of particular significance.

It should not be forgotten that the advent of a new material or
method has been the signal for a number of failures. It is the
awareness of this fact that has thankfully instilled a cautious
attitude.

A common, and often reasonable criticism of limit state
procedures is that use is made of the techniques of statistics which
are not justified either from the knowledge currently available on
strength distribution, on loading or on and structural behaviour.
Proposals to develop codes using statistical procedures on even
higher levels have met with a decidedly cool response in the UK but
there are types of structure where the probabilistic approach is
appropriate.

8 Fire Resistance and Durability

It is now usual in the early stages of design of normal reinforced
concrete building structures to determine the fire resistance
required and to choose a desired durability. Here the fire
resistence is as established in the Building Regulations for
buildings of various types of occupancy, and durability is in terms
of resistence to given exposure conditions. Only when these
parameters have been considered it is possible to decide such matters
as the member sizes and appropriate cover to the reinforcement.
These aspects are not considered for bridge design despite the
regular occurence of fire damage to bridges.

9 Cement and Concrete Technology

A further consideration is the changes in the properties of cement
that have taken place since earlier codes of practice were drafted.
Nowadays a higher early strength is achieved almost to the point of a
flash set in some cases; hence a modern code of practice may
recommend a higher concrete strength than necessary structurally, but
which is part of the concrete properties relating to durability.
    The selection of a concrete mix is now something of a juggling
act. Some of their properties were not as well understood as they
were thought to be, but the engineer had the confidence that came
with familiarity - and perhaps ignorance. He could assume that the
strength of concrete would continue to rise indefinitely: the
twenty-eight day strength was 50% above that at seven days and a
further 25% at a year - with the probability that there was still
more to come! Things are much more complicated now. It is no longer
reasonable to assume that a twenty eight day strength will give a fat
safety margin. On the contrary, BS 8110 gives no age allowance
unless there is evidence to justify a higher strength for a
particular structure. Especially of relevance to water-retaining and
- excluding structures, if the concrete cracks it may no longer heal
itself, since all the hydration would have taken place. Again,
calcium chloride and high aluminous cement have bitten the dust - or
contributed to it. Alkali-silica reaction and carbonation have
reared their ugly heads (although the suggestion has been made by at
least one eminent engineer that the one may counteract the other).
    The availability and use of pulvarised fuel ash (pfa), ground
granulated blastfurnace slag (ggbfs), silica fume and
super-plasticisers present engineers with additional, if confusing
choices.
    In the face of all these complications a welcome simplifying
assumption is that for a given kind of mix, the stronger the
concrete, the more durable. It is this assumption that has made it
possible, in BS 8110, to tabulate durability with strength and
water/cement ratio. It cannot be claimed that it has eradicated all
controversy: there are numerous, and vocal, objections, some but not
all of them mutually contradictory. Fruitful topics of argument are
myriad and include the need to limit the cement concrete to reduce
cracking, the means of ensuring an adequate concrete cover to
reinforcement and the equivalent cementitious value of additives.
Champions of pfa have been heard to claim that it is the cement which
should be regarded as the additive! Bearing in mind what is
happening in some countries in Europe, where cement may contain 25%
of anything, perhaps the claim is not ridiculous.

10 Quality Assurance

There have been significant advances in the productivity and
reliability of high technology process engineering in the last 20
years or so which have been attributed in no small measure to the
adoption of a formal "systems" approach to operational and production
management which has become known as Quality Assurance (QA). It

should be noted that the definitions below, taken from the relevant
British Standard (Ref 2) and the I.S.O. Standard on which it is
directly based, apply:-

quality management: That aspect of the overall management
function that determines and implements the quality policy.
quality system: The organizational structure, responsibilities,
procedures, processes and resources for implementating quality
management.
quality control: The operational techniques and activities that
are used to fulfil requirements for quality.
quality assurance: All those planned and systematic actions
necessary to provide adequate confidence that a product or
service will satisfy given requirements for quality.

The success of QA in the top echelons of manufacturing industry,
particularly those companies involved with the defence procurement
programme, led to a campaign by the UK Central Government
administration to promote wider adoption of QA throughout all sectors
of industry. It has also been suggested that a recognised QA scheme
can be useful to combat protectionism in exporting to certain
states. This was certainly the case with at least one major sector
of the construction materials industry. Some professional quality
specialists see BS 5750 as relevant to construction even without
adjustment, but it is quite obvious that BS 5750 was primarily
drafted around the requirements of manufacturing industry. For this
reason early in 1985 a group of consultants consisting of G Maunsell
& Partners and three other firms developed a coordinated quality
system with BSI with the intention of launching it as a sector scheme
approved by the National Accreditation Council to provide the
framework for a working system.
    The group therefore participated in the production, with BSI, of
an interpretative document called "The Quality Assessment Schedule"
(QAS) which stands alongside BS 5750 to describe the fundamental
requirements of the system for civil and structural engineering
design. This has been adopted by BSI and issued as one of their
official publications (Ref 3).
    In the years to come the place of quality assurance in the
design and construction of bridges will in turn influence codes of
practice.

11 Bridge Inspection, Assessmemt and Repair

The traditional bridge form in Britain before the industrial
revolution was been the brick and masonry arch. It is interesting
that current research in Britain is addressing the analysis of such
structures as there are quite a few things that are not fully
understood. As an arch bridge relies on its springings and
abutments, spandrel walls and filling materials, all of which may
vary, it is apparent that the codification of design of new bridges
or old ones of this form is not the subject of an established Code of
Practice. The British Department of Transport method of assessing

the strength of existing arches is subject to current review and research and doubtless will be modified in the light of research and experience.

More generally bridge inspection and assessment is a major specialist task. Routine inspections can be superficial owing to the lack of skilled men and resources at the right time.

12   Britain and Europe

It is necessary to refer now to Britain's participation in the European Community. From a technological standpoint, at least, it is logical that the best information available in all the member countries should be brought together into one pool. Whether the way chosen to do this is the best remains to be seen. It will not do to pretend that the technical advantages are all there is to it; the political context is extremely important.

It is intended that by the end of 1991 the "internal market" will be complete. This is when there will be free access throughout the European Community of products and services. A Construction Products Directive sets out the rules. In implementation there will be European Standards which will cover every aspect of construction.

In the first instance national standards and codes will continue to be acceptable in parallel with European Standards (or ISO, International Standardisation Organisation Standards where they do not exist). Eventually only European Standards will be acceptable - but it will take many years before the other documents are superseded.

A number of British Standards have been granted the status of European Standards, or Euronorms, and given dual numbering. In some instances British Standards have been drafted with an eye to becoming Euronorms; alternatively, ISO Standards to whose drafting Britain has contributed may also carry both British Standard and Euronorm numbers.

13   Eurocodes

More complicated is the series of Eurocodes, being drafted by engineers of various European Nationality. The list of Eurocodes is as follows:-

    EC 1   General Principles
    EC 2   Concrete
    EC 3   Steel
    EC 4   Composite
           Steel/Concrete
    EC 5   Timber
    EC 6   Masonry
    EC 7   Foundations
    EC 8   Seismic
    EC 9   Actions/Loadings

Wherever possible the drafting Committees have made use of work
done in appropriate international organisations: EC 1 uses much of
the work of ISO together with that of other International bodies; EC
2 is very closely based upon the work of CEB; EC 3 on that of
European Commission for Structural Steelwork and so on. Much of the
work on the first three Eurocodes has been done although if a
National Code is a compromise than an International Code must be a
second order compromise. The draft addresses building design and
bridge design will follow later. The strategy of Eurocodes requires
that national standards work is limited to a care and maintenance
basis, that any new work is done in the European Standards framework
and that the European Standards should accommodate UK requirements -
and in addition structured for ready application.

Any pessimism as to the early realisation of the Eurocode
strategy may be taken as justification for the argument over just one
word: Actions, eventually to be used as the title for EC 9. (The
word is intended to mean a wide range of manifestations in different
units, including forces, restraints, deformations, shrinkage,
moisture and thermal movements, etc.) In a footnote to one draft of
EC 2 there is the statement that the term 'effects of the actions'
covers a wider range of effects than the term 'action effects', which
only refers to bending moments, torsional moments, longitudinal
forces, and tangential forces.

There are numerous other examples in the Eurocodes of familiar
words which have assumed unfamiliar guises. To be fair it is not a
one-way traffic since quite a few English words have been accepted to
which continental purists would take exception.

14 Practical Use of BS 5400

BS 5400 (Ref.4) represents a considerable achievement in the field of
limit state codes of practice and its use is necessary in the UK when
designing bridges for the Department of Transport and for other
Client Authorities. As the author was not concerned in its drafting
directly other than with Part 9 Bridge Bearings it is perhaps
appropriate to emphasise the advances it has made in bringing limit
state thinking into a bridge code.

15 BS 5400 for Steel Bridge Design

BS5400: Part 3 deals with steel structures and much of the research
work carried out in the last twenty years has been incorporated
particularly the Merison Committee work on steel plate panels and box
girders. Such aspects mean that the code is a complex one and
generally a great deal of cross referencing is required from one
clause to another. One part of the code is not always consistent
with another part. An example is the use of the design factor which
in Part 3 is included in the resistance side of the calculation
whereas in Part 4 it is included as a load factor.

Part 3 in particular provides basic formulae for the various
tables and graphs used in design which in turn promotes the

development of computer software for number crunching. Maunsell has
developed a programme caused STRIDE which has considerably assisted
the design process since it frees the engineer and gives him time to
develop an optimum design. The corollary is that for steel design the
document enables an engineer to carry out the necessary checks for
each element of a proposed or existing structure but gives little
guidance as to the initial sizing of members or layout of stiffeners
etc. There are no short cut rules to obtain optimum panel sizes such
as for webs. For this reason the design process can be quite lengthy
until the engineer becomes suitably experienced.

For the design of steelwork in buildings the BS 449 is now
joined by a limit state code BS 5950. This complex document will
require further consideration and amendment in due course. The same
comments can be applied to BS 5400 Part 3 for the design of steel
bridges but happily to a lesser extent. Bridge engineers are more
attuned to complex calculations in recent years.

It is interesting to note that Eurocode 3 for steel structures
deals with building type structures and there is no immediate
intention to enlarge the document to deal with bridges to the
sophistication of BS 5400. Nevertheless in the future both bridges
and offshore structures will be brought into the Eurocode net.

16 BS 5400 for Concrete Bridge Design

Before the issue of BS 5400 concrete bridges were designed similar to
the building codes and as modified by memoranda issued by the
Government Department. Like the concrete code for buildings BS 8100,
BS 5400 does not facilitate the design of simple structures and the
relevant clauses are dotted around. An important aspect is that the
design is limit state and whilst much that may be made of the change
to ultimate limit state format this is not necessarily the most
drastic change since much of the bridge design is still governed by
serviceability criteria using elastic analysis. For example global
analysis of structures is still performed elastically even if
ultimate section design is used. For bending of reinforced concrete
the serviceability criteria of crack control governs not ultimate
strength.

Prestressed concrete sections in bending are still designed
elastically for the Class 1 and Class 2 criteria. The application of
partial prestressing to Class 3 is not yet permitted.

Special details such as half joints, deep beams, end blocks and
the like require extended analysis. In the circumstances either
elastic or ultimate design is employed depending on the results of
research.

17 Improvement in Design Information

Over the last twenty years there has been a dramatic improvement in
the quality and quantity of advice readily available to bridge
engineers including wind and temperature loadings, design for shear
in both prestressed and reinforced concrete, design of partially

prestressed concrete and design of end blocks and deep beams. Similarly box girders have been extensively reported together with the estimation of crack and shrinkage effects. With the close relationship of concrete bridge design to the methods of erection such as incremental launching, balanced cantilevering and precast segmental construction it is difficult to visualise such progress having been made so comprehensively without the use of computers.

As the Eurocodes loom over the horizon the British Standards Institute and all the other bodies involved with Codes of Practice will be dedicating more and more effort into the European dimension, consequently work on British Standards will inevitably take a lower priority. Although BS 5400 as a bridge code will be generally improved it is not the current intention to do a drastic rewriting. On the other hand the Eurocodes will concentrate at least in the initial period on dealing with building structures and it is therefore likely that BS 5400 will have some considerable further life in it. Perhaps much of it will be incorporated in the future Eurocode for bridge construction.

18 References

Practical Design of Reinforced and Prestressed Concrete Structures
    FIP Recommendations, Thomas Telford Ltd, London 1984.

British Standards Institution BS 5750 : 1987 Quality Systems Part
    1, Specification for design/development, production,
    installation and servicing. (ISO 9001)

QAS 8370/243 : Quality Assessment Schedule to BS 5750 Part 1,
    relating to civil and/or structural engineering led
    multi-disciplinary engineering project design.

BS 5400: Steel, concrete and composite bridges.

    Part 1: 1978. General Statement

    Part 2:   1978. Specification for Loads

    Part 3:   1982 Code of practice for design of steel bridges

    Part 4:

1984. Code of practice for design of concrete bridges

    Part 5:   1979. Code of practice for design of composite
                bridges
    Part 6:   1980. Specification for materials and workmanship
                steel

    Part 7:   1978. Specification for materials and workmanship,
                concrete, reinforcement and prestressing tendons

Part 8:   1978. Recommendations for materials and workmanship, concrete, reinforcement and prestressing tendons

Part 9:   Bridge Bearings:
Section 9.1 1983. Code of practice for design of bridge bearing
Section 9.2 1983. Specification for material, manufacture and installation of bridge bearing.

Part 10:   1980.Code of practice for fatigue

THE FIXED TRAFFIC LINK ACROSS THE GREAT BELT IN DENMARK

N.J. GIMSING
Professor, Dept Structural Engineering, Technical University of
Denmark, Lyngby, Denmark
(Technical Adviser for the East Bridge of the Great Belt Link)

Abstract
The paper presents the project for a fixed traffic link across the
Great Belt in Denmark, consisting of a bored railway tunnel, a multi-
span concrete box girder bridge for road and railway, and a major
suspension bridge for road. The link is at present under construction
and will be completed in 1996.
Keywords: Bored Tunnel, Box Girder Bridge, Cable-Stayed Bridge, Suspension Bridge, Ship Collision.

1 Introduction

After having seriously discussed a fixed traffic link across the Great
Belt in Denmark for more than 5o years, the project has finally
reached the construction stage.

When completed in 1996 the Great Belt Link will comprise:

- the second longest underwater railway tunnel in Europe, as it will
  be surpassed only by the Channel Tunnel, which is under construction in the same period as the Great Belt Link.

- the longest road bridge in Europe, as the East Bridge will be
  approximately 10% longer than the Oland Bridge in Sweden.

- the longest combined road and railway bridge in Europe, as the
  West Bridge will be approximately twice as long as the present
  record holder, the Storstrøm Bridge in Denmark

The total cost of the link is estimated at approximately 18.5
billion Danish Crowns (DDK) corresponding to 1.7 billion Pounds (GBP).

2 Geographical Conditions

Denmark consists of the peninsula Jutland, forming a part of the
European continent, and a considerable number of islands among which
Zealand (with the capital of Copenhagen) and Funen are the largest
(Figure 1).

Figure 1. The location of the major Danish bridges.

The population of the country is split among the different parts with approximately 45% in Jutland, 10% on Funen, and 45% on Zealand.

Due to the country's location on the European map, the Danish highways and railways have to carry not only the domestic traffic between the national provinces, but also to form an important part of the international traffic routes from Sweden, Norway and Finland to the European continent.

The decision to start construction of fixed links across the many straits separating the different parts of Denmark was taken already in the 1920.es, and in 1935 the first Little Belt Bridge (Figure 2) joined Jutland and Funen. Two years later, the Storstrøm Bridge south of Zealand was added, so that the country at the end of the 1930.es was united in two traffical units separated only by the Great Belt.

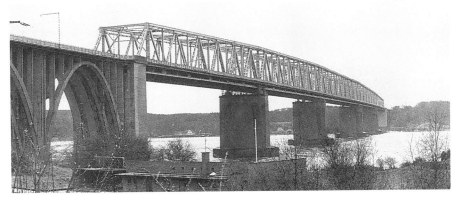

Figure 2. The first Little Belt Bridge from 1935.

The main reason for not adding the final stone, a bridge or tunnel across the Great Belt, in direct continuation of the other prewar bridges was that the task of crossing this strait is of a quite different magnitude. Thus, the widths of the Little Belt is less than 1km, the Storstrøm approximately 3.5km, but the Great Belt 18km.

As the main ferry routes across the Great Belt operate between the towns of Korsør on Zealand and Nyborg on Funen, it was natural also to consider this location for the alignment of the fixed link. Here, the Great Belt is divided into two channels, the East Channel and the West Channel, separated by the small island of Sprogø so that the fixed link could be composed of two separate bridges or tunnels.

The two channels are of almost equal widths, approximately 8km, but with a quite different bottom profile (Figure 3). Thus, the East Channel has shallow water depths of less than 10m extending almost 3km out from each shore, whereas the central deep water channel reaches a water depth of 55m. In the West Channel the water depth is more constant and between 18m and 28m over a width of almost 5km.

GREAT BELT

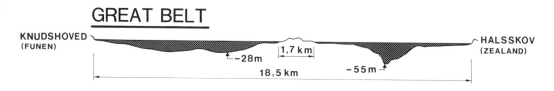

Figure 3. Cross section of the Great Belt at the location of the fixed link.

Also, the two channels differ by their importance for navigation, as the international navigation route from the Baltic Sea to the North Sea runs through the East Channel, whereas the West Channel is used only by local ship traffic to ports on Funen and the smaller islands to the south.

3 The 1978 design for a fixed link

The first realistic designs for a bridge across the Great Belt were worked out already in 1936, but due to the outbreak of the Second World War in 1939 all plans of starting actual construction had to be postponed to more peaceful times.

After the war, renewed investigations into the possibilities of constructing a bridge across the Great Belt were initiated in 1948, and during the following 25 years a large variety of technical solutions were studied by a number of committees and working groups.

Then, in 1973, the long investigation period came to a preliminary end when the Danish Parliament approved the first Act on the construction of a fixed traffic link across the Great Belt.

In this Act it was stated that the link should be constructed as a pure bridge link with bridges across both the East and the West Channels.

The actual design of the Great Belt Bridge according to the Act of 1973 was carried out in the period from 1976 to 1978, and it resulted in a choice of two solutions for the large spans across the navigation channel (in the East Channel), either a cable stayed bridge with a main span of 780m or a suspension bridge with a main span of 1416m (Figure 4).

Figure 4. Suspension bridge across the East Channel
according to the design from 1978.

Across the West Channel a low level bridge with a vertical clearance of maximum 14m was planned to be built.

In the summer of 1978 the bridge authority, named Statsbroen Store Bælt, was ready to sign the first construction contract (comprising preparation of work site areas on Zealand), but before this was actually accomplished a newly formed coalition government decided to postpone the construction of the link for another 5 years.

At the end of the postponement period, in 1983, it was decided to consider once more other technical solutions than the combined road and railway bridge. Thus, in the following 3 years a large number of combinations of bridges, immersed tunnels and bored tunnels were evaluated, not only from a constructional and traffical viewpoint but also to a large extent by considering the environmental aspects.

4 Formation of the Great Belt Link Ltd.

In parallel with the technical investigations, political discussions started in order to clarify whether it would be possible to find a technical and financial solution that could attract a majority in the Parliament. Immediately, this did not seem very easy to accomplish, as the Government in power wanted to start construction of the combined road and railway bridge (according to the 1973 Act), whereas the largest opposition party, the Socialdemocrats, would support only the construction of a pure railway link.

However, in the summer of 1986 a political compromise was reached by the two political groups, and the solution was to construct the link in two stages so that the railway link would be opened first, but be followed by the road link after a period of less than 4 years.

With the requirement that the railway link should come first, it became indispensable to let the train cross the East Channel in a tunnel, either immersed or bored, whereas the crossing of the West Channel requiring only moderate navigation openings most favourably could be accomplished by the construction of a low level bridge, initially opened to train traffic only, but later extended to also allowing a passage of vehicular traffic. Thus, the total link would consist of three major crossings: The East Railway Tunnel, the East Road Bridge and the West Road and Railway Bridge (Figure 5).

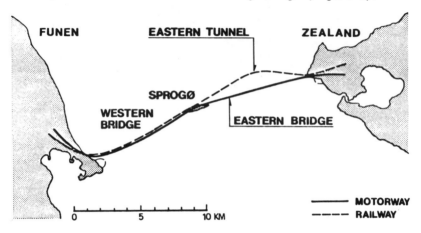

Figure 5. Bridges and tunnel of the Great Belt Link
(courtesy of K.H.Ostenfeld)

For the roadway link across the East Channel two technical solutions seemed possible, either the construction of a high level bridge with a large navigation span or an immersed tunnel.

Based on the political compromise from 1986, the Danish Parliament approved in 1987 the Second Act on the construction of a fixed traffic link across the Great Belt.

In this Act it was specified that a limited company with the Danish state as the only shareholder should be formed to coordinate the design, construction and operation of the Great Belt Link. The financing should only to a very limited extent be based on a direct funding by the state but mainly on domestic and foreign loans. However, it was stated that these loans would be guaranteed by the Danish state.

The formation of the Great Belt Link Ltd., or just STOREBÆLT as it is generally called, was initiated in the spring of 1987, and in the following months conceptual designs for the following structures were prepared by the consulting engineers (COWIconsult):

Railway tunnel under the East Channel constructed as a bored tunnel, an immersed steel tunnel or an immersed concrete tunnel.

Combined road and railway bridge across the West Channel based on either a single deck steel structure, a double deck composite structure or a single deck concrete structure.

Road bridge across the East Channel with either a cable stayed or a suspension span across the navigation channel.

Road tunnel under the East Channel to be constructed as an immersed steel tube or an immersed concrete tube.

Based on these conceptual designs, the master plan for the construction of the total fixed link was worked out (Figure 6). According to this plan the railway link shall be opened to traffic in the spring of 1993 and the road link in the summer of 1996.

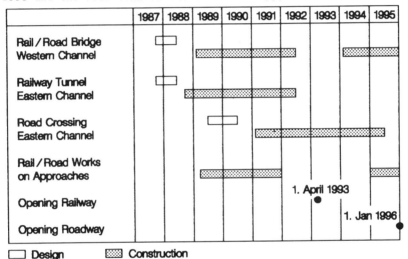

Figure 6. Master plan for the construction of the Great Belt Link
(courtesy of K.H.Ostenfeld)

Immediately after completion of the conceptual designs the work was started on preparing the tender designs for the East Railway Tunnel and the West Bridge, whereas the tender designs for the East Road Link were planned to be worked out two years later due to the later completion date of this part of the link.

5 East Railway Tunnel

For the East Railway Tunnel three tender designs were prepared, as the conceptual design phase had shown that the difference in cost between the three basic solutions was so small that none of them could be ruled out immediately.

This assumption was to a large extent verified when the bids were received on the different options in June 1988, as it is seen from the following list of construction costs offered:

Bored tunnel   3.2 - 4.3 billion DKK
                          (approx. 290 - 390 million GBP)

Immersed concrete tunnel   3.8 - 4.1 billion DKK
                          (approx. 350 - 370 million GBP)

Immersed steel tunnel   3.7 - 4.0 billion DKK
                          (approx. 340 - 360 million GBP)

It appears that there was a considerable overlap between the bids on the different tunnel types, although the immediate indication was that the cheapest solution was to be found for a bored tunnel. However, the fact that the bids generally were accompanied by a number of reservations and that STOREBÆLT had to carry a larger risk if choosing a bored tunnel with its inherent geological uncertainty implied that the bids had to be carefully investigated and evaluated.

In October 1988 STOREBÆLT was ready to start the contract negotiation with the consortium that had presented the most attractive bid, and on November 28, 1988 the contract to construct a bored railway tunnel under the East Channel was awarded to the MT Group, a consortium comprising the Danish contractor Monberg & Thorsen and three foreign contractors (German, French and American).

The double track railway tunnel will consist of two single track tubes each with an internal diameter of 7.7m and interconnected by cross passages at 250m distance (Figure 7). The tunnel lining will generally be made of 400mm thick precast concrete segments bolted together. Only at the locations of the cross passages the lining is going to be made of cast iron. All tunnel lining is to be waterproof for the full hydrostatic pressure of up to 8 bars.

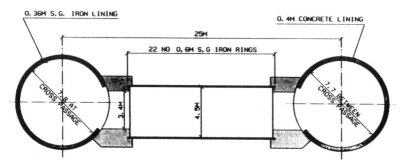

Figure 7. Cross section of the East Railway Tunnel
(courtesy of K.H.Ostenfeld)

The total tunnel length will be approximately 8km of which 7.3km will be constructed by boring, whereas the remaining lengths will be made by the cut and cover method.

The four full face tunnel boring machines (TBMs) are designed
to work in the upper till layers with large granitic boulders as well
as in the lower marl layers, which are expected to be highly permeable
in a number of fissured regions. It is therefore necessary to incorporate into the TBMs an earth pressure balanced shield.

In the beginning of 1989 the consortium began to construct the
tunnel ramps on Zealand (Figure 8) and off the coast of Sprogø, and
at the same time the fabrication of the tunnel boring machines was
initiated.

Figure 8. Aerial photo of the tunnel ramp on Zealand.

At the end of 1989 the fabrication of the tunnel lining elements
began at a special plant erected on the work site area on the Zealand
side.

The TBMs are to be delivered at the site in the early spring of
1990, and immediately afterwards the tunnel boring process will start
(Figure 9).

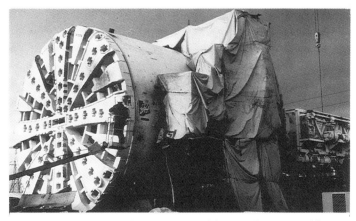

Figure 9. The tunnel boring machine under assembly
at James Howden Co., Ltd. in Glasgow.

6 West Bridge

The preparation of the tender designs for the West Bridge was started in the autumn of 1987, shortly after completion of the conceptual designs for this bridge.

In the conceptual design phase the following three proposals had been investigated (Figure 10):

CONCRETE GIRDERS

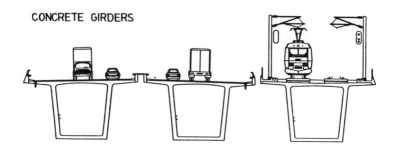

STEEL GIRDER

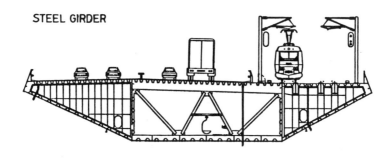

COMPOSITE GIRDER

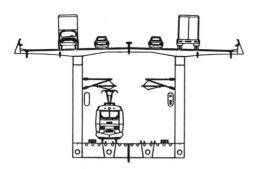

Figure 10. Cross sections of the three bridges included in the tender designs for the West Bridge (courtesy of K.H.Ostenfeld).

- a composite proposal comprising a double deck structure with an upper concrete slab under the roadway, two vertical steel trusses, and a lower railway deck in steel.

- a concrete proposal comprising three individual box girders one carrying the two railway tracks, and each of the other two a carriageway for the motorway.

- a steel proposal comprising one 37m wide box girder with the railway tracks on one side and the motorway on the other.

The investigations had shown that the optimum span for the three options differed substantially, and it was therefore decided to base the composite design on a span length of 144m, the concrete design on a span length of 105m, and the steel design on a span length of 120m. In all cases, the span lengths were maintained along the total bridge length as no special navigation spans were to be provided.

The preliminary cost estimates from the conceptual design phase indicated that the composite proposal and the concrete proposal would be very close with regards to construction cost, whereas the steel proposal appeared to be somewhat more expensive.

Consequently, at the start of the tender design phase it had to be decided whether all three options or only two should be further elaborated to the level of tender designs. The final decision was to continue with all three options, as the total cost of the steel option was very sensitive to the cost of fabricating the steel boxes. Therefore, the cost could be noticeably reduced if a major ship yard decided to undertake this production (as it had been the case a few years earlier at the construction of the Farø Bridges, which were based on a similar design).

For the substructures two different designs were investigated during the conceptual design phase, either a concrete caisson with open cells or a concrete caisson with a solid slab at the bottom. In cost these two options seemed to be very close, but as it was felt that the open caisson to a larger degree would assure a uniform contact with the soil, this option was chosen for the tender design.

The total bridge length varied slightly (between 6048m and 6120m) with the superstructure design due to the difference in span length between the three options.

Irrespectively of the superstructure design the alignment was chosen to have a constant horizontal curvature with a radius of 2okm and also a constant vertical curvature.

For tendering the West Bridge, five consortia were prequalified in March 1988, and in the following month the tender documents were distributed to these consortia.

On November 30, 1989 the bids were received indicating the following construction costs for the tender designs:

Composite Bridge   3.0 - 4.2 billion DKK
                   (approx. 270 - 380 million GBP)

Concrete Bridge   3.0 - 4.6 billion DKK
                   (approx. 270 - 420 million GBP)

Steel Bridge   4.0 - 4.7 billion DKK
                   (approx. 360 - 430 million GBP)

As it can be seen, these bids to a large extent confirmed the conclusion drawn after the conceptual design phase regarding the relative cost of the different options. Thus, the lower bids for the composite bridge and the concrete bridge options were very close indeed, whereas the lower bids for the steel bridge option were substantially higher. The reason for this was that the larger shipyards had lost interest in bridge girder fabrication after having experienced a boom in the shipbuilding activities.

Apart from a number of bids on bridges in accordance with the tender designs, a relatively large number of bids based on alternatives or variants were submitted by the participating consortia. This was in contrast to the situation at the tendering of the East Railway Tunnel, where no alternatives had been presented.

The alternatives for the West Bridge comprised a number of concepts based on substantially different concepts, such as a multispan arch bridge, a double decked concrete truss of high strength concrete, a haunched concrete girder, and a pure steel truss with both the upper roadway deck and the lower railway deck comprising stiffened steel plates. Also for the substructures alternatives were received, mainly characterized by application of caissons closed at the bottom by a solid slab and by using prefabricated pier shafts to reduce the amount of work at the exposed site.

The bids on alternatives ranged from 2.4 billion DKK to 4.0 billion DKK (approx. 220 million GBP to 360 million GBP).

The lowest bid was received on an alternative design comprising a double decked concrete truss made of high strength concrete and with external cables. This alternative was carefully studied, but the final conclusion was that it did in too many aspects involve an extrapolation of existing technology and that it would be impossible to predict the long term behaviour under the dynamic train load and in the hostile environment of the Great Belt. This alternative was therefore abandoned.

Figure 11. Alternative design for the West Bridge as presented
by the European Storebælt Group.

The second lowest bid was received on a concrete box girder bridge
(Figure 11) characterized by the following features:

- superstructure composed of two box girders, a narrow one under
  the railway and a wide one carrying the entire roadway.

- haunched girders with a parabolic bottom slab to be erected
  as a double cantilever having the full length of the span
  and reaching from midspan to midspan.

- piers to be made of multicellular caissons with closed
  bottoms.

- pier shafts made of precast concrete units.

After evaluation of all bids in the lower end of the interval,
it was found that the contract should be negotiated either with the
consortium having presented the lowest bid on the tender design for
the composite bridge or with the consortium having presented the above
mentioned alternative bid based on a haunched concrete box girder.

After a preliminary clarification round, STOREBÆLT decided in
March 1989 to start contract negotiations with "European Storebaelt
Group"(ESG), the consortium that had presented the alternative with
haunched concrete girders.

During these negotiations it also became necessary to update the
alternative design technically, e.g. by increasing the depth of the
railway girder substantially and relocate the expansion joints. Furthermore, it was decided to increase the total bridge length to 6611m
as it was disclosed that the transition point between bridge and
embankment at Sprogø should be displaced approximately 500m towards
the East to arrive at the optimum solution.

In comparison with the tender design for a concrete bridge, the ESG design was characterized by a somewhat larger span in the continuous section (110.4m), whereas the spans adjacent to the expansion joints were reduced to 81.75m.

On June 26, 1989 the contract was finally signed with the ESG consortium that originally comprised a Dutch, a Swiss, a British and a Danish contractor, but later during the contract negotiations was extended to incorporate two more Danish contractors.

During the autumn of 1989, the contractors prepared the work site area on Funen where the bridge components shall be fabricated in very large units weighing up to 6000 tons. For the 110.4m bridge length reaching from the center of one span to the center of the next the total structure will consist of only five elements:

- a multicellular caisson with a closed bottom
- two pier shafts
- a double cantilever box girder element for railway
- a double cantilever box girder element for roadway

For the transportation and erection of these elements a special floating crane (Figure 12) is at present being built in Holland for delivery in the autumn of 1990.

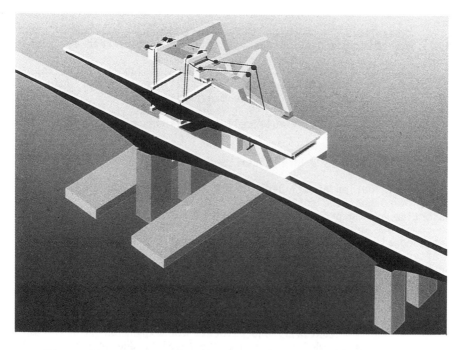

Figure 12. Computer plot of the special floating crane
to be used for transport and erection of the West Bridge.

7 East Road Bridge

During the conceptual design phase in the summer of 1987 both a bridge
and an immersed tunnel were investigated for the road link across
the East Channel. However, as the cost estimates had shown that the
tunnel would be characterized by substantially higher construction
costs, as well as higher operational costs, it was decided in October
1988 that only the bridge option should be further elaborated in the
tender design phase.

During the conceptual design phase, it was shown that the optimum
spans for both a cable stayed bridge solution and a suspension bridge
solution were close to those found for the combined road and railway
bridge during the 1977-78 design period. Consequently, the conceptual
designs for the main spans of the East Road Bridge were prepared
simply by transforming the double deck truss girder from 1978 to a
single deck box girder (Figure 13).

For the approach spans the changes were somewhat larger, as it
was found that the spans should be increased to arrive at an optimal
solution. Thus, the conceptual designs for the road bridge differed
not only by the change from a double deck truss to a single deck box
but also by having a larger span (164m instead of 144m).

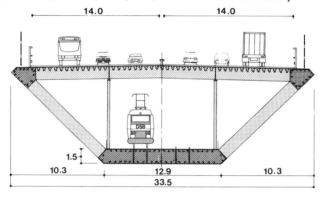

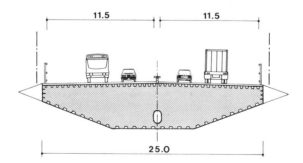

Figure 13. Cross section of the designs for a combined
road and rail bridge from 1978 and a pure road bridge
from 1986.

In the conceptual design from 1987 the length of the East Bridge was assumed to be the same as chosen in 1978 (approx. 5.5km). This length was well justified in case of selecting an immersed railway tunnel, as this required a 1200m ramp extending out from the coast of Zealand, thus allowing an embankment of similar length for the bridge. However, after having decided in November 1988 to construct a bored railway tunnel without a ramp off the coast, it was decided to abandon the offshore embankment by extending the bridge to a length of 6.8km.

The conceptual designs from 1987 were prepared by the consulting engineers without assistance from architects, but as it was considered of great importance that especially this part of the total link would be aesthetically pleasing the architects were asked to update the engineering design during the summer of 1988.

This resulted in a number of architectural recommendations regarding the shape of the girders, the pier shafts, the abutments and the pylons. Thus, the pylons were proposed to be constructed without the traditional cross beam below the bridge girder and with a more monolithic transition between the pylon legs and the pier structure (Figure 14).

Figure 14. Architect's proposal from 1988 for the
East Bridge main span.

During the spring of 1989, a number of supplementary investigations were carried out as part of a general updating exercise of the conceptual designs for the East Road Bridge.

For the main spans it was decided not only to determine the optimum span for a cable stayed and a suspension bridge, but to make cost estimates for quite large intervals of main spans for both bridge types. Thus, for the cable stayed bridge option a main span interval from 800m to 1400m and for the suspension bridge a main span interval from 1000m to 2000m were studied. The cost estimates included all relevant items within a reference length of 2360m. Consequently, not only the cost of superstructure and substructure but also the required amount of compensation dredging (that differed from case to case) were included.

From the cost estimates it could be concluded that the theoretical optimum for a cable stayed bridge was achieved for a main span of 900m, whereas the optimum main span was around 1200m for a suspension bridge. For both bridge types the well-known flatness of the cost minimum was found, so that main spans could be chosen somewhat higher than the theoretical optimum without seriously affecting the total construction cost. However, to go to the upper end of the span intervals would give a significant cost increase beyond the budget limits.

Decisive for the final choice of main span length and type of bridge were to a large extent considerations regarding the navigational safety.

At the bridge site the navigation is regulated by a separation scheme comprising two shipping lanes, each with unidirectional traffic. The total width of the separation scheme is 1600m and the angle between the direction of navigation and the bridge axis is 67.5 degrees. Thus, a main span of 1800m would be required if the existing separation scheme should be kept free of bridge piers.

However, the navigation authorities did not initially rule out the possibility of having the two pylon piers positioned in the existing separation scheme, provided that the side spans could be used by smaller ships leaving the main span open for the largest ships, which are the most difficult to manoeuvre. Also, it would be possible by dredging at the shoulders of the existing shipping lanes to improve navigation by relocating the separation scheme in such a way that a more perpendicular intersection with the bridge axis was achieved, and this should also make it more acceptable to reduce the widths of the shipping lanes.

To study the influence of the main span length on the navigational safety, a number of real-time simulations were conducted at the Danish Maritime Institute and, initially, tests were run for a 780m cable stayed bridge and a 1400m suspension bridge. The result of these simulations clearly indicated that the smaller span was unacceptable, whereas the larger span did not have the same adverse effect on the navigational safety, provided that the ships were fully functional and were maneuvered by competent navigators (such as pilots accustomed to the conditions in the Great Belt).

The updating exercise performed in the spring of 1989 also included the approach spans, as more elaborate analyses of the hydrological conditions had shown that the number of approach span piers was not so critical as earlier anticipated. At the same time, the bidding of the West Bridge had clearly indicated that a concrete superstructure was competitive to a steel superstructure.

The investigations related to the approach spans indicated that the optimum solution was achieved by having approach spans composed of continuous concrete girders with a constant span length of approximately 120m and a single box cross section (Figure 15), or continuous steel girders with span lengths of approximately 140m.

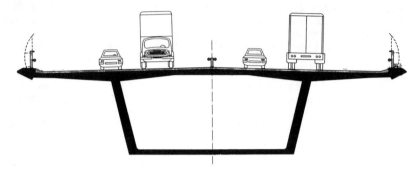

Figure 15. Cross section of the concrete proposal for the approach spans of the East Bridge.

The tender design phase for the East Bridge began in September 1989 and it was initiated by preparation of four outline designs, comprising cable stayed bridges with spans of 916m and 1204m as well as suspension bridges with spans of 1448m and 1688m.

The main purpose of the outline design phase was to determine the feasibility of the different options in relation to the navigational conditions, to the constructability, and to the cost of construction. Consequently, the outline designs were not only investigated to determine the quantities required, but were also followed by further navigational simulations, by wind tunnel tests, and by thorough evaluations of constructional methods.

The outline design investigations clearly indicated a benefit for navigation safety by choosing the larger span and, at the same time, it was found that construction costs of the different options were more equal than earlier expected - due to the fact that the bridges with smaller spans required several costly measures to be taken to secure them against ship collision and to improve navigational conditions.

With this result, the Board of the Great Belt Link Ltd. decided in February 1990 that the tender design of the East Bridge should be based on a main span in the upper range, i.e. 1600-1700m.

So, besides becoming the largest bridge in Europe when considering the total length, the East Bridge will also get the longest free span.

8 References

Gimsing, N.J. (1984) **Storebæltsbroen/-tunnelen** (in Danish). Teknisk Forlag, Copenhagen

K.H.Ostenfeld: Denmark's Great Belt Link, COWIconsult, 1989.

CORROSION RESISTANCE IN STEELS FOR MOTORWAY BRIDGES

K. JOHNSON
British Steel Technical, Swinden Laboratories, Rotherham,
England, UK
W. RAMSAY
British Steel, General Steels Commercial, Motherwell, Scotland, UK

Abstract
During the 1980s steel has shown a steady increase in share of the highway bridge market and one of the factors that has led to this growth has been improvements in corrosion protection. The principal corrosion protection options that are available to the bridge engineer are discussed.

The majority of steel bridges are protected by the application of painting systems. Typical Department of Transport specified systems involve the application of 5 or 6 coats to produce a total dry-film thickness of 200-300 microns and require minor maintenance after 6 years and major maintenance after 15 years.

Recently a high-build elastomeric urethane coating has been used on a new steel bridge over the River Ythan. Plate girders were blast cleaned in the shops and treated with a single, spray-applied coat to a minimum dry-film thickness of 1000 microns. A period of at least 20 years to first maintenance is anticipated.

Weathering steels account for about 10% of the steel bridge market. They form a stable protective oxide coating and achieve a low terminal corrosion rate. Their limitations are described and suggestions for their future use are made.

Finally, the enclosure of structural steelwork on composite bridges is described as a method of corrosion protection which will receive much greater emphasis in the future.
Keywords: Corrosion, Costs, Coatings, Weathering steels, Enclosures.

1   Introduction

Until recently, the commonly held view was that steel bridges suffer from the inherent disadvantage of requiring corrosion maintenance. Conversely, concrete bridges were considered to be essentially maintenance free. It is now known that is not the case and, in a sense, the position has been reversed.

In March 1989, following a great deal of public concern, a report by G. Maunsell and Partners on the performance of concrete bridges was published. The report covered a survey of 200 representative structures drawn from the 5900 concrete bridges for which the Department of Transport is responsible in England. It concluded that approximately three-quarters of these bridges are affected by chloride attack and severe corrosion will inevitably occur unless urgent action is taken.

In contrast, it has always been recognised that steel bridges require corrosion protection and the materials, methods and procedures for providing this are well established and effective. An appropriate way of illustrating the advances that have been made in corrosion protection of steel bridges may be to compare the practices which were used on the old Forth Rail Bridge and the relatively new Forth Road Bridge.

The Rail Bridge was completed in 1890 and at that time little thought was given to the maintenance of the structure. Paradoxically that subject has since passed into British mythology (a 'Forth Bridge Job' is one that is never ending). Even more oddly, there is no truth to the story that a team of painters starts at one end of the bridge, works its way to the far bank and then goes back and starts again. The reality is more complicated but, even so maintenance painting is a continuous effort.

The protection that was originally applied to the external surfaces of the bridge steelwork involved wire brushing to remove rust and scale, applying hot boiled-linseed oil and then a number of coats of red lead based paints. The insides of the large tubular members were treated with a white lead based paint, to show up any rusting. The total area which was painted amounted to approximately 145 acres! Little consideration was given to gaining access for maintenance; painters were expected to have the climbing skills of Alpine mountaineers. Consequently, since the advent of the Health and Safety laws in 1974, access to some of the more difficult areas of the structure is at present not possible.

In contrast, when the Forth Road Bridge was built, 26 years ago, much more thought was given to the protection and maintenance of the structure. The fabricated steelwork was treated using purpose-designed plant installed in disused hangars at Drem Airfield, about 5 miles from the site. Automatic centrifugal blast cleaning was followed by automatic zinc metal spraying and then all external surfaces were treated with an etch primer, a primer, and two finishing coats. Provision was made for access for maintenance painting by the installation of two hand-driven gantries. In recent years these have been replaced by power-driven gantries and a new walkway has been added to give direct access to the upper parts of the superstructure.

These two bridges, standing side by side over the Firth of Forth, therefore symbolise some of the changes in attitude and advances in materials and techniques that have been made in corrosion protection of steel bridges. In the 26 years since the Forth Road Bridge was built, further advances have been made and some of these are described in this paper.

2   The cost of corrosion protection on steel bridges

Figure 1 shows a breakdown of costs in typical bridgework tenders. Although the absolute costs are clearly subject to change, it can nevertheless be seen that the cost of initial

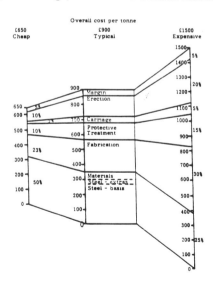

Fig. 1.   Breakdown of costs in a typical bridgework tender.

corrosion protection varies between 10-15% of the total cost of the erected steelwork. These costs do not take account of subsequent maintenance, which is discussed later in this paper.

Various methods are used to protect steel bridgework and obviously the choice of method will reflect on the cost of protection. The principal methods used are outlined below.

3 Conventional painting systems

The majority of steel bridges are protected from corrosion by the application of paint coatings. For many years the steelwork for new bridges has been shop-treated by blast-cleaning, followed by the application of primer and intermediate coats. After transportation to site and subsequent erection the steelwork is then further treated with finishing coats.

In the UK, much of this work is covered by the Department of Transport's 'Specification for Highway Works, Part 6, Series 1900, Protection of Steelwork Against Corrosion' and the equivalent SDD document. This describes methods of surface preparation, metal coatings, multicoat painting systems and procedures for testing of paints, application of paints, etc.

The accompanying 'Notes for Guidance on the Specification for Highway Works' then goes on to describe 11 protective systems. Selection of a protective system for a particular bridge is dependent upon environment, accessibility and required durability.

The painting systems described in the document are based upon the use of conventional blast-primers, primers, intermediate coats and finishing coats. Typical systems involve the application of 5 or 6 coats to produce a total dry-film thickness of 200-300 microns. For example, in a marine environment, with 'ready' access, painting after blast cleaning would involve the application of:-

Shop treatment -
1 coat Zinc Phosphate Chlor-Rubber Alkyd Blast Primer
2 coats Zinc Phosphate Chlor-Rubber Undercoat
1 coat MIO Chlor-Rubber Undercoat
Despatch to site and erect

Site treatment -
1 coat Chlor-Rubber Undercoat
1 coat Chlor-Rubber Finish

Overnight drying intervals would normally be required between each coat and the total minimum dry-film thickness would be 300 microns. Bridge beams treated in this manner would be held in the fabrication/painting shop for at least 4 days and would then require extensive site treatment, which is expensive, difficult to control, subject to the weather and introduces substantial delays to the site programme.

In the opinion of the writers this approach to bridge painting is outdated and could be greatly improved by the use of modern high-build painting systems.

4 High-build painting systems

In recent years protective coatings technology has made significant advances, partly as a result of the challenge of protecting North Sea offshore structures. In extreme environments high demands are made of corrosion protection systems and subsequent maintenance is very difficult and expensive. The coatings which have been developed for use on such structures rely heavily on 2-pack chemically-cured resin systems which produce highly durable coatings and can be applied as very thick or 'high-build' films.

On the initiative of Grampian Regional Council, a new steel bridge over the River Ythan at Newburgh, about 10 miles north of Aberdeen, has been treated with a high-build elastomeric urethane coating. This was a composite design plate-girder bridge in 3 spans and the complete protective system was applied as a single coat of 1000-1250 microns, in the shops, after fabrication.

Sophisticated spray-equipment is required to apply these coatings since they usually have a very short pot-life. In this case a twin-feed airless spray unit was used, involving a proportioning pump, an in-line mixer and thermostatically controlled in-line fluid heaters. Early apprehension about the use of such equipment under fabricating shop conditions proved to be unfounded and in general the spraying operation proceeded smoothly without any major difficulties. Problems relating to masking of areas to be left uncoated, spraying procedures, inspection and quality control, treatment of site weld areas, treatment of damage areas etc. were resolved during the course of the work.

A large number of dry-film thickness readings were taken during the course of the work and these are shown in Fig. 2 as a frequency distribution. This shows that

- very few measurements were below the minimum requirement of the specification (i.e. 1000 microns or 1 mm)
- more than 70% of measurements were in excess of the specification maximum
- the mean film thickness on all of the measurements taken on the bridge beams was 1.71 mm (with a standard deviation of 0.63 mm)

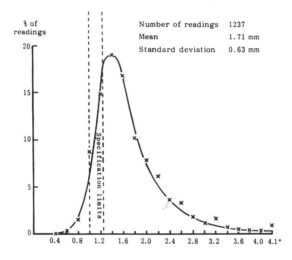

Fig. 2. Frequency distribution of all film thickness measurements on 6 beams, treated with the high-build elastomeric urethane system.

Laboratory tests on plate panels which were coated during the course of the contract gave very encouraging results. Direct pull-off adhesion tests produced values between 1000-1700 psi, a level of adhesion which is considered excellent for such a coating. Salt spray tests on thick coatings such as this are normally limited to about 2000 h duration. In this case tests were continued to 9000 h, giving excellent results, both in terms of spread of corrosion from a scribe-mark and the general durability of the coating. These tests and other field experience with high-build coatings suggest that a period of at least 20 years to first maintenance can reasonably be expected from a protective coating of this type.

At the time of the contract the cost of the High Build System was found to be slightly higher than the cost of the equivalent 6 coat DoT system, though life cycle costs would be expected to show a considerable advantage.

Following their experience with the Waterside Bridge, Grampian Regional Council have used this type of high-build paint system on further bridges and it is understood that the Scottish Development Dept. (SDD) have now approved this system as an alternative which may be considered alongside the conventional DoT systems for future steel bridges.

5   Weathering steel bridges

The first use of weathering steels in bridges in the UK was a small footbridge over an artificial lake at York University which was erected in 1967. During the ensuing 23 years approximately 100 weathering steel bridges have been built in the UK. Current use of weathering steels accounts for about 10% of the total steel bridge market. The majority of these have been in the medium span range and are of plate-girder construction.

Weathering steels, of which Cor-Ten is the most common, are low alloy steels typically containing up to 3% of alloying elements such as phosphorus, chromium, nickel, copper and vanadium. On exposure to air under suitable conditions they rust to form an adherent oxide coating. This acts as a protective layer which, with time and under appropriate conditions, causes the corrosion rate to reduce until it reaches a negligible level. Conventional coatings are therefore unnecessary since the steel provides its own protection.

These steels have properties comparable with Grade 50 steels to BS4360 - Weldable Structural Steels and this standard also includes weathering steel grades (WR50).

WR50A or Cor-Ten A can be used for many structural applications, including footbridges, but for bridgework generally, considerations of notch ductility require the use of WR50B or Cor-Ten B.

During the early part of their life weathering steels corrode in a similar manner and at a similar rate to mild steels. As the protective oxide layer develops the corrosion rate falls to a low terminal value, Fig. 3. The time required for a weathering steel to form a stable protective coating depends upon its orientation, the degree of atmospheric pollution and the frequency with which the surface is wetted and dried. For this reason, each proposed bridge structure should be considered individually with respect to its local environment as

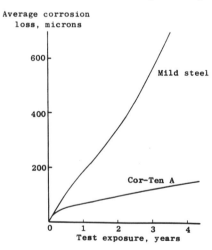

Fig. 3.   Industrial exposure - Port Talbot.

there can be large differences in corrosion rates, dependent upon geographical location and positional effects.

The Department of Transport Standard BD/7/81 and the equivalent SDD document refer to the use of 'Weathering Steels for Highway Structures' and states that they should not be used when:-

- the atmospheric chloride ion concentration is greater than 0.1 mg/100 cm² per day, average
- the atmospheric SO_2 concentration is greater than 2.1 mg/100 cm² per day, average
- the bridge is over a road subject to de-icing salts and the headroom is less than 7.5 m

A British Steel study of airborne salinity at eighteen sites in the UK concluded that although very high chloride levels are encountered at the coastline, these fall very rapidly on moving inland. The values obtained suggest that only a very narrow coastal strip (approximately 1-1½ km) has a sufficiently high airborne salinity to adversely affect the performance of weathering steels.

Usually it is in industrial locations that weathering steels show their greatest advantage over mild steel. Sulphur dioxide in the air reacts with the alloying elements to produce insoluble sulphates which plug the rust pores and the corrosion rate falls rapidly.

It has been observed in the UK that bridges and buildings in Cor-Ten weather to different extents, as measured by both appearance and metal loss, depending upon their compass orientation. Controlled atmospheric corrosion tests have shown that greatest corrosion occurs on north-westerly facing steel surfaces, almost certainly because they are wetted for longer due to rain carried by the prevailing wind and they take longer to dry, since they are most shaded from the sun (Fig. 4).

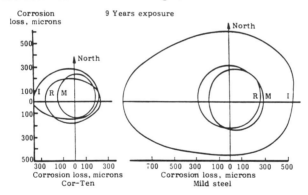

Fig. 4. Corrosion losses of mild and Cor-Ten A steels exposed vertically, facing different compass directions.

Weathering steels should be blast-cleaned, to remove millscale, before exposure to provide a sound uniform surface for the formation of the oxide coating. Under most conditions a stable rust patina will be established in about 2-3 years, changing in colour from brown through to almost black, though the ultimate colour will depend upon the conditions at the site.

Drainage of corrosion products can be expected during the first 2 or 3 years exposure and can stain or streak adjacent materials, e.g. concrete piers. Provision should be made to divert this from vulnerable surfaces.

A continually wet surface will cause weathering steels to corrode at approximately the same rate as mild steel. Detailing should therefore avoid crevices or other water retention areas. Drainage and ventilation should be provided. All interior surfaces (e.g. box girders) and faying surfaces should be painted or sealed to prevent entry of moisture. Structures should be periodically inspected to ensure that all joints and surfaces are performing satisfactorily.

It is generally considered that bridge steelwork in bare unprotected weathering steels, with a 1 mm corrosion allowance, approximates in cost to bridge steelwork in Grade 50 steel, protected with a conventional multicoat painting system.

It has also been claimed that paint coatings perform significantly better when applied on weathering steel, as compared to a mild steel substrate. These claims have come predominantly from the United States where it is suggested that improved paint performance on weathering steels will double the time to first maintenance of a typical bridge painting system. A recently conducted survey in USA indicated that 300 000 t of bare Cor-Ten will be used on bridges over the next five years and that 130 000 tons of painted Cor-Ten will be used over the same period.

Painting of weathering steel bridges has not been used as a standard practice in the UK, though occasionally painting has been used as a remedial measure on weathering steel bridges which have inadvertently been built in an inappropriate environment. Indeed the DoT Standard BD/7/81 states that 'paint does not have an increased life on weathering steel'.

The underlying reasons for any possible improved paint performance are unclear and will probably be to some extent dependent upon prevailing climatic conditions. Consequently, in view of the major cost consequences of this possibility, a test programme has been commenced by British Steel to evaluate the performance of various bridge painting systems on weathering steels and on mild steel in a number of environments. Clear indications from this test programme are not expected for another 4-5 years.

6   Enclosure of steel bridges

This method of protecting structural steelwork on composite bridges was proposed in 1980 by TRRL. It relies upon the principle that clean steel does not corrode significantly at relative humidities as high as 99%, provided certain environmental contaminants, such as sulphur dioxide and/or chlorides, are absent or are present only at low levels. The concept therefore is to enclose steel bridge beams (which are already sheltered by a concrete deck) with plastic or other sheeting, thereby reducing the corrosivity of the environment to which the steel is exposed.

After preliminary experiments at a number of bridges to establish the feasibility and validity of the method, the first full scale enclosure was carried out on the 3 span composite river bridge at Conon, near Inverness. The bridge had been completed in 1982 and the painted beams were totally enclosed in 1984, with stiffened anodised aluminium sheeting.

Measurements have since been made periodically of humidity, temperature, time-of-wetness, atmospheric chlorides and sulphur dioxide and corrosion rates have been measured on bare steel test panels. The results of these tests, carried out both inside and immediately outside the enclosure have confirmed that the method produces an environment of low corrosivity and suggests that the painted steel beams will remain maintenance free for decades. The enclosure method may also be applicable to unpainted steel beams and could also be used to extend the life of weathering steel bridges which have inadvertently been constructed in unfavourable environments.

A further advantage of the method is that the enclosure panels can provide a permanent access platform for future inspection and maintenance, providing the cladding is designed for such loading. This built-in access must be considered as a major benefit for bridges over rivers, motorways etc. A major use of the enclosure method is currently being undertaken on the Tees Viaduct where GRP sheeting is being used.

The method is applicable to both new and existing bridges though beam sizes and spacings will obviously affect the economics. Deep, closely spaced beams, which allow a large surface area of structural steelwork to be enclosed with a small amount of cladding will give the most economic enclosures. Beam depths of less than about 2 m make access and inspection more difficult, as well as being economically less favourable. Beam spacings of more than 2.75 m represent an upper limit imposed by the strength and stiffness of many potential enclosure cladding materials.

The Steel Construction Institute are currently involved in writing, on behalf of the Department of Transport, a draft Departmental Standard and Advice Note on enclosures. This will include design guides and advice on cladding materials.

7 Maintenance of steel bridges

7.1 Maintenance of conventional painting systems

The majority of steel bridges are protected from corrosion by the application of paint coatings. Consequently the mechanisms by which paint films fail are of relevance.

All paint films are permeable to atmospheric gases and in particular oxygen, water vapour and water. The degree of permeability varies from one coating type to another and is also dependent upon film thickness; however, in all cases water and oxygen reach the steel surface. The reasons for corrosion not subsequently occurring are complex and unresolved but are almost certainly to a large extent due to the paint film imposing a large electrical resistance in the path of any possible corrosion current. This protection mechanism is only effective while the paint adheres to the steel surface. If adhesion is lost or markedly reduced, osmotic pressures are developed under the paint film causing blistering, general loss of adhesion and, eventually, corrosion of the steel surface. This degradation occurs, in a sense, through the film.

A second form of degradation occurs as a result of spread of corrosion from the discontinuities in the coating. These may be 'holidays' (i.e. omissions during application), mechanically damaged areas, localised film weaknesses (e.g. air or solvent voids), or localised disruptions of the coating by the above 'through film' mechanism. In all of these cases corrosion occurs at the discontinuity and spreads under the film, forming a crevice in which corrosion can proceed at an increased rate, leading to an underfilm degradation of the paint coating.

The third mechanism by which paint coatings degrade is concerned with erosion from the surface of the film. This may be the result of UV degradation or, more simply, mechanical erosion due to wind-borne sands etc.

In the wet, temperate climate of the UK through-film and underfilm degradation are the common modes of failure; UV degradation is less significant and mechanical erosion is specific to a small number of extreme sites.

It is expected that new steel bridgework which has been treated with one of the Department of Transport conventional multicoat painting systems will require minor maintenance after 6 years and major maintenance after 15 years.

The procedures employed by the Department for the maintenance of conventionally painted steel bridgework involve:-

- defining the extent of the failure; local, general etc.
- carrying out a site survey
- writing a specification for the maintenance work
- carrying out a site feasibility trial
- execution of the contract
- testing of materials during the course of the contract

all of which is described in the Departmental Standard BD/18/83 and Advice Note BA/13/83.

The systems used are, of course, conventional multicoat systems, similar to those used on new steel bridgework.

7.2 Maintenance of high-build painting systems

Experience with these systems on bridgework is limited so that predictions of maintenance periods can not be made with certainty. However, accelerated tests and general field experience suggest a period of at least 20 years before major maintenance is required, and possibly much longer.

The thick, highly cross-linked films produced by these materials pose their own problems when the need for that maintenance does arise. Most experience of this kind has been with high-build (usually solvent-free) epoxy coatings which have proved to be difficult to maintain due to lack of adhesion between the original coatings and the maintenance coats. This difficulty has been overcome, at a cost, by insisting that the surface of the original coating is abraded before the maintenance coats are applied. At the present moment, this requirement would also be made for the high-build elastomeric urethane coatings even though these are much less susceptible to intercoat adhesion problems. In practice this would be achieved by lightly blast-cleaning the overall surface and, since blast-cleaning is a well established practice on bridge maintenance sites, would not be expected to cause significant problems in terms of either practicability or cost.

7.3 Maintenance of weathering steel bridges

Weathering steel bridges are not entirely maintenance free. Structures should be periodically inspected to ensure that decks and expansion joints are not leaking, that the weathering steel surface is free from debris and that all surfaces are performing satisfactorily.

Departmental Standard BD/7/81 requires that visual examinations of critical areas should be carried out at least every 2 years and steel thickness measurements should be made at intervals of approximately 6 years during the life of the structure.

7.4 Maintenance of enclosures

The enclosure sheeting which protects the bridge beams will, of course, be fully exposed and will eventually require maintenance or replacement. The form this takes and the period involved will be dependent upon the local environment and the cladding material used.

7.5 Commuted maintenance costs

The evaluation of future maintenance costs for highway bridges is a continuing controversial topic. Until recently, commuted maintenance costs were applied to steel bridges but not to concrete bridges.

Recent DoT Standards BD36/88 and BA28/88, which are intended for the purpose of comparing alternative designs, include costs for inspection and maintenance of concrete bridges. The steel industry considers that this is a step in the right direction but that the cost rates given in the new standard still do not truly reflect the real situation and that steel is still at a disadvantage.

8   The market for steel bridges

During the 1980s steel has shown a steady increase in its share of the highway bridge market. This has been due to a combination of several factors:-

- steel costs have fallen relative to concrete costs
- fabrication costs have reduced
- the introduction of limit state design codes
- improved corrosion protection systems for steel and a growing awareness that concrete bridges are not maintenance free

A market survey has been carried out on all highway bridges in England and Wales, completed during the period April 1983 to April 1988, whose maximum span was in the range 15 m - 50 m. An excellent response, in excess of 90%, gave a base of 612 structures.

Figure 5 shows the changing market share of steel and concrete over the 5 year period, both in terms of numbers of bridges and deck area. Steel has increased its share both in terms of numbers of structures built (10%→28%) and in terms of deck area (11%→53%). Deck area is considered to be the most realistic method of making this comparison.

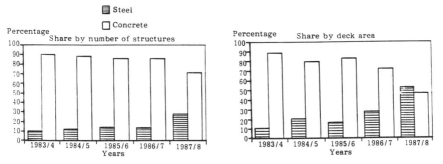

Fig. 5. Highway bridges in England and Wales.
1983-1988   15 m - 50 m span

The data were also analysed by span range, Fig. 6, which shows that steel has increased its share in all 3 span categories, but particularly in the intermediate 25-35 m range.

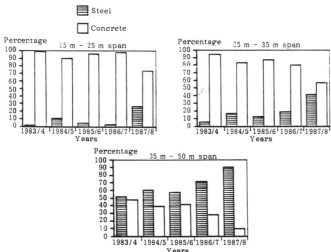

Fig. 6. Highway bridges in England and Wales.
1983 - 1988   Share by deck area

9 Prospects for the future

Corrosion protection remains a critical part of the overall package which will ensure that steel maintains and increases its share of the total bridge market.

In the USA there are no Federal procedures or practices dealing with bridge corrosion protection as there are in the UK. Moreover there is much greater emphasis placed on air pollution by the organic solvents in paints. The net result is that the developments in bridge corrosion protection in the USA are heavily concerned with weathering steels:-

- the growing use of weathering steels for bridges. The bad experience of the Michigan Department of Transportation with weathering steel bridges during the late 1970s has generally been attributed to leaking deck-joints, salt-spraying and bad detailing and a major growth in use is now expected
- the recent introduction of Cor-Ten B-QT, a high yield strength steel which is being promoted for plate-girder bridgework
- the growing use of painted Cor-Ten, to increase life to first maintenance

In mainland Europe, weathering steels are used for bridges, but not extensively and the principal developments seem to be concerned with reducing the cost of conventional painting systems. This is being achieved by reducing the number of coats applied, e.g. a current French chlorinated rubber specification for rural, urban and industrial environments involves the application of only 3 coats to produce a film thickness of 180 microns and is claimed to give a life of 15 years to first major maintenance. Possibly the French climate is less aggressive than in the UK?

In the UK, the authors expect a continuing growth in market share for steel bridges. This will involve more weathering steel bridges, a growing use of the enclosure method, particularly for river bridges and possibly motorway bridges, and a move away from conventional multicoat painting systems towards the very high-build systems as used on the Waterside bridge, possibly applied in specially equipped, purpose built shops.

10 Acknowledgements

The authors wish to thank Dr. R. Baker, Director of Research, British Steel for permission to publish this paper.

11 References

Bishop, R. Enclosure - An alternative to bridge painting. TRRL research report No. 83.
Johnson, K.E. and Stanners, J.F. The characterisation of corrosion test sites, EUR 7433 EN.
Wallbank, E.J. (1989) The performance of concrete in bridges. G. Maunsell & Partners for Department of Transport.

FIBER REINFORCED PLASTIC BRIDGES IN CHONGQING 1983–1988

R.N. BRUCE, Jr
Dept Civil Engineering, Tulane University, New Orleans,
Louisiana, USA

Abstract
This paper is a brief report describing the physical characteristics
of two fiber reinforced plastic bridges built in Chongqing, Peoples
Republic of China, during the decade of the 1980's. The two bridges
that are described are the GRP Pedestrian Cable-Stayed Bridge, and the
Guanyinqiao Pedestrian Bridge. The GRP Pedestrian Cable-Stayed Bridge
was completed in 1986, and the Guanyinqiao Pedestrian Bridge was completed in 1988. Both bridges are a unique combination of high
strength steel and reinforced concrete with glass reinforced plastic
(GRP). It is noted that this paper pertains to work done by others -
the paper is a "trip report" which provides information on an innovative use of FRP (GRP) materials in bridge structures.
Keywords: Fiber Reinforced Plastic Bridges, Glass Reinforced Plastic
Bridges, Plastic Bridges, Cable-Stayed Bridges, Bridges.

1 Introduction

Through the efforts of Mr. Chen Kesheng, Senior Engineer, and Madame
Cheng Liping, Deputy Chief Engineer of the China Highway and Transportation Society; the author was invited to inspect two of the three
fiber reinforced plastic bridges in Chongqing. Both of the bridges
were conceived and built by The Research Institute of Composite
Material Bridges of the Chongqing Institute of Transportation.

It is noted that this paper pertains to work done by others, and
that the author did not participate in the projects reported. The
paper is a "trip report" which provides information on an innovative
use of FRP (GRP) material in bridges. Much information pertaining to
mechanical properties of the materials, technical details of the
design, and methods of fabrication and erection, were not available to
the author; and thus the paper is limited to the brief descriptions
provided. In the case of the first bridge described, some technical
information was provided by Mr. Tang Guodong of the Compostite
Material Bridge Institute in Chongqing.

2 GRP Pedestrian Cable-Stayed Bridge

The first bridge visited was the GRP Pedestrian Cable-Stayed Bridge
joining two parts of the campus of the Chongqing Institute of

Transportation. The bridge was completed in 1986, and is shown in
Fig. 1.

Fig. 1 GRP Pedestrian Cable Stayed Bridge

The layout of the bridge is an unsymmetrical system having a single
tower and a single harped array of cable stays. The side spans and
the tower are of reinforced concrete, with the tower having a height
of 11m and an inclination of 15° from the vertical. Each of the seven
cable stays consists of 19 No. 5 steel wires encased in a polyethylene
tube.

The entire length of the bridge is 50m which includes a non-
continuous main span and two side spans. The main span consists of a
single FRP box girder with cover plates. The FRP girder is 27.4m
long, and 4.3m wide. The weight of the FRP girder is approximately 8
tons. The cost of the single FRP box girder was 120,000 yuan, or approximately 45% of the total cost of the project. The total cost of
the completed structure, including a test program, was 260,000 yuan
for a unit cost of 1,000 yuan per square meter. To translate the unit
cost into dollars would result in a range between $25 per square foot
and $13 per square foot, depending on the exchange rate used.

Complete technical information regarding bridge design, materials,
and construction, was not available at the time of inspection;
however, limited information has been provided related to the glass
fiber reinforced plastic, and related to design criteria.

The GRP is composed of high strength plastic reinforced with glass fiber. The bond between the fiber and the plastic is enhanced by the use of a resin. The high strength plastic is referred to as a plastic resin. Some of the physical properties of the GRP were provided.

The structural stiffness (EI and GI) is dependent on material parameters and member cross-section. Although the physical stiffness of the material (E, G) is less than that for concrete or steel, the stiffness of individual structural members can be increased by increasing the geometric stiffness in terms of moment of inertia of the cross-section. Values of elastic modulus (E, G) can be increased by the proper choice of fiber orientation.

The FRP box girder used in the GRP Pedestrian Cable-Stayed Bridge consisted of a top and bottom flange, and five vertical webs. The arrangement of fiber reinforcement in the webs could be placed at an angle of 45° in order to provide increased shear resistance to shear stresses, thus improving the strength and stiffness of the laminated sections used in the box girder. This predetermined arrangement of the fiber reinforcement can result in a material having anisotropic properties. These anisotropic properties must be considered in design. In the case of fiber reinforcement placed in a longitudinal direction only, physical properties may be determined.

The elastic modulus of the GRP in tension and compression are given, respectively, in equation (1) and (2).

$$E_o = E_f \cdot V_f + E_m \cdot V_m \tag{1}$$

$$E_o = E_f \cdot V_f + E_m (1 - V_f) \tag{2}$$

where

E_o = modulus of GRP

E_m = modulus of plastic resin

E_f = modulus of glass fiber

V_m = percentage content of resin

V_f = percentage content of fiber

The specific gravity of the GFP is between 1.4 - 2.2. For the GRP reinforced in a longitudinal direction only, the tensile strength is 10,000 kg/cm^2.

Design criteria included a design live load of 350 kg/cm^2; with a factor of safety of 10 for direct stress, and a F.S. between 3 and 6 for shear stress. The allowable deflection is L/600. The highest design temperature under service conditions is 70°C.

The design procedure involved a sequence of three steps. The first step was the design of a single laminate, involving the material composition, quantity, and arrangement. The second step was the design of the laminated plates, involving the direction and building sequence of the single laminates into a laminated plate or sandwich structure.

The third step was the overall design of the structure, involving volume, size, and geometric proportions. It is pointed out that the three steps in the design procedure are interrelated and interdependent.

Special commentary may be made with respect to the design of the FRP box girder. The depth of the girder is variable so as to form the deck with a longitudinal slope of 2%, sufficient for draining water on the deck. All of the laminated plates in the box girder have a thickness of 10 cm. except the center longitudinal web panel which has a thickness of 16cm.

The FRP box girder was fabricated in the New Material Factory of Wuhan Industry University. The box girder was transported and erected in one piece. The bridge was tested under combinations of dead load and live load. The tests included static and dynamic tests to determine the structural behavior of the completed GRP bridge. The static test indicated compatible results when observed values and calculated values were compared. Maximum mid span deflection was measured to be 3.5 cm., less than the allowable deflection of L/600.

The dynamic tests included an impact test and a "People Passing Bridge" vibration test. In the impact test, a weight of 1.47kN was dropped from a height of 1.5m, impacting the mid-span of the GRP bridge. As a result of the impact test, the frequency of the GRP box girder was determined to be equal to 3.15Hz, the strain amplitude was 41.75$\mu\varepsilon$, and the resistance coefficient was 0.107.

In the "People Passing Bridge" tests, three procedures were used. When a group of 40 persons walked randomly through the bridge, the total strain amplitude of the GRP grider was less than 26$\mu\varepsilon$, approximating the relevant static strain. When two groups of persons, each group having 15 persons, quickly marched reversely along the two sides of the bridge, the vibration was greater than the first case, and the measured frequency was 2.5 Hz.

In the third procedure, 4 persons ran across the GRP girder with 3Hz frequency, aggravating the vibration. The maximum strain amplitude was measured to be 50$\mu\varepsilon$, significantly larger than the first procedure involving 40 persons. The running frequency of the people approximated the natural frequency of the GRP girder, thus causing the significant amplification.

The completed bridge was reviewed by the PRC Ministry of Communications, the agency responsible for transportation. About 40 experts of highway, railway, material and urban construction took part in the review.

The review indicated that the cost of the FRP bridge is less than the cost of the same bridge in steel, and that the FRP bridge is maintenance free. This GRP Pedestrian Cable-Stayed Bridge is the first in the world, and the design may be used for the pedestrian bridges of cities in China.

3 Guanyinqiao Pedestrian Bridge In Chong Qing

The second bridge visited was the Guanyinqiao Pedestrian Bridge of Jiangbei District in Chongqing. The bridge was completed in May of 1988. The bridge is described as a space frame, with FRP deck griders

suspended from reinforced concrete rigid frames.   The model of the
bridge is shown in Fig. 2.

Fig. 2   Model of Guanyinqiao Pedestrian Bridge

The bridge crosses an intersection of two main streets in downtown
Chongqing.   The structural layout is in the shape of an ancient
Chinese coin.   The span of the rigid frame is 70m, with the overall
total length of the bridge equal to 157 m.   The four long FRP girders
are 19 m long.   The four FRP girders forming the central geometry are
9m long.   All eight FRP girders are 4.3m wide, and 0.9m deep.   All
girders are hung from the space frames by high strength wire.   Total
weight of the FRP girders was 43 tons, with the 19m girders each
weighing 7.29 tons and 9m girders each weighing 3.46 tons.   The cost
of the FRP girders in place represented 42% of the total cost of the
completed project.   The total cost of the project was 1,700,000 yuan.
This amount converts to $485,000 to $243,000 depending on the exchange
rate used.

The FRP girders were fabricated in Chongqing Glass Fiber Product
Factory then transported and erected in single pieces.   Those responsible for the design and construction of the FRP bridges in Chongqing
have indicated certain characteristics of FRP bridges as follows:

1.   Cost of FRP bridges is less than steel bridges of same type.
2.   There is no rusting problem, and maintenance is minimized.

3. Tests indicate that the FRP structure is maintenance free for 40 years, except in especially bad environment. Tests indicate that is it unnecessary to worry about the problem of life span for FRP bridges.
4. FRP has good properties for resisting fatigue and low energy impact.
5. FRP is easy to form and to color.

The final appearance of the Guanyinqiao Pedestrian Bridge in service is shown in Fig. 3.

Fig. 3 Guanyingiao Bridge in Service

Attempts are being made to obtain additional technical information on the GRP bridges in China, including the vehicular bridge in Beijing. When such information is received, it will be incorporated into an expanded version of this report.

References

1. Guodong, Tang, Analytical and Experimental Work of GRP Cable-Stayed Pedestrian Bridge. Composite Material Bridge Institute, Chongqing, China.

THE TEST PERFORMANCE OF A FULL-SCALE PRESTRESSED BRICKWORK BRIDGE ABUTMENT

S.W. GARRITY, T.G. GARWOOD
Dept Civil Engineering and Building, Bolton Institute of Higher
Education, Bolton, England, UK

Abstract
A prestressed clay brickwork diaphragm wall, representing part of a
6.6m high bridge abutment, was built in the laboratory. The wall was
test loaded by a system of hydraulic rams fixed to a similarly
constructed reaction wall. The rams were arranged to simulate both
the longitudinal load applied to the abutment at carriageway level
and the earth pressure forces acting on the back face. In the test,
no cracking was observed in the brickwork until the applied loading
was well in excess of that corresponding to the design service
condition. At the final stage of the test there was no indication of
impending shear or bending failure. The bending moment and shear
force resisted by the abutment at this stage were greater than those
produced by the design ultimate loads as defined in BS 5400.
Keywords: Brickwork, Prestressing, Bridges, Full-Scale Testing.

1 Introduction

Concrete has generally been regarded by engineers as a durable
material requiring little or no maintenance. However, an alarming
number of Britain's concrete highway bridges, many of which were
constructed during the major road building programmes of the 1960s
and early 1970s, are now showing signs of deterioration. Although
there are many likely causes of such deterioration, the most serious
and widespread problem is reinforcement corrosion arising from the
use of de-icing salts in winter. Bridge piers and abutments are often
the worst affected, in some cases requiring major repairs. Based on
the results of a survey of 200 bridges in England by Wallbank (1989),
it has been estimated that it would cost about £600 million over a
ten year period to repair all the concrete bridges owned by the
Department of Transport. This figure does not include the cost of
disruption to the road users.
    In the past, masonry was used extensively in Britain for arch
bridges and for the piers and abutments of beam and slab bridges.
Such masonry structures have generally performed well over long
periods of time, in many cases requiring only minimal maintenance.
Inevitably, masonry bridge abutments were usually of relatively
massive construction because of masonry's inherently low flexural
strength. Reinforced concrete construction was considered to be

stronger in flexure and less costly than masonry. Consequently, in modern construction, masonry has been used primarily as an aesthetically pleasing cladding for reinforced concrete abutments rather than as a structural material.

Curtin et al (1982, 1982a, 1982b) have shown that the prestressed masonry "diaphragm" or cellular wall is an efficient structural form which can be used economically to resist large lateral loads. Prestressed clay brickwork diaphragm wall construction may therefore be a cost-effective alternative to reinforced concrete for bridge abutments, particularly if whole life costs and aesthetic appeal are considered. As far as the authors are aware, the first instance of where this form of bridge abutment construction was used in practice was the Glinton - Northborough by-pass in Cambridgeshire, as reported by Bell (1989a).

Curtin (1986), Ambrose et al (1988) and Curtin and Howard (1988a) have carried out tests on prestressed clay brickwork diaphragm walls. However, the wall sections tested and the magnitude and distribution of the test loads were not appropriate for a bridge abutment.

The work reported in this paper is the first of a series of tests on a full scale prestressed clay brickwork bridge abutment built in the laboratory. The headroom available limited the height of brickwork to 4.275 m. This would be the height of brickwork, from the top of foundation to the capping beam soffit, in an abutment of 6.6m overall height as shown in Figure 1.

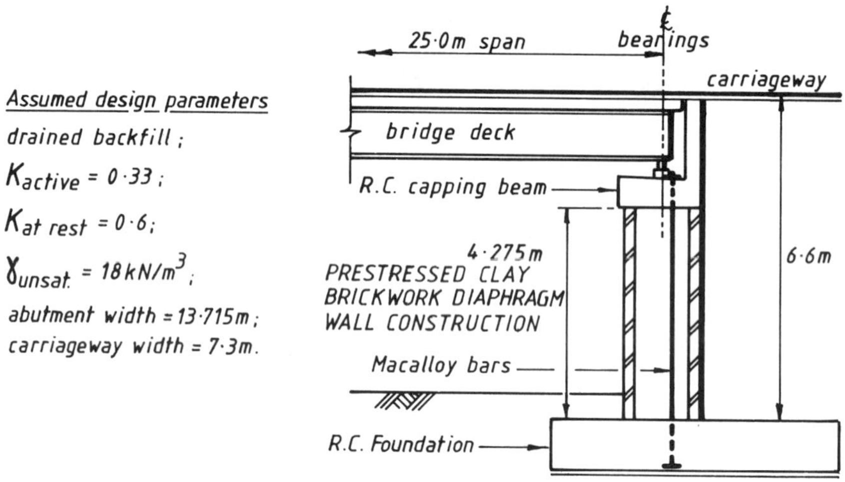

Assumed design parameters

drained backfill ;

$K_{active} = 0.33$;

$K_{at\ rest} = 0.6$;

$\gamma_{unsat} = 18\,kN/m^3$;

abutment width = 13.715m ;
carriageway width = 7.3m.

25.0m span   bearings

carriageway

bridge deck

R.C. capping beam

4.275m
PRESTRESSED CLAY
BRICKWORK DIAPHRAGM
WALL CONSTRUCTION

6.6m

Macalloy bars

R.C. Foundation

Fig.1. Highway bridge with prestressed brickwork
diaphragm wall abutments

2 Test details

2.1 Test arrangement
Details of the test arrangement are shown in Figure 2. Engineering bricks having an average crushing strength of 103 N/mm² and 5.8%

water absorption were used with a 1:¼:3 (cement:lime:sand) volume batched mortar. The brickwork bonding pattern shown in Figure 3 was developed to obviate the need for steel ties which have often proved to be a maintenance liability in building construction. Minimum use is made of cut or special bricks and the bonding pattern is versatile in that the web spacing and the overall depth can be altered in multiples of 225mm. The test abutment construction is described in greater detail elsewhere by Garrity and Garwood (1989b).

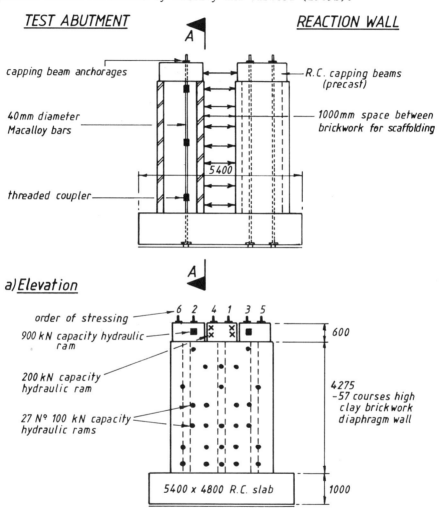

Fig.2. Details of test arrangement

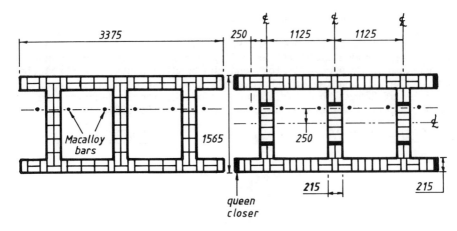

All dimensions are in millimetres

Fig.3. Brickwork bonding pattern
(alternate courses shown)

2.2 Prestressing

The prestressing force was provided by 6 No. 40mm diameter Macalloy
bars at an eccentricity of 250mm. The prestressing force in each bar
was 910 kN; the prestress in the brickwork was 1.02 N/mm² and 3.74
N/mm² in the front and back flanges respectively. In the 5 month
period between prestressing and testing, the losses were as shown in
Table 1.

Table 1. Summary of losses of prestress

Source	Loss (%)
Relaxation of Macalloy Bars (Manufacturers data)	3.5
Elastic Shortening of Brickwork (after topping up of prestress)	0.1
Creep of Brickwork	0.8
7°C Ambient Temperature Rise	0.7
TOTAL LOSS OF PRESTRESS = 5.1%	

2.3 Test loading

Most bridge abutments are designed to withstand the combination of forces shown in Figure 4. However, force P1 was not simulated in the test as it would have a small beneficial effect. Forces P2 and P3 were combined and applied using six large capacity hydraulic rams. Twenty seven smaller rams, each with a load capacity of 100kN and connected to a single electrically controlled pump, were used to provide force P4. These rams were arranged in seven levels, as shown in Figure 2, to produce the trapezoidal distribution of earth pressure loading that would be applied to the full 6.6m height of the abutment. The shear force and bending moment diagrams shown in Figure 4 correspond to the active earth pressure condition with no longitudinal load (i.e. P2 = 0).

The forces produced by the rams were measured using load cells connected to a data logger. Proving ring measurements were also taken to check the load cell readings.

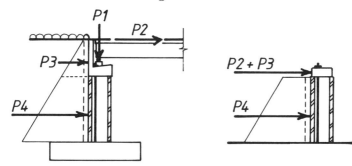

a) *Abutment design loading.*   b) *Simulated loading*

P1 = *vertical reaction from deck due to self weight, superimposed dead and live load*

P2 = *longitudinal load*

P3 = *earth pressure force above brickwork*

P4 = "   "   "   "   "   on   "   "

Fig.4. Design and simulated loading

3 Behaviour of the abutment under test

The test was carried out over two days. On the first day, only the earth pressure forces (P3 and P4) were applied. The loads were increased in eight equal increments so that by load stage 8 the total earth pressure force was twice that produced by active earth pressure based on Rankine's theory. At this stage no cracking was observed. The horizontal deflection at the top of the wall was approximately 1.5mm.

On the second day of the test the earth pressure force was maintained at a fairly constant level of twice the active value while the longitudinal load (P2) was increased in ten equal increments. At the end of the test, load stage 18, the longitudinal load was 750 kN. The applied loads and the corresponding bending moments and shear forces, for the significant load stages of the test, are given in Table 2.

Table 2. Details of significant load stages

Load stage	Total earth pressure load [P3 + P4] (kN)	Longitudinal load [P2] (kN)	Shear force (kN)	Bending moment (kNm)
1	138.9	0	139	309
4	500.7	0	501	1139
8	1014.5	0	1015	2314
9	1027.3	75	1102	2648
10	966.0	150	1146	2886
11	985.3	225	1210	3200
12	997.1	300	1297	3647
15	1055.8	525	1581	4791
18	1075.4	750	1825	5924

At load stage 11, small vertical cracks appeared in the webs at the foot of the abutment. This cracking was probably caused by the hogging curvature of the base which produced large horizontal tensile strains in the top surface of the concrete; these strains were transmitted through the bottom bed joint to the brickwork. The cracks did not develop significantly with increased loading.

At load stage 12, cracks became visible in the concrete base; the position of some of these cracks coincided with the aforementioned vertical cracks in the brickwork webs.

At load stage 15, horizontal cracking developed in the bottom bed joint of the rear flange of the abutment. Decompression had occurred, i.e. the prestress had been annulled, and furthermore the flexural strength of the brickwork had been reached.

At load stage 18, the horizontal crack at the foot of the rear flange had opened to a width of approximately 5mm. Additionally, the crack had propagated along the web bed joint to within approximately 550mm of the front face of the abutment. The abutment was, in effect, rotating bodily about the foot of the front flange. However, at this stage there was no indication that either shear or flexural failure of the abutment was imminent.

It is interesting to note that there was some diagonal cracking in the base resulting from the vertical shearing action caused by the downward line load from the front flange and the upward forces from the prestressing bars.

On removal of the load, the abutment returned to its original position and all the cracks closed up.

The development of cracks in the brickwork abutment and reinforced concrete base is summarised in Figure 5.

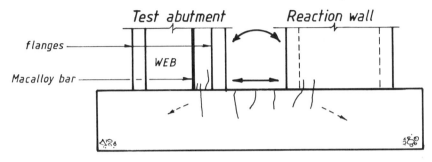

a) *Load stage 12*

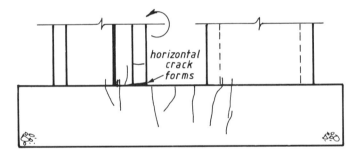

b) *Load stage 15*

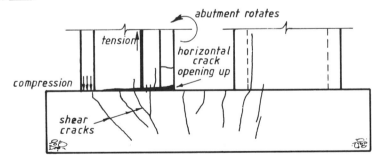

c) *Load stage 18*

Fig.5. Development of cracks during testing

4 Discussion

4.1 Flexural cracking of the base during testing

The cracks in the reinforced concrete base noted at load stage 12 of the test were probably caused by the combination of the tensile force F1 and the hogging bending moment M1 at section A-A, as shown in Figure 6a. In practice however, the concrete base would not have to resist force F1. Also, if the equilibrium of the zone of concrete below the brickwork is considered, as shown in Figure 6b, the bending moment at section A-A would be considerably less than M1. Thus, in practice, the tensile strain in the concrete at the top of the base would be much less than that produced in the test; thus flexural cracking of the base and subsequent cracking of the brickwork webs would probably not occur.

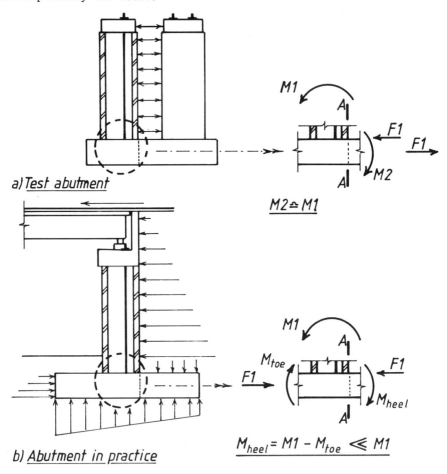

Fig.6. Flexural cracking of the base

172

4.2 Horizontal bending of abutment flanges

In design, the spacing of the webs of a diaphragm wall is often governed by the low flexural strength of the flanges. In fact, calculations based on the characteristic tensile strength of the brickwork indicate that the flanges would not be strong enough to resist the earth pressure loading. To investigate this possible limitation, 10 of the 27 No. 100kN capacity hydraulic rams were positioned close to the centre of the flanges, as shown in Figure 2. As a consequence, between load stages 8 and 18, the flanges were subjected to local shear forces and bending moments which were greater than those which would be generated by at-rest earth pressure. However, the difference in the horizontal deflections measured at the abutment web and at the flange mid-span position was negligible; this indicates that there was no cracking in the flanges as a result of horizontal flexure. The high apparent flexural resistance of the flanges is attributed to horizontal arching of the brickwork between the abutment webs.

4.3 Shear

There was no sign of diagonal cracking in the webs of the abutment either before or after flexural cracking had occurred. Furthermore, when the maximum load was applied, at load stage 18, there was no evidence that shear failure was imminent. It is also interesting to note that the vertical cracks at the bottom of the abutment, which appeared at load stage 11 as a result of bending in the concrete base, did not move any further up the web and did not trigger off a shear failure in the subsequent loading.

Roumani and Phipps (1988b) have shown that the major factors which affect the shear strength are the amount of prestress, the shear span/depth ratio and the principal tensile strength of the brickwork. Using the elastic shear stress distribution for an I-section, the principal tensile stresses were calculated for load stage 14. This was the final load stage before flexural cracking occurred. The principal tensile stresses were 1.54 N/mm^2 at the web-flange interface and 1.02 N/mm^2 at the centroid of the section. No attempt was made to determine the principal tensile stresses after flexural cracking.

4.4 Design service condition

In order to assess the structural performance of the test abutment under working or service conditions, it is necessary to determine realistic values for the horizontal longitudinal and earth pressure forces shown in Figure 4. Using the recommendations of BS 5400(1978) for the bridge shown in Figure 1, the total longitudinal load applied over a single notional carriageway width is 408kN. Assuming that the bridge deck would be sufficiently rigid to distribute this load uniformly through the bearings and capping beam over the full width of the abutment, the longitudinal load appropriate for the 3.375m wide test section would be 100kN. Related to an abutment of 6.6m total height this would produce a bending moment of 660kNm at foundation level.

For the service condition, given the small value of the abutment deflection at load stage 8, it is considered that earth pressure forces should be based on the backfill in the at-rest state. Taking the at-rest pressure coefficient to be 0.6, the total earth pressure force acting on the abutment, caused by drained backfill of weight 18kN/m³ and a surcharge of 10kN/m², would be 928kN. The corresponding bending moment is 2188kNm.

Combining the effects of the longitudinal load and the earth pressure forces means that, under service conditions, the abutment would be subjected to a maximum shear force of 1028kN and a maximum bending moment of 2848kNm. As can be seen from Table 2, these values are very close to the shear force and bending moment resisted by the abutment at load stage 10, when no cracking had occurred. However, although minor cracking was noted at load stage 11, it was not until after load stage 14, when the shear force and bending moment were 1504kN and 4437kNm respectively, that horizontal flexural cracks were observed.

It is relevant to note that the design service bending moment is less than 3665 kNm, which is the magnitude of the decompression moment.

4.5 Design ultimate condition
Using an effective partial safety factor of 1.375 for the longitudinal load and 1.65 for the earth pressure forces, the design ultimate shear force for the abutment would be 1669kN and the design ultimate bending moment would be 4518kNm. However, in the test, the abutment resisted a shear force of 1825kN and a bending moment of 5924kNm without failure occurring.

Hence, although the actual strength of the abutment was not determined experimentally, it has been demonstrated that the abutment was strong enough to resist the design ultimate shear force and bending moment.

5 Conclusions and principal findings

5.1 Losses
Over the five month period between prestressing and testing, the loss of prestress in the abutment was between 5% and 6%. Allowing for a wider range of ambient temperature than that encountered in the laboratory, it is recommended that a loss of prestress of between 10% and 15% is used in design.

5.2 Cracking of the abutment foundation
Although small vertical cracks were induced at the foot of the brickwork webs by the hogging curvature of the concrete base, it is unlikely that such cracks would occur in practice. However, as an interim measure, it is suggested that the flexural tensile stress in the top surface of the concrete base, under service conditions, is limited to 2 N/mm².

The authors also recommend that the abutment foundation should be reinforced to cater for vertical shear beneath the brickwork produced by the upward tensile force in the Macalloy bars and the vertical compressive stress in the front flange of the abutment.

5.3 Horizontal flexural strength of the flanges

The effective horizontal flexural strength of the 215mm thick abutment flanges, spanning between webs of the same thickness at 1125mm centres, was adequate to resist the maximum earth pressure loading.

5.4 Resistance to BS 5400 design service and ultimate loads

Under simulated design service loading, as defined in BS5400, there was no cracking in the brickwork. Also, the abutment was able to resist bending moments and shear forces greater than those produced by the design ultimate loads, and there was no indication that either shear or flexural failure was imminent.

6 Acknowledgements

The authors would like to thank the following:- Armitage Brick Limited for providing the bricks for the project; McCalls Special Products, for providing the Macalloy bars and prestressing accessories and Tilcon Limited for providing the premixed lime and sand for the mortar.

Funding for the project was provided mainly from research grants awarded by the former National Advisory Body for Public Sector Higher Education.

7 References

Ambrose, R.J., Hulse, R. and Mohajery, S. (1988) Cantilevered prestressed diaphragm walling subjected to lateral loading. **Proceedings of the Eighth International Brick/Block Masonry Conference** (ed J.W. De Courcy), Elsevier Applied Science, Dublin, pp. 583-594.

Bell, S.E. (1989a) Development of prestressed clay brickwork in the United Kingdom. **Proceedings of the Fifth Canadian Masonry Symposium**, Vancouver, pp. 155-163.

British Standards Institution (1978) **BS5400 : Steel, Concrete and Composite Bridges**. London.

Curtin, W.G. (1986) An investigation of the structural behaviour of post-tensioned brick diaphragm walls. **The Structural Engineer**, 64B, 4, pp. 77-84.

Curtin, W.G. and Howard, J. (1988a) Lateral loading tests on tall post-tensioned brick diaphragm walls. **Proceedings of the Eighth International Brick/Block Masonry Conference** (ed. J.W. De Courcy), Elsevier Applied Science, pp. 595-605.

Curtin, W.G. and Phipps, M.E. (1982) Prestressed masonry diaphragm walls. **Proceedings of the Sixth International Brick/Block Masonry Conference** (ed Laterconsult s.v.l.), Andil, Rome, pp. 971-980.

Curtin, W.G., Shaw, G., Beck, J.K. and Bray, W. (1982a) Post-tensioned brickwork. **Proceedings of the Sixth International Brick/Block Masonry Conference** (ed Laterconsult s.v.l.), Andil, Rome pp. 961-970.

Curtin, W.G., Shaw, G., Beck, J.K. and Pope, L.S. (1982b) Post-tensioned, free cantilever diaphragm wall project. **Proceedings of the Institution of Civil Engineers Conference - Reinforced and Prestressed Masonry,** Thomas Telford Limited, London, pp. 79-88.

Garrity, S.W. and Garwood, T.G. (1989b) The construction and testing of a full-scale prestressed clay brickwork diaphragm wall bridge abutment. **Proceedings of the 2nd International Masonry Conference** (to be published), London.

Roumani, N.A. and Phipps, M.E. (1988b) The ultimate shear strength of unbonded prestressed brickwork I and T section simply supported beams. **Proceedings of the British Masonry Society,** 2, pp. 82-84.

Wallbank, E.J. (1989) **The Performance of Concrete in Bridges: A Survey of 200 Highway Bridges.** H.M.S.O., London.

OPTIMUM COST DESIGN OF COMPOSITE AND NONCOMPOSITE STEEL BRIDGES

M. MAFI
Civil Engineering Dept, Union College, Schenectady, New York, USA

Abstract
Bridges with composite and noncomposite cast-in-place concrete slabs
on steel stringers are considered for optimization. A quick review
of relevant concepts in optimization is done. Design variables,
constraints, and objective functions are defined. Using a
univariate method, a program was written in FORTRAN that searches
through possible system configurations and finds the lowest cost
design for a particular locale. In three to five seconds after
inputting the required information, the user will be informed of
complete descriptions of the optimum design and few near optimum
designs if he wishes to. This includes thickness of concrete slab,
slab reinforcement and its distribution, type and number of rolled
steel sections and their spacings. The output can also be tied to a
CADD program and a drawing be generated for the optimum design with
the dimensions shown.
Keywords: Bridge Designs, Bridge Optimization, Short-Span Bridges,
Composite Bridges, Noncomposite Bridges.

1 Introduction

The number and percentage of deficient bridges in he United States
is alarmingly high. The result of a survey of bridges in the
United States (Better Roads 1986) shows that nationwide 42% of
bridges are substandard. This percentage for some states is as high
at 69%. The available funds are hardly sufficient for the
rehabilitation or replacement of all deficient bridges (Hegarty
1986). Many structurally deficient and functionally obsolete
bridges were built prior to 1950, and have exceeded their useful
life and should be replaced. The above facts make it imperative
that new bridge construction be as economical as possible.

Efforts have been made on several fronts to decrease the cost
of bridges. Standardization of components, new selection
techniques, development of new materials, and optimization of cross
sections are among the efforts that can be named.

In a separate publication (Mafi, 1988), the above mentioned
studies have been cited and development of an expert system for
bridge selection is reported. Using the expert system developed in
that study, the user can determine which bridge type is the most

economical for a given site while also providing structural
integrity, adequate serviceability, and sufficient safety. After a
bridge type is selected, however, further economy can still be
achieved by refining component proportioning, spacing, and
interconnection of members in such a way as to produce the minimum
cost.   In other words, optimization techniques can be used to
minimize the cost of bridges with respect to the dimensions of
different components, arrangement of members, and geometry of
structure within a selected bridge type.

   In this paper two widely used bridge types were selected for
optimization - namely, composite and noncomposite cast-in-place
concrete deck on steel stringers. These two bridge types will be
described and their overall design consideration will be
highlighted. The optimization problem will be defined. The general
definitions of design variables, constraints, objective function,
method of programming, and solution approach used in the
optimization technique will be followed by the specifics regarding
the bridge problem at hand.

2 The Bridge Types Selected and Their Design Considerations

The brief description of this type of bridge superstructure in the
National Cooperative Highway Research program report 222 (1980)
states:

>   "This is a very common type of bridge. A wide range
>   of standard wide flange and I-Beam shapes are
>   available. The size and spacing of beams may be
>   adjusted from site to site to optimize the use of
>   material. The standard rolled shapes are typically
>   used for spans which are less than 90 ft. This
>   bridge system is widely used all over the world.
>   Studs are generally used to achieve composite
>   action except for very short spans where it may be
>   economical to omit the studs" (p.97).

   A typical cross section of this type of bridge is shown in
Figure 1.

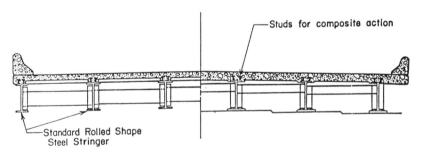

PERMANENT FORM WITH SITE-
CAST DECK AT MIDSPAN   SITE-CAST DECK AT ABUTMENT

Fig. 1. Typical cross sections of bridges with cast-in-place
        concrete deck on steel stringers

In the present study, both noncomposite and composite arrangements were considered. Due to the high cost of providing shoring most composite bridges are constructed without shoring, and hence this construction method has been considered in this study. No cover plates are considered for composite sections; since, according to the manual of steel construction (1980:2-97), "high labor cost has made their use rare."

Dead loads, truck and lane loads, impact load and the proper combination of these are used according to AASHTO specifications. The load factor design method is used for analysis of the bridge. In this method three load levels are considered, i.e., maximum design load, overload, and service load.

For short-span bridges, the use of rolled wide-flange beams is always more economical than the use of plate girders, except for areas where fabrication costs are fairly low. In this study wide flanges ranging from W5 to W36 were used. The properties for these wide flanges are the ones listed in the manual of steel construction (American Institute of Steel Construction, 1980). Today the carbon steel used for bridges is A36, which as a minimum yield point of 36 ksi. When higher strength is desired, A588 steel with minimum yield point of 50 ksi is usually specified. A few bridges were analyzed and designed to identify all the factors involved. All the requirements of AASHTO were considered in the designs.

3 The Bridge Optimization Problem

The bridge under consideration consists of several beams that span in the direction of the traffic, with a reinforced concrete deck placed on top of the beams. The concrete deck is designed as a continuous slab spanning transversely over the several interior beams, which act as simple supports. The beams are then designed to carry longitudinal bending moments. If adequate mechanical shear connectors are used, the slab can be considered as a cover plate for the beams and thus help them to carry the longitudinal moments compositely. The use of shear connectors allows for the use of smaller steel beams, but the cost of the shear connectors must be added.

Within this concept, many designs satisfy the functional requirements. The problem at hand is that of developing an automated design procedure which, by varying the factors involved, seeks the optimum geometry and member selection that yield the minimum costs.

4 Design Variables

4.1 Design Variables in General
The set of quantities that describes a structural system is comprised of two groups. These quantities that are fixed before analysis and are not varied during optimization, are called pre-assigned parameters. Design variables, the second group, are varied during optimization. The combination of pre-assigned parameters and

design variables completely defines a design.

4.2 Design Variables For The Bridge Problem

The variables of the problem can be grouped into the following categories:

Geometrical variables
    Dimensions of the rolled wide flange beam sections
    Concrete slab thickness
    Area of main reinforcing steel and distribution steel in
        concrete slab
    Degree of beam-slab interaction (composite, noncomposite)
    Number and spacing of beams
    Span of the bridge
    Width of the bridge
Material variables
    Concrete strength
    Yield point of reinforcing steel
    Yield point of structural steel
Cost variables
    Unit cost of reinforcing bars
    Unit cost of structural steel
    Unit cost of concrete
Loading variables
    Type of loading (AASHTO truck and alternate military
        loading, 125 percent of AASHTO loads, and PaDOT permit
        load)
    Type of wearing surface
    Curb and parapet types
Additional variables for composite design
    Type and size of shear connectors
    Transverse and longitudinal spacing of shear connectors
    Modulus of elasticity of concrete
    Cost of single stud
    Range of longitudinal shear
    Anticipated cycles of stress reversal

5 Constraints

5.1 Constraints in General

As noted previously, any set of values assigned to the design variables, collectively referred to as the design vector {X}, is called a design. Only those designs that satisfy all the requirements are acceptable. These are called feasible designs. The rest of the possible designs are unacceptable, and hence are called infeasible. The restrictions that must be satisfied in order to produce a feasible design are called constraints. Constraints can usually be expressed in the form of equalities and/or inequalities such as:

$$g_i (\{X\}) < 0 \qquad i = 1, \ldots K$$

$$h_j (\{X\}) = 0 \qquad j = 1, \ldots L$$

where {X} is the vector of N design variables

K is the number of inequalities

L is the number of equalities

5.2 Constraints For The Bridge Problem
The constraints considered are those relating to slab design, noncomposite stringer design, and composite stringer design.

5.2.1 Slab Design Constraints
The constraints for the concrete slab are the same for composite and noncomposite designs. They are as follows:

a. A minimum slab thickness of $(S + 10)/30$ is required, but not less than 6.5 inches, where S is the distance between the edges of the flanges plus half the beam flange width. In some state departments of transportation this minimum is set at higher values.
b. Half an inch will be considered as an integral wearing surface.
c. The top cover for the reinforcement in the slab should be at least 2.5 inches.
d. The bottom cover for the reinforcement in the slab should be at least 1 inch.
e. The overhang of the deck slab outside of the fascia stringer should not exceed 4.25 feet, or half the stringer spacing, whichever is smaller.
f. The maximum steel ratio is 0.75 ρ_b.
g. A minimum steel ratio of $200/f_y$, or enough to develop a factored moment at least 1.2 times the cracking moment is required, unless the reinforcement is at least one-third greater than that required by the analysis.
h. The same main reinforcement is to be used for the negative and positive moments of the continuous span.

5.2.2 Stringer Design Constraints for Composite Systems
For the sake of brevity, the constraints on stringer design for only composite systems are given. They are as follows:

a. There should be at least three stringers.
b. The number of stringers should not exceed the maximum number of beams designated by the user.
c. The fascia beam should be at least as strong as the intermediate ones.
d. The depth of the stringer should be less than the maximum depth input by the user.
e. If $d/t_w \overset{<}{=} 13300/ (F_y)^{1/2}$ and $V_u \overset{<}{=} 0.35\ F_y dt_w$, then the section is compact, and the ultimate strength of the section must be greater than the maximum moment, which is defined as
$$M_{max} = 1.3\ \{M_{DS} + M_{DC} + 5/3\ (M_L + I)\}$$
where M_{DS} = moment due to dead load on stringer
    M_{DC} = moment due to dead load on concrete

$M_{L + I}$ moment due to live load plus impact

f. If the above criteria are violated, but $D/t_w \overset{<}{=} 150$, and $V_u \overset{<}{=} 0.58\ F_y Dt_w$, then the section is noncompact, and the stresses acting on steel due to 1.3 M_{DS} acting on steel alone, and 1.3 { M_{DC} + 5/3 ($M_{L + I}$)} acting on the composite section, should not exceed F_y.

g. The stresses on the steel beam due to M_{DS} acting on steel beam alone and those due to {M_{DC} + 5/3 ($M_{L + I}$)} acting on the composite section should not exceed 0.95 F_y.

h. The deflection of the composite section due to live load plus impact should not exceed 1/800 times the span.

i. The height-over-diameter ratio of the studs must be greater than or equal to 4.

j. Eighty-five percent of the ultimate strength of all the studs placed between the points of zero moment and maximum moment must be greater than the smaller value of the formulas

$$P_1 = A_s F_y$$
$$P_2 = 0.85\ f'_c\ bc$$

where A_s = total area of steel section (in²)

F_y = minimum specified yield point for the steel section (psi)

f'_c = specified 28-day compressive strength of concrete (psi)

b = effective flange width (in)

c = thickness of the concrete slab (in)

k. The ultimate strength of one stud with a height-over-diameter ratio of more than 4 is equal to

$$S_u = 0.4\ d^2\quad (f'_c\ E_c)^{1/2}$$

where d = diameter of stud (in)

E_c = modulus of elasticity of concrete (psi)

l. To satisfy fatigue requirements for studs, the following inequality must hold:

$$Z_r n \overset{>}{=} S_r p$$

with $Z_r = \alpha d^2$ and $S_r = \dfrac{V_r Q}{I}$

where α = 13,000 for 100,000 cycles
        10,600 for 500,000 cycles
         7,850 for 2,000,000 cycle
         5,500 for over 2,000,000 cycles
     d = diameter of studs (in)

V_r = range of shear due to live loads and impact. At any section, the range of shear shall be taken as the difference between the minimum and maximum shear envelopes (lb)

Q = static moment about the neutral axis of the composite section of the transformed compressive concrete area (in³)

I = moment of inertia of the transformed composite stringer (in⁴)

p = center-to-center distance between rows of studs (in)

n = ratio of modulus of elasticity of steel, E_s, to that of concrete, E_c.

6 Objective Function

6.1 Objective Function in General
Among the many feasible solutions, some are better than others. Such a conclusion is possible only if there is a quality of which more or less can be found in the better solution. Mathematically speaking, that desired quality must be expressed in the form of a function of the design variables, F ((X)). This function, usually called the "objective function" or "merit function," can be evaluated, and its value is then used to compare feasible solutions. Finding the best alternative can then be looked upon as finding the maximum nor minimum value of the objective function.

6.2 Objective Function for the Bridge Problem
The objective function is the total cost of the structural steel, concrete, reinforcing steel, and studs (if composite). The unit costs associated with these items are input by the user. The unit costs of the structural steel and the reinforcing steel are expressed in dollars per pound, whereas the unit price for concrete is usually given in dollars per cubic yard. Assuming that the reinforcing bars in the positive and negative areas extend across the bridge, the weight of the main reinforcing bar, W_R, is

$$W_R = 2 \ (A_{SM}/144.0)(B)(L)(\gamma_R)$$

where: L = span length (ft)

B = bridge width (ft)

A_{SM} = area of main reinforcement per foot of slab along

the span of the bridge (in²)

γ_R = unit weight of reinforcing bars (lb/ft³)

The weight of distribution reinforcement, W_D, is
$$W_D = (A_{SD}/144.0)(B)(L)(\gamma_R)$$
where: A_{SD} = area of distribution reinforcement per foot across the bridge (in²)
and the weight of temperature reinforcement, W_T, is

$$W_T = (A_{ST}/144.0)(B)(L)(\gamma_R)$$

where: A_{ST} = area of temperature reinforcement per foot along the span (in^2)

The cost of all types of reinforcement, C_R, is thus

$$C_R = (W_R + W_D + W_T)(U_R)$$

where U_R is the unit cost of reinforcing bars provided by the user.

The cost of the concrete, C_C, is

$$C_C = (L)(B)(t_s/12.0)(U_C)$$

where t_s is the thickness of slab in inches, and U_C is the user-provided unit cost of concrete.

The cost of steel beams, C_B is

$$C_B = (W)(L)(N_B)(U_B)$$

where W = weight per unit length of steel beam under consideration

N_B = number of steel beams used

U_B = unit cost of steel beam given by the user.

Therefore, the value of the objective function, or the total cost for the noncomposite superstructure under consideration, C_T, is

$$C_T = C_R + C_C + C_B$$

where C_R , C_C , and C_B are as defined earlier.

To calculate the total cost for the composite design, the cost of the studs C_S, must be added to the cost equation:

$$C_S = (N_S)(N_B)(U_S)$$

where N_S = number of necessary studs for one stringer

N_B = number of composite steel stringers in the superstructure

U_S = cost of one stud.

The total cost for a composite design, C_{CT}, is

$$C_{CT} = C_R + C_C + C_B + C_S$$

where C_R, C_C, and C_B are defined in the same way as for the noncomposite design.

7   Problem Formulation

7.1 Formulation of an Optimization Problems in General

After design variables, constraints, and the objective function have been identified, any problem of optimal structural design can be stated in the following form: choose values of the design variables, subject to the given constraints, that will minimize the objective function; that is, choose the components of vector {X} satisfying

$$g_i\,(\{X\}) \overset{<}{=} \qquad i = 1,\ \dots K$$

$$h_j\,(\{X\}) = 0 \qquad j = 1,\ \dots L$$

so that

$$F\,(\{X\} \rightarrow \min.$$

Each design variable or component of {X} can be considered as one dimension in a design space. Since the vector {X} is composed of design variables, and the behavior of the structure can be computed by analysis equations for any {X}, these equations can be separated from the mathematical formulation and used to evaluate the response behaviors which are needed in the constraint expressions. This formulation, which focuses attention on the design variable space, is used in this study.

7.2 Search Method
The method used in this study is called the univariate method in which the minimum of the function is approached in a succession of approximations by adding a scalar multiple of a unit vector along one of the coordinate axes to the previous approximation. This method proceeds according to the following formula:

$$\{X\}_{i+1} = \{X\}_i + \alpha_i \{S\}_i$$

where $\{X\}_i$ = the vector representing the current design

$\{X\}_{i+1}$ = the new design

$\{S\}_i$ = the cyclic ordering of the unit vectors,
$S_1 = (1, 0, 0, \ldots, 0)$, $S_2 = (0, 1, 0, \ldots, 0)$, $\ldots$ and $S_n = (0, 0, 0, \ldots, 1)$

α_i = the positive or negative step chosen so that

$$F(\{X\})_{i+1} < F(\{X\}_i).$$

After a starting point, $\{X\}_0$, has been chosen, a multiple of the unit vector along the first coordinate axis in the design variable space is added to this initial design to obtain a new point. The objective function is calculated for the new point and compared with the value of the objective function for the starting point. If movement in the direction of the coordinate axis does not decrease the objective function, the sign of the step is changed to evaluate movement in the opposite direction. Movement along the first coordinate axis continues until no further improvement in the value of the objective function takes place. The next move will then be made along the second coordinate axis. Here again, a probe is made in both directions to determine the direction along which the objective function decreases.

This procedure continues until a move along all the coordinate axes has been attempted, at which point a new cycle starts by using the first coordinate axis again.

A number of variations of the method are possible, depending on the rationale used for implementing this exploitation (Fox 1971).

The starting point, the minimum step length, and the stop criterion are the important factors to be considered in the univariate method. A judicious selection of these three factors adds to the efficiency of the optimization program.

In structural design problems, it is usually not difficult to select an initial design, since any over-design configuration that satisfies the constraints can serve as the initial point in the design variable space. The other two factors, the minimum step

185

length and the stop criterion, should be customized for any
particular problem.

7.3 Particulars of the optimization program
In order to increase the efficiency of the optimization program, the
number of variables had to decrease. An effort was made to include
as many variables as possible in the category of pre-assigned
variables defined in 4.1, above. Span, bridge width, the yield
stress of steel, the ultimate strength of concrete, and whether the
structure is composite or note are all pre-assigned.

Some variables, like spacing and number of stringers, are not
independent of each other. Discrete values for moments of inertia
and other cross sectional properties of structural steel were used.
The slab thickness was also treated as a discrete-valued variable
with half-inch increments in order to obtain a more practical value.

A number of modifications had to be made to the universal search
method to accommodate the above points. After each move along the
axis representing the number of stringers, another move is made
along the axis corresponding to the deck thickness. For each
combination of number of stringers and thickness of slab, coupled
with the values for pre-assigned variables, the slab design and the
selection of the rolled wide-flange stringer are carried out
considering all the relevant constraints.

In essence, the program searches through possible system
configurations and rejects all that are structurally infeasible.
Only the designs satisfying all of the constrains will be retained
for calculation of cost and for eventual comparison to find the one
with the lowest cost.

7.4 Assumptions For The Bridge Problems
The following assumptions have been made in developing the
optimization algorithm:
   a. If sufficient bracing is indicated by the user, all the beams
      are assumed to deflect equally. Otherwise, each beam is
      assumed to deflect due to its share of live load.
   b. The weight of the concrete is assumed to be 145 lb/ft^3.
   c. The modulus of elasticity of steel is assumed to be
      29,000 ksi.
   d. The modulus of elasticity of concrete is obtained from the
      following formula:

$$E_c = \omega^{1.5} \, (33) \, (f'_c)^{1/2}$$

   where E_c = modulus of elasticity of concrete (psi)

   ω = unit weight of concrete, here 145 lb/ft^3

   f'_c = ultimate strength of concrete (psi)

   e. The maximum deflection of a simply supported stringer is
      assumed to occur at midspan.

f. Bond and shear are assumed to be satisfactory for concrete slab.
g. The weight of the curbs and parapets is assumed to be carried by all the stringers.
h. The concrete slab is assumed to provide full lateral support for the stringer.
i. Two studs in a row are used on the upper flange of the stringer.

8 Program Description

The program is composed of four parts. The first part is the input section. The program gives the user two choices for inputting information. The user can provide information interactively or store all the information needed in a file which will be read by the program. The input file contains questions and provides sample responses to those questions. The user will have to edit only those numerical values that are different from the default values.

In the second part, the information is processed and some data generation takes place. At this point, the optimization procedure starts and other information stored on separate files is accessed by the control program. During the optimization procedure the objective function has to be calculated repeatedly. The third part of the program does the objective function calculations. The final part of the program deals with the output when the user is given several options about the amount, kind, and format of results.

9 Utilization of the program, an example

9.1 Example input

This example is intended to show the results of one run of the program. The best way to provide the input is to call a special input file and type the numerical values that relate to the problem at hand. Any editor that handles ASCI files can be used. The user input can be typed anywhere on the appropriate lines. For this example problem, some important pieces of information from the input file are as follows:

Span: 65 ft; width: 47.5 ft; loading: HS20 and military load; Fy for structural steel: 50 ksi; Fy for reinforcing bars: 60 ksi; f'_c for concrete: 4 ksi; No. of traffic lanes: 4

Information on the size of studs used, limits on depth of stringers, minimum and maximum slab thickness, parapet specification, unit prices, and a few other pieces of information from the input file are not mentioned here for the sake of brevity.

9.2 Example Output

The program with the above input was run on a VAX 11/785 at Union College and it took 6.7 seconds to run. At the completion of a run, the program informs the user of several options for the form, extent, and type of output. At the request of the user the program can create a file which can be read by a CADD program to create the cross section of the optimum design. Figure 2 shows the computer generated cross section of the minimum cost composite design for this problem.

Information about the optimum design and near optimum designs can be requested in tabular form. Table 1 shows that information for the composite bridge in the example.

Table 1.   Information on the optimum and the next-best design for the composite bridge.

DESIGN	NO. OF BEAMS	BEAM SPACING (ft)	STEEL SECTION	SLAB THICKNESS (in.)	MAIN REINFORCEMENT (in.²/ft)	DISTRIBUTION REINFORCEMENT (in.²/ft)	NO. OF STUDS	COST ($)
Optimum	5	10.75	W36X135	9.00	0.61	0.41	156	144829
Next Best	6	8.60	W30X124	8.00	0:58	0.39	142	151259

10 Other advantages of the program

In addition to providing optimum and near optimum solutions, the program enables the user to see the effects of changes of several variables on the optimum design. For example, many of the following questions can be answered by changing the appropriate number in the input file and then noting the effect on the optimum design in three seconds.
- What happens to overall cost if I use the more expensive $F_y=50$ ksi steel instead of the regular $F_y=36$ ksi?
- What is the difference in design and cost if I make my system composite?
- How do changes in concrete strength and price affect optimum design?
- What is the effect on costs of minimum slab thickness mandated by several states?
- What is the optimum space between stringers for different span lengths to provide lowest cost?

11   Appendix A. Conversion Table

$$1 \text{ inch} = 0.0254 \text{ m}$$
$$1 \text{ foot} = 0.3048 \text{m}$$
$$1 \text{ lb/ft}^3 = 157.086 \text{ Newton/m}^3$$
$$1 \text{ Ksi} = 6.89 \times 10^6 \text{ Newton/m}^2$$

12 References

American Association of State Highway & Transportation Officials (AASHTO), **Standard Specification for Highway Bridges**, (1989).
American Institute of Steel Construction. **Manual of Steel Construction**, 8th ed., 1980.

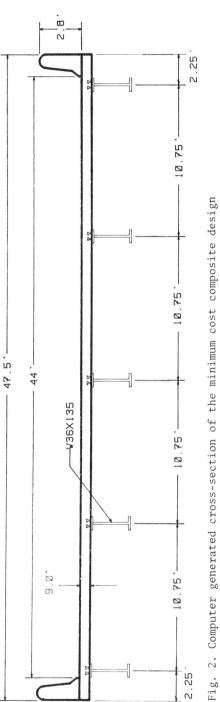

Fig. 2. Computer generated cross-section of the minimum cost composite design for the example problem.

Better Road, (1986), Nov., p. 42.

Fox, R.L. (1971) **Optimization Methods for Engineering Design,** Addison-Wesley, Reading, Ma.

Hegarty, M.J. (1986), Design variables and systems scenario of low-volume bridges, **Problem Report,** West Virginia Univ., Morgantown, WV.

Mafi, M. (1985) Cost-effective, Short-span Bridge System. Ph.D. Thesis, Pennsylvania State University, State College, Pa.

Mafi, M. (1988) A basic expert system for type selection of bridges (BEST Bridge), in **Proceedings of ASEE Conference,** Portland, Oregon,

National Cooperative Highway Research Program, **Bridges of Secondary Highways and Local Roads: Rehabilitation and Replacement.** NCHRP Report 222, May 1980.

A COMPUTER-BASED METHODOLOGY FOR INVESTIGATION OF BRIDGE VEHICLE/SUPERSTRUCTURE INTERACTION

T.E. FENSKE, S.M. FENSKE
Dept Civil Engineering, University of Louisville, Louisville,
Kentucky, USA

Abstract
This paper presents a summary of our current efforts in the
development of an analytical methodology for the investigation of
vehicle-superstructure interaction problems which result from a
vehicle traversing a highway bridge system. This investigation is
prompted by uncertainties associated with current design practices
which have been shown to be either overly conservative or unsafe under
general traversing vehicular loading.
    The analytical methodology considers the vehicle-superstructure
interaction problem based upon a moving mass, moving force, or
pseudo-moving static live load formulation. The kinematic coupling,
due to the velocity of the traversing vehicle mass, is included in the
moving mass formulation. The analytical methodology is based upon the
finite element analysis method, and necessary algorithms have been
implemented. The direct integration solution procedure is employed.
Keywords: Bridges, Bridge Dynamics, Computer-Aided Analysis/Design,
Kinematic Coupling, Bridge Superstructure/Vehicle Dynamic Interaction.

1 Introduction

Highway bridges are very large, intricate structures in which the
means of transferring vehicular loading through the bridge and into
the supporting substructure is not completely understood. This is
because the extent of participation of each structural component of
the bridge is not fully known. For example, in a highway girder
bridge there are typically three major structural components which can
be considered to compose a bridge superstructure. They include (1)
the deck slab, (2) the girders, and (3) the diaphragms. The vehicular
loading is passed from the deck into the girders and diaphragms and
then to the supporting substructure. There is also a fourth component
for girder bridges. This fourth component is the edge stiffness
resulting from a guardrail (barrier) or sidewalk which, if present,
may have a significant effect on the behavior of the bridge.
    The complexity of a bridge makes an accurate structural analysis
difficult to perform. As a vehicle traverses a highway bridge
superstructure initially at rest, the superstructure deflects from the
original equilibrium position. Forces acting upon the superstructure
are a combination of forces due to vertical acceleration of the
vehicle and bridge masses, the horizontal effect of vehicle velocity,
plus the static force due to the weight of the combined vehicle/bridge
system. For girder bridges, these forces combine to give maximum
static and dynamic effects that are distributed to the various
supporting girders. The traversing vehicle, which causes the combined

191

static and dynamic effects, is termed "live loading" because the loading effect is transient.

2 Current Solution Approaches

Many analytical methods exist which can be used to predict the complex response of the vehicle-superstructure interaction problem. However, current bridge design practice, as recommended by the American Association of State Highway and Transportation Officials (AASHTO, 1983), avoids a direct treatment of the dynamic behavior of vehicle-bridge interaction. Instead, the dynamic behavior is addressed by increasing the static design live load by an "impact factor" taken as a function of span length and having a maximum value of 0.3. Again considering a girder bridge, the static design live loads used are evaluated based upon wheel load distribution factors. These distribution factors represent how the loading effect of the traversing vehicle is distributed to the various supporting girders. The applicable parameters for using the AASHTO truck wheel load distribution are based upon the bridge type and girder spacing.

These standard design procedures for evaluating live load distribution and impact (dynamic effect) are easy to apply and require only minimal computational effort. However, the need to achieve greater economy in bridge construction and the need to insure safe design standards mandates the development of more accurate and reliable procedures. Refinement of the AASHTO empirical factors is necessitated by the need to avoid unconservative applications under certain limited conditions of span and bridge continuity. Application of AASHTO distribution factors have been shown to lead to unsafe conditions for design under some situations (Hays and Hachey, 1984). Also, because of the simplified nature of the AASHTO empirical distribution factor applied to various design conditions, it is overconservative for most applications. Under these conditions, design practice leads to excessive costs. Another study (Dorton, et. al., 1977) illustrated that, an accurate lateral live load distribution factor would result in a design moment that is 27 percent less than the AASHTO recommended value.

Using the AASHTO impact formula of $50/(L+125) \leq 0.3$ to account for the dynamic effect of the traversing vehicle by increasing the static design live load moment is obviously unreliable in the cases where the lateral load distribution is unreliable (since the design live load is directly proportional to the distribution factor). However, even in cases where the distribution factor is sufficiently accurate, the use of the AASHTO empirical impact formula is highly questionable because of its limited consideration of the dynamic parameters of the bridge system. The parameters normally used to describe the dynamic behavior of bridges are the natural frequencies, mode shapes, and the displacement and acceleration response under forced vibration. It is recognized that a substantial magnification of the static live load effect can occur in a continuous bridge system when a vehicle traverses the superstructure at a speed coinciding with the natural frequency of the structure. This effect requires either rapid vehicle transit or a relatively large vehicle-superstructure mass ratio, which generally has not been the case. However, as the trend in new designs continues toward lighter and more flexible superstructures, the problem of the dynamic behavior is becoming increasingly more prevalent (Brown, 1977). The dynamic effect has been shown to be both considerably greater and considerably less than the AASHTO recommended value of 30 percent (Green, 1977).

These significant variations from the impact formula can be related to the superstructure natural frequency, the effect of which is not included in the AASHTO empirical impact formula. When the frequency of the bridge is between 2 Hz and 5 Hz, it is in the range of most large trucks and, thus, the dynamic effect of the vehicle-superstructure becomes significant. On the other hand, if the bridge natural frequency is far removed from this truck frequency range, the dynamic effect is considerably less. The primary reason that a more rigorous dynamic effect is not incorporated into the AASHTO impact formula is the difficulty associated with an accurate and reliable analysis procedure. A central difficulty in all analysis approaches to the vehicle-superstructure interaction problem involves finding a suitable method for treating the kinematic coupling term which arises in the mathematical formulation of the problem. Existing analytical solution techniques make use of simplifying assumptions so as not to treat the dynamic coupling effect.

The kinematic coupling effect is due to the traversing vehicle position change with respect to time, i.e., the velocity of the vehicle. To illustrate the kinematic relationship, a simplified vehicle-superstructure interaction model represented by a simple beam and single mass particle is shown in Figure 1. The equation of motion for the superstructure can be expressed in generic form as

$$m\ddot{y} + c\dot{y} + ky = F(x,t) \tag{1}$$

in which the superstructure properties are represented by m (mass),

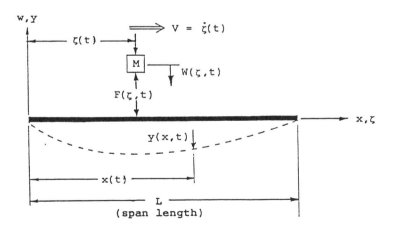

Fig.1. Vehicle-Superstructure Interaction

193

c (damping), k (stiffness), and y (=y(x,t), the displacement function representing the vertical displacement at position x and time t). $\dot{y}$ and $\ddot{y}$ represent the corresponding velocity and acceleration. The coupling effect comes from examining the forcing function F(x,t). F(x,t) is the reaction force exerted by the moving particle mass on the beam superstructure. When the mass is at position ζ(t), the forcing function F(x,t) can be related to the vertical acceleration of the particle by Newton's second law, i.e.,

$$F(x,t) = -m[g + \frac{d^2w}{dt^2}(\zeta,t)] \; \delta(x-\zeta) \qquad (2)$$

In Equation 2, mg represents the weight of the particle and $\delta(x)$ is the Dirac Delta function having value of 1 when ζ(t) = x and zero elsewhere. Since the particle mass position ζ(t) is a function of time, the explicit form of the transverse acceleration is obtained as

$$\frac{d^2w(\zeta,t)}{dt^2} = \frac{\partial^2 w}{\partial x^2}(v^2) + (2v)\frac{\partial^2 w}{\partial x \partial t}$$

$$+ \frac{\partial w}{\partial x}(\dot{v}) + \frac{\partial^2 w}{\partial t^2} \qquad (3)$$

in which v = $\dot{\zeta}$(t), i.e., longitudinal mass velocity. The kinematic coupling is from the mixed derivative and neglecting this effect restricts the formulation to small velocities. The formulation of this problem type is called a "moving-mass" problem. However, neglecting the velocity effects in Equation 3 reduces the analysis problem to a "moving force" approximation. In either case, the variation from the AASHTO impact formula can be very significant.

These bridge design difficulties may be avoided or controlled by developing an accurate and reliable bridge analysis procedure. Obtaining accurate and reliable live load distribution factors and related dynamic contributions requires a reliable methodology that accurately considers the significant parameters that influence the outcome of the analysis. Unquestionably, when considering large redundant systems for which closed-form solutions are nonexistent, the most reliable and comprehensive method of analysis available today is the finite element method.

3 Analytical Methodology

As previously stated, the finite element analysis procedure is utilized in the investigation of the transient response of general bridge systems subjected to traversing vehicular loading. For example, slab bridges have been modeled using either isotropic or orthotropic plate finite elements. These elements can be either 4-node Melosh plate elements or 4-, 8-, or 9-node Mindlin plate elements. The Mindlin plate elements account for shear effect, and either full or reduced integration can be employed. On the other hand, girder type bridges have been modeled using a combination of beam and plate elements.

The equivalent nodal force vector, used in the solution of the equation of motion, is formulated for the various traversing vehicle locations. These vehicle locations can be specified by using a constant acceleration, linear acceleration, or constant velocity approach. When the effect of the mass of the moving vehicle is considered, the analysis is coupled, and it is termed "moving mass" analysis. However, when only the force of the moving vehicle is considered, the analysis is linear and is termed "moving force" analysis. A pseudo-moving static effect also can be examined.

Direct integration is utilized in the solution of the dynamic interaction problem. Several schemes have been implemented: Newmark's Beta method, Hilbert's Alpha method, and a modified Hilbert scheme.

3.1 Basic Algorithms
The general equation of motion for the bridge system is given by

$$M\ddot{Y} + C\dot{Y} + KY = F(x,t) \tag{4}$$

where Y is the nodal displacement vector and $\dot{Y}$ and $\ddot{Y}$ represent nodal velocities and accelerations, respectively. M, C, and K represent the global mass, damping, and stiffness matrices. $F(x,t)$ represents the nodal forcing function.

If the mass of the moving vehicles is included in the nodal forcing function, $F(x,t)$, then the contribution of the kinematic coupling effect can be expressed as

$$F(x,t) = -mg - m\left[\frac{\partial^2 W(\zeta,t)}{\partial t^2}\right] = F_S + F_D{}^* \tag{5}$$

This equation is applicable for the loaded elements only. "m" is the mass of the traversing vehicle. In this equation, $F_D{}^*$ can be defined as the equivalent dynamic nodal loading and can be expressed as

$$F_D{}^* = -M^*\ddot{Y} - C^*\dot{Y} - K^*Y \tag{6}$$

where M^*, C^*, and K^* can be considered as the equivalent mass, damping, and stiffness matrices. Substituting the forcing function expression into the equation of motion and rearranging yields

$$(M + M^*)\ddot{Y} + (C + C^*)\dot{Y} + (K + K^*)Y = F_S. \tag{7}$$

This equation represents the moving mass problem that includes the kinematic coupling effect. Neglecting the kinematic effect would allow for the moving force formulation of

$$(M + M^*)\ddot{Y} + C\dot{Y} + KY = F_S. \tag{8}$$

Similarly, the pseudo-moving static formulation can be expressed as

$$KY = F_S. \tag{9}$$

The described formulation of the analytic methodology has compared favorably in extensive comparisons to closed-form solutions and experimental results.

4 Application of Analytical Methodology

The first step in applying the analytical methodology is modeling the bridge system. For example, the highway girder bridge is modeled using a combination of beam and plate finite elements. A typical two-span highway girder bridge is modeled as shown in Figure 2. Figure 2(a) represents the overall bridge geometry and Figure 2(b) shows the generated finite element mesh. The beam and plate elements selected

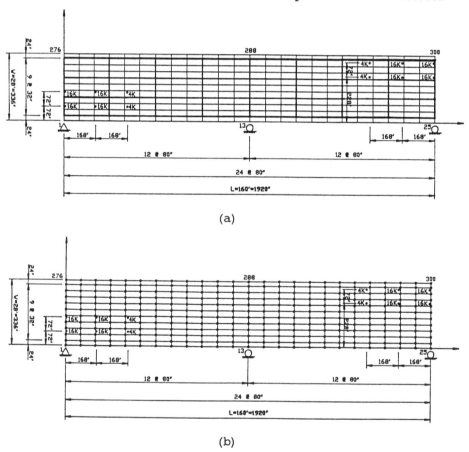

(a)

(b)

Fig.2. Highway Bridge Model

196

satisfy requirements for nodal compatibility. The bridge supporting
girders, guardrails, and transverse diaphragms are modeled
as beam elements. The roadway deck is modeled using plate bending
elements.

The dynamic interaction analysis results are obtained upon
formulating the analysis model by employing the described analytical
algorithms. The results of the analytical methodology compare very
favorably to closed-form solutions. For example, consider the simply
supported plate shown in Figure 3 (without being given specific data
regarding vehicle force, mass, velocity, plate geometry, etc.) A
comparison of results for the moving mass (MM), moving force (MF), and
static (S) analysis for both an isotropic (I) and orthotropic (O)
plate is given in Table 1.

A more general slab bridge is illustrated in Figure 4. For this
structure, the use of orthotropic plate elements is required. The
plate element bending rigidities can be evaluated using the shaded
portion shown on the bridge cross-section. Again, without giving the
specifics of the interaction problem shown in Figure 4, a comparison
of results is presented in Table 2.

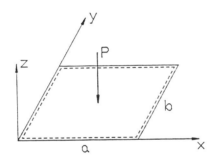

Fig.3. Simply Supported Plate

Table 1.   Center Node Comparison - Simply Supported Plate

Plate Type	Analysis Procedure	Deflection (in.) AM	CF	% Diff
I	S	-0.81006	-0.81000	0.0074
O	S	-0.20188	-0.21200	4.9400
I	MF	-0.03360	-0.03170	5.8600
O	MF	-0.00670	-0.00660	0.9700
I	MM	-0.03440	-0.03180	7.6400
O	MM	-0.00680	-0.00700	3.5200

AM-Analytical Methodology, CF-Closed Form

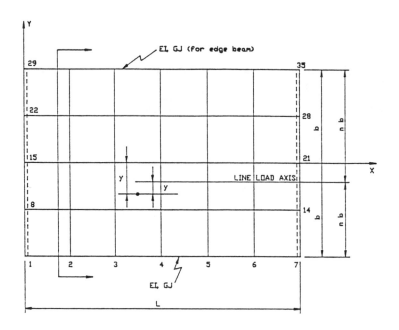

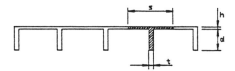

Fig.4. General Slab Bridge

Table 2. Center Node Comparison - General Slab Bridge

Analysis Procedure	Deflection (in.) AM	CF	% Diff
MF	−0.00619	−0.00610	1.3800
MM	−0.00652	−0.00671	2.8600

AM-Analytical Methodology, CF-Closed Form

5 Conclusion

The developed analytical methodology has been demonstrated to be an accurate and reliable analysis tool for the investigation of the dynamic interaction problem.

6 Acknowledgement

This work has been sponsored by the National Science Foundation under Grant No. MSM-8708691.

7 References

American Association of State Highway and Transportation Officials, Standard Specifications for Highway Bridges, Thirteenth Edition, 1983.

Brown, C. W., "An Engineer's Approach to Dynamic Aspects of Bridge Design," Proceedings, Symposium on Dynamic Behavior of Bridges, Crowthorne, Berkshire, England, 1977.

Dorton, R. A., Holowka, M., and King, J. P. C., "The Conestogo River Bridge -- design and testing," Canadian Journal of Civil Engineering, Vol. 4, 1977.

Green R., "Dynamic Response of Bridge Superstructures - Ontario Observations," Proceedings, Symposium on Dynamic Behavior of Bridges, Crowthorne, Berkshire, England, 1977.

Hays, C. O., and Hachey, J. E., "Lateral Distribution of Wheel Loads on Highway Bridges Using the Finite Element Method," Structures and Materials Research Report No. 84-3, University of Florida, Gainsville, Florida, 1984.

A RELIABILITY-BASED CAPACITY RATING OF EXISTING BRIDGES BY INCORPORATING SYSTEM INDENTIFICATION

H-N. CHO
Dept Civil Engineering, Hanyang University, Seoul, Korea
C-B. YUN
Dept Civil Engineering, Korean Advanced Institute of Science and
Technology, Seoul, Korea
A.H-S. ANG
Dept Civil Engineering, University of California, Irvine,
California, USA
M. SHINOZUKA
Dept Civil Engineering, Princeton University, Princeton,
New Jersey, USA

Abstract
This paper develops practical models and methods for the assessment of safety and rating of damaged and/or deteriorated bridges by incorporating a system identification technique for the explicit inclusion of the degree of deterioration or damage and of the actual bridge response. And, based on the proposed model, reliability-based rating methods are proposed as LRFR(Load and Resistance Factor Rating) and system reliability-index rating criteria. The proposed limit state model explicitly accounts for the degree of deterioration or damage in terms of the damage and response factors. The damage factor in the paper is proposed as the ratio of the current stiffness to the intact stiffness.

Based on the observation and the results of applications to existing bridges, it may be concluded that the proposed rating models, which explicitly account for the uncertainties and the effects of degree of deterioration or damage based on the system identification technique, provide more realistic and consistent safety-assessment and capacity- rating.

1 Introduction

In spite of the remarkable advances in structural modeling, numerical analysis and nondestructive testing, it is still difficult to predict realistic failure behavior and capacity-rating of existing road bridges, especially when those bridges are deteriorated or damaged to a significant degree. Recently, this has led to an increasing attention to the problems of safety assessment and rating of existing bridges, as well as the identification of the degree of deterioration or damage of those bridges in connection with the maintenance and rehabilitation problems.

This paper develops practical models and methods for the assessment of safety and rating of damaged and/or deteriorated bridges by incorporating such an elaborate technique as system identification or more approximately FFT analysis for the explicit inclusion of the degree of deterioration or damage and of the actual bridge response. And, based on the proposed model, reliability-based rating methods are proposed as LRFR(Load and Resistance Factor Rating) and system reliability-index rating criteria.

2 Limit State Model for Deteriorated or Damaged Bridges

For the development of a probability-based LRFR criteria[Cho and Ang 1989] a linear limit state function for a structure may be used and stated as

$$g(\cdot) = R - \sum_i S_i \qquad (1)$$

where R is the structural resistance, and S_i is the ith load effect.

A realistic safety assessment or rating of existing bridges requires a rational determination of the degree of deterioration or damage. Therefore, the limit state function, Eq.1, must incorporates some random variates to reflect such deterioration/damage and the underlying uncertainties. Various approaches, such as system identification [Shinozuka *et al.* 1978,1982], fuzzy set theory[Yao 1986] or probabilistic measure of structural redundancy have been suggested[Frangopol and Curley 1986]. A rational approach utilizing the system identification[Hoshiya and Saito 1984] for the damage assessment of structural members are proposed herein for practical application. However, in lieu of such elaborate techniques as system identification, the resistances of the deteriorated or damaged members are made to be estimated on the basis of visual inspection and/or simple in-situ tests with a practical analysis utilizing the FFT technique, supplemented with the engineering judgement.

The true resistance R may be modelled as,

$$R = R_n \cdot D_F \cdot N_R \qquad (2)$$

where R_n = the nominal resistance of the intact member in which the material strengths of the resistance are assumed as the real nominal values estimated on the basis of the NDT test results if available; D_F = the damage factor, which is the ratio of the current stiffness, K_D, to the intact stiffness, K_I, i.e. K_D / K_I ($\approx \omega_D^2 / \omega_I^2$) in which ω_D and ω_I are the fundamental natural frequencies of the damaged and the intact structures; and N_R = the correction factor for adjusting any bias and incorporating the uncertainties involved in the assessment of R_n and D_F.

The mean-value of N_R represents the bias correction, and its C.O.V. represents the uncertainties in the material strength, fabrication, construction, structural modelling, and damage assessment.

For the rating purpose, the proposed load model needs to consider only the dead load and truck loads as the primary loadings for short span bridges. Thus, the load effects may be expressed as:

$$S = C_D D_n N_D + C_L L_n K N_L \qquad (3)$$

where C_D, C_L = the influence coefficients for dead and live load effects, which are estimated based on the current stiffness parameters of the deteriorated or damaged bridges ; D_n, L_n = the nominal dead and truck loads, respectively; K = the response ratio equal to $K_S(1+I)$ in which K_S is the ratio of the measured stress to the calculated stress and I is the impact factor, either measured or calculated; and N_D, N_L = the correction factors for adjusting the bias and uncertainties in the estimated D_n and L_n, respectively.

In Eq.3 the uncertainties in the load effect model should include those associated with the analysis, load distribution, traffic load model and load test. Also, the random live-load effect should be evaluated on the basis of the current stiffness parameters of the deteriorated or damaged bridges, and the actual or estimated response ratio of the measured stress to the calculated one.

3 Damage Assessment by System Identification

For the purpose of capacity rating of existing bridge structures, which might have experienced significant structural damage through years of service, the structural damage may be approximately modelled as a reduction of the stiffness of a critical element[Yun and Min 1989] as

$$D^{(e)} = K_o^{(e)} - K^{(e)} = \sum \alpha_e K_o^{(e)} \tag{4}$$

where $K^{(e)}$ and $K_o^{(e)}$ are the element stiffness matrices with and without damage, respectively; and α_e is the element coefficient($-1 < \alpha_e < 0$). For the cases of reinforced concrete bridges, it may be represented by a reduction of the stiffness of a critical element as

$$D_F = \frac{(EI)^{(e)}}{(EI)_o^{(e)}} \tag{5}$$

In this study, the identification of the reduction of the element stiffness due to structural damage is carried out by two steps as follows;

(1) Determination of modal properties : The first few natural frequencies and the corresponding vibrational modes of the damaged structure are evaluated based on the dynamic response records measured at several locations of the structure. The extended Kalman filtering algorithm [Hoshiya and Saito 1984; Yun, et al. 1988] has been applied to the estimation of the modal parameters. For this purpose, a state vector is defined by including the unknown parameters as the augmented state variables and nonlinear state equation is constructed from the equation of motion.

(2) Determination of structural damage : Comparing the estimated modal quantities with those of the undamaged structure, the location and degree of damage are determined with an aid of the inverse modal perturbation technique[Yun and Min 1989] :

$$[T][P^{(k)}]\{\alpha\} = \left\{ \begin{array}{c} \mu_{ok}(\omega_k^2 - \omega_{ok}^2) \\ \beta\{\phi\}_k \end{array} \right\} \tag{6}$$

where $\{\alpha\}$ is the vector for the unknown element damage coefficient; $[T]$ and $[P^{(k)}]$ are the transformation and modal perturbation matrices associated with the k-th mode, which are obtained in terms of the known or measured quantities; ω_{ok} and μ_{ok} are the k-th natural frequency and generalized mass of the undamaged structure; ω_k and $\{\phi\}_k$ are the k-th natural frequency and mode shape of the damaged structure which are measured; and β is the unknown scale factor of the measurement value of $\{\phi\}_k$. The unknown damage coefficient vector $\{\alpha\}$ is evaluated by way of minimizing the estimation error.

4 Reliability Assessment of Existing Bridges

An existing bridge structure may be rated on the basis of a specified target reliability. This requires the assessment of the reliability of a bridge in either element-reliability or system-reliability level. Obviously, system reliability is more desirable for the rating purpose regardless of the degree of the modelling accuracy or numerical approximations involved in the assessment. For this reason, the stable configuration approach(SCA)[Quek and Ang 1986] is used to provide a practical and approximate method for estimating the system reliability against ultimate collapse of R.C. bridges, which may be considered to be a brittle mode of failure for which the SCA is particularly effective. Of course, for steel girder bridges that fail in ductile mode, the failure mode approach (FMA) considering only a few dominant failure modes may be more desirable and can be used without any further numerical difficulties. For the purpose of bridge rating, a second order bound estimate based on the approximate formulation of SCA or FMA is preferred to more elaborate theoretical formulation. A computer code developed based on the approximate SCA formulation of system reliability and the AFOSM method incorporating the equivalent-normal transformation and Haosfer-Lind's iterative algorithm are used for a practical evaluation of the system reliability index $\beta_s = -\Phi^{-1}(P_F)$. Then, using the reliability index as a requirement for structural safety, the values of β_s are suggested as a guide for developing rating criteria of deteriorated and/or damaged bridges.

5 Reliability-Based Rating Criterion

A load and resistance factor rating(LRFR) criterion may be developed corresponding to a specified target reliability index. Based on Eqs.2 and 3, the following rating criterion expressed in terms of the rating factor, RF, which is the ratio of the nominal load carrying capacity, P_n, to the standard rating load, P_L, specified in the code may be obtained:

$$RF = \frac{P_n}{P_L} = \frac{\phi' D_F R_n - \gamma'_D C_D D_n}{\gamma'_L C_L K P_L} \tag{7}$$

For the rating of an existing deteriorated bridges, the following two load levels of capacity rating of Eq.7 may have to be provided. At the service or lower level, the capacity rating may be referred to as the Service Load Rating(SLR) which corresponds to the allowable safe load level for the normal operation of the bridge. At the over-load or higher load level, the capacity rating may be referred to as the Maximum Over-Load Rating(MOR) which corresponds to the absolute permissible load level for special operation(over-load permit) of the bridge. The nominal safety parameters, ϕ', γ'_D and γ'_L, of the LRFR criterion of Eq.7 for each load level, SLR or MOR, corresponding to a specified target reliability (tentatively, $\beta_o = 3.0$ for SLR, $\beta_o = 2.0$ for MOR) may be calibrated by using the well established procedure for code calibration[Ravindra and Lind 1983].

For the assessment of reliability of existing bridges and the calibration of rating criterion, the statistical uncertainties of resistance and load effects are estimated from the data available in Korea [Shin et al. 1988] and partly from the engineering judgement. As the results of the calibration, the proposed LRFR criteria are shown in Table 1, which is given as in the form of the rating provisions for R.C. T-beam and slab bridges.

Table 1. LRFR-Rating Criteria

Bridge	SLR	MOR
R.C. T-beam	$\dfrac{0.80\,D_F\,R_n - 1.20\,C_D\,D_n}{1.85\,C_L\,K\,P_L}$	$\dfrac{0.95\,D_F\,R_n - 1.20\,C_D D_n}{1.65\,C_L\,K\,P_L}$
R.C. Slab	$\dfrac{0.85\,D_F\,R_n - 1.20\,C_D\,D_n}{1.85\,C_L\,K\,P_L}$	$\dfrac{1.00\,D_F\,R_n - 1.20\,C_D\,D_n}{1.65\,C_L\,K\,P_L}$

6 Application Examples

6.1 Damage Assessment

At first, an example analysis is carried out by using records generated by the simulation technique in order to verify the present method of the damage estimation. The structural model used is a R.C. beam with a uniform T-section as in Fig 1. Four different cases with damages at single or multiple locations are investigated. It is assumed that response time histories due to an impact load applied at Node 7 are measured at four nodes,i.e. Nodes 4,5,6 and 7. The first three natural frequencies and the first mode shape for each case are determined from the response time histories by using the extended Kalman filtering technique. Then, the element damage coefficients are estimated by the inverse modal perturbation. Table 2 summaries the assumed exact and the estimated values for the element damage coefficients. The results indicate that the present method can identify the locations of the damages very precisely. The accuracy of the estimated degree of damage has been founded to be somewhat deteriorated, but still remains in a reasonable range. Currently, experimental studies are being carried out to verify the applicability of the method to real structures.

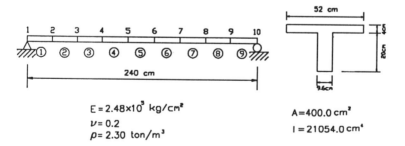

$$E = 2.48 \times 10^5 \ \text{kg/cm}^2$$
$$\nu = 0.2$$
$$\rho = 2.30 \ \text{ton/m}^3$$

$$A = 400.0 \ \text{cm}^2$$
$$I = 21054.0 \ \text{cm}^4$$

Fig.1. Structural model of R.C. beam with a uniform T-section.

Table 2. Exact and Estimated Damage Coefficient(α_e)

Cases	Element No.	Assumed Exact Damage	Estimated Damage	Natural Frequencies ($\omega_1, \omega_2, \omega_3$ in Hz)
damage I	3		0.0	
	4		0.0007	60.1
	5	0.4	0.5810	257.0
	6		0.0006	547.0
	7		0.0001	
damage II	3	0.4	0.5380	
	4		0.0030	58.0
	5	0.4	0.5361	242.0
	6		0.0037	536.0
	7		0.0	
damage III	3	0.4	0.5210	
	4		0.0	56.8
	5	0.4	0.5206	238.0
	6	0.2	0.1956	533.0
	7		0.0021	
damage IV	1	0.4	0.5970	
	4		0.0016	58.7
	5	0.4	0.5549	250.0
	6	0.2	0.2093	534.0
	7		0.0	
damage V	3	0.2	0.2202	
	4		0.0	60.4
	5	0.2	0.2202	247.0
	6	0.2	0.2205	557.0
	7		0.0	
damage VI	1	0.3	0.4178	
	4		0.0010	61.1
	5	0.2	0.2257	252.0
	6	0.2	0.2266	554.0
	7		0.0	

Note: Natural Frequencies of the undamaged structure (in Hz)
$\omega_{o1} = 64.3$, $\omega_{o2} = 257.5$, $\omega_{o3} = 578.8$

6.2 Reliability Assessment

The models and methods of reliability analysis using the stable configuration approach and the AFOSM algorithm proposed in the paper are applied for the reliability-assessment of three existing R.C. T-beam bridges which were field-tested for the safety-evaluation and capacity rating. Table 3 summarizes the bridge data obtained from the measurements and calculations performed for the safety-assessment of these bridges.

Table 3. Bridge Data

Bridge	Jinjeon	Wolcheon	Seosang
Bridge Type	R.C. T-beam	R.C. T-beam	R.C. T-beam
Design Load	D-9	D-13.5	D-13
Construction Year	1932	1956	1957
Rating Load	DB-24	DB-24	DB-24
No. of Girder	6	3	4
Girder Space(cm)	127	195	110
Span Length(cm)	1000	1400	1100
M_N (t-m)	64.2	226	123
M_D (t-m)	24.5	87.5	29.8
M_L (t-m)	10.4	35.9	37.5
D_F	0.75	0.89	0.86
$\sigma_{cal} / \sigma_{test}$	3.35	1.64	1.22
$(1+i_c)/(1+i_t)$	0.90	1.17	1.05
δ_c / δ_t	1.22	4.48	5.28
K	1.06	0.69	1.03
S_{Dn}(t-m)	24.5	87.5	29.8
S_{Ln}(t-m)	11.02	24.59	38.63
1+I	1.3	1.12	1.26

The results of reliability evaluation in terms of the element-reliability and system-reliability indices of the example bridges are shown in Table 4.

The data summarized in Table 3 as well as the visual inspections of these bridges indicate that the bridges are significantly deteriorated and/or damaged. Accordingly, the estimated reliability indices are found to be remarkably low for the bridges as shown in Table 4. It has to be noted that the low reliability indices of the bridges result also from the fact that the current rating load used for safety assessment is higher than the original design load. It may be also observed that the estimated system-reliability indices are consistently higher than the results for the element-reliability in spite of the fact that the system reliability evaluated based on the SCA method is an approximate conservative estimate. It may be argued that the system reliability index may have to be used, if possible, rather than the element-reliability index in order to take into account some reserved safety related to the redundancy of the bridge system. Furthermore, the results of the reliability evaluation as compared with the rating factors discussed in the following section, indicated that the reliability index may be used as a preliminary measure of the rating of existing bridges with only visual inspection before performing extensive field measurement.

Table 4. Reliability Indices of Bridges

Bridge	Jinjeon	Wolcheon	Seosang
Element-Rel.	1.42 (Int. Girder)	2.87 (Int. Girder)	1.64 (Ext. Girder)
System-Rel.	1.81	3.70	2.19

Table 5. Rating of Bridges

Bridge		Jinjeon	Wolcheon	Seosang
LRFR	SLR	0.45	1.24	0.68
	MOR	0.90	2.13	1.01
WSR		0.48	1.02	0.54
LFR	0.52	1.37	0.72	
AASHTO	I.RF	0.48	1.27	0.67
	O.RF	0.80	2.12	1.12

6.3 Reliability-Based LRFR Criterion

The recommended provisions of LRFR criterion as in Table 1 are applied for the rating of the bridges discussed in the previous section. As the rating vehicle for the LRFR criterion, the DB-24 which is the standard design-truck of the KSCE(Korea Society of Civil Engineers) bridge design code is used. The DB-standard trucks are similar to the AASHTO's standard design trucks. Note that, for example, the configurations of D-18 and DB-18 of the KSCE correspond to those of H-20 and HS-20 of the AASHTO. The combined weight on the first two axles of DB-18 is 18 tons which corresponds to 20 short-tons of HS-20. A comparison of the results between the proposed LRFR rating and the conventional rating such as WSR, LFR and the AASHTO rating is presented in Table 4. There are strong coherences between the estimated values for the SLR-rating factors and the reliability indices, which vary depending on the state of damage and/or deterioration of the bridges. It may be observed from Table 5 that, in general, the results of the conventional rating methods are not considerably different for those of the LRFR method for these particular bridges, although there are still some significant differences between the results of the proposed LRFR rating and those of the conventional ratings. However, considering that the current WSR or LFR criterion could result in rather irrational ratings because of inherent shortcomings in the conventional criteria which do not account for all the underlying uncertainties and the degree of the deterioration and damage, and thus are not reliability based, the results of the LRFR may be regarded as more rational ratings. It is interesting to observe that the rating of each example bridge at SLR and MOR levels are somewhat similar to those of the AASHTO rating although the AASHTO rating specification is not reliability-based. However, as stated above, the differences between the proposed and the conventional ratings may, mainly, be attributed to the fact that the proposed LRFR do explicitly account for the degree of the deterioration and damage in terms of the damage factor and the safety factors, ϕ, γ_i , which are derived on the basis of the reliability methods. Furthermore, it may be noted that the results of rating of each example bridge reasonably well correspond to the reliability indices presented in Table 4.

7 Conclusions

Based on the observation and the results of the application, it may be concluded that the proposed rating models, which explicitly account for the uncertainties and the effect of the degree of deterioration or damage based on the system identification or FFT analysis, provide more realistic and consistent safety-assessment and capacity-rating. Thus, it is strongly recommended that the system reliability index β_s and LRFR rating, rather than the conventional WSR or LFR rating, be preferably used in practice for the realistic assessment of safety and remaining reserved capacity of deteriorated and/or damaged bridges.

8 References

Cho, H-N. and Ang, A.H-S., "Reliability Assessment and Reliability-Based Rating of Existing Road Bridges", 5th International Conference on Structural Safety and Reliability(ICOSSAR'89), USA, Aug. 7-11, 1989

Frangopol, D.M. and Curley, J.P., "Effects of Redundancy Deterioration on the Reliability of Truss Systems and Bridges", Proc. of a Session, Structural Div. ASCE Convention, Seattle, Washington, Apr. 1986, pp.30-45.

Hoshiya, M. and Saito,E., "Structural Identification by Extended Kalman Filter", J. of Eng. Mechanics, Vol 110, No.12, Dec.1984, pp1757-1770

Quek, S-T. and Ang, A.H-S., "Structural System Reliability by the Method of Stable Configuration", SR Series No. 529, Dept of Civil Eng., Univ.of Ill., Urbana-Champaign, Nov. 1986.

Ravindra, M.K. and Lind, N.C., "Trends in Safety Factor Optimization,"Beams and Beam Columns, ed. R. Narayanan, Applied Science Publishers, Barking, Essex, UK, 1983, pp. 207-236

Shin, J-C., Cho, H-N. and Chang, D-I., "A Practical Reliability-Based Capacity Rating of Existing Road Bridges," J. of Structural Eng./Earthquake Eng., JSCE, Vol.5, No.2, Oct.1988, pp. 245-254.

Shinozuka M., Yun,C.B and Imai,H., "Identification of Linear Structural Systems", Technical Report NSF-ENG-76-12257-2, Columbia University, NewYork N.Y. May.1978

Shinozuka, M., Yun,C.B. and Imai,H., "Identification of Linear Structural Dynamic System", J. of ASCE, Vol 108, No.EM6, Dec.1982 , pp1371-1389

Yao, J.T.P., "An Unified Approach to Safety Evaluation of Existing Structures", Proc. of a Session, Structural Div., ASCE Convention,Seattle, Washington, Apr. 1986, pp. 22-29.

Yun,C.B., Kim,W.J., and Ang,A.H-S., "Damage Assessment of Bridge Structures by System Identification", Proc.of Korea. Japan Joint Seminar of Emerging Technologies in Structural Engineering and Mechanics Nov. 23-25,1988, Seoul Korea, pp182-193

Yun,C.B, and Min,J.K., Estimation of Structural damages by Inverse Modal Perturbation Method", M.S. thesis, KAIST 1989

BRIDGE RAIL DESIGN AND CRASH WORTHINESS

M. ELGAALY, H. DAGHER, S. KULENDRAN
Civil Engineering Dept, University of Maine, Orono, Maine, USA

Abstract
This paper address the issues of bridge rail design and
crash worthiness. It gives brief description of research
work including full-scale crash tests conducted in the
United States. It briefly discusses the recent AASHTO Guide
Specifications for bridge railings. It describes geometry
and strength design for standard steel and aluminum open
bridge railings developed for a consortium of the
transportation departments in the New England states.
Finally, it presents results of computer simulations of
crash tests and address plans for full-scale tests.
Keywords: Bridge Railing, Geometric Design, Strength
Design, Full-Scale Crash Tests, Computer Simulation.

1 Introduction

Although the primary purpose of bridge traffic railings is
to contain vehicles, consideration is also given to
protection of the occupants of the vehicle in collision
with the railing, protection of other vehicles near the
collision, and protection of persons and property on
roadways or other areas underneath the bridge. Recently the
American Association of State Highway and Transportation
Officials (AASHTO) have issued guide specifications for
bridge railing which require that railings and transitions
be subjected to full-scale crash testing, to insure safe
performance. Other considerations, which are not directly
safety related, include railing cost effectiveness,
appearance and freedom of view, snow removal, and
maintenance of the railing such as painting and accident
damage repair.

The authors of this paper have conducted research work
sponsored by a consortium of the transportation departments
of the New England states to review their bridge railings
and transitions standards in the light of the crash
worthiness question. The objective of the work was to
develop standard open bridge rail and transition designs

which meet, to the extent possible, the geometric and strength requirements in the AASHTO guide specifications and pass the full-scale crash testing.

During the course of their studies, the authors reviewed national and international technical literature, including reports and videotapes of several full-scale crash tests, conducted mostly in the United States. Based on these reviews and the guidelines outlined in the AASHTO Guide Specifications, two open bridge railing and transition designs, one is made of steel and the other is made of aluminum, were proposed. Computer simulations of an 18,000 lb truck crash into the designed rails and transitions at a speed of 50 mph and an angle of 15 degrees were performed using a computer program, BARRIER VII. These designs passed the computer simulated crash tests, and will be subjected to actual full-scale crash tests in the near future.

In this paper research work conducted by others will be reviewed, and geometry and strength design considerations will be discussed. Requirements for full-scale crash tests, and computer simulation of vehicle crash into the railing will be described. Emphasis will be placed on the open bridge railings developed for the transportaion departments of the New England states.

2 Review Of Research Work

Most of the research work related to bridge railing was conducted in the United States. This work includes full-scale crash testing as well as developing and testing energy absorbing devices. In 1970, the California Department of Transportation reported the results of their first full-scale vehicle impact test, Nordlin et al. (1970). Later, they reported the results of five tests on concrete parapet with structural steel tube placed 12 in. above the parapet, Nordlin et al. (1971), and the results of two full-scale vehicle impact tests on steel bridge rail of relatively open profile, Nordlin et al. (1972). Between 1978 and 1980 a total of 30 full-scale crash tests were conducted by the Texas Transportation Institute (TTI), Buth et al. (1984). Twenty-one of these tests were performed on five in-service railing designs and nine were performed on an instrumented wall. The five railing designs tested are steel and aluminum and had relatively open profiles. The tests employed compact (1,800 lb), medium-size (2,250 lb), and full-size (4,500 lb) sedans, and 20,000 lb and 32,000 lb buses. The speed at impact varied between 55 and 60 mph and the angle of impact varied between 15 and 25 degrees. In order to develop comprehensive performance standards for bridge rails and crashworthy designs that could be recommended for use, the Federal Highway Administration

(FHWA) contracted with the Southwest Research Institute to conduct another series of bridge rail tests. These tests begun in 1983, and were performed on six different types of bridge rail designs selected from a list of 166 approved State standard plans, Bronstad et al. (1987). The bridge rail types include concrete parapets, steel railing of relatively open profiles and combinations thereof. All bridge rail systems with few exceptions were tested at 60 mph with an 1800 lb minicar impacting at 20 degrees and a 4500 lb full-size sedan impacting at 25 degrees. One system was evaluated using a 40,000 lb inter-city bus impacting at 60 mph and 15-degree angle. In 1985, the results of twelve full-scale crash tests performed on a railing system consisting of 10-gage thrie-beam steel rail attached to W6x9 steel posts spaced at 8 ft. 4 in. were reported, Bryden and Phillips (1985). The system was tested with and without a 6 in. curb using 4500 and 1800 lb vehicles.   In 1986, a Highway Planning and Research (HPR) program pooled-fund study on bridge rails was initiated. This study called for the Texas Transportation Institute to Conduct 38 bridge rail and 12 transition tests. Results from crash testing a relatively open profile steel bridge rail were reported, Buth et al. (1988). The bridge rail was tested using an 18,000 lb single-unit truck. The speed of the vehicle just prior to impact was 50.8 mph and the angle of impact was 15.1 degrees. A metal rail 18 in. high mounted on top of a 32 in. concrete parapet was crash tested using 80,000 lb van-type truck, Hirsch et al. (1986). The bridge rail was struck by such a truck at 48.4 mph at an angle of 14.5 degrees. Brief description of some of the aforementioned crash tests and the lessons learned from these tests are given by McDevitt (1987).

The use of energy-absorbing devices in bridge railing design was considered. In 1969 an article which describes the applications of a NASA-developed fragmenting tube energy absorber in the construction of highway bridge rails was published, Woolam (1969). Four full-scale crash tests were conducted to determine the capabilities of this system, Stocker et al. (1970), and a design method was developed, Woolam and Garza(1970). An energy-absorbing bridge rail system that uses the plastic deformation of steel rings as the primary impact energy absorber has been developed through full-scale crash testing and the use of BARRIER VII computer program, Kimball et al. (1976). The system design is capable of withstanding impacts by large vehicles such as buses and trucks without imparting high accelerations to impacting smaller vehicles. Ten full-scale crash tests were performed with vehicles ranging from 2,000 to 40,000 lb impacting with a speed of 55mph and at an angle of 19 degrees. A bridge rail which incorporates structural steel tube railings and posts and high strength rubber energy absorbers was developed and tested, Beason et

al. (1986). Results of the crash tests show that this rail
can smoothly redirect a 4500 lb automobile impacting with a
velocity of 60 mph and an angle of 25 degrees and remain in
service with no maintenance.

In Great Britain six full-scale impact tests were made
on designs of aluminum bridge rails, Jehu et al. (1969),
and three types of steel bridge rails were subjected to
vehicle impact tests, Blamey (1972). Clark (1972), traced
the development of bridge rails in G.B. which lead to the
issuing of the Department of the Environment's Technical
Memorandum No. BE5. He discussed this memorandum and
compared the aluminum and steel alternatives. In the
Netherlands the Institute for Road Safety Research Carried
out research into the most suitable bridge rail, Paar
(1973). Based on this research the Netherland authorities
issued guidelines for bridge rail design. In Japan,
Shimogami (1987), the Ministry of Construction has
collected a lot of accident data throughout 20,000 km of
designated sections of the national highways. The Public
Works Research Institute of the Japanese Ministry of
Construction has analyzed these data.

3 AASHTO Guide Specifications

Based on a nationwide data averaged over the years 1984 and
1985, there are about two fatal accidents involving bridge
rails every day in the United States.

In 1988, AASHTO issued Guide Specifications for Bridge
Railings as an alternative to existing bridge railing
specifications in the AASHTO Standard Specification for
Highway Bridges. These new Guide Specifications define
three bridge railing performance levels and associated
crash tests and performance requirements, along with
guidance for determining the appropriate railing
performance level for a given bridge site. The performance
levels and selection procedures given in the Guide
Specifications are based on considerations of the
probabilities that a bridge railing will be subjected to
given impact conditions, the consequence of those impacts
given that performance level railings are in place, and the
cost of providing the various performance level railings.

The crash testing and evaluation of the results are in
accordance with the crash test procedure given in the NCHRP
Report 230 (1987). The test vehicles include 1,800 lbs
small automobile, 5,400 lbs pickup truck, 18,000 lbs medium
single-unit truck, and 50000 lbs van-type tractor-trailer.
The impact speed varies between 45 and 65 mph and the angle
of impact varies between 15 and 20 degrees.

These Guide Specifications are applicable to railings
for new bridges and for bridges being rehabilitated to the
extent that railing replacement is obviously appropriate.

The Specifications are not applicable to determine the
adequacy of existing railings.

4 Geometric Design

At all performance levels, the geometry of the railing
should guard against bumper, wheel, and hood snagging and
should provide stable post-impact trajectories for the
impacting vehicle.

4.1 Curbs and Sidewalks
Curbs and sidewalks are considered as integral parts of the
bridge railing. A brush curb will have an influence on the
capability of a railing system to contain and redirect an
errant vehicle and hence it is required to crash test the
railing with the curb along with it. The width of a brush
curb shall not exceed 9 inches and desirably 6.
Furthermore, it is recommended that the distance between
the face of the curb and the face of the rail be as small
as possible.
    Of particular concern is the potential for the brush
curb to produce a ramping effect that might allow a vehicle
to climb over a railing or a tripping effect that would
increase the likelihood for high-bodied vehicles to roll
over a railing. Furthermore, in order to insure a safe
post-impact trajectory, the body of an errant vehicle
should impact the rail before the wheels impact the curb.
    A sidewalk is, also, an integral part of a railing
system and should be crash tested along with the railing.
The minimum recommended width of a sidewalk is 4 feet.

4.2 Railings
The acceptability of railing geometry shall be verified
through crash testing. The Guide Specifications
recommendations for railing geometry do not insure that
when crash tested, a rail will pass the acceptance
criteria. On the other hand, if the recommendations are not
followed, there is a little chance a rail will pass the
crash test. As recommended in the Guide Specifications, the
minimum height of the railing measured from the top of the
roadway shall not be less than 27 inches for performance
level 1, and 32 inches for performance level 2. The
distance between the bottom of the bottom rail and the top
surface of the road or curb shall not exceed 10 inches.
    The traffic face of all railings should be smooth and
continuous. Rail splices should be designed to take the
forces and the deformations resulting from the impact,
without separation of the two spliced parts of the rail.
The clearance from the face of the rail to the face of the
post must be big enough to preclude contact by substantial
vehicle parts that might penetrate an opening in the

railing. In the Guide Specifications a minimum clearance of 10 inches is suggested. This is an absolute minimum for any railing configuration where snagging is an obvious possibility and it is recommended that this clearance should be a minimum of 12 inches. The spacing between the posts in an open railing is dictated by the design loads and should not exceed 10 feet.

The aforementioned recommended dimensions are intended to guard against bumper, wheel and hood snagging for automobiles and to provide stable post-impact trajectories for automobiles, light trucks, and passenger vans.

4.3 Transitions

Traffic barriers on bridge approaches must be properly transitioned to traffic railings on bridges. Inadequate guardrail stiffening at the transition may result in substantial deflection of the guardrail and subsequent snagging or poor redirection of the vehicle. The bridge rail and the approach rail should have similar structural characteristics. Based on computer simulation and full-scale crash tests, the use of larger posts near the bridge is not as effective as reducing the spacing of standard posts.

5 Strength Design

The Guide Specifications contains no instruction on the structural analysis of bridge railings. Instead, acceptability of the railing is to be determined through crash testing. However, the expense of crash testing makes a cut-and-try approach to design impractical, and guidance is needed for the structural design of a railing system.

The Guide Specifications give magnitudes, distributions, and locations of railing design loads. For each performance level three components of the design load are given; namely horizontal (or transverse), longitudinal, and vertical. For example the magnitudes of the transverse design loads are 30, 80, and 140 kips; for performance levels 1, 2, and 3, respectively. The load for performance level 2 should be applied over an area which is equal to 28 inches along the length of the railing and 14 inches in the perpendicular direction in the plane of the railing. The location of the center of this patch is 17 inches from the surface of the road. Dimensions of the loaded patch and its location from the surface of the road for performance levels 1 and 3 are given in the Guide Specifications. Applied simultaneously with these transverse loads, are the vertical and longitudinal loads in the plane of the railing. The magnitude of the in-plane loads for performance level 2 are $\pm$ 24 kips in the longitudinal direction of the railing and 15 kips downward or 5 kips upward. These in-plane loads are applied over the same patch area as the lateral loads. The magnitude of the in-plane loads for performance levels 1 and 3 are given in the Guide Specifications. The in-plane loads, longitudinal and vertical, are to be applied at the traffic faces of the longitudinal rail elements. Loads on the posts are assumed to be transmitted through the longitudinal rail elements. Longitudinal loads are to be distributed to no more than three posts.

In the Guide Specifications it is assumed that a railing will be near its ultimate strength when subjected to the aforementioned design loads. Thus, railing analysis and design should be based on ultimate strength approaches, for example, through the application of plastic analysis to steel bridge railings. It is further suggested that anchorages and splices provide the full moment and shear capacity of the rails and at least 50 percent of the tensile strength of the rail.

For maintenance purposes, it is highly desirable that bridge decks, railing attachments to bridge decks, and bridge railings be designed so that impact damage to railing does not carry into the bridge deck or other bridge elements. Railing loads applied to the bridge deck slab shall be based on the ultimate strength of the railing used.

In the course of developing a proposed open rail design for New England strict adherence to the Guide Specifications was not possible without producing very conservative designs which are not recommended from both economical and performance points of view. A flexible but strong railing system can absorb more energy and minimize the damage which an impacting vehicle will sustain. It appears that the Specifications need modifications at least with respect to open bridge rail design.

6 Dynamic Analysis

The interaction of a vehicle with a barrier is difficult to simulate on a computer. Dynamic effects, large displacements , and inelastic behavior must all be considered. The dynamic loads are not explicitly specified, but must be determined by satisfying force equilibrium and displacement compatibility between the vehicle and the barrier.

A computer program BARRIER VII, which takes into consideration the aforementioned effects, was developed at the University of California, Powell (1970 and 1973). A dynamic step-by-step analysis is carried out by the program. At the beginning and the end of each step, the following parameters are known and determined, respectively:

216

1. The positions, velocities, and accelerations of all points on the vehicle and barrier,
2. The magnitudes and directions of the normal and friction forces exerted between the vehicle and the barrier, and the positions at which they act, and
3. The axial forces and bending moments in the barrier members.

To run the program the barrier is idealized as a framework. The range of member types available in the program permits considerable flexibility in the idealization procedure. The program also has features which permit the analysis of barriers which do not lie in a single plane. For example, bridge rails in which a low level barrier makes contact with the vehicle wheels while a higher level barrier makes contact only with the body can be analyzed.

The vehicle is idealized as a body of arbitrary shape which possesses mass and rotational inertia. The part of the vehicle boundary which interact with the barrier is defined by specifying a number of points at which contact with the barrier may be made. A discrete non-linear spring is then associated with each point. The springs are assumed to have no mass and no damping. Wheel positions can also be defined, and brakes can be specified to be either on or off during the analysis.

The agreement between the results from the program and available experimental data is very encouraging and the solution technique is surprisingly reliable in view of the complexity of the problem. For some complex barriers, however, experimentation in the selection of time steps, damping values, and other parameters may be necessary to obtain stable solutions.

BARRIER VII was used to evaluate the standard bridge rails and transitions developed for New England.

7 Proposed Standard Designs

A steel and aluminum bridge railing designs were developed which meet, to the extent possible, the geometric and strength requirements of the Guide Specifications; the geometric design is shown in figure (1). Alternate designs were also developed, the alternate steel design is shown in figure (2). Computer simulations of an 18,000 lb truck crash into the designed rails and transitions (steel and aluminum) at a speed of 50 mph and an angle of 15 degrees were conducted using the computer program BARRIER VII.

The bridge rail model used in the analysis of the proposed steel design is shown in figure (3). The vehicle trajectories after impacting at the rail mid-span are shown in figure (4). The transverse and longitudinal shears and moments time-histories in the post are shown in figure (5).

217

Finally, time-histories of axial tension and bending moment in the top rail are shown in figure (6).

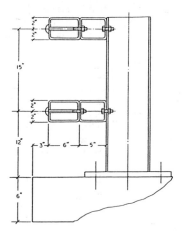

Fig.1 Geometric design

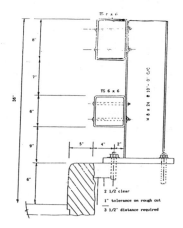

Fig.2 Alternate design

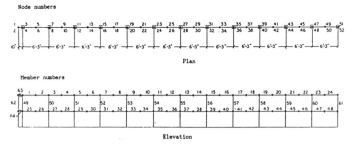

Fig.3 Bridge rail model

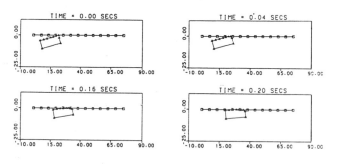

Fig.4 Vehicle trajectories after impact

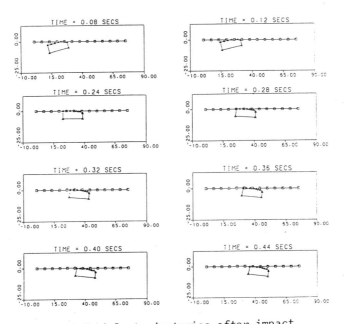

Fig.4 Vehicle trajectories after impact

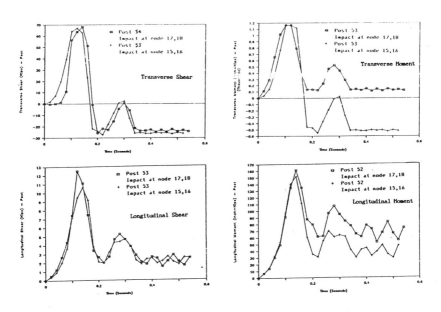

Fig.5 Transverse and longitudinal shears and moments in post

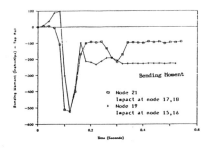 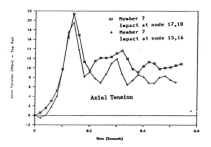

Fig.6 Bending moment and axial tension in top rail

8 Full-scale Crash Tests

As required by the Guide Specifications any new bridge rail
design should be crash tested before it is used. There are
only a few testing facilities in the United States which
are specialized in full-scale crash testing. The required
test length of a barrier system is 75 feet. Two crash tests
on the transition and three on the bridge rail will be
required.

9 Acknowledgment

The funds received from the consortium of the New England
Transportation Departments made this study possible. The
technical support of the consortium bridge rail technical
committee is appreciated.

10 References

Beason, W.L. et al. (1986) A low-maintenance, energy-
    absorbing bridge rail, in **Texas Transportation
    Institute Research Report 417-1F.**
Blamey, C. (1972) Bridge parapet tests carried out in
    collaboration with the British Steel Corporation, in
    Transport and Road Research Laboratory Report LR 492.
Bronstad, M.E. et al. (1987) Bridge rail designs and
    performance standards, in **Report No. FHWA/RD-87/049,
    Federal Highway Administration, Washington, DC.**
Bryden, J.E. and Phillips, R.G. (1985) Performance of a
    thrie-beam steel-post bridge rail system, in
    Transportation Research Record 1024.

Buth, E.F. et al. (1984) Safer bridge railing, in **Report No. FHWA/RD-82/072, Federal Highway Administration, Washington, DC.**

Buth, E.F. et al. (1988) Test report 7069-15 - Illinois 2399 bridge rail, in **Texas Transportation Institute Draft Report.**

Clark, M.N. (1972) Steel parapets to BE5, in **Journal of the Institution of Highway Engineers, Vol.19, No.10.**

Hirsch, T.J. et al. (1986) Concrete safety shape with metal rail on top to redirect 80,000 lb trucks, in **Transportation Research Record 1065.**

Jehu, V.J. et al. (1969) Bridge parapet tests carried out in collaboration with British Aluminum Company, in **Road Research Laboratory Report LR 281.**

Kimball, C.E. et al. (1976) Development of a new collapsing-ring bridge rail system, in **Transportation Research Record No. 566.**

McDevitt, C.F. (1987) Crash testing bridge rails and transitions, in **Public Roads, Vol.51, No.3.**

Nordlin, E.F. et al. (1970) Dynamic tests of California type 9 bridge barrier and type 8 bridge approach guardrail, in **Highway Research Record No. 302.**

Nordlin, E.F. et al. (1971) Dynamic tests of the California type 20 bridge barrier rail, in **Highway Research Record No. 343.**

Nordlin, E.F. et al. (1972) Dynamic tests of the California type 15 bridge rail, in **Highway Research Record No. 386.**

National Cooperative Highway Research Program Report 230 (1987) Recommended procedures for the safety performance evaluation of highway appurtenances.

Paar, H.G. (1973) Crash-barrier research and application in the Netherlands, in **Highway Research Record No. 460.**

Powell, G.H. (1970) Computer evaluation of automobile barrier systems, in **Federal Highway Administration Report No. FHWA-RD-73-73.**

Powell, G.H. (1973) BARRIER VII: A computer program for evaluation of automobile barrier systems, in **Federal Highway Administration Report No. FHWA-RD-73-51.**

Shimogami, T. (1987) Efficiency analysis of traffic safety countermeasures, in **Annual Report Roads (Japan).**

Stocker, A.J. et al. (1970) Full-scale crash tests of the fragmenting-tube-type energy-absorbing bridge rail, in **Highway Research Record No. 302.**

Woolam, W.E. (1969) Aerospace energy absorber applied to highway safety, in **U. S. Naval research Laboratory, Shock Vibration Bulletin 40, Part 5.**

Woolam, W.E. and Garza, L.R. (1970) Design fabrication and installation of a fragmenting-tube-type energy absorber in conjunction with a bridge rail, in **Highway Research Record No. 302.**

PART FOUR
REINFORCED CONCRETE BRIDGES

A SIMPLE METHOD FOR DESIGNING BRIDGE DECK SLABS

B. deV. BATCHELOR
Professor of Civil Engineering, Queen's University, Kingston,
Ontario, Canada
P. SAVIDES
Research Assistant, Queen's University, Kingston, Ontario, Canada

Abstract
This paper presents a summary of studies conducted over the
past two decades on the effects of compressive membrane
action on the behaviour of bridge deck slabs. Analytical
and experimental models used in the investigations are
described. The work has resulted in the development of a
very simple, economical method for designing bridge deck
slabs. This method takes into account the beneficial
effects of compressive membrane action and has been
incorporated into the Ontario Highway Bridge Design Code
since 1979, and into the most recent edition of the
Canadian Code for the design of highway bridges. The
simplicity and economy of the method have aroused much
interest in other jurisdictions.
Keywords: Bridge Deck Slabs, Compressive Membrane Action,
Empirical Design Method.

1 Introduction

The traditional approach to concrete bridge deck slab
design has been to consider bending effects only, assuming
that the resulting shear capacity is adequate. The methods
of analysis are usually elastic and any effects of membrane
action are ignored. In reality, however, typical bridge
deck slabs fail in punching at much higher loads than their
bending capacities.

The main reason for this strength enhancement is the
existence of in-plane compressive forces in the slabs which
are referred to as "compressive membrane action" or
"arching action". Thus the slab panels tend to act as
arches between supporting girders.

Theoretical and laboratory investigations at Queen's
University, supplemented by field studies carried out by
the Ontario Ministry of Transportation, have established
that the effect of compressive membrane action in bridge
deck slabs is considerable, and can be predicted quite

closely in certain systems. The results of these investigations have been confirmed by independent studies both in the U.S.A. and the United Kingdom.

These studies have formed the basis for a simple, economical, empirical deck slab design method prescribed in the Ontario Highway Bridge Design Code (OHBDC) (1979, 1983). The inclusion of the effects of compressive membrane action in design results in much lower reinforcement percentages than would be required with conventional design approaches. The method has also been effectively used in the evaluation of existing deck slabs.

The Canadian Standards Association has adopted the empirical design method in the CSA Standard S6-M88, "Design of Highway Bridges" (1988) and there is indication that the American Association of State Highway Officials (AASHTO) will do the same in the next edition of the "Standard Specifications for Highway Bridges".

2 Studies of compressive membrane action in slabs

The fact that bridge deck slabs with relatively high span/thickness ratios tend to fail in the punching mode rather than in flexure under concentrated loads, has been known for some time. Harris (1957) reported that this was observed by Freyssinet in 1945 during tests on prestressed concrete slabs. This was later confirmed by Guyon (1960) in tests on prestressed concrete bridge deck slabs. The considerable strength enhancement observed by Ockleston (1955, 1958) in tests on the concrete floor of a three-storey building is well known. He attributed this enhancement to compressive membrane action, and researchers have since concentrated in determining the exact contribution of this action to the flexural strength of slabs, particularly the flat slabs in buildings. A comprehensive review of these studies has been presented by Batchelor (1987), therefore only a few of these will be elaborated on below.

A good explanation (Figure 1) of compressive membrane action was presented by Liebenberg (1960). Figure 1(a) shows that compressive membrane action can develop in a cracked unreinforced slab, while Figure 1(b) shows that in a cracked reinforced slab the applied load is resisted by flexure and compressive membrane action. However, it can be seen from Figure 1(c) that compressive membrane action will not develop in a plate constructed from a material having the same stress strain relationship in tension and compression.

Typically, compressive membrane effects develop very rapidly, and reach a maximum value at small slab deflections (Park 1964)). Taylor and Hayes (1965) reported increases of up to 60% in the punching strength of laterally restrained slabs. Similar findings were reported

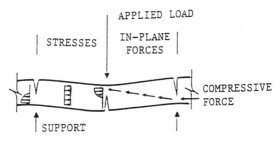

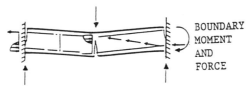

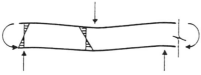

a) COMPRESSIVE MEMBRANE ACTION

b) BOUNDARY ACTION

c) STRESS RESULTANTS IN SLAB HAVING
   SAME STRESS-STRAIN RELATIONSHIP
   IN COMPRESSION AND TENSION

Fig. 1.   Compressive membrane action in slabs.

by Brothcie and Holley (1971). Since the 1960's, a number
of studies on the   strength of bridge deck slabs have been
conducted on direct models at Queen's University at
Kingston, Ontario, Canada with a view to determining the
degree of compressive membrane action.   The results of the
studies on square bridge deck slabs are reported by Tong
and Batchelor (1971) and Batchelor and Tissington (1976).
They show that compressive membrane action does develop in
these slabs, and suggest methods for determining the
magnitude of this effect.

   A number of studies have been conducted on rectangular
deck slabs of composite I-beam bridges as reported by
Newmark et al. (1946, 1958). Siess and Viest (1953)
reported that the design of slabs of such bridges was much
too conservative. Similar findings were reported by Thomas
and Short (1952). A methodical study of the behaviour of
the deck slabs of composite I-beam bridges was presented by
Hewitt and Batchelor (1975). They proposed methods   for

determining the magnitude of compressive membrane action in such systems, which later formed the basis for the design specifications in the Ontario Highway Bridge Design Code.

Hewitt (1972) extended a model used by Kinnunen and Nylander (1960) to study the punching strength of simply supported slabs. Hewitt successfully introduced the effects of compressive membrane action and boundary moments into this model in studying the punching strength of deck slabs of I-beam bridges. Figure 2 shows details of the modified model in which the outer portion of a segment of the slab, bounded by the shear crack and two radial cracks subtending the angle ß, is considered to be loaded through a compressed conical shell that develops from the perimeter of the loaded area to the root of the shear crack. The conical shell of varying thickness is assumed to have

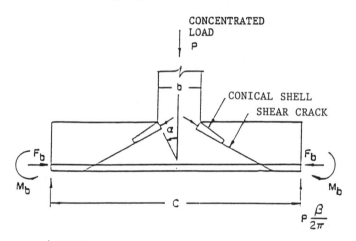

a) BOUNDARY FORCES IN SLAB

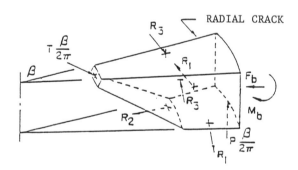

b) FORCES ON ELEMENT

Fig. 2. Model of slab at punching.

dimensions shown in Figure 2(a) such that the compressive stresses at the intersection with the loaded area and at the root of the crack are approximately equal.

The element shown in Figure 2(b) is acted upon by the external force $P\beta/2\pi$ and by the stress resultants described by Batchelor and Hewitt (1975). They have explained how this model can be used to determine the values of the boundary forces from tests of actual bridge deck systems.

3 Selected laboratory and field studies

3.1 Laboratory studies

In a major study of the punching strength of slabs of steel/concrete composite bridges, Hewitt (1972) used 1/8 scale models of medium span bridges. Some deck slab panels contained conventional orthotropic reinforcement, while others contained isotropic reinforcement varying from zero to 0.6%. Reinforcement location and concrete strength were varied. The slab panels were tested under static and repeated concentrated loading as described by Hewitt (1972), Hewitt and Batchelor (1975) and Batchelor (1987).

It was shown that punching strength of a slab panel was not significantly influenced by the position of the load, previous failures in adjacent panels, strength of concrete or dead load stresses. The studies confirmed that the conventional deck slab design method was too conservative. The results of these studies led Hewitt and Batchelor (1975) to propose that 0.2% isotropic reinforcement, top and bottom, is adequate from considerations of strength and serviceability. From repeated loads tests, Batchelor et al. (1978) concluded that the 0.2% isotropic reinforcement has considerable strength against fatigue failure. The 1979 edition of the OHBDC adopted this empirical design method with 0.3% isotropic reinforcement instead of 0.2%.

The first edition of the OHBDC in 1979, created wide interest in the empirical design method. Beal (1982) and Kirkpatrick et al. (1984), confirmed the findings of the earlier studies. In 1985, general confirmation of the Ontario empirical deck slab design method was provided by Fang et al., Elling et al. and Tsui et al.

Recently, Perdikaris and Beim (1988) confirmed that isotropically reinforced bridge decks designed according to the OHBDC (1983) outperform the ones reinforced orthotropically in accordance with the AASHTO (1983) provisions. However, they also concluded that a constant moving load is more critical than either a fixed pulsating load or a static load. Studies are also in progress in the United Kingdom as reported by Jackson (1988).

3.2 Transversely prestressed bridge decks

An investigation of transverse prestressing in bridge decks was carried out at McMaster University, Ontario, from which

Moll (1985) proposed that the thickness of bridge decks be reduced, and the two layers of normal reinforcement in each direction be replaced by one layer of reinforcement prestressed only in the transverse direction. Tests on a composite bridge deck designed accordingly, proved to be satisfactory in terms of the ultimate limit state. The prestressed slab also performed satisfactorily under service loads with respect to both deflection and cracking.

Savides (1989) reported on static loading tests performed at Queen's University, Ontario, on a three-girder model bridge constructed to approximately 1/4 scale, with a 43 mm thick deck, having a span/depth ratio of 13.2. A view of the bridge model is shown in Figure 3. The deck was prestressed only in the transverse direction. The deck slab panels failed in punching at an average safety factor of 7.4. The transverse prestress enhanced the development of compressive membrane action, as well as the behaviour of the deck slab under service loading. From the experimental results, there is every indication that the prestressed deck slab could be an acceptable if not improved alternative in composite bridge deck design.

While the investigation at Queen's University, was taking place, Poston et al. (1988) published a study on the effects of transverse prestress in bridge decks. After tests on a 0.45 scale bridge model, they concluded that a transversely prestressed deck can be expected to fail in punching shear at a minimum safety factor of seven, confirming the results of the Queen's investigation.

3.3 Field Studies

Dorton et al. (1977) have reported the results of tests on a full scale experimental bridge in Conestogo, Ontario. Some of the deck slab panels were provided with conventional orthotropic reinforcement, while others contained isotropic reinforcement ranging from 0.1% to 0.6%. The behaviour of the bridge under extensive testing, confirmed the validity of the empirical design method and justified its inclusion in the 1979 edition of the OHBDC. Future tests on the Conestogo bridge (Bakht and Markovic (1985)) have shown that the deck slab continues to perform satisfactorily. The Ontario Mininstry of Transportation and Communications has also conducted extensive field testing on deck slabs in service. The test procedure and results have been reported by Bakht and Csagoly (1979) and also confirm the validity of the empirical method of deck slab design.

4 Design applications

The first (1979) edition of the Ontario Highway Bridge Design Code specified an acceptable empirical deck slab design method based on the studies conducted in Ontario.

Fig. 3. View of the model bridge deck.

The code reduced the design of most deck slabs simply to a
prescription of the provision of two layers of 0.3%
isotropic reinforcement top and bottom for a total of four
layers of reinforcement, provided that all the following
conditions were met:

Slab span (girder spacing) does not exceed 3.7 m.
The slab extends at least 1.0 m beyond the exterior
girder, or a kerb of equivalent cross section is
provided.
The spacing of reinforcement in each layer does not
exceed 300 mm.
Intermediate diaphragms or cross frames are provided at

a maximum spacing of 8.0 m.
Diaphragms or cross frames are provided at all supports.
The skew angle of the slab system does not exceed 20.
The ratio of girder spacing to slab thickness does not
exceed 15, with the minimum slab thickness being 190 mm.

The empirical method was made mandatory in the second
edition of the OHBDC (1983), provided that certain less
stringent conditions are met. For example, the condition
regarding skew is relaxed as shown in Figure 4, with a
doubling of reinforcement in the support region. In
addition, the empirical method is allowed for slabs
supported on concrete girders without intermediate
diaphragms. The minimum slab thickness has been increased
to 225 mm for durability purposes and thereby providing an
increased level of compressive membrane enhancement. In
addition, charts are provided for use in evaluating deck
slabs which do not conform to the specified criteria.

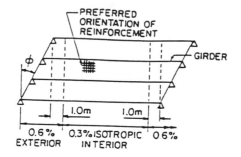

a) SUMMARY OF REINFORCEMENT
   SPECIFICATIONS

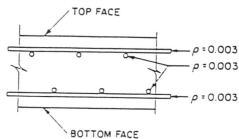

b) REINFORCEMENT IN INTERIOR PANEL

Fig. 4. Slab reinforcement in the Ontario Highway Bridge
        Design Code (1983).

These charts were derived from the theoretical models and computer programs described by Batchelor et al. (1985), and have proven extremely useful in rehabilitation.

The 1988 edition of the Canadian Standards Association's "Design of Highway Bridges" (CSA/CAN-S6-M88) has incorporated the empirical method for designing bridge deck slabs as an alternative to the yield line theory. There is every indication that the next AASHTO edition of the "Standard Specifications for Highway Bridges" will include the empirical design method specified by the Canadian bridge codes.

5 Conclusions

The studies described in this paper have led to the conclusion that the use of elastic methods for analysing bridge deck slabs gives results that are wasteful. This is compounded by the practice of ignoring compressive membrane action in a system that has been shown to develop high membrane forces, and which typically fails in punching rather than the bending mode assumed in conventional design.

The empirical method used in Canada for designing reinforced concrete deck slabs is simple to apply, and results in considerable savings in reinforcement without compromising safety. The significant saving involved will provide the incentive for the adoption of this empirical method in other jurisdictions in the near future.

Recent research has indicated the feasibility of developing a similar method for designing transversely prestressed bridge deck slabs.

6 References

American Association of State Highway Officials, (1983) **Standard Specifications for Highway Bridges.** AASHTO, Washington, D.C.

Bakht, B. and Csagoly, R.P. (1979) **Bridge Testing.** Ontario Ministry of Transportation and Communications, Res. Rep. SRR-79-10.

Bakht, B. and Markovic, S. (1985) Reinforcement savings in deck slab by a new design method, in **Proc. Int. Colloq. on Concrete in Developing Countries**, Lahore, Pakistan, Vol. 2, pp.595-614.

Batchelor, B.deV. (1987) Membrane enhancement in top slabs of concrete bridges, in **Concrete Bridge Engineering: Performance and Advances** (ed. R.J. Cope), Elsevier Applied Science, Ch. 6, pp. 189-213.

Batchelor, B.deV. Hewitt, B.E. and Csagoly, P.F. (1978) An investigation of the fatigue strength of deck slabs of composite steel/concrete bridges, in **Transportation**

Research Record 664-MTC, pp. 153-161.

Batchelor, B.deV. and Tissington, I.R. (1976) Shear strength of two-way bridge slabs. **J. Str. Div., Amer. Soc. Civ. Engs**, 102(ST12), pp. 2315-2331.

Batchelor, B.deV. and Tong, P.Y. (1970) **An Investigation of the Ultimate Shear Strength of two-way continuous Bridge Slabs Subjected to Concentrated Loads.** Ontario Ministry of Transportation and Communications, Res. Rep. No. RR167.

Beal, D. (1982) Load capacity of concrete bridge decks. **J Str. Div., Amer. Soc. Civ. Engs**, 108(ST4, pp. 814-832.

Brotchie, J.F. and Holley, M.J. (1971) Membrane action in slabs, in **Cracking, Deflection, and Ultimate Load of Concrete Slab Systems**, Amer. Conc. Inst. Publication SP-30, pp. 345-377.

Canadian Standards Association (1988) **Design of Highway Bridges.** CSA Standard S6-M88, Rexdale, Ontario, Canada.

Csagoly, P.F. Holowka, M. and Dorton, R.A. (1978) The true behaviour of thin concrete bridge slabs, in **Transportation Research Record 664-MTC**, pp. 171-179.

Dorton, R.A. Holowka, M. and King, J.P.C. (1977) The Conestoga river bridge - design and testing. **Can. J. Civ. Eng.**, 4(No. 1), pp. 18-39.

Elling, C.W. Burns, N.H. and Klingner, R.L. (1985) **Distribution of Girder Loads in A Composite Highway Bridge.** Center for Transportation Research, University of Texas at Austin, Res. Rep. CTR 350-2.

Fang, I.K. Worley, J. and Burns, N.H. (1985) **Behavior of Ontario-type Bridge Decks on Steel Girders.** Center for Transportation Research, University of Texas at Austin, Res. Rep. CTR 350-1.

Guyon, Y. (1960) **Prestressed Concrete.** John Wiley and Sons, Vol. 2.

Harris, A.J. (1957) Prestressed concrete runways: history, practice and theory. **Proc. of the Inst. of Civ. Engs**, 6, pp. 45-66.

Hewitt, B.E. and Batchelor, B.deV. (1975) Punching shear strength of restrained slabs. **J. Str. Div., Amer. Soc. Civ. Engs**, 101(ST9), pp. 1837-1853.

Jackson, P. (1988) **Membrane Action in Bridge Deck Slabs.** B.C.A. Bulletin, Issue No. 1, pp. 2-3.

Kinnunen, S. and Nylander, H. (1960) **Punching of Concrete Slabs without Shear Reinforcement.** Trans. Royal Inst. Technology, Stockholm, No. 158.

Kirkpatrick, J. Rankin, G.I. and Long, A.E. (1984) Strength evaluation of M-beam bridge deck slabs. **The Structural Engineer**, 62B(No. 3), pp. 60-68.

Liebenberg, A.C. (1966) **Arch Action in Concrete Slabs.** South African Council for Sci. and Ind. Res., R234, Nat. Bldg Res. Inst., Bull. 40.

Moll, E.L. (1985) **Investigation of Transverse Stressing in Bridge Decks.** M.Sc. Thesis, McMaster University, Hamilton, Ontario, Canada.

Newmark, N.M. Siess, C.P. and Peckham, W.M. (1948) **Studies of Slab and Beam Highway Bridges, Pt II.** University of Illinois, Eng. Exp. Stn, Bull. 375.

Newmark, N.M. Siess, C.P. and Penman, R.R. (1946) **Studies of Slab and Beam Highway Bridges, Pt I.** University of Illinois, Eng. Exp. Stn, Bull. 363.

Ockleston, A.J. (1955) Load tests on a three storey reinforced concrete building in Johannesburg. **The Structural Engineer,** 33, pp. 304-322.

Ockleston, A.J. (1958) Arching action in reinforced concrete slabs. **The Structural Engineer,** 36, pp. 197-201.

Ontario Ministry of Transportation and Communications (1979) **Ontario Highway Bridge Design Code, 1st edn.** Downsview, Ontario, Canada.

Ontario Ministry of Transportation and Communications (1983) **Ontario Highway Bridge Design Code, 2nd edn.** Downsview, Ontario, Canada.

Park, R. (1964) Ultimate strength of rectangular concrete slabs under short-term uniform loading with edges restrained against lateral movement. **Proc. Inst. Civ. Engs,** 28, pp. 125-150.

Perdikaris, P.C. and Beim, S. (1988) Reinforced concrete bridge decks under pulsating and moving load. **J. Str. Eng.,** 114(No. 3), pp. 591-607.

Poston, R.W. Phipps, A.R. Almustafa, R.A. Breen, J.E. and Carrasquillo, R.L. (1988) Effects of Transverse Prestressing in Bridge Decks. **J. Str. Eng.,** 114(No. 4), pp. 743-764.

Savides, P. (1989) **Punching Strength of Transversely Prestressed Deck Slabs of Composite I-Beam Bridges.** M.Sc. Thesis, Queen's University at Kingston, Ontario, Canada.

Siess, C.P. and Viest, I.M. (1953) **Studies of Slab and Beam Highway Bridges, Part II - Tests of Continuous Right I-beam Bridges.** University of Illinois, Eng. Exp. Stn, Bull. 416.

Taylor, R. and Hayes, B. (1965) Some tests on the effect of edge restraint on punching shear in reinforced concrete slabs. **Mag. Conc. Res.,** 17, pp. 39-44.

Thomas, F.G. and Short, A. (1952) A laboratory investigation of some bridge deck systems. **Proc. Inst. Civ. Engs,** 1 (pt. 1), pp. 125-187.

Tong, P.Y. and Batchelor, B.deV. (1971) Compressive membrane enhancement in two-way bridge slabs, in **Cracking, Deflection, and Ultimate Load of Concrete Slab Systems,** Amer. Conc. Inst. Publication SP-30, pp. 271-286.

Tsui, C. Burns, N.H. and Klingner, R.L. (1985) **Behavior of Ontario-type Bridge Decks on Steel Girders: Negative Moment Region and Load Capacity.** Center for Transportation Research, University of Texas at Austin, Res. Rep. CTR 350-3.

BEHAVIOUR OF INTEGRAL REINFORCED CONCRETE SKEWED SLAB BRIDGES

H. DAGHER, M. ELGAALY, J. KANKAM
Dept Civil Engineering, University of Maine, Orono, Maine, USA

Abstract
The design of a rigid-frame single-span bridge is
straightforward when the bridge has no skew. For skew-
bridges, the design is complicated by higher shears and
moments that develop near the obtuse angle. The finite
element method can be used to compute these moments and
shears but it is too time consuming for everyday design.
This paper outlines work being done to simplify the design
of such bridges.
Keywords: Design, Rigid-frame, Concrete, Skew-slab, Bridge,
          Shear, Moment, Finite Element.

1 Introduction

Many investigations have been conducted on the behaviour of
skew slabs under load: Bakht (1988), Cheung et. al.
(1968), Clark (1984), Cope (1977, 1980, 1983a, 1983b,
1984a, 1984b, 1985, 1989), Greimann et. al. (1983),
Mahmoudzadeh et. al. (1984), Mehrein (1967), and Nilson
(1980). However, most of this work relates to skew slabs
which are supported on columns or on nonintegral supports.

This paper describes the development of a simplified
method for the economical design of rigid frame reinforced
concrete skewed slab bridges. A recent nationwide survey
conducted as part of this research indicated that several
States had bridges of this type either already existing or
projected for the future. Besides being less expensive,
rigid-frame bridges are less affected by de-icing chemicals
and function for extended periods without appreciable
maintenance or repair.

These bridges have high shears and moments near the
obtuse angles of the slab, which can be computed with
finite element modelling. However, the finite element
method is too time consuming for everyday design. In an
ongoing research program at the University of Maine,
special purpose pre- and post-processors were developed to
expedite the finite element analysis. Parametric studies

are being conducted, leading to a simplified design method for such bridges. Later on in the research program, reinforced concrete models of these bridges will be tested for experimental verification of the analytical results. This will eventually lead to the production of a design manual for such bridges.

2   Finite Element Modelling

The finite element models of the structure comprise the following elements:

Flat Plate elements with five degrees of freedom per node are used in the walls and skew. In general, these elements are to be restrained against in-plane rotation (drilling) when the elements intersecting at one node are in the same plane.

Beam elements with six degrees of freedom are used in the beams at the edges of the skew slab.

Boundary elements are used to restrain the 'drilling' degrees of freedom in plate elements when these are not in the direction of a global coordinate axis. Boundary elements are also used to obtain the reactive moments and forces at the base of the walls.

3   Preprocessor

A special preprocessor has been developed for rigid frame reinforced concrete skewed slab bridges which are to be analysed using the SAP4 Finite Element Structural Analysis Program, described in Bathe, Wilson and Peterson (1974). A modified version of this program which had been adapted for the University of Maine computer system was used for the analysis.

With the preprocessor, the user interactively specifies only the overall dimensions, material properties and loading conditions of the bridge. The preprocessor itself generates the numbers and coordinates of the nodes of the structure, the numbers and connectivities of the plate and beam elements, and other necessary data in the format required by the SAP4 program.

The preprocessor consists of a main program and seven subroutines as described below:

Program Jpreskew : The Main Program; reads in design data and calls the subroutines to perform their assigned functions.

Subroutine Jnodpd: Generates nodal points and their coordinates.

Subroutine Jpenuc: Generates plate elememts and their connectivities.

Subroutine Jbelc : Generates beam elements and their connectivities.

Subroutine Jbedir: Generates boundary elements and their directions.

Subroutine Jbccod: Generates the general restraint condition codes for all nodes.

Subroutine Jcdrbe: Generates codes for displacement and rotation for boundary elements.

Subroutine Jsapdf: Creates the SAP4 data file.

An example of a finite element mesh generated by the preprocessor is shown in Figure 1.

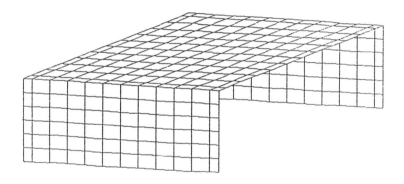

Fig.1.   Typical finite element mesh for rigid frame skew slab bridge

4   Convergence Studies

Using the preprocessor, convergence studies have been carried out to determine the optimum size of element to be used to ensure maximum accuracy of the results of the analysis without taking unnecessarily large amount of computer time.

For the structure shown in Fig. 2, vertical displacements caused by the self-weight of the structure at

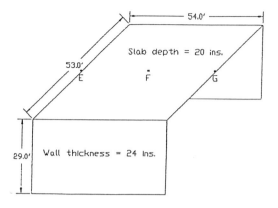

(a)   The whole frame

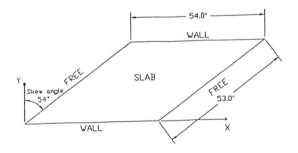

(b)   Top view of the slab

Fig.2.   Structure used in convergence study

239

points E and G, where maximum values occur in the slab, as well as at point F, the center of the slab, are given in Table I for three different models. It can be seen that very good convergence occurs as the number of nodes in the structure is increased from 495 in model II to 1078 in model III.

Table 1. Results of convergence studies

	Model I	Model II	Model III
No. of nodes	136	495	1078
No. of elements*	112	448	1008
Max. dimension of element (ft.)	9'	4.2'	2.7'
V.D.** at E and G	0.222"(86.0%)	0.252"(97.7%)	0.258"(100%)
V.D.** at F	0.138"(93.2%)	0.145"(98.0%)	0.148"(100%)

*: Total number of plate elements in the slab and two walls
**: Vertical displacement.

5 Postprocessor

A postprocessor is in the course of development to put the results of the analysis into the form of contour plots, curves, and tables showing the distribution of moments and shearing forces in the loaded structure. At the time of preparing this paper, the following components of the postprocessor had been completed:

Program Jposkew1: This program operates on the output from the SAP4 Finite Element Structural Analysis Program. It resolves the moments and membrane forces of plate elements into the required components and puts them into appropriate files.

Program Jposkew2: This program matches the centroidal coordinates of plate elements with the corresponding moments and membrane forces produced by JPOSKEW1. The output from JPOSKEW2 is used to generate contour plots such as those shown in Figs. 3a and 3b. It can be seen from these two plots that the bending moments near the obtuse angles are much higher than those elsewhere. For

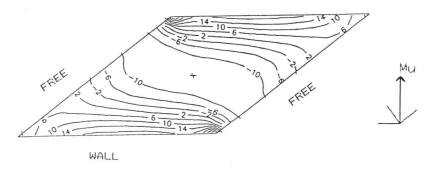

(a). Mu, bending moments per unit length in the vertical
plane containing the normal span. (in.kips./in.)

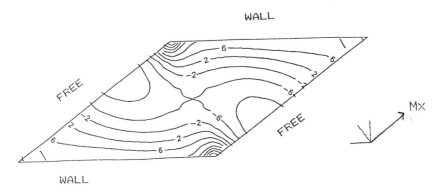

(b). Mx, bending moments per unit length in the vertical
plane containing the skew span. (in.kips./in.)

Fig.3. Typical bending moments per unit length in the skew slab of
the structure shown in Fig.2., subjected to its own weight.

sections of the skew slab near either wall, the bending
moment per unit length, Mx, in the vertical plane
containing the skew span increases from 1.6 in.kips./in.
in the acute corner to 35.5 in.kips./in. in the obtuse
corner, with an average value of 10.4 in.kips./in.

Program Jposkew3:   This program matches the nodal
coordinates of the structure with the corresponding
vertical displcacements. The output from the program is
used to plot the deflected shape of the structure in the
vertical direction, such as that shown in Fig. 4.

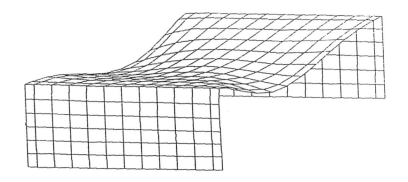

Fig. 4. Deformed shape of structure in Fig.2 subjected to
its own weight.

6 Design Guide

The first phase of the analytical work is expected to be
completed by the end of June 1990. Preliminary design aids
in the form of tables and contour plots giving the
distribution of moments and shearing forces per unit length
in the structure will be ready by then. These will be
produced for unit concentrated, line, and distributed loads
in various locations in the structure; the effects of
actual design loads will be obtained by superposition. The
design guide will include examples describing the use of
the proposed simplified design method.

7 Experimental Verification

To verify the analytical results obtained in the first year
(1989-90) of the research, model testing of one-quarter
scale reinforced concrete specimens will be carried out in
the Structural Engineering Laboratory at the University of

Maine in the period 1990-92. The results of the tests will be used to refine the analytical design method, if required.

8 Conclusions

Rigid frame reinforced concrete skewed slab bridges have high shears and moments near the obtuse angle of the slab. These shears and moments can be computed with finite element modelling but the method is too time consuming for everyday design. An outline has been given of work being done to produce a design guide which will simplify the design of such bridges.

9 Acknowledgements

The authors wish to acknowledge the Maine Department of Transportation, and particularly Mr. Jame Chandler, Bridge Engineer, for funding this research.

10 References

Bakht, B. (1988) Analysis of some skew bridges as right bridges. **ASCE Jr. of Str. Eng.**, Vol. 114, No. 10, pp. 2307-2322.

Bathe, K.J., Wilson, E.L. and Peterson, F.E. (1974) SAP4: Structural analysis program for static and dynamic response of linear systems, **Univ. of Calif.**, Berkeley.

Cheung, Y.K., King, I.P. and Zienkiewicz, O.C. (1968) Slab bridges with arbitrary shape and support conditions: a general method of analysis based on finite elements. **Proc. Inst. Civ. Engrs.**, 40, pp. 9-36.

Clark, L.A. (1984) Punching shear near the free edges of slabs. **Concrete**, Aug., pp. 15-17.

Cope, R.J. and Rao, P.V. (1977) Non-linear finite element analysis of concrete slab structures. **Proc. Inst. Civ. Engrs.**, Part 2, pp. 159-179.

Cope, R.J., Rao, P.V. and Edwards, K.R. (1980) Non-linear finite element analysis techniques for concrete slabs, in **Proc. Int. Conf. Numerical Methods for Non-linear problems**, Swansea, pp. 445-456.

Cope, R.J. and Rao, P.V. (1983a) A two-stage procedure for the non-linear analysis of slab bridges. **Proc. Inst. Civ. Engrs.**, Part 2, 75, Dec., pp. 671-688.

Cope, R.J. and Rao, P.V. (1983b) Moment distribution in skewed slab bridges. **Proc Inst. Civ. Engrs.**, Part 2, 75, Sept., pp. 419-451.

Cope, R.J. (1984a) Material modelling of real reinforced concrete slabs, in **Proc. Int. Conf. Computer-Aided Analysis and Design of Structures,** Part I, Split, Yugoslavia, Sept., pp. 85-117.

Cope, R.J. and Rao, P.V. (1984b) Shear force in edge zones of concrete slabs. **The Struct. Engr.,** Vol. 62A, No. 3, Mar., pp. 87-92.

Cope, R.J. (1985) Flexural shear failure of reinforced concrete slab bridges. **Proc. Inst. Civ. Engrs.,** Part 2, 79, Sept., pp. 559-583.

Cope, R.J. and Clark, L.A. (1989) **Concrete slabs: analysis and design.** Elsevier Appl. Sc. Publishers, London and NY.

Greimann, L.F., Wolde-Tinsae, A.M. and Yang, P.S. (1983) Skewed bridges with integral abutments. **Transportation Research Record 903,** Bridges and Culverts, National Academy of Sciences.

Mahmoudzadeh, M., Davis, R.E. and Semans, F.M. (1984) Modification of slab design standards for effects of skew. **USDOT, FHWA and Calif. DOT,** Sacramento.

Mehrain, M. and Bouwkamp, J.G. (1967) Finite element analysis of skew composite girder bridges. **Dept. of Civil Eng., Univ. of Calif.,** Berkeley.

Nilson, A.H. et al. (1980) Finite element analysis of reinforced concrete. **ASCE.**

OPTIMIZATION OF PRESTRESSED CONCRETE SINGLE CELL BOX GIRDER BRIDGES

D.N. TRIKHA, P.C. JAIN, N.M. BHANDARI, B.R. BADAWE
Civil Engineering Dept, University of Roorkee, Roorkee, India

Abstract
The paper presents an optimization study of single cell prestressed concrete box girder bridge superstructures by the application of a FEM-SLP technique developed by integrating the finite element analysis with optimization procedure using sequential linear programming method. Eight noded isoparametric flat shell elements have been adopted for the finite element analysis of the superstructure. The powerful simplex method is used for the solution of linear programming problem. The analysis procedure interacts with the optimization technique at each design iteration stage. Optimization is carried out with respect to the size variables which define thicknesses of various components of the box girder and the shape variables identified as the ratios of the width of the box to the overall roadway width (B/C) and the depth of the girder at a suitable location to the bridge span (D/L). Minimization of the weight of the superstructure defines the objective. Suitable safety criteria have been devised to keep the stresses within the safe limits throughout the continuum.

Six bridge problems, three each of the cantilever and the simply supported types have been studied and the results presented in tabular form. Graphs showing the variation of the optimum weight with D/L ratios and the contours of equal weight have been presented for a typical structure.
Keywords: Optimization, Box Girder Bridge, Superstructure, Finite Element Method, Sequential Linear Programming.

1   Introduction

In the last 3-4 decades, a tremendous increase in bridge construction activity has been witnessed throughout the world. Of the various possible cross sectional shapes, single cell prestressed concrete box girders have often been preferred over other shapes for medium span ranges of 40 m -100 m for reasons of economy, pleasing aesthetics, high torsional stiffness and amenability to the modern methods of construction. It thus becomes imperative to minimize the requirement of materials by minimizing the total weight of such structures. Since the analysis of box girder bridges requires considerable computational effort, it is essential that analysis and optimization proceed simultaneously starting from an initial choice of the cross-sectional dimensions.

Box girders have complex cross-sectional configuration, and in case of thin walled sections, they develop distortional stresses of comparable magnitudes as the primary stresses. For true estimate of these stresses, use of rather intricate finite element procedure employing suitable shell elements becomes essential, and as such, the optimization procedure must necessary remain simple to be integrated with the analysis procedure to save computational effort.

The efficiency and accuracy of the finite element analysis are dependent on the type of the element used. In the present study, eight noded isoparametric flat shell elements have been chosen by combining bending action with membrane action, as described briefly in a subsequent paragraph. The effect of prestressing force is simulated by equivalent load concept.

Optimization of box girder bridges has received considerably less attention. Lacey and Breen [1] carried out an optimization study of segmentally constructed box girder bridges considering the minimization of the cost as the objective function. Datta et al.[2] have treated the problem as a constrained non-linear programming problem and considered thicknesses of the bottom flange, area of prestressing steel and eccentricities of the prestressing force as design variables. Sinha [3] considered thicknesses of the web and the soffit slab as design variables and achieved the optimum weight of box girder bridges by FEM-SLP approach.

In the present study, the thicknesses of the web, the soffit slab, the deck slab between the webs and the cantilever slab have been treated as the design variables and minimisation of the weight of the bridge as the objective function. Sequential linear programming technique is adopted for optimization. The powerful simplex algorithm is used for solving the linear programming equations. Shape variables viz. B/C and D/L ratios are varied externally to obtain the global optimum weight.

2   Finite Element Analysis

Since the finite element method is well documented, it is not neccessary to describe the same. In the present study, an eight noded isoparametric parabolic shell elements have been used as obtained by superposition of the characteristics of the plane stress element developed by Ergatoudis [4] and those of the plate bending element reported by Hinton et al. [5,6]. The element is based on the Mindlin's plate theory which differs from the conventional plate theory in the assumption that the normals to the mid surface before deformation remain straight but not necessarily normal to the mid surface after deformation (Fig. 1). Thus effect of transverse shear deformations is accounted for. The chosen shell element has five degrees of freedom at each node with an additional sixth fictitious degree of freedom, $\{\theta z'_i\}$ associated with fictitious couple $\{M'_{zi}\}$ to facilitate the assembly of the overall stiffness matrix of the structure.

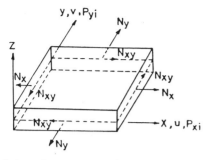

(a) PLANE STRESS (MEMBRANE)
ACTION

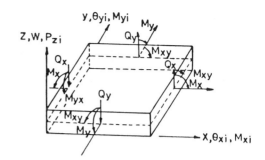

(b) PLATE BENDING ACTION

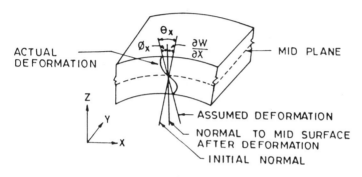

(c) DEFORMATION OF THE CROSS SECTION OF PLATE

FIG.1–GENERALIZED PLANE STRESS AND PLATE BENDING
ACTIONS AND DEFORMATIONS OF PLATE SECTION

Thus the force-displacement relations for the element may be
stated as

$$\{F'\}^e = [K']^e \{\varepsilon\}^e \tag{1}$$

where $[K]^e$ is the stiffness matrix of order 48x48 for the chosen
element.

For small deformations, the membrane action and the bending
action remain uncoupled and as such, the above relationship may
be expressed as

$$
\begin{Bmatrix} F'_p \\ F'_b \\ M'_z \end{Bmatrix}
=
\begin{bmatrix} [K'_p] & [0] & [0] \\ [0] & [K'_b] & [0] \\ [0] & [0] & [K'_z] \end{bmatrix}
\begin{Bmatrix} \delta'_p \\ \delta'_b \\ \theta'_z \end{Bmatrix}
\tag{2}
$$

where the subscripts p and b stand for the plane stress and the plate bending actions respectively, and θ'_z is the rotation about z axis. The terms kij in $[K'_{\theta z}]$ which relate θ'_z rotations at the eight nodes to M'_z moments have been defined using a fudge factor α_f so that

$$kij = \alpha_f \; EtA \quad for \quad i = j,$$

and $$kij = -\frac{1}{7} \alpha_f \; EtA \quad for \quad i \neq j,$$

where   A   =   Area of the element
        t   =   Thickness of the element
        E   =   Modulus of elasticity of the material
        α_f =   Fudge factor the value of which is taken   0.001.

After due trials, 56 number of the chosen 8-noded elements have been used to discretize the cantilever type of box girders, whereas for simply supported bridges, 52 such elements have been found to be sufficient for discretization (Fig. 2).

Figure 2 also shows simulation of prestressing force by equivalen nodal loads. Losses in the prestressing force have been calculated both at transfer and at service stages by conventional methods at different locations coinciding with the vertical mesh lines in order to determine the forces to be applied at these locations so that these forces when added to the prestressing force applied at the ends result in achieving the net prestressing force acting at a given location.

3   Criteria For Safe Design

The component plates of a box girder are essentially in a bi-axial state of stress. Several experimental investigations have been reported as regards the strength of concrete subjected to a bi-axial state of stress under different combinations such as compression-compression, compression-tension and tension-tension. In the present study, the failure envelope based on the experimental investigations of Kufer et al. [8,9] has been adopted. For devising the safe design criteria for the above stress combinations, the same envelope is assumed to be valid at the working stress stage. The safe design criterion thus obtained for different stress combinations.

(i) Compression–compression state

$$\frac{(\sigma_1 + \sigma_2)^2}{(\sigma_1 + 3.65\sigma_2)} \leqslant \sigma_0 \qquad (3a)$$

or   F $\leqslant \sigma_0$ \qquad (3b)

Where F given by the left hand expression of equation 3a has been

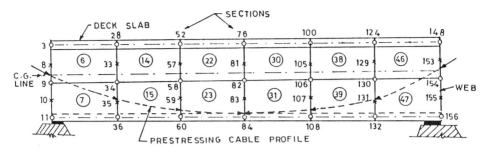

(i) DISCRETIZATION AND SYMBOLIC REPRESENTATION OF PRESTRESSING FORCE

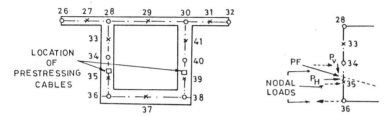

(ii) LOCATION OF PRESTRESSING   (iii) EQUIVALENT NODAL LOADS DUE
CABLES AT ANY SECTION   TO APPLIED PRESTRESSING FORCE PF

(a) SIMPLY SUPPORTED BOX GIRDER BRIDGE

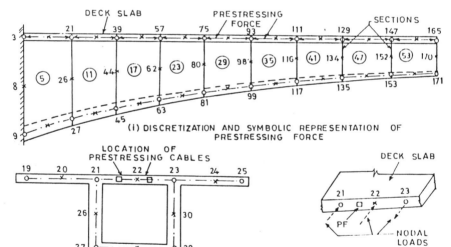

(i) DISCRETIZATION AND SYMBOLIC REPRESENTATION OF
PRESTRESSING FORCE

(ii) LOCATION OF PRESTRESSING CABLES   (iii) EQUIVALENT NODAL LOADS DUE
AT ANY SECTION   TO APPLIED PRESTRESING FORCE PF

(b) CANTILEVER BOX GIRDER BRIDGE

FIG.2 — DISCRETIZATION AND SIMULATION OF PRESTRESSING FORCES
AS EQUIVALENT NODAL LOADS

249

called the 'reference stress', σ_o is the uniaxial permissible compressive stress in concrete and σ_1 and σ_2 are the principal stresses in directions 1 and 2 respectively.

(ii) Tension-compression state

For a tension-compression biaxial state of stress at a point, the criterion for safe design may be written as devised

$$\frac{1}{\mu} \, \sigma_1 - 0.625x \, \sigma_2 \leqslant \sigma_o \qquad (4a)$$

$$\text{or} \quad F \leqslant \sigma_o \qquad (4b)$$

Where left hand side of equation 4a represents the reference stress 'F' and μ is the ratio of the permissible stresses in tension and compression for concrete.

(iii) Tension-tension state

In case both the principal stresses are tensile, the safe design criterion may be written as

$$F \leqslant \mu \, \sigma_o \qquad (5)$$

Where 'F' is the reference stress for this state of stress.

4   Optimization Strategy

The stress analysis by the finite element method for given thicknesses and the optimization of these thicknesses by linear programming have been integrated to obtain minimum weight of a bridge superstructure. Suitable constraints on the variation of the thicknesses which are treated as the design variables have been imposed by specifying the maximum and minimum values. Since the sensitivities of each design variable are different at each design stage, a stagewise linear programming has been adopted to develop an economical solution algorithm. The sensitivities of the stresses and the displacements, which help in deciding the best directions in which to implement changes, are determined at the stage of the finite element analysis. Thus starting from some initial chosen values, the problem of optimization has been formulated as a sequence of linear programming problem by imposing restrictions on move limits of the design variables at each iteration stage.

4.1 Identification of design variables

Two types of design variables have been identified as given below.

(A) Size variables which comprise the thicknesses of the component plates of the box section. They are designated as t_1, t_2, t_3 and t_4 as the thicknesses of the web, the soffit slab, the deck slab between the webs and the cantilever slab respectively.

(B) Geometric variables which define the configuration of the box section.. These are

(i)   B/C ratio i.e. the ratio of the box width   B   to the total
        width   C   of the deck inclusive of the cantilever portions at
        the deck slab.

(ii)   D/L ratio i.e. the ratio of the depth   D   of the box girder
        to the span length L.   D   is the maximum box girder depth
        at the pier end for cantilever bridges, and the constant
        girder depth for simply supported bridges.

In case of cantilever bridges, the depth of the cell is assumed
to vary parabolically along the span. If D is the maximum depth of
the box girder at the pier section, and   Do   is the depth of the
girder adopted at the free end, the depth   'D$_x$' of the cell at a
distance   x   from the pier is obtained as (Fig. 3).

$$D_x \; = \; \; Do + (D \; - \; Do) \, (1 \; - \; \frac{x}{L} \,)^2 \tag{6}$$

In case of simply supported bridges, the depth   D   of the box
girder is assumed to be constant for the entire span, as is normally
the case.

4.2 To achieve a global optimum
The local optima are obtained first by carrying out the parametric
studies by applying the FEM-SLP process as below.

(i) For a given span length L and keeping the value of D/L ratio
    constant, the local optimum weights for different   B/C ratios
    are obtained, from which the optimum value of   B/C ratio is
    determined.

(ii) Keeping the   B/C ratio constant, the local optimum weights for
     different   D/L ratios are obtained from which the optimum value
     of   D/L ratio is determined.

The ranges of the variations of the   D/L and B/C ratios are
suitably determined using past experience.

5   Formulation of Linear Programming Problem

5.1 Move limits
Certain move limits have been imposed on the variation of the design
variables so that starting from the initial values of $\{ T \} = \{ t_1, t_2, t_3, t_4 \}^T$.
These variables can assume any suitable values within the $\{ T \}^{max}$
and $\{ T \}^{min}$.   Thus

$$\{ T \}^{min} \; \leq \; \{ T \} \; \leq \{ T \}^{max} \tag{7}$$

The maximum and minimum changes in the design variables are thus
given by

$$\{dT\}^{max} = \{T\}^{max} - \{T\}$$
and
$$\{dT\}^{min} = \{T\}^{min} - \{T\}$$

(8)

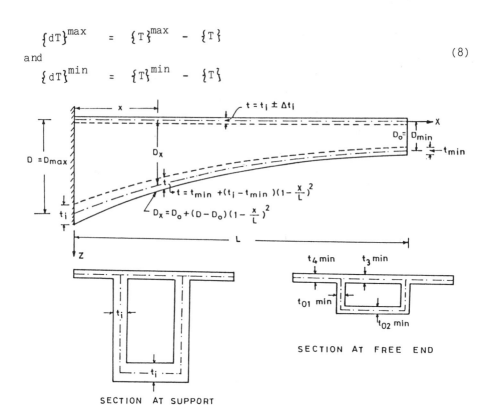

$t = t_i \pm \Delta t_i$

$D_o = D_{min}$

t_{min}

$D = D_{max}$

D_x

$t = t_{min} + (t_i - t_{min})(1 - \frac{x}{L})^2$

$D_x = D_o + (D - D_o)(1 - \frac{x}{L})^2$

t_i

L

$t_4 min$ $t_3 min$

$t_{01} min$

$t_{02} min$

SECTION AT FREE END

t_i

t_i

SECTION AT SUPPORT

(a) CANTILEVER BRIDGE (DEPTH VARYING ALONG SPAN)

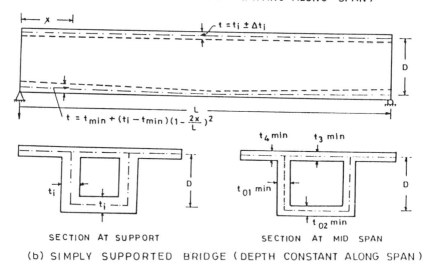

$t = t_i \pm \Delta t_i$

D

$t = t_{min} + (t_i - t_{min})(1 - \frac{2x}{L})^2$

L

$t_4 min$ $t_3 min$

t_i

t_i

D

$t_{01} min$

$t_{02} min$

D

SECTION AT SUPPORT

SECTION AT MID SPAN

(b) SIMPLY SUPPORTED BRIDGE (DEPTH CONSTANT ALONG SPAN)

FIG. 3 — VARIATION OF DESIGN VARIABLES

The changes in the design variables may be either positive or negative. If they are negative, there are transformed to non-negative values through the following side constraint :

$$\{dT^+\} \quad = \quad \{dT\}^{max} - \{dT\}^{min} \tag{9}$$

5.2 Objective function

Minimization of the total weight of the superstructure has been adopted as the objective function for the optimization problem. The total weight of the superstructure may be expressed as the sum of the weights of all its elements 'e' as given below :

$$W \quad = \quad \sum_{i=1}^{e} \quad A_i \times t_i \times \gamma \tag{10}$$

where A_i, t_i and γ are the area, the thickness and the material density of the element i. Since for a given bridge superstructure of specified geometric configuration, the density of the material and the areas of the elements are constant, the variation of the element thicknesses determine the weight change. The spanwise variation of these thicknesses have been assumed as below, Fig. 3. The deck slab thickness (t_3) and the cantilever slab thickness (t_4) are assumed constant over the span for both the cantilevered and simply supported bridges. their thickness t at any section is given by

$$t \quad = \quad t_i \tag{11}$$

where, i = 3, 4 and t_i is the specified thickness

The web and the soffit slab thicknesses, t_i (i = 1,2) are assumed to vary parabolically along the span as given below :

(i) for cantilever bridges

$$t \quad = \quad t_{oi}[1 - (1 - \frac{x}{L})^2] + t_i(1 - \frac{x}{L})^2 \tag{12}$$

and (ii) for simply supported bridges

$$t \quad = \quad t_{oi}[1 - (1 - \frac{2x}{L})^2] + t_i (1 - \frac{2x}{L})^2 \tag{13}$$

Where L   = span length of a girder
       t_{oi} = the minimum thickness at the free end/mid-span for the cantilever/simply supported bridges
       t_i = the thickness at the support section
       t = thickness at distance x from the support section.

Since the chosen design variables are the thicknesses t_i at predefined locations, and t_{oi} remains constant during any variation

of t, the first term in each of the above equations makes zero contribution to the variation in total weight 'W' of the superstructure.
Hence if the change in the thicknesses during the optimization process
be {dT} which minimizes the increase in total weight, the change in
the total weight dW may then be written as

$$dW = W_1 dt_1 + W_2 dt_2 + W_3 dt_3 + W_4 dt_4 \qquad (14)$$

Where $W_i = A_i \gamma$ = constant for an element

The change in thicknesses can be either positive or negative, the
objective function may be written by using the transformed thicknesses, as below :

$$\text{To minimize} \quad W_1 dt_1^+ + W_2 dt_2^+ + W_3 dt_3^+ + W_4 dt_4^+ \qquad (15)$$

5.3 Stress constraints

To ensure safety of different component plates during optimization,
the constraints on stresses such that their combination does not
affect the safety of structure any where, have been applied. At any
predefined critical locations K', the reference stress F', defined
above for different combinations of stresses and after considering
the effect of changes in the design variables, should be less than
the permissible value. The equation for stress constraints is thus
written as

$$F_m(K) + \sum_{i=1}^{4} \frac{\partial F_m(K)}{\partial t_i} dt_i - \sigma_o \leq 0 \qquad (16)$$

where $F_m(K)$ is the reference stress F at K^{th} critical point for
m^{th} loading condition,

$\dfrac{\partial F_m(K)}{\partial t_i}$ is the change in the reference stress F with respect to
thickness t_i for an increase in the thickness by dt_i,

and σ_o is the uniaxial permissible compressive stress in concrete, at transfer/service stage.

To these stress constraints, the side constraints as stated
in equation 9 are added to form the complete linear programming
problem equations which are solved by the simplex method.
The stress sensitives $\partial F/\partial t_i$ of reference stress F with respect
to the design variables t_i are evaluated within the finite element
analysis as described below in brief.

5.3.1 Stress sensitivity

The stress resultants N_x, N_y and N_{xy} for plane stress action and
M_x, M_y, M_{xy}, Q_x and Q_y for the plate bending action at the element
gauss point are determined as forces per unit width of the plate.
From these, the extreme fibre stresses σ_x, σ_y and τ_{xy}, principle

stresses σ_1 and σ_2 and reference stress F in terms of σ_1 and σ_2 are easily evaluated.

Let F and $\{\delta\}$ be the reference stress and the displacement vector at a gause point for a particular design value t_i. Using comma notation for differentiation, for a change of Δt_i in the design variable, the displacements and reference stresses corresponding to the new design variables $(t_i \pm 1/2 \Delta t_i)$ may be expressed as $(\{\delta\} \pm 1/2 \{\delta\}_{,t_i} \Delta t_i)$ and $F[\{\varepsilon\} \pm 1/2 \{\varepsilon\}_{,t_i} \Delta t_i]$. The derivative of the reference stress F, corresponding to the change dt_i in the design variable t_i may then be written using central difference equation as below :

$$F_{,t_i} = \frac{F[\{\delta\} + \frac{1}{2}\{\delta\}_{,ti}\, \Delta t_i] - F[\{\varepsilon\} - \frac{1}{2}\{\delta\}_{,ti}\, \Delta t_i]}{\Delta t_i} \qquad (17)$$

The change in variable Δt_i is decided by experience and degree of accuracy desired. Evaluation of derivatives of the nodal displacement with respect to the design variables t_i, $\{\varepsilon\}_{,ti}$ required in the above equation has been done as explained below :

5.3.2 Derivatives of displacements

The equations of equilibrium, equation 1, obtained during the finite element analysis relate the nodal load vector $\{R\}$ to the nodal displacement vector $\{\varepsilon\}$. Differentiating both sides of this equation with respect to t_i, and expression for the derivatives of the displacements $\{\varepsilon\}_{,ti}$ is obtained as

$$[K]\{\varepsilon\}_{,ti} = \{R\}_{,ti} - \{\varepsilon\}[K]_{,ti} \qquad (18)$$

in which the right hand side may be termed as the force derivative requiring evaluation of the derivatives of the nodal load vector and stiffness matrix with respect to the design variables.

The derivative of the nodal load vector $\{R\}_{,ti}$ is formed in the stiffness subroutine at the same time as $\{R\}$ is set up [10].

The element stiffness matrix $[K^e]$ consists of $[K^e]^p$ and $[K^e]^b$ representing the plane stress and plate bending actions respectively and a fictitious stiffness term $[K^e_{\theta z}]$ corresponding to the sixth degree of freedom.

The plane stress stiffness submatrix is given by

$$[K^e]^p = \iint [B]^T[D][B]\, dv \qquad (19)$$

in which [B] and [D] do not contain any term depending on thickness t. Hence the derivative of $[K^e]^p$ with respect to t is obtained simply by dividing every term in $[K^e]^p$ by t. The differentiation of $[K^e_{\theta z}]$ is similarly obtained by dividing it by t.

The plate bending stiffness submatrix connecting nodes i to j is given by

$$[K^e]^b = \int_{-1}^{1} \int_{-1}^{1} [B_i]^T[D][B_j]\, |J|\, dg\, d\eta \qquad (20)$$

255

Here the submatrix [D] of the elastic rigidities consists of terms containing t^3 and t. Hence the differentiation of the terms in $[K^e]_b$ is done by considering individual terms separately and dividing them by 3/t and 1/t respectively.

The matrix $[K]_{ti}$ is formed simultaneously with the formation of $[K^e]$ in the stiffness subroutine. Thus all the quantities required for the formation of equations in the linear programming problem have been evaluated during the finite element analysis. These equations are processed by simplex algorithm. The slack variables are added whenever necessary to change the inequalities into equalities.

The variation in the initially chosen values of the design variables satisfying the stress constraints and minimizing the weight of the superstructure are evaluated in each iteration till the specified number of iterations are over or the optimum is achieved. The optimum variations in the initial values, the optimum values of the design variables and the corresponding optimum weight of the structure are thus obtained.

6 Computer Coding of FEM–SLP Technique

The integrated FEM-SLP procedure formulated above has been suitably codified as Program RMBPT in FORTRAN IV language to fcilitate Computer aided optimization of both cantilever and simply supported types of bridges. Manual effort required for the preparation of input data for the program has been minimized by writing a subroutine for a preliminary design of the structure including determination of the prestressing force required at selected sections. The live loads and prestressing forces are then automatically transferred to the relevant nodes of the sections, as the desired equivalent nodal loads.

The program RMBPT comprises of one main and seventeen subroutines. Besides the usual data to be given in a finite element analysis program, the present program RMBPT requires additional input data consisting of the initial as well as the maximum and minimum values of the thicknesses of the four components of the box girder viz. the web, the soffit slab, the deck slab between the webs and the cantilever slab, permissible stress values at transfer and service stages, ratio of the permissible tensile strength to the compressive strength of concrete, loads and their positions with respect to the sections along the span, node numbers of the sections at which the loads and the prestressing forces are to be transformed and the patch/wheel loads and x and y co-ordinates of the corners of each patch load on the deck.

The output from the program consists of all the input data unless suppressed, statical shear forces and bending moments due to self weight and imposed live loads and required prestressing forces at different sections, the displacements u, v, w, θ_x, θ_y and θ_z at each node and their derivatives with respect to each design variable, the stress resultants, principle stresses and reference stress at the specified critical gauss points, the constraint equations of the linear programming problem, the optimum values of the design variables viz. all the four component thicknesses and thicknesses of all the elements and finally the optimum weight of the super-structure.

7   Problems Studied

To establish the validity of the FEM-SLP approach. Six bridge super-structures, three each of the cantilever and the simply supported types, as described in Table 1 below, have been studied [11].

Table 1.

Bridge identification	Span (m)	Type	No. of lanes	Remarks
CB_1	46	Cantilever, no foot paths	two	Varying depth
CB_2	46	Cantilever, with foot paths	two	Varying depth
CB_3	75	Cantilever, with foot paths	two	Varying depth
SSB_1	60	Simply supported, no foot paths	two	Constant depth
SSB_2	60	Simply supported, with foot paths	two	Constant depth
SSB_3	90	Simply supported, with foot paths	two	Constant depth

The wheel loads and the materials have been considered as per the Indian Roads Congress Specifications [12]. Concrete of M40 grade with permissible compressive stress of 15.20 N/mm^2 at transfer of prestress stage and 11.85 N/mm^2 at the service stage have been considered. The prestressing cables, 12T13 of Freyssinet group, with a nominal ultimate tensile strength of 1800 N/mm^2 per cable have been used. The cables are located in the deck slab in in case of cantilever bridges and in the two webs in case of simply supported bridges.

To arrive at the global optimal value for each of the above problems, the local optima have been obtained by varying the shape variables in the following range :

B/C ratio :   0.4 to 0.55 for both types of bridges

D/L ratio :   Cantilever bridges - 0.08 to 0.15 at the support, and constant depth at the free end

Simply supported bridges - 0.025 to 0.10

The intervals in the above ranges have been chosen as appropriate to each problem to locate the optimum values.

8 Results

The optimum values of the thicknesses and the corresponding optimum weights obtained and their initial values for each of the above bridges are given in Table 2 below :

Table 2.

Identification of bridge	Span m	Optimum thickness (mm) t_1	t_2	t_3	t_4	Optimum values of B/C	D/L	Optimum weight (kN)
CB_1	46	155.23 (250.0)	150.0 +(250.0)	192.94 (250.0)	150.0 (200.0)	0.5	0.09	3790.5 (4430.5)
CB_2	46	150.0 (300.0)	150.0 (250.0)	176.41 (250.0)	150.0 (200.0)	0.4	0.15	4378.4 (5680.0)
CB_3	75	200.0 (350.0)	200.0 (350.0)	272.10 (350.0)	150.0 (300.0)	0.4	0.15	10075.3 (14498.5)
SSB_1	60	150.0 (250.0)	150.0 (250.0)	226.76 (250.0)	150.0 (170.0)	0.45	0.025	3659.2 (4418.3)
SSB_2	60	150.0 (300.0)	150.0 (300.0)	150.0 (250.0)	169.42 (200.0)	0.55	0.04	5242.2 (7403.0)
SSB_3	90	150.0 (350.0)	200.16 (350.0)	150.0 (300.0)	251.44 (200.0)	0.5	0.05	10361.2 (12509.8)

+( ) Figures in the brackets indicate the initial design values specified.

The minimum thickness of 150 mm has been specified for any plate for the above structures.

Graphs which show that the variation of the optimum weights with B/C and D/L ratios for a representative bridge CB_3 are given in Fig. 4 and Fig. 5 respectively. The local optima for each value of these shape variables can be seen in these figures.

With the local optima obtained for various values of B/C and D/L ratios, the contours of equal weight have been drawn as shown in Fig. 6. From this figure, it can be seen that the global optimum weight of 10075 kN is obtained at B/C and D/L ratio of 0.4 and 0.15 respectively, as against the initial design value of 14498.5 kN indicated in Table 2. The optimum weight obtained by the application of the proposed FEM-SLP Process is about 43.9 percent less than the initial design value. It may be mentioned that the initial design was based on statical analysis of the bridge.

9 Conclusions

It can be claimed from the results of the various bridge examples

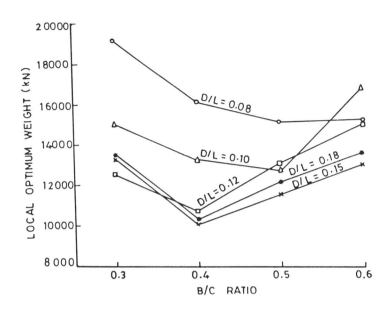

FIG.4 – VARIATION OF THE OPTIMUM WEIGHT WITH
B/C RATIO-CANTILEVER BRIDGE CB 3
(SPAN = 75m)

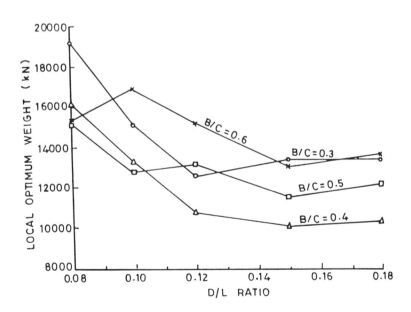

FIG.5 – VARIATION OF THE OPTIMUM WEIGHT WITH D/L
RATIO- CANTILEVER BRIDGE CB 3 (SPAN =75m)

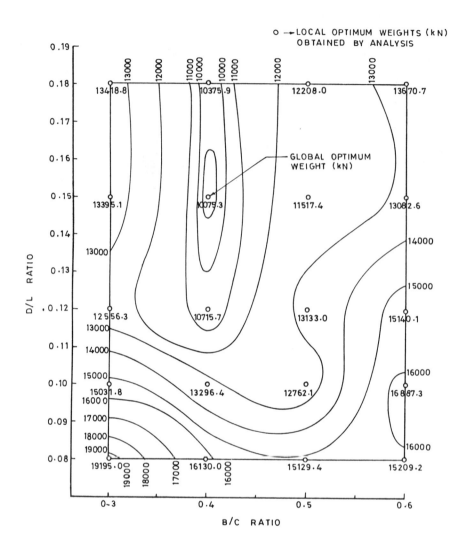

FIG. 6 — LOCAL OPTIMUM SOLUTIONS AND CONTOURS OF EQUAL
WEIGHT FOR DIFFERENT VALUES OF SHAPE VARIABLES
—CANTILEVER BRIDGE CB3
(SPAN = 75m)

presented above that the integrated FEM-SLP approach developed in the present study offers a valuable tool for the interactive analysis and optimization procedures of prestressed concrete box girder bridges and results in considerable saving of the computational effort if finite element analysis and optimization were to proceed independently. It achieves safe design satisfying several constraints. The results obtained for the six bridge problems considering several variables like span lengths, nature of supports and presence or absence of foot paths indicate that considerable economy can be achieved by the application of this technique, in the design of prestressed concrete box girder bridges.

10 References

1 Lacey, G.C. and Breen, J.E. (1975) Design and optimization of segmentally precast box girder bridges. Research Report No.121-3, Centre for Highway Research, The University of Texas, Austin.
2 Datta, A.B. and Dayaratham, P. (1978) Optimal sizes of segmental prestressed concrete cantilever bridge girder. **The Bridge and Structural Engineer,** ING/IABSE, Vol.8, No.3.
3 Sinha, S.K. (1985) Segmental prestressed concrete box girder bridges. Ph.D. Thesis, University of Roorkee, Roorkee, India.
4 Ergatoudis, J.G., Irons, B.M. and Zienkiewicz, O.C. (1968) Curved, isoparametric quadrilateral elements for finite element analysis. **International Journal of Solids and Structures,** 4, 31-42.
5 Hinton, E., Razzaque, A. and Zienkiewicz, O.C. (1975) A simple finite element solution for plates of homogeneous, sandwitch and cellular construction. Proc. Institution of Civil Engineers, London, Part 2, 59.
6 Ahmad, S., Irons, B.M. and Zienkiewicz, O.C. (1970) Analysis of thick and thin shell structures by curved finite elements. **Int. Jour. of Numerical Methods in Engg.,** Vol.2, 419-451.
7 Mindlin, R.D. (1961) Influence of rotatory inertia and shear on flexural motions of isotropic elastic plates. **Journal of Applied Mechanics,** 18, 31-38.
8 Kupfer, H.B., Hilsdorf, H.K. and Rusch, H. (1969) Behaviour of concrete under biaxial stresses. **ACI Journal Proceedings,** 6, No. 8, 656-666.
9 Kupfer, H.B. and Cerstle, R.H. (1973) Behaviour of concrete under biaxial stresses. **Journal of the Engineering Mechanics Division,** ACI Proceedings, 99, No. EM4, 852-856.
10 Trikha, D.N., Jain, P.C., Bhandari, N.M. and Badawe, B.R. (1990) Integrated FEM-SLP technique for optimization of prestressed concrete box girder bridges. Sent to Institution of Civil Engineers London.
11 Badawe, B.R. (1989) Cross sectional optimization of prestressed concrete box girder bridge superstructures. Ph.D. Thesis, University of Roorkee, Roorkee, India.
12 IRC-6-1966. (1985) Standard specifications and code of practice for road bridges. Section II, Indian Roads Congress.

STUDY OF CAST-IN-SITU CEMENT CONCRETE MULTI-CIRCULAR VENTED HIGHWAY BRIDGES

K. LAKSHMINARAYANA, M.K.L.N. SASTRY
Faculty of Civil Engineering, Bangalore University, Bangalore, India

Abstract
Cast-in-situ cement concrete multi-circular vented highway bridges, under certain circumstances, are advantageous and economical in cost. These bridges are in usage since 1962. Bridges having skew angle upto 35° are not uncommon. These bridges are now being used as "Multi-purpose high level bridges." The main structure of the existing bridge is of cast-in-situ mass concrete of 1:3:6 with 20% plums and having vents upto about fortysix in a series without an expansion joint. The primary objectives are to assess the load factor of the bridge and to evolve a suitable 'typical design', which forms the design with respect to intended function, economy of structure and conformity with appropriate codes. Undistorted structural models of concrete M10 are used without and with nominal reinforcement surrounding the vent, having scale factor of 1/10 of the prototype, are tested to various position of live load. Ultimately each model will be tested for assessing the ultimate capacity of the model and for locating the absolute critical section. Salient features regarding model behaviour are observed and also increase in value of the load factor of the model for nominal reinforcement is assessed. For designing these bridges the 'typical design' evolved by the author based on treating 'half portion of one vent' as sub-components of fixed beam having variable moment of inertia is found to be satisfactory and favourable for the Bridge Designer's.
Keywords: Concrete M10, 70R track load, Standardised normal distribution curve, Undistorted structural model, Absolute critical section, Auxiliary specimen, Effective width of deck.

1 Introduction

Cast-in-situ cement concrete multi-circular vented highway bridges (designated as CCMV bridges) are provided in places where the river is shallow or the bridge site is sandy, alluvial or clayey soil; hard strata is not met with even below a reasonable depth. In such circumstances, by adopting these bridges as against slab and beam bridges with well foundation, there will be a substantial saving in cost. These bridges(5) are also called "cast-in-situ pipe bridges." Recently these bridges are constructed as a "Multi-purpose high level bridges" which help communication, water supply and irrigation.

Fig.1 is the photograph of a two lane bridge of this type, having eight vents of 2.5m diameter, constructed in Bangalore(India) in the year 1967. The main structure is mass concrete of 1:3:6 with 20% plums (Fig.14) having no reinforcement and diameter of vents ranging from 1.5m to 3.4m. In case of deep sandy soils, the main structure is supported on a bed of 0.3m thick 1:5:10 cement concrete (pad) which inturn rests on boulder packing in sand of 0.3m thick. If the bed of the river is either alluvial or clayey, the usual practice is that the soil is excavated to a depth of 1.2m below the boulder packing and filled up with sand (Fig.14). There will be a horizontal construction joint at a height corresponding to the centre of vents and a vertical construction joint in between every third and fourth vent.

1.1 Review of literature
Most of the information available for these bridges, are linear waterway calculations; no structural design has been attempted which satisfies the appropriate codes of practice.

1.2 Object of investigation
The aim of the work presented in this paper is to assess the load factor of the bridge and to evolve a suitable 'typical design' based on a simplified approach which forms the design with respect to intended function, economical structure, and conformity with the appropriate codes (1 to 4).

It is aimed to investigate these bridges within the scope of certain parameters, that is, (i) two lane bridge of 7.5m roadway (ii) end to end of parapet of 8.4m (iii) 76mm thick wearing coat (iv) main structure is of concrete M10. Undistorted concrete models of M10 are tested to check the validity of the 'typical design' and to assess the load factor of the model and in turn the load factor of the selected Prototype (5). Hence undistorted concrete models of M10 without (5 models namely, L1 to L5 (Fig.6) and L9, and with nominal reinforcement (NR) surrounding the vents (3 models namely, L6R, L7R, L8R (Fig.7) having a scale factor of 1/10 of the Prototype (Fig.14), are tested to various position of live load (10,12) of different intensities and upto 12t (service load including impact allowance is about 9t). Fig.3 shows the live load unit on the model and it will be the corresponding scaled-down load of 70R track load. (Fig.10). Fig.3 illustrates loading unit, instrumentation for vertical deflection and surface strains. Thus, the loading position is studied for an eccentric position of the model(10,11) in order to obtain maximum longitudinal moment at the top (Fig.4) and bottom (Fig.7) of circular vents and its transverse moment on the top of deck. Live load at B (Figs.3, 9) is one such position. Initial crack load, formation and development of cracks are studied. Further tests will be conducted on each model to assess the reserve strength of the model by increasing the loads (0t, 11t, 13t, 15t) and the loading on the model continued until the failure of the model.

Necessary instrumentation such as Deflection gauges and Demec gauges are employed to measure the deflections and strains at various points (Figs.3, 4) of the model.

2 Concrete model M10

While designing the models, certain components like cut-off walls, railings, parapets which do not affect the response (10,12) of the model are completely eliminated. However, for clarity dummy kerbs and parapets (Fig.3) are introduced at the time of taking the photographs of crack pattern for the models (Fig.4). The loading frame (Fig.2) fabricated for the purpose of testing the models, provides an elastic foundation to the structural model within the loading frame itself. The simulated depth of elastic foundation (compacted sand) for the model corresponds to a foundation depth of 4.2m in the prototype (Fig.14).

The structural model automatically takes into account (6, 11) dead loads, thrust and shear, the varying moment of inertia, distribution of loads on the corners and other secondary stresses which have been neglected in the theoretical calculations.

Mix design (8) by combined fineness modulus of combined aggregate and fineness modulus of cement has been considered to obtain cement concrete of M10; cement concrete of 1:2.82:3.63 by weight and with water/cement of 0.67 is adopted as suitable mix for the cement (portland pozzolana) used. Characteristic works cube strength (1) of concrete model M10 at 28 days is found to be 110.11 kg./sq.cm; however its value should not be less than 102 kg./sq.cm (1).

Any structural model which lies "within acceptance range" of the standardized normal distribution curve has to be treated as "under acceptance"(9). Each model after testing, is investigated whether it is within the "accepted region" of the standardized normal distribution curve using test results of auxiliary specimen (Fig.8).

3 Typical design

Recently the classification of vehicles has been revised and the latest codal provision (2) states that "Class 70R loading should be used on National highways". Studies have shown that this loading is comparatively heavier than the loadings in advanced countries (13).

However, 70R track load (Fig.10) being the critical load for these bridges, the same loading is considered in the design. In the proposed investigation, loading aspects such as i) Dead loads ii) Impact load, iii) Breaking force iv) Effect of distribution of track load (1, 7) and effective width of slab for overlapping effective widths(1) have also been given due consideration (Fig.10).

From the preliminary investigation, cracks due to failure of these bridges are supposed to develop on the top most point of the first circular vent from the end of the prismatic structure or the end of an expansion joint. This aspect has been considered as one of the important criteria of failure of models and verified (Fig.9) while testing the concrete models (Figs.5 to 7) and found to be true.

Sum of dead load bending moment (statical moment) and absolute live load bending moment(SM) is design longitudinal bending moment (statical moment), and this diagram is shown in Fig.11. The fixed end moments are assessed by column-analogy method (14). Fig.12 is the final longitudinal bending moment diagram for the longitudinal strip per metre width of deck, by working stress method. Design code (1) permit the support moments calculated by elastic methods, to be redistributed (6) upto 15% provided that the mid-span moment is decreased by a similar amount. Fig.12 shows the variation of longitudinal design bending moment along the deck of the roadway. Fig.13 shows the contours of tensile flexural stresses that are based on elastic theory method and redistribution of moments at various sections corresponding to the extreme fibres of deck and vent.

4 Test results

The behaviour of the concrete model M10 without and with nominal reinforcement (0.157% of vented area) around the vents is studied with respect to deflections, surface strains, formation and development of cracks. Table 1 is the summary of initial crack load and ultimate resisting capacity of concrete models M10 without and with nominal reinforcement around the vents. Thus load factor of prototype is 2.06 and its capacity is increased to 2.4 by providing nominal reinforcement around the vents.

Fig.9 shows the response of the model towards 70R track load; loading position refers to section B (Fig.9) and centre of vent number 1 is the critical section, and stressed to the maximum value (curve numbers 1 and 6 in Fig.9). Figs.5 to 7 shows the crack pattern for a models L5 and L8R respectively.

5 Analysis of Prototype

The permissible tensile flexural stress (1,4) in 'tension zone' for mass concrete 1:3:6 with 20% plums and concrete M10 is 'zero' and 3.4 kg. per sq.cm respectively. Keeping the thickness to 0.6m above, below and in between the vents, for the CCMV bridge of concrete M10 (Fig.14), the 'typical design' was carried out for different vent diameters; it is observed that 3.4m diameter vent is the 'optimum vent' and the critical sections are checked for the principal stresses also. Special attention has been given to the aesthetic aspects of the structure.

6 Conclusions

A close observation of Fig.13 reveals that sufficient area of tensile zone exist in the structure. Since 84.35% being the concrete work of M10 for the main structure, it should be constructed using controlled concrete M10 rather than mass concrete 1:3:6 with or without plums, and the coarse aggregate for concrete work of M10 should be 20mm and down size only.

In order to withstand secondary stresses, a nominal temperature reinforcement (8mm ϕ Tor rods) of quantity about 0.15% of minimum area of cross section of the deck should be provided around the vent both in circumferential and transverse direction.

For the CCMV bridge constructed in concrete M10 of thickness 0.6m, it is noticed that 3.4m diameter vent is the 'optimum vent'. For further increase in the diameter of the vent from 3.4m to 3.7m of the CCMV bridge constructed in concrete M10, the thickness in between the vents, at the top and the bottom of the vents should be 0.9m instead of 0.6m.

It is better to provide a bell-mouthed entry and exit for the circular vents and it refers to an arc of quadrant of a circle having a radius of about 0.044 times the diameter of vent (Fig.14). This modification should be treated as essential from an aesthetic point of view, reduces the afflux and increases the coefficient of discharge, and will relieve the 'stress concentration' at the corners of the vent.

For every quarter portion of the circular vent, for a bell-mouthed entry and exit, the form work consists of 3 subunits instead of a single unit.

If the length of the bridge exceeds 45m (1) it would be desirable to provide an expansion joint. Since the embankments may be vulnerable to attack by floods, Reinforced Cement Concrete Cantilever type wing walls are economical rather than Return type wing walls. Fig.14 shows the cross sectional elevation of CCMV bridge prototype having a vent diameter of 3.4m with suggested modifications. Details of wearing coat, kerbs, parapets, drainage arrangements and expansion joint are similar to as other bridges.

Thus the design of CCMV bridges by the 'typical design' using a simplified approach has important characteristics such as, safety of structure with economy, satisfies appropriate codes, follows the current practice of determining the liveload moment, the design procedure is easy to follow, saves the designer's time and moreover, the design is based on the structural model behaviour.

6 Acknowledgement

This investigation, research work forms a part of long term research project leading to Ph.D degree and carried out at the Civil Engineering Department (U.V.C.E.) of Bangalore University. The financial support of the University Grants Commission, New Delhi, for this research work is gratefully acknowledged.

Table 1   Initial crack load and ultimate load of CCMV models of M10 without and with nominal reinforcement around the vents

Sl No	Model number, particulars	Position of load and its designation	Initial crack load (t)	Ultimate load (t)	Average ultimate load (t)	Load factor and remarks
1	L4	Centre of vent (Section B)	12.00	18.00	18.00	18/8.75=2.06
2	L5	---"---	16.50	18.00		
3	L6R	---"---	no cracks	no cracks Max.applied load= 22.35t	21.00	21.00/3.75=2.40
4	L7R	---"---	17.27	18.29		
5	L8R	---"---	15.00	22.35		
6	L1	Section A	20.00	20.00		Section A is not critical
7	L3	Section C	17.00	21.00		Section C is not critical
8	L9	Section D	15.00	22.35		Section D is not critical

Appendix. - References

1 -,(1979) **Code of Practice for Plain and Reinforced Concrete,** IS:456, Indian Standards Institution, New Delhi, 147 pp.

2 -,(1974) **Standard Specifications and Code of Practice for Road Bridges, Section-II - Loads and Stresses,** IRC:6, The Indian Roads Congress, New Delhi, 34 pp.

3 -,(1975) **Standard Specifications and Code of Practice for Road Bridges,** Section I - General Features of Design, IRC:5, The Indian Roads Congress, New Delhi, 30 pp.

4 -,(1974) **Standard Specifications and Code of Practice for Road Bridges,** Section III - Cement Concrete (Plain and Reinforced), IRC:21, The Indian Roads Congress, New Delhi, pp.3-21, 26-36, 32-34, 52-59.

5 -,(1971) **Construction of a cast-in-situ Pipe Bridge across Begur River at mile 7/8 of Harihara - Ballayanmandur Road,** Report, Karnataka Public Works Department, Karnataka, 26 pp.

6 -,Davies, J.D.(1964) **Structural Concrete,** The MacMillan Company, New York, pp.7-9, 55-66, 99-107, 116-127, 149-159.

7 Johnson Victor, D.(1973) **Essentials of Brdige Engineering,** Oxford and IBH Publishing Co., New Delhi, pp.13-53, 66-97, 102-132, 299-303.

8 Jaikrishna, and Jain, O.P.(1966) **Plain and Reinforced Concrete,** Vol.I, Nemchand and Bros., Roorkee, pp.49, 58-59, 63-65, 79, 121.

9 Richard B. Ellis, (1975) **Statistical Inference,** Basic Concepts, Prentice-Hall, Inc., Englewood Cliffs, N.J., pp.48-59.

10 Ricardo P.Pama, and Anthony R. Cusens,(1969), **Load distribution in multibeam concrete bridges,** Ist. International Symposium on Concrete Bridge Design, ACI, Detroit, Sp.Pub.No.23-7.

11 Sterling Kinney,J.(1962) **Indeterminate Structural Analysis,** Addision-Wesley Publishing Company, Inc, Massachusetts, USA, pp.431-441, 584-636.

12 Stevens, L.K. and Gosbell, K.B.(1969), **Model studies on a beam and slab concrete bridge, and concentrated loads,** Ist. International Symposium on Concrete Bridge Design, ACI, Detroit, Sp.pub.No.23-5.

13 Thomas, P.K. (1975) **A Comparative study of Highway Bridge Loadings in Different Countries,** U.K. Transport and Road Research Laboratory, Supplementary Report 135 UC, 47 pp.

14 Wang, C.K.(1953) **Statically Indeterminate Structures,** McGraw-Hill Book Company, Inc., pp.289-292, 298-303.

Fig.1 CCMV Bridge

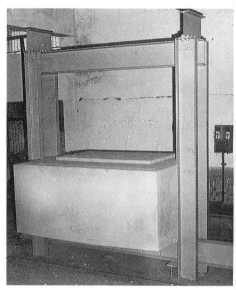

Fig.2 Loading Frame

Fig.3 Instrumentation for the Concrete Model M10.
Fig. shows Loading Unit for IRC 70R Track
Load and Observation of the 'Surface Strain'

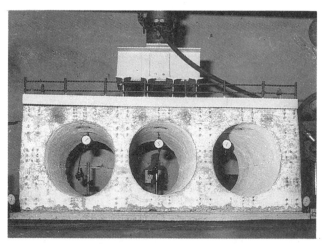

Fig.4 Front Face of Model
showing Instrumentation.
Fig. shows the Loading
Unit at Section C

Fig.5   Crack Pattern for
        the Front Face of
        Model L5 (Plain
        concrete M10)

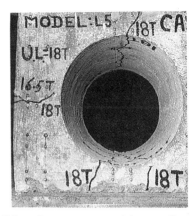

Fig.6   Crack Pattern for
        the End Face and
        Rear Face of
        Model L5 (Plain
        concrete M10)

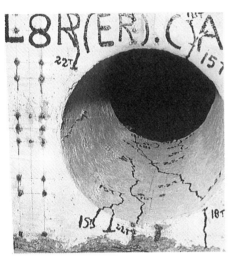

Fig.7   Crack Pattern for
        the Bottom of Vent
        (V1) and Rear Face
        for the Model L8R
        (Model of M10 with
        nominal reinforce-
        ment around the
        vents)

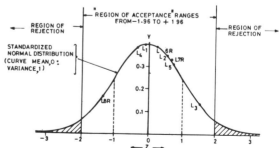

FIG. 8 – COMPRESSIVE WORKS CUBE STRENGTH (BASED ON STATISTICAL STUDY)
        OF CONCRETE M10 (AGE = AVERAGE 32 WEEKS) FOR STRUCTURAL
        MODELS.

269

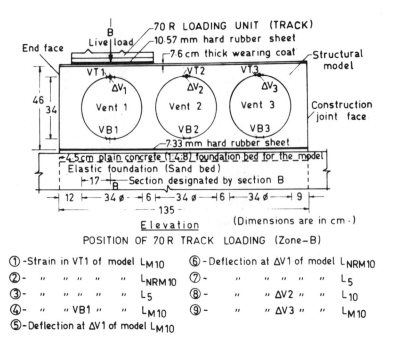

POSITION OF 70R TRACK LOADING (Zone-B)

① -Strain in VT1 of model L_{M10} ⑥-Deflection at ΔV1 of model L_{NRM10}
② -   "   "   "   "   " L_{NRM10} ⑦-   "   "   "   "   " L_5
③ -   "   "   "   "   " L_5 ⑧-   "   " ΔV2 "   " L_{10}
④ -   "   " VB1 "   " L_{M10} ⑨-   "   " ΔV3 "   " L_{M10}
⑤-Deflection at ΔV1 of model L_{M10}

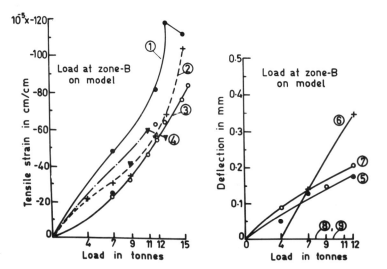

FIG. 9 -FOR THE LOAD AT 'ZONE-B' ON THE MODEL (L_5,L_{M10} and L_{NRM10}) EXPERIMENTAL VALUES OF STRAIN (at VT1, VB1) AND DEFLECTION (ΔV1, ΔV2 and ΔV3) DIAGRAM (Vent diameter = 0.34m, Scale factor = $1/10$)

270

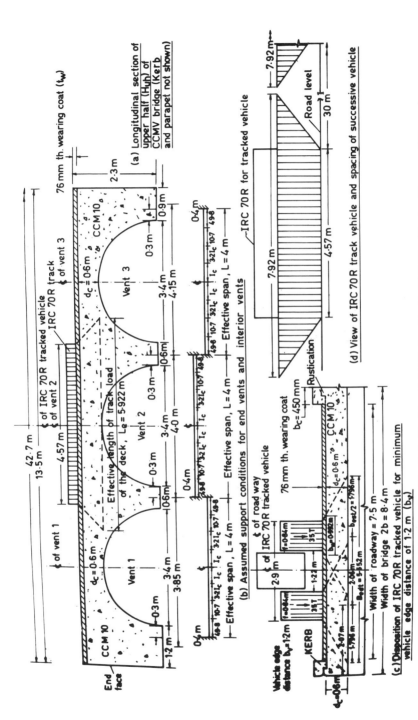

(a) Longitudinal section of upper half (H_uh) of CCMV bridge (Kerb and parapet not shown)

(b) Assumed support conditions for end vents and interior vents

(c) Disposition of IRC 70R tracked vehicle for minimum vehicle edge distance of 1·2 m (b_vℓ)

(d) View of IRC 70R track vehicle and spacing of successive vehicle

FIG.10 - DETAILS OF POSITION OF IRC 70R TRACK VEHICLE ON CCMV BRIDGE   (Scale : 10 mm = 500 mm)

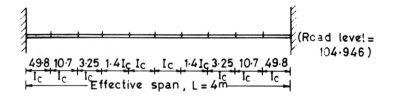

49·8 10·7 3·25 1·4I_c I_c I_c 1·4I_c 3·25 10·7 49·8
I_c I_c I_c I_c I_c

Effective span, L = 4m

FIG. 11 (a)—ASSUMED BOUNDARY CONDITION
(Scale: 10 mm = 400 mm)

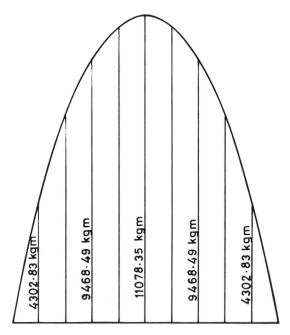

4302·83 kgm   9468·49 kgm   11078·35 kgm   9468·49 kgm   4302·83 kgm

FIG. 11 (b)—DESIGN LONGITUDINAL BMD (Statical moment)
FOR LONGITUDINAL STRIP (LS)

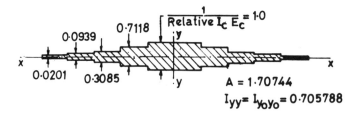

$\dfrac{1}{\text{Relative } I_c\, E_c} = 1\cdot0$

0·7118

0·0939

0·0201   0·3085

A = 1·70744

$I_{yy} = I_{y_0 y_0} = 0·705788$

FIG. 11 (d)—ANALOGOUS-COLUMN SECTION

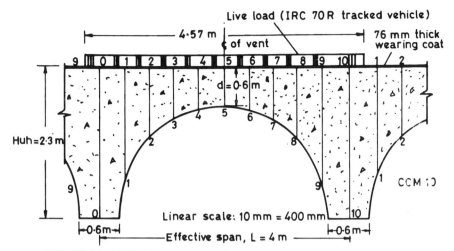

Live load (IRC 70R tracked vehicle)

4·57 m
$\mathcal{C}$ of vent

76 mm thick
wearing coat

d=0·6 m

Huh=2·3 m

CCM ID

Linear scale: 10 mm = 400 mm

|←0·6m→| ←————— Effective span, L = 4 m —————→ |←0·6m→|

FIG. 12 (a)–SECTIONAL ELEVATION OF LONGI. STRIP FOR
CCMV BRIDGE DECK

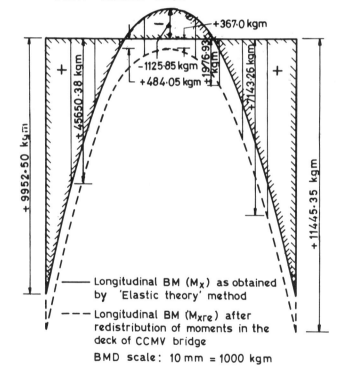

+367·0 kgm

−1125·85 kgm

+484·05 kgm

+1976·93 kgm

+45650·38 kgm

+7143·26 kgm

+9952·50 kgm

+11445·35 kgm

——— Longitudinal BM (M_x) as obtained
by 'Elastic theory' method

– – – Longitudinal BM (M_{xre}) after
redistribution of moments in the
deck of CCMV bridge

BMD scale: 10 mm = 1000 kgm

FIG. 12 (b)–VARIATION OF LONGITUDINAL BENDING MOMENT
(M_x) IN CCMV BRIDGE

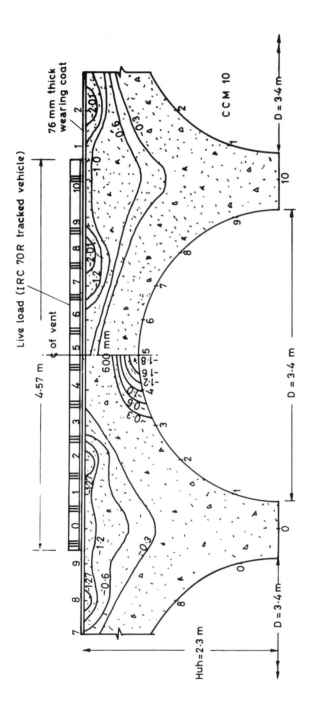

Live load (IRC 70R tracked vehicle)

76 mm thick wearing coat

¢ of vent

4·57 m

Huh=2·3 m

600 mm

CCM 10

D = 3·4 m

D = 3·4 m

D = 3·4 m

FIG. 13 (a)-TOTAL TENSILE FLEXURAL STRESS (kg/cm²) CONTOURS DEVELOPED IN THE LONGI. STRIP OF CCMV BRIDGE BASED ON 'ELASTIC THEORY' METHOD

FIG. 13 (b)-TOTAL TENSILE FLEXURAL STRESS (kg/cm²) CONTOURS DEVELOPED IN THE LONGI. STRIP OF CCMV BRIDGE CONSIDERING 'REDISTRIBUTION OF MOMENTS'

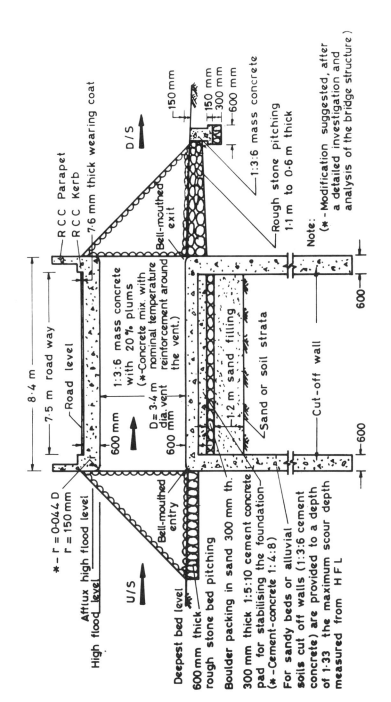

FIG. 14 —CROSS SECTIONAL ELEVATION OF (SELECTED PROTOTYPE) CAST-IN-SITU
CEMENT CONCRETE MULTI-CIRCULAR VENTED HIGHWAY BRIDGE

THE USE OF OPEN THIN-WALLED CONCRETE STRUCTURES AS RAIL BRIDGES

S.A. EL-HAMMASI
College of Technological Studies, Shuwaikh, Kuwait

Abstract
The behaviour of open thin-walled, restrained concrete
beams of channel cross-section have been investigated in
this paper for the post-cracking stages of loading. A
mathematical model has been developed for the computation
of the deformations and geometrical properties at any stage
of loading, to analyze any shape of open cross-section
structures that are subjected to interaction of bending and
torsion or any of them separately. To confirm the mathematical method approach, five identical concrete beams of
channel cross-section which were restrained against warping
at the ends by adequate diaphragms, have been tested under
different ratios of loads. The loading takes place from
zero up to failure. The results of the experimental work
and theory were put in comparison. They are in very agreeable sense.
Keywords: Torsion, Bending moment, Warping torsion,
Bimoment, St.Venant's torsion, Post-cracking, Open cross-
section, Thin-walled, Beam, Angle of twist, Deflection.

1 Introduction

In practice it is very rare to find structures subjected to
pure torsion particularly those designed to resist gravity
loads. Torsion is nearly always accompanied by bending
moment and shear. Examples of structures subjected to
combined loading are curved beams, spandrel beams and all
structures that carrying eccentric loads.

Many experimental and theoretical studies have been
conducted on pure torsion, as well as its interaction with
bending moment and shear, in reinforced concrete. Torsion
and its interaction with bending in thick-walled cross-
sections is moderately well understood, since the applied
torsion is totally resisted by pure St.Venant's torsion.

Also, the behaviour of thin-walled structures of open
cross-section under pure torsion have been well established, in the elastic range for homogeneous materials.
Both St.Venant's torsion and warping torsion are taken into

account for this loading. However,torsion in thin-walled
reinforced concrete structures of open restrained cross-
section, restrained against warping, has not been developed
yet in the post-cracking range.

The primary aim of this paper, therefore, is to investigate the behaviour of such structures. A theoretical method
has been developed and its accuracy has been confirmed by
comparing results with those obtained experimentally under
different ratios of bending to torsion. The study includes
the all stages of loading from the point of cracking up to
failure.

Secondary warping of the elements that form the cross-
section is ignored in these stages of loading due to its
little effect at post-cracking range.

2 Research significance

The behaviour of homogeneous elastic members with thin-
walled open section under the interaction of bending and
torsion has been investigated to a limited extent Refs.11,
12. However, if concrete beams are used , it is inevitable
that they will crack under service loading. After initial
cracking all the geometrical properties of the cross-section change significantly and become a function of the
various forces acting on the cross-section and the condition of the end supports. It is therefore necessary to
study the behaviour of open thin-walled concrete beams in
the post-cracking range.

2.1 Experimental programme

Five identical specimens were constructed in reinforced
concrete and tested under different ratios of bending and
torsion. The shape of the cross-section was that of a
channel, 0.5m deep and 0.5m wide with a uniform wall thickness of 50 mm. The overall length of each beam was 5.0 m.
Detailed information of the cross-sectional dimensions and
the amount of , longitudinal and transverse reinforcement,
are shown in Fig.(1). The aim of the experimental work
described here was to investigate the behaviour of concrete
thin-walled beams of open cross-section under the interaction of bending and torsion. Results obtained from this
work have been compared with those obtained theoretically.

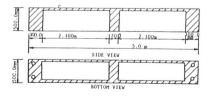

Fig.1a. General arrangement of test specimen

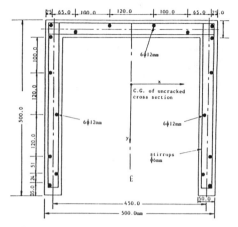

Fig.1b. Details of reinforcement in X-section

A diaphragm was provided at each end of the beams and were adequately reinforced to restrain warping. Also each had additional diaphragm at midspan for fixing the loading rig and for preventing local damage in this area, as shown in Fig.(2) and Fig.(3) respectively.

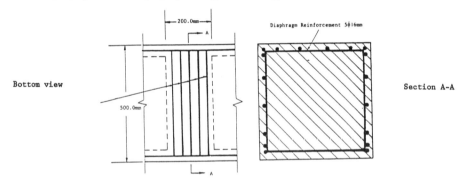

Fig.2. Reinforcement details of middle diaphragm

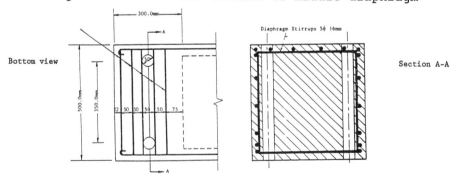

Fig.3. Reinforcement details of end diaphragms

278

The test procedure has been taken the following arrangement:

(a)Rig for interaction of bending and torsion was composed
as follows. A lever-arm assembly which is a cantilever
supporting a hydraulic jack for applying pressure,
through a loading bar, which was resting on a steel box.
The load applied through the loading bar which passes
the loading lever-arm assembly and an other similar
lever which was fixed in the two bottom beams of the
testing frame. The testing frame is a big steel frame
composed of big steel elements; two top beams; two
bottom beams and two columns at each end. the beams and
columns are connected together by big adjustable bolts,
Fig.(4) illustrates the mechanism of loading and
supporting.

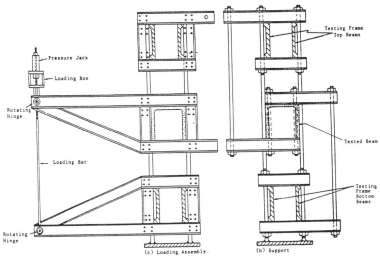

Fig.4. Loading/supporting mechanism for interaction of bending/torsion

(b)Rig for pure torsion consisted of two loading jacks
acting simultaneously under the same pressure, but at
opposite sides of the tested beam and also in opposite
direction for the purpose of creation of pure couple as
shown in Fig.(5). The supports are the same as those
used in rig (4).
(c)Rig for pure bending the load was applied by a hydraulic
jack, at the midspan, which was fixed on the top beams
of the testing frame. The test beam was supported on
rounded steel bars which are welded to steel plates
which rests on concrete blocks. Strains are measured by
using a Demec gauge. The strains were read both in
longitudinal and transverse reinforcements. Also angle
of twist and deflections are recorded for each stage of
loading.

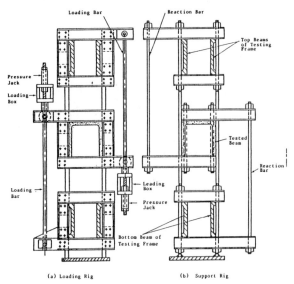

Fig.5. Testing rig for pure torsion

2.2 Mathematical approach

A secant modulus method of equilibrium has been used in this work to formulate and calculate the deformations, stresses and geometrical constants for every cross-section along the beam. An iteration method was used for which the effective use of a computer was essential. Some assumptions have been made for the sake of simplicity without unduly affecting the results obtained by the method:

(a) cross-section remains rigid, i.e., does not distort.
(b) the effect of shear deformation is negligible and ignored.
(c) the steel in the tension and compression zones and concrete in compression may be transferred to an equivalent area of concrete which has a modulus of elasticity (Ec) in the following way:

$$A_{cs} = n.A_s \qquad \text{steel in tension} \qquad (1)$$

$$A_{csc} = (n-1).A_{sc} \qquad \text{steel in compression} \qquad (2)$$

$$t_{eq} = r.t \qquad \text{concrete in compression} \qquad (3)$$

where $n = E_{ssec}/E_c$; $r = E_{csec}/E_c$

The geometrical properties of the uncracked cross section, including reinforcement, were calculated first before entering cracked stage. Strains may then be calculated for the cracking load as given by Eq.(15).

The cross-section is divided into strips of concrete, the coordinates x, y and w of each strip being calculated at its center. The coordinates of the reinforcement bars are defined by their centers. At first cracking the strips of concrete that lie in the tension zone are omitted by putting t(i)=0.

In all stages of loading the secant modulus is used for both concrete and steel. The following equation Ref.12 gives the secant modulus of concrete at every level of stresses.

$$Ecsec(\epsilon_c) = f'_c(\frac{2}{\epsilon_o} - \frac{\epsilon_c)}{\epsilon_o^2)}) \qquad (4)$$

The equivalent thickness of each strip in the compression zone is given by Eq. (3). Also, for steel either in tension or in compression, the equivalent concrete is calculated according to Eqns.(1) and (2) respectively. Fig.(6) and Fig.(7) show the stress-strain curves for concrete and steel respectively. The cycles of iteration take place until satisfactory convergence occurs, characterized by

$$(I^{cr}_{ww(n-1)} - I^{cr}_{wwn}) < 0.001\ I^{cr}_{wwn}$$

which means that strains in two successive iterations are nearly equal.

$$I^{cr}_{ww} = \Sigma t_{eq}(i) \cdot \Delta w^2_c(i) + \Sigma A_{cseq}(j) \cdot w^2_s(j)$$

$$I^{cr}_{wy} = \Sigma t_{eq}(i) \cdot \Delta \cdot w \cdot c(i) \cdot y_c(i) + \Sigma A_{cseq}(j) \cdot w_s(j) \cdot y_s(j)$$

$$I^{cr}_{wx} = \Sigma t_{eq}(i) \cdot \Delta \cdot w_c(i) \cdot x_c(i) + \Sigma A_{cseq}(j) \cdot w_s(j) \cdot x_s(j) \qquad (5)$$

$$I^{cr}_{xx} = \Sigma t_{eq}(i) \cdot \Delta \cdot x^2_c(i) + \Sigma A_{sceq}(j) \cdot x^2_s(j)$$

$$I^{cr}_{yy} = \Sigma t_{eq}(i) \cdot \Delta \cdot y^2_c(i) + \Sigma A_{sceq}(j) \cdot y^2_s(j)$$

$$I^{cr}_{xy} = \Sigma t_{eq}(i) \cdot \Delta \cdot y_c(i) \cdot x_c(i) + \Sigma A_{sceq}(j) \cdot y_s(j) \cdot x_s(j)$$

$$S_w = \Sigma t_{eq}(i) \cdot \Delta_{wc}(i) + \Sigma A_{sceq}(j) \cdot w_s(j)$$

$$S_x = \Sigma t_{eq}(i) \cdot \Delta \cdot x_c(i) + \Sigma A_{sceq}(j) \cdot x_s(j) \qquad (6)$$

$$S_y = \Sigma t_{eq}(i) \cdot \Delta \cdot y_c(i) + \Sigma A_{sceq}(j) \cdot y_s(j)$$

$$\alpha x = (Ixx.Ixy-Ixy.Ixy)/(Ixx.Iyy -Ixy^2) \qquad (7a)$$

$$\alpha y = -(Iyy.Ixx-Ixy.Ixx)/(Ixx.Iyy -Ixy^2) \qquad (7b)$$

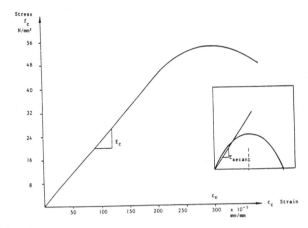

Fig.6. Stress-strain curve for concrete

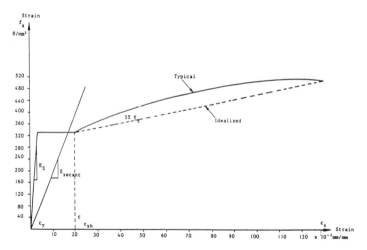

Fig.7. Idealized stress-strain curve for steel

Eqns. (7a) and (7b) give the components of the shift of the shear centre. As the shear centre changes the amount of bimoment will be changed as well, and be given by the following expression

$$Mww = Mww + Mxx.ax + Myy.ay \qquad (8)$$

2.3 Mixed torsion

It is well known that, after cracking, the flexural and torsional stiffness of concrete decreases and the degree of this reduction depends significantly on the level of stress over the cross-section. However, in this particular type of structure, the St.Venant's torsional stiffness which is the most affected, decreases sharply and is given by the following expression Ref.(8)

$$GCcr = E_s(Ao)^2A(1+m)/(u.s) \tag{9}$$

where $m=Alt.s/(A.u)$

A common measure of the significance of warping torsion is given by k, the characteristic length which, after cracking, may be defined as

$$kcr = L \sqrt{GCcr/(E_c I_{ww}^{cr})} \tag{10}$$

This term tends to be very small since the reduction in St.Venant's torsional stiffness is relatively greater than that of the torsional warping stiffness. If the St.Venant's torsional stiffness is neglected, then the bimoment, due to the restraint of the warping deformation at any cross-section, is given by:

$$M_{ww} = -T(L-4z)/8 \qquad 0 < z < L/2 \tag{11a}$$

$$M_{ww} = T(3L-4z)/8 \qquad L/2 < z < L \tag{11b}$$

Making this assumption the equation for bimoment is analogous to that for bending moment due to a concentrated load at midspan. This analogy may be extended to any type of loading.

The curvatures due to the bimoment and bending moments are given by the following expressions:

$$\phi'' = -Mww/(EcI_{ww}^{cr}) \tag{12a}$$

$$n'' = -Mxx/(EcI_{yy}^{cr}) \tag{12b}$$

$$\zeta'' = Myy/(EcI_{xx}^{cr}) \tag{12c}$$

These expressions may be numerically integrated by, Simpson's Rule, to give the angle of twist and deflections at any cross-section along the beam.

The shear stresses are functions of the section statical moments Swxcr, Sxcr and Sycr and are given by the following equation:

$$q = \frac{Tw.S_w^{cr}}{I_{ww}^{cr}} \quad \frac{Vy.S_y^{cr}}{I_{yy}^{cr}} \quad \frac{Vx.S_x^{cr}}{I_{xx}^{cr}} \tag{13}$$

where the contribution of St.Venant's torsion has been neglected, as mentioned before .

The strains in the stirrups, concrete struts and longitudinal reinforcement which are mainly due to shear stresses, are given by the following expression

$$\epsilon_n = q.\tan\alpha/(Esh.th) \qquad (14a)$$

$$\epsilon_c = q/(Ec.tc.\sin 2\alpha) \qquad (14b)$$

$$\epsilon_l = q/(Esl.tl.\tan\alpha) \qquad (14c)$$

3 Evaluation and presentation of results

3.1 Interaction of loading

For applications of loading within the elastic range of the steel, the interaction relationship between bending moment and warping moment is linear up to yielding of the first longitudinal steel bar, Fig.(8). This implies that the geometrical properties are not greatly affected by the relative amounts of bending moment and warping moment during these stages of loading. Since the mechanical properties of the steel are everywhere within the elastic limit, the areas in the compression and tension zones do not change considerably. The following expression gives the value of strain under any ratio of loading:

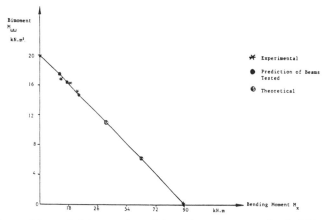

Fig.8. Interaction of bimoment and bending moment of yielding of steel

$$\epsilon = \frac{M_{ww}.w_{cr}}{I_{ww}^{cr}} + \frac{M_{xx}(I_{xx}^{cr}.ycr - I_{xy}^{cr}.xcr)}{I_{xx}^{cr}.I_{yy}^{cr} - I_{xy}^{cr} {}^{2})} / Ec \qquad (15)$$

It is clear from the above expression that, for a particular value of strain, any increment in the level of one of the forces must result in a corresponding decrement in the other. If the geometrical properties are constant which is nearly the case. Experimental results confirms this point see Fig.(8). In the final stages of loading, after the steel has started to yield, the torsional warping stiffness and flexural stiffness of every cross-section of the beam are noticeably affected.

The predicted loads and actual loads that cause failure do not differ considerably. This is apparent from Table 1. For example, Beam 1, the ultimate torques obtained experimentally and theoretically differ by 8.8%. For Beam4, which is under pure torsion, the difference was only 3.3%.

Table 1.   Failure loads for each beam

		Experimental results				Theoretical prediction		
B	e	P_a	$P+P_a$	T	M	P	T	M
	m	kN	kN	kN.m	kN.m	kN	kN	kN.m
1	2.350	23.00	24.40	57.37	31.24	26.78	62.92	31.46
2	1.560	31.00	33.06	51.60	40.60	40.96	63.90	48.13
3	1.175	40.00	42.73	50.21	52.38	54.40	63.92	63.68
4	1.000	60.00	60.00	60.00	0.00	62.00	62.00	0.00
5	0.000	222.00	222.00	0.00	260.85	265.00	0.00	311.37

At these stages the relationship between bending and warping moment is no longer linear, but it takes the shape shown in Fig. (9).

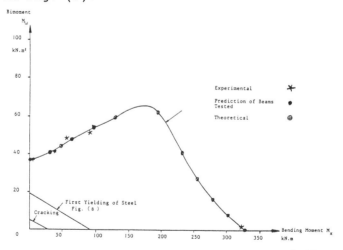

Fig.9. Interaction of bimoment at failure

From this figure it is noticed that at ultimate load, the application of bending moment enhances the torsional

capacity by up to 80%in the case of Beam 1 (M/T=1.0). This
enhancement,which extends over a considerable part of the
range, was not apparent before yield. These predictions are
confirmed experimentally, as shown in Table 1 and Fig.(9).
    The longitudinal strain in the longitudinal reinforce-
ment is obtained theoretically and recorded experimentally
in Figs.(10-13). These show strains in two longitudinal
bars at different positions around the cross-section over
half the length of the beams. The longitudinal strains are
composed of two components. First component is due to
warping moment and bending moment; the second component is
due the change in the mechanism of resisting shear in the
post-cracking stage, given by Eq.(15a). From Figs.(10-13)
the following may be observed:

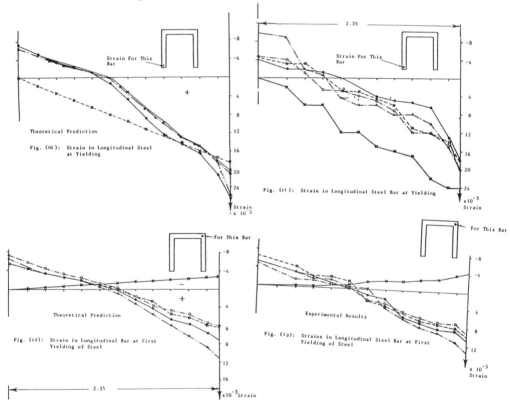

Fig. (10): Strain in Longitudinal Steel at Yielding

Fig. (11): Strain in Longitudinal Steel Bar at Yielding

Fig. (12): Strain in Longitudinal Bar at First Yielding of Steel

Fig. (13): Strains in Longitudinal Steel Bar at First Yielding of Steel

(a) There are two distinct regions around any cross-section
    two compression and two tension ,and along the beams for
    the chosen bars for comparison of the experimental and
    theoretical,which are consistent in both.
(b) The maximum values predicted by theory and obtained

experimentally are very close. The maximum range of difference is approximately 20%.

The greatest strains appear at midspan and end cross-sections of the beams. The failures are compressive, i.e., $\epsilon_c > \epsilon_o$, and dominates all beams except Beam 5 where failure occurred simultaneously in concrete and steel. From the angle of twist point of view, in the elastic range, i.e., before yielding of the first bar of steel, it was noticed that the angle of twist could be a function of the torsion only, because the geometrical properties, from the point of cracking until first yielding of steel bar, are constant.

The predicted angle of twist obtained by secant modulus method are nearly the same as those obtained by using theory of elasticity for warping of homogeneous materials of thin-walled structures of open cross-sections.

Figs.(14) and (15) refer to theoretical and experimental results respectively and illustrate the angle of twist at midspan of the beams at all loads up to failure.

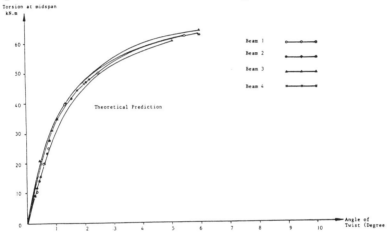

Fig.14. Torque-angle of twist at midspan for each beam up to failure

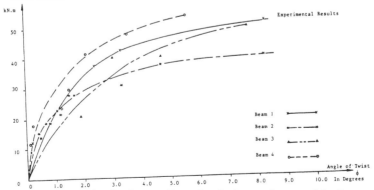

Fig.14. Torque-angle of twist at midspan for the beam each state of loading

287

The results are quite agreeable. For example Beam4, which is subjected to pure torsion. In this beam the difference is approximately 0.2°. It also apparent that the curves are almost coincident until first yielding takes place. This proves that the contribution of shear strains to the angle of twist may be neglected.

After yielding the angle of twist increases rapidly, because the stiffness reduces very significantly as will be observed from the Figs.(14) and (15). Fig.(16) shows the ultimate angle of twist at midspan of the beam for different ratios of bending moment to torsion, both theoretically and experimentally. The shape of curves are similar but their peaks differ by approximately 1.5° and occurred at Beam1. In Beam4 which is under pure torsion, the difference is very small, approximately 0.3°.

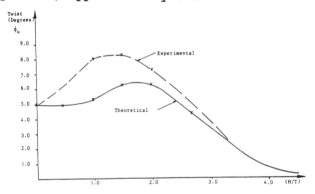

Fig.16. The effect of bending on the angle of twist of midspan

Figs.(17) and (18) represent the theoretically and experimentally obtained deflections of the test specimens at midspan at each stage of loading.

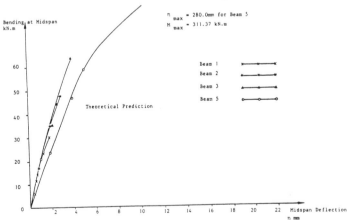

Fig.17. Bending deflection curve at midspan of each beam for each stage of loading

288

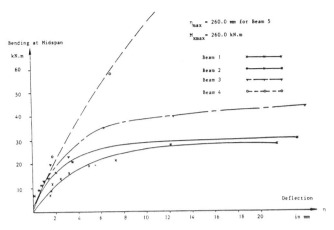

Fig.18. Bending deflection curve at the midspan of the beam for each stage of loading

The deflections are generally small with the exception of Beam5 which is under pure bending.

The maximum deflections recorded for beams subjected to torsion and bending simultaneously range from 5-17 mm. These values are relatively small for proving the validity of the theoretical method used. However, Beam 5 which is under pure bending, the maximum deflections obtained experimentally and theoretically are 265mm and 280mm respectively. The difference, only 5.3%, confirms the reliability of the method used for theoretical prediction. It is further evidence that the existence of torsion with bending, even in small amounts, reduces midspan deflection to a negligible degree.

The strains in the stirrups have been studied carefully in this work. The strains in the stirrups are caused mainly by shear flow due to the combined loading. The strains in the stirrups are given by Eq. (14a).

Figs.(19) show the experimental and theoretical results of strains in the stirrups at a particular cross-section for each beam at every stage of loading.

At first yielding of the steel, the theoretical and experimental results are in fair agreement. Also the existence of flexural shear does not affect the level of strains or extent of cracking significantly. This is even the case in Beam3 which has the highest level of bending moment in the beams subjected to combined loading.

Cracking in the beams was observed and recorded in detail during the test. In Beams 1-4, where torsion exists, the first cracks were observed to be inclined; this is due to the domination of St.Venant's stresses around the cross-section. In Beam5, which is under pure bending, they were vertical.

In Beams 1-4, cracks did not appear everywhere on the surface of the beams at the early stages of loading. The cracks at these early stages, before yielding of the

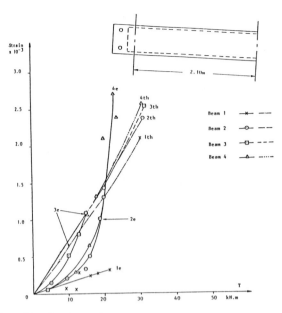

Fig.19a. Strain in stirrup 2.10m from midspan in the right hand web

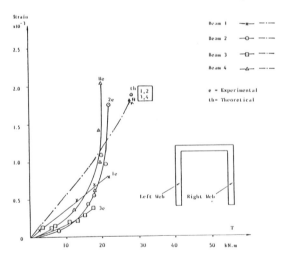

Fig.19b. Strain in stirrup 2.10m from midspan in the left hand web

longitudinal steel, appeared at the quarter points. At later stages of loading this cracking extended towards the supports and midspan. Cracking in these beams was primarily inclined and covered most surfaces by spiraling around the axis of the beam. This is shown in Plates(1-4) where the angle of inclination is approximately 45°. The crack pattern is very similar in all these beams although some of them are under the action of bending moment in addition to torsion.

The cracks in Beam5 are typical of those that occur in beams under pure bending moment and flexural shear.

Cracks in diaphragms were observed. Until the longitudinal steel started to yield, the diaphragms of Beams 1-4 at both ends and at the middle were uncracked. After yielding, cracks started to appear in the end diaphragms. These may be classified as either main cracks or secondary cracks.

The secondary cracks appeared on the sides and on the top and bottom faces of the end diaphragms and had different patterns and directions. The change of direction occurred largely as a result of stress concentrations in the diaphragms due to the mechanism of the support system. The main cracks propagated on the outside and inside faces of the diaphragms, perpendicular to the axis of the beam. These were basically shear cracks arising from transverse torques which are equal to the warping torsion in the end diaphragms.

4 Conclusions

a) The portion of applied torque resisted by St.Venant's torsion is small for beams of practical length and de creases sharply after cracking.

b) Geometrical properties change significantly after cracking but remain constant until first yield of the steel takes place. After yielding, the geometrical properties change very rapidly as the load increases.

c) The effect of torsion is very significant in comparison to bending. Accordingly designers must consider torsion seriously at the design stage even under service loads. In the case of moderate ratios of torsion to bending, bending has been found to be of secondary importance with respect to warping torsion.

d) The angle of twist is not affected by the existence of bending before first yielding of the steel. However, it is affected considerably after yielding up to failure.

e) Before yielding, deflections are functions of bending only. Even after yielding, with the existence of moderate ratios of torsion to bending, observed deflections are very small. This is because torsion is the most influential action determining the ultimate capacity of the structure.

f) The existence of bending moments with torsion after yielding of the steel helps significantly in raising the ultimate warping torsional capacity of the structure.

g) The contribution of shear flow to the axial strains in longitudinal reinforcement is negligible.

h) The effect of flexural shear on axial strains in the stirrups is very small in comparison with that due to warping torsion.

5 References

ACI Committee 318. (1983) **Building Code Requirements for Reinforced Concrete** (ACI 318-83), American Concrete Institute, Detroit.

Desayi, P. and Krishnan, S. (1964) Equation for the Stress-Strain Curve of Concrete. **J. American Concrete Institute**. Proc. Vol.61, No.3, 345-350.

El-Hammasi, S.A. (1986) The Behaviour of Open Thin-walled, Reinforced Concrete Beams Under Interaction of Bending and Torsion, **Ph.D. Thesis**, Department of Civil Engineering, University of Bristol, England.

Hsu, T.C. (1973) Post-Cracking Torsional Rigidity of Reinforced Concrete Sections, **J. American Concrete Institute**, Vol.70, No.5, 352-360.

Hwang, C.S. and Hsu, T.T.C. (1983) Mixed Torsion Analysis of Reinforced Concrete Channel Beams - A Fourier Series Approach, **J. American Concrete Institute**, No.80-36, 377-220.

Kollbrunner, C.F. and Basler, K. (1969) **Torsion in Structures: An Engineering Approach**. Springer-Verlag, New York. (Originally published in German, 1966).

Krpan, P. and Collin, M. (1981) Predicting Torsional Response of Thin-walled Open RC Members, **J. Structural Division**, ASCE, Vol.107, No.ST6. Proc. Paper 1633, pp.1107-1127.

Lampert, P. (1973) Post-Cracking Stiffness of Reinforced Concrete Beams in Torsion and Bending. Analysis of structural systems for torsion, SP-35, **J. American Concrete Institute**, Detroit. 385-433.

Roberts, T.M. and Azizian, Z.G. (1983) Instability of Thin-walled Bars, **J. Engineering Mechanics**, ASCE, Vol.109, No.3, 781-794.

St.Venant. (1855) **De la Torsion de Prismes**, Tome XIV de l'Academie de Science, Paris, France. (in French).

Timoshenko, S.P. (1945) Theory of Bending, Torsion and Buckling of Thin-walled Members of Open Cross-section, **J. The Franklin Institute**, Vol.239, No.5, 343-361.

Vlassov, V.Z. (1935) **Thin-walled Elastic Beams** (in Russian, 1st Edn.) (The second, revised and augmented edition published in 1958 was translated into English by the National Science Foundation, Washington, USA.)

Yoo, C.H. (1980) Biomoment Contribution to Stability of

Thin-walled Assemblage, **Computers and Structures**, 11, 465-471.

Zbirohowski-Koscia, K.F. (1968) Stress Analysis of Cracked Reinforced and Pre-stressed Concrete Thin-walled Beams and Shells, **Magazine of Concrete Research**, London, Vol.20, No.65, 213-220.

PART FIVE
CABLE BRIDGES

SOME THOUGHTS ON CABLE BRIDGE DESIGN

J.M. SCHLAICH
University of Stuttgart, West Germany

Abstract
Nowadays, the suspension bridge is considered inferior to
the girder bridge for short spans and to the cable-stayed
bridge up to spans around 1,000 meters. Thus, variety in
bridge design may further dry out, what a pity. This paper
wants to show that this need not be accepted.

Fig. 1: Pedestrian suspension bridge over the Neckar near
Stuttgart, span 114 m, built 1988

1 The interrelation of structure and span in bridge building

Watch-making and model-making teach us that the more we **reduce** the scale of an object, the more we will boost its cost. The fate of the dinosaur and the postulations of early men like Galileo Galilei warn us that **enlarging** a structure true to size or proportion will ultimately cause it to collapse under its own weight. This is the picture of the dilemma which the designer and builder of bridges has faced from time immemorial, namely to be torn between the economical considerations and structural demands of cost and dead load, respectively - they being the criteria under which the required span of a bridge dictates its structural form. However, the factors governing these criteria have not always been the same. Their reorientation parallels the **steady** - sometimes even **radical** - process of evolution of the art of building.

The development of advanced production methods and new construction techniques has been instrumental, above all, in influencing the first criterion, namely the cost-size relationship. Introducing erection by free cantilevering, for instance, enabled the limit of 200 m of span to be surpassed in the construction of prestressed concrete box girder bridges, an achievement that would otherwise, that is applying the conventional method of erection from falsework, have been prohibitively expensive. Inversely, when mass-production provided us with prallel wire cable at a relatively low price, it opened the door for the cable-stayed bridge to set foot in the short span range below 100 m.

Similarly, the improvements in material quality and design practices have provided ways to circumvent the second criterion, namely load-bearing capacity being voraciously devoured by dead load. This can be done in two ways: either directly by increasing the breaking length β/γ, that is using materials of higher strength β and lower density γ, or indirectly by abandoning redundancy, that is avoiding bending in favour of pure compression and, above all, tension. Thus, the fact that natural stone, by its very nature, offers little room for its properties being refined explains why the masonry arch bridge, after a notably **steady** progress reached an early climax at a modest 60 m span. It was not until the emergence of concrete, an artificial stone, and steel that the span of arch bridges could be made longer to reach 300 m and more. It is notable enough that although steel and concrete are wide apart in terms of their respective β / γ ratios, they do share much the same limit in length of individual span, as a result of the compressive action.

The long truss and cantilever bridges are formidable
examples of the successful application of a very skillful,
that is to say simple fabrication technique. Devoid of any
abundance and making every single member work to maximum
efficiency, they are still a serious rival of the cable
stayed bridge for spans up to 500 m, particularly so in the
USA. As we look at the specimens this type of bridge has
produced, some of them beautiful and others so very ugly,
it becomes evident that the application of technical codes
and rules, perfectly valid as they may be, does not automatically translate into aesthetic quality, but that such
quality is something we must wilfully seek and strive for
in every instance anew. While in the past, economic
considerations and the absence of reliable methods of stress
analysis dictated cross sections standardized in shape and
members uniform in length, making for dull and uninteresting
shapes, today's computer-controlled production facilities
once again place variety into reach and inventive faculty
in demand.

The most striking example for a **sudden radical step** in
its history of evolution as a consequence of one single
technical innovation is the suspension bridge at the point
of time when mass-produced drawn wire became available and
Röbling developed the basics of cable spinning in the middle
of the last century. By then, French and British engineers
had of course already demonstrated that even mechanization
and industrialization could not shake the fact that the
largest distances can only be spanned with structures that
work in tension. Thus, in 1826 Telford used forged chain
links to build the suspension bridge over Menai Strait with
a span of 177 m, outdistancing all other bridges in
existence at that time. Parallel to this development, there
were also plans for building cast-iron arch bridges that
would span similar distances. But the Brooklyn cable suspension bridge built by Röbling shortly after (between 1869
and 1883) had a span of 486 m, which was three times
superior to what was then technically feasible with any
other structural form. And in fact, the suspension bridge
has lost nothing of its superiority as regards potential
length of individual span. Although the cable-stayed bridge
is under way of almost doubling its present record spans
of 457 m (Hooghly River Bridge, Calcutta) and 462 m
(Annacis Bridge, Vancouver), to reach 857 m with the
Normandy Bridge, Le Havre, it is still far from matching the
suspension bridge. The Humber Bridge, which was opened to
traffic in 1981, freely spans 1,410 m, the Akashi Bridge
and the bridge over the Straits of Messina will have spans
near 2,000 m and 3,000 m, respectively. Today, the potential
limit is at about 18,000 m, this being the maximum distance
over which we could rely on the self-supporting capability

of high-tensile steel cable. Synthetic fibres, like **Kevlar** with its extremely favourable β/γ would permit even larger lengths to be spanned, larger than necessary. But then, of course, let us not forget that there are still some challenges waiting for us, like the Straits of Gibraltar...

Hence, where large spans are concerned, obviously the structural form is dictated by what is technically feasible. And the beautiful suspension bridge is and remains the queen.

For everyday life, however, the small spans are of much more interest. In view of the formidable technological progress over the past decades, particularly so in the field of concrete construction, including prestressed concrete, erection by free cantilevering, self-supporting travelling formwork, incremental launching and similar, as well as in steel construction with the introduction of high-tensile steels, new welding techniques and composite construction etc., one would expect a noble beauty contest to be in full swing. But, alas, there is no such thing! The borderlines are staked off, fast and sound: for spans up to 200 m, take a slab or a deep beam or a box girder, distinguishable only at close range, whether you made them out of concrete or steel or composite, beyond that the cable stayed bridge and that's it.

2 The small spans

Confining the variety of forms to the uniformity of beam bridges for the small spans and to cable-stayed bridges for the larger spans (surely the few suspension bridges that have been built for the very large crossings do not really count) cannot be satisfying neither from a cultural point of view nor for the engineer himself and his professional reputation. This holds true particularly for urban bridge building, including the elevated roadways and, above all, the pedestrian bridges. It would seem just too simple and paradoxical at the same time to put the blame for this impoverishment on wealth alone - in other words to say that where wages are high and materials relatively cheap, the notions of imaginative faculty, of diversity and filigree, of forms that reflect the flow of forces and of the lovingly elaborated detail no longer stand a chance. Holding responsible the authorities on the grounds that they tend to choose the cheapest bid, would be unfair as long as they are not offered a true choice of alternatives: from among several equivalent designs, it is only natural to pick the cheapest. Much rather, the question comes up of whether it is not ourselves who are to blame, for our keeping to the beaten track, whether out of convenience or lack of discrimination, instead of availing ourselves of the array of technological and scientific possibilities of our time.

As a matter of fact, there are first signs of what pro-
mises a welcome change, chiefly coming from abroad. French
engineers excel in showing us how to build interesting
hollow box bridges out of concrete, some of them with per-
forated steel webs. Swiss engineers have built cable-stayed
bridges with spans shorter than 100 m, using thin slabs
as girds thus entering a span range where the box girder
was thought to be unbeatable. In a project near Stuttgart,
a cable-supported slab is to be built thus expanding the
idea of external prestress. Even the arch bridge experiences
an occasional revival.

It would be beyond the scope of the present paper to
discuss all these and similar efforts and developments.
Hence, the following chapters will only be concerned with
the question of **whether and if so, under what circumstances
does the suspension bridge still stand a chance to be
adopted for the smaller span range,** like for instance in
the design of pedestrian bridges? If diversity and light-
ness are to be valid means towards aesthetic quality, this
question should not remain unasked.

3 Comparison of cable-stayed bridges and suspension bridges

How can the suspension bridge ever come to be generally
accepted as a valid substitute for the cable-stayed bridge
within the span range of up to 100 m when the latter is
generally preferred even for the much larger spans, that
is up to 1,000 m? The idea that lies behind comparing the
suspension bridge to the cable-stayed bridge and no other
is that beam bridges and arch bridges are subject to entirely different considerations in terms of how they blend
and integrate into their environment. At the most, we could
compare the cable bridge to the slab suspended from an arch,
as is, for instance, the Langer's beam, because the bridge
with a thin deck slab offers the immediate advantage of
a short ramp length (Figure 2) which is often a decisive
factor in densely built areas.

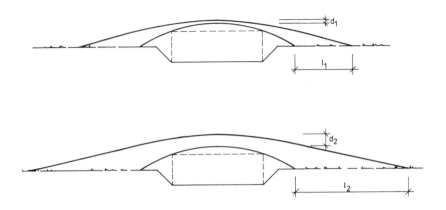

Fig.2. A slender deck makes for short ramps

It may come as a surprise - and yet it is postulated
herein and will be substantiated in the following that there
is, **in principle**, no length of span where the suspension
bridge could be considered inferior. There is, in fact, just
one "gap" - let us place it roughly between 100 m and
1,000 m - where certain drawbacks of the suspension bridge
may become more prominent. But below that range, both are
certainly at least equivalent choices.

And speaking of drawbacks - it should not be overlooked
that the cable-stayed bridge, too, exhibits a number of inherent disadvantages as against the suspension bridge and
which recede into the background only when its advantages
as regards construction (which will be described later on)
come to the fore. The asset of greater stiffness attributed
to the cable-stayed bridge, like its higher resistance to
concentrated loads is, in reality, of no interest at all,
except maybe to carry railroad loads. What is more, it comes
accompanied by alternating stresses in the stay cables, which
can reach magnitudes that by far exceed those occurring in
the main cables of the suspension bridge. It is true that
the suspenders of the suspension bridge do suffer much the
same fate as the stay cables of the cable-stayed bridge but
then, the former can be exchanged easily enough, even under
traffic load. By the way, even the main supporting cables
of a suspension bridge could be designed for replacement
under traffic loading, as had been demonstrated on the
Williamsburg Bridge competition design in New York, which
will be discussed later (Figures 15 and 16).

In those cases, where the deformations of the suspension bridge under concentrated loads do constitute a problem, it can be effectively strengthened by providing additional stay cables. This will retain the low tower height of the suspension bridge (h/l approx. 1/10 as against approx. 1/7 for the fan and 1/5 for the harp cable-stayed bridge). This solution results into varying patterns for the side spans (Figure 3).

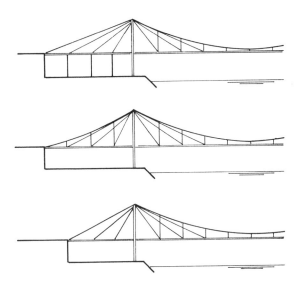

Fig.3. Different side span configurations
in a combined suspension/cable-stayed bridge

Surprisingly enough, the Brooklyn Bridge as designed by Röbling remained the only combined suspension/cable-stayed bridge. The most plausible explanation seems to lie in the change of attitude towards statical behaviour and the progressive improvement in analytical methods. In his time, Röbling designed along the lines of what he deemed right. The fact that his design was indeed uncalculable and unpredictable for him did not seem to worry him - but then, he had no competitor. So, he regarded all loads to be carried by the main cables and added the stay cables. Later, as scientific methods of stress analysis became more reliable, engineers would only build what they could calculate. This was the first step towards simplifying the system - and making it uninterestingly dull: with vertical suspenders and side spans propped up to keep the tower top as rigid as possible. This approach was intended to minimize the risk of strainless deformations or changes in the natural curvature of the main cables under concentrated loads, which were difficult to analyze. Accordingly, the new concept of the

303

inclined hanger which was introduced with the Severn Bridge
in 1965 B.C. (B.C. meaning Before Computers) was still a
chancy enterprise, though in the author's opinion through
no flaw inherent to the concept itself, but rather because
the design underestimated the magnitude of the alternating
stresses with regard to their detailing. Another reason why
Leonhardt's beautiful and brilliant monocable bridge never
became popular may have been the fact that the concept was
carried to the very limits of what was susceptible of
analysis at the time (Figure 4). It is a true blessing that
today, this subject is no longer a topic, thanks to the progressive improvement in methods of stress analysis and cable
anchorage techniques.

Fig.4. Monocable bridge by F. Leonhardt, 1961

Also with respect to their structural details, the peculiar problems associated with the cable-stayed bridge are
quite a few: The fan arrangement of the cables presupposes
a complex pattern for their connection to the tower. On the
other hand, distributing the cable anchors at equal distances along the height of the tower will cause the bending
moments to increase which in turn calls for more cables.
Also, as the angle of inclination of the cables varies with
height, anchoring them along the edges of the deck turns
into a tedious and expensive job. Yet, the alternative
choice, which is central suspension, is all the worse.
Finally, the builder faces the dilemma of whether he should
provide for a vertical (and therefore stiff) support of the
deck at the tower and thus make it subject to high bending
moments or suspend it from the cables which will get him into
conflict with the steep cable connections at the tower.

Of course, none of these drawbacks are such that they
could not be overcome or make cable-stayed bridges generally
prohibitive. And besides, the purpose of the present paper
is not to speak out against any particular type of bridge,
but rather to speak up for the suspension bridge, in the
interest and name of greater variety. Also, the shortcomings
of a given structure also lead to enhanced efforts for
improvement. We have, for instance, developed a modified

scheme to overcome the one problem typical of the cable-stayed bridge, namely the connection of the cables to the deck and to the tower head. The solution consists in using fixed anchors for the deck and adjustable ones for the tower. The concept uses simple details and a particular geometry for the tower head. The second problem, namely the support of the deck at the tower can be solved with a very slender and flexible deck. Provided sufficient length of the tower legs to permit satisfactory response to temperature differences, the deck can be monolithically joined to the tower, without causing the moments to increase.

The true asset of the cable-stayed bridge lies in that it permits erection by free cantilevering **while simultaneously** remaining self-anchored in respect of the horizontal component of the cable forces. As a consequence, it permits the mass of the abutments to be kept relatively low (Figure 5).

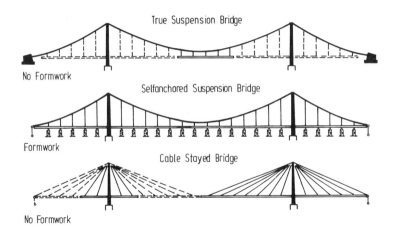

Fig.5. Deck structure erection
(suspension bridge and cable-stayed bridge compared)

Now, with the suspension bridge, it's **either or.** When back-anchored, it may be erected without falsework, as its supporting cables will be anchored in the abutments, freeing the deck from any bearing function. The abutments, however are of necessity heavy (and therefore expensive) if their sliding is to be safely prevented, in view of the very flat angle at which the cable ends meet the abutment.

By way of contrast, in a self-anchored suspension bridge, the cables are connected to the ends of the deck slab and the horizontal component of their load transmitted to the deck will be canceled by compression, leaving only the vertical components of the cable forces to be transmitted to the foundations - a task that can be accomplished by slender piers or ties. On the other hand, the deck of the self-anchored suspension bridge must be erected entirely from falsework because it cannot be suspended from the main cables until the construction of the deck is completed.

4 Suspension bridges for the small spans

It hence follows that, economically speaking, the suspension bridge stands perfectly up to the cable-stayed bridge whenever conditions are such that the conventional mode of construction from falsework is neither prohibitively expensive nor unacceptable because of undue interference with traffic. These conditions are typically given in the span range smaller 100 m, as for instance in pedestrian bridges over streets or railways where falsework does not constitute a serious problem. Where large distances are involved, as for instance in the case of bridges over waterways, rivers or straits and in which, by their very nature (or else such long lengths of individual span would not be necessary) it would be too unsafe and costly to instal temporary piers, the cable-stayed bridge may fully exhibit its inherent virtues in terms of construction.

Yet, and as we all know, it is the exception that proves the rule: In Osaka (Japan), a self-anchored suspension bridge with a span of no less than 300 m was recently completed. To erect the deck, two auxiliary piers were set up in the shallow waters, temporarily obstructing navigation. For the small spans and in particular in those cases where the deck is narrow and therefore light, as would typically be the case with a pedestrian bridge, the cost involved for back-staying is not prohibitive. The back-anchored suspension bridge will therefore be the preferred choice whenever the of necessity **straight** deck of the self-anchored suspension bridge, as dictated by compression, would leave too little a margin of freedom for the designer. The back-anchored suspension bridge knows nothing of such geometrical constraints, thus giving free rein to the designer's inventive creativity.

It is at this point that the **structural analysis** becomes so **complex a problem** and so formidable a task that it is, in fact, not until quite recently that such bridges can be built with a clear conscience and free of the preoccupation of not being able to master the structure geometrically, statically and/or dynamically. Suspension bridges are essentially kinematic systems which are stable only under a specific load condition. They are normally not prestressed, meaning that they are stress-less once the dead load is removed. Given a non-plane but three-dimensional geometry of the suspension system, as a consequence of a non-straight deck, the mere calculation of the initial geometry under load is already rendered impossible unless geometrically non-linear methods are applied that will cover large and strainless deformation. To this end, the geometric stiffness matrix (large strainless deformations) must be introduced in the load displacement equation, in addition to the elastic stiffness matrix (small strains), so as to cover those forces that will be generated from the displacement of the members. It is a feature peculiar of the suspension system that it renders notably realistic results, which is attributable to the fact that the finite elements represent the real cable sections between clamps. Apart from the initial geometry as such, the purpose of this analysis is to calculate the (unstrained) fabrication lengths of the components. Precise cutting to size and fabrication to the fraction of an inch, is the key to a successful construction of such bridges. This holds true particularly in those cases where they are additionally prestressed by stays. Adjustment by means of turnbuckles, which are again and again installed out of lack of confidence, is a painfully tedious process. Although primarily a means of defining the initial geometry, the same methods or programs are of course also used at a later stage to calculate load and erection conditions. Al- though with this type of bridge, the superstructure - even if very slender - is not considered critical in terms of buckling stability because of its elastic bedding, it is nonetheless felt to be a huge progress that today, the in- stability problem as a whole can be treated along the lines of the Theory of Second Order because the assumption of reasonable imperfections causes no problems.

Hence, it is no exaggeration to say that **modern struc- tural design** can pride itself to having cleared the way to- wards **absolute freedom in the design of suspension bridges**, including any such geometries which only a few years ago nobody would have dared to think of.

Some practical examples from the author himself together with R. Bergermann may render evidence of this. Their development basically followed the same scheme: freehand sketches, often a simple working model, finalizing form and design, detailing, **comparison with the rough calculation**, demonstration models or montage photographs.

A second purpose of these examples is to demonstrate
that where pedestrian bridges are concerned, the attributes
of light or heavy, filigree or awkward need not be a matter
of material choice. With these bridges, concrete lends itself to be used for the decks and of course for the
foundations; steel for the cables and towers. But this is
by no means a hard-and-fast rule: timber is definitely also
a choice.

5 Some examples

Footbridge in the Rosenstein Park in Stuttgart (Figure 6)
The Rosensteinbrücke in Stuttgart (W-Germany) is a suspension bridge of the asymmetric, self-anchored type. The slab
is made of reinforced concrete 35 cm thick and is "prestressed" by the horizontal component of the cable forces.
The inclined hangers enliven the appearance and are at the
same time a helpful tool in distributing concentrated loads,
although they come accompanied by higher alternating
stresses than the vertical hanger design. The two main
cables with a diameter of 75 mm are passed over a cast-iron
saddle at the top of the tower and tied together by means
of clamps to make allowance for the action of the differential forces. There is no material contact between tower and
deck slab. The tower was jacked up to strip the deck and
tension the cables. (Built in 1977, in consultation with
H. Luz, M. Bächer, architects.)

6

7

Footbridge in the Rosenstein Park in Stuttgart (Figure 7)
The Small Rosensteinbrücke is separated by a hill from the
bridge just described. The footbridge is designed as a
back-anchored suspension bridge. The path leads downhill
through the small valley which is to be spanned. This topographic situation lent itself to rest the deck slabs
directly on the two supporting cables between two abutments.
The two main cables are cable-supported to provide for
additional stability. The cable support consists of an upward curved cable which is tied to the two main cables by
diagonal ropes to form a three-dimensional cable truss in
which the upper flange is formed by the concrete slabs
which are joined to the supporting cables by means of
clamps. The deck slabs are made of concrete 10 cm thick,
with open expansion joints. (Built in 1977, in consultation
with H. Luz, M. Bächer, architects.)

Bridge over the river Neckar in Stuttgart (Figures 1 and 8)
The Max-Eyth-Brücke in Stuttgart is designed as a back-
anchored suspension bridge with a span of 114 m between
towers. The structure, with its two main cables splayed out
from the top of the towers, its network of inclined hangers
and a concrete deck slab gives the impression of sober
symmetry in the main section across the river, while the
side sections are tailored to react perfectly to the very
different topographic conditions prevailing on either side
of the river.
   One river bank is perfectly flat with a beautiful stock
of trees in which the bridge ends and branches out, in its
whole length up to the abutments supported from the network
of hangers. The other bank is steep and the path comes down
from behind a valley. Here, the deck takes an outward curve
short of the tower and merges with the path. The walker
becomes aware of the structure before crossing it. The
tower is stayed by means of straight ropes.
   The undeniable extra cost that accrues from backstaying
and which was necessary because sinking auxiliary piers into
the navigable river bed was not acceptable here and therefore leaving no other choice but suspend the prefabricated
deck slabs directly from the cables, finds a further justification in offering the designer perfect geometrical freedom in the shaping of the slab - a freedom that has been
fully exhausted here. (Built 1988, with B. Schlaich-Peter-
hans, architect.)

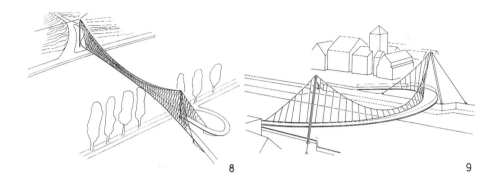

8

9

Pedestrian bridge in Kelheim (Figure 9)
The deck of the pedestrian bridge over the Rhein-Main-Donau-
Kanal (major waterway between the rivers Rhine and Danube)
in Kehlheim (W-Germany) spans the channel in a near semi-
circular plan, between two straight approaches running
parallel to the channel. The approaches were necessary in
order to reach the required elevation to suit the headroom
requirements for navigation. It would, of course, have been
possible to do it the shorter way and provide for a break
at the upper ends of the approaches, routing the bridge
at a right angle over the channel. But the continuous form
of the deck not only seemed much more beautiful but it could
also be used for statical purposes in that the structural
demands of the circular slab could be satisfied by
suspending the concrete deck but along one single side.
The supporting cable is born by two inclined towers
and forms a beautiful geometric pattern in combination with
the hangers. The slab is prestressed to compensate for
warping moments. The prestressing steel is arranged in a
ring and over the gravity center of the cross section. The
traditional classification into back-anchored and self-
anchored does not work with this bridge: it is, in fact,
both at the same time. The leaning towers require staying
while the circular slab acts as an arch within the plan,
the compressive forces of which are offset by the horizontal component of the supporting cables. The construction
work posed no problem because the flat topography permitted
the erection of the deck from falsework and the excavation
work for the channel was to start only after the bridge
was completed. Stripping and cable tensioning was done by
jacking up the towers. (Built in 1987, with Ackermann und
Partner, architects.)

Pedestrian street overpass over the Neckarstrasse in Stuttgart (Figure 10)
The pedestrian bridge spanning the Neckarstrasse in
Stuttgart is a back-anchored suspension bridge, asymmetri-
cally suspended from a new hotel building for which it
serves as a canopy over the entrance and as a link to the
nearby park. The concrete slab widens towards the building
to which it is monolithically tied. The deck is movably
supported, firstly by the cables which are socketed to the
deck slab in an exactly opposite arrangement and, secondly,
on a hill. The fittings, mountings, anchorages and sockets
are made of stainless steel so as to be able to leave the
beautiful structure unspoiled from anti-corrosion coating.
(Built in 1988, with Kammerer, Belz und Partner, architects.)

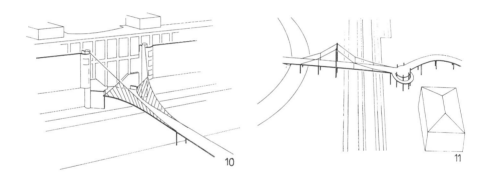

Pedestrian bridge in Bad Windsheim (Figure 11)
The pedestrian bridge in Bad Windsheim (W-Germany) has a
span of no more than 19.5 m. It crosses the railway and
links the town to a park. The approach is a timber-planked
steel grid, resting on piers at a spacing of 6.5 m, which
are arranged in pairs. The design of the approach, including the spacing pattern was also to be used for the
bridge itself and therefore repeats and reflects itself
in the deck and in the suspension. The result is a self-
anchored structure suspended from chains. The very small
sizes, as for instance in the present design, are best and
most economically served by means of eyebars for bolting
instead of ropes and steel mountings or fittings. (Built
in 1988, with Schunck und Partner, architects.)

Pedestrian bridge over a road in Stuttgart (Figure 12)
This bridge, spanning the main road "Am Kochenhof" uses a
concrete slab which is only 13 cm thick and rests on a prestressed asymmetric rope truss. The design sprang from the
desire of creating a structure that would combine and convey
both the idea of elegant weightlessness and technical sober-
ness to blend in with the nearby Stuttgart fair and exposition buildings. The bridge was tensioned at the time
of writing (December 1989).

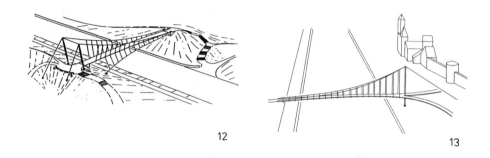

12

13

Pedestrian bridge in Berching (Figure 13)
The pedestrian bridge crossing the Rhein-Main-Donau-Kanal
outside the town of Berching (W-Germany) is designed as
a self-anchored asymmetric suspension bridge. The main
emphasis was placed on creating a buoyant and weightless
structure. The thin concrete slab rests on cross girders
made of steel that are cantilevering on both sides and
feature undisguised (metal-to-metal) suspension. (To be
placed under construction in 1990, with Ackermann und
Partner, architects.)

Pedestrian bridge for the IGA 1993 in Stuttgart (Figure 14)
The two bridges as projected for the international horti-
cultural show (IGA) in Stuttgart are very much alike and
will lie close together. Each bridge has three branches,
to provide for a crossing over the railway and an access
down to the railroad platforms. They are partly self-
anchored, partly back-anchored. In the one case it was
possible to stabilize the tower by the suspension system
alone, while the other case requires additional backstaying.
(To be placed under construction in 1990, with H. Luz and
H. Egenhofer, architects.)

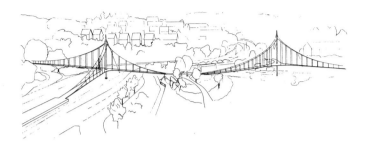

14

Williamsburg Bridge, New York (Figures 15 and 16)
Finally, the author wants to acquaint the reader with the
design of a larger suspension bridges with a span of 487 m.
Here, too, the cable-stayed bridge would have been the
obvious choice, following the conventional rules. But with
the existing abutments being well preserved and taking into
account urbanistic and aesthetic considerations, it seemed
both technically feasible and desirable to propose a suspension bridge.

Within an international design competition, the City of
New York invited tenders for the rehabilitation of the old
Williamsburg Bridge, including its approaches. The Williamsburg suspension bridge was built in 1903 and provides a
link between Manhattan and Brooklyn. It handles a traffic
load of approximately 240,000 people a day, carried on eight
lanes and two streetcar rails. The new bridge is to have
only six, but wider, lanes and three streetcar rails. The
tender specifications called for a proposal in which the
new bridge could take the place of the old one with the
least possible interference with traffic and would blend
neatly into the existing surroundings without requiring
any buildings to be pulled down.

The proposal described in the following was awarded the
first prize (in collaboration with Prof. R. Walther, Mory,
Maier, engineers, Basel, and M. Goldsmith, Chicago, Ackermann und Partner, München, architects) and contained a
number of distinctions:
- it proposed a suspension bridge instead of a cable-stayed
bridge which would have been the conventional choice for
the length of span concerned. The proposed structure would
thus harmonize with the two existing neighbouring suspension
bridges under the aspects of aesthetic appeal and scale;

313

- it provided for additional stay cables in the main span as a means of preventing deformations, similar the practice used for the old Brooklyn Bridge, now 105 years old;
- it suggested an orthotropic deck made of steel and steel towers filled with concrete;
- it would make use of the existing, well-preserved abutments for backstaying the main cables;
- it proposed that erection be accomplished by constructing the new bridge in two halves on either side of the old bridge and which, upon completion, would carry the traffic load while the old bridge was pulled down, whereupon the two halves would be joined locating the axis of the new bridge to coincide exactly with the axis of the old one. The bridge would have to be closed down for no more than approximately two weeks and no existing building would be touched in the process.

15

It is common practice today, to make cables replaceable. In cable-stayed bridges, this is no problem. The present proposal shows, for the first time, a method of how the main cables of a suspension bridge can be replaced under traffic load. To this end, the main cables would not be made to form one single cable unit, as is normally the case, but would be made up of 12 individual strands or ropes to be passed over the cast-iron saddle and tied together by means of clamps. When required, the individual cables would be exchanged one at a time, thus making the bridge loose but one twelvth of its load-bearing capacity while replacing

one strand. Reducing the mass of the cable into several
separate cables makes it transparent and conveys the idea
of weightlessness.

Moreover, it enables the use of the same rope or strand
type for the main cables, the suspenders and the stays.
In order to provide for satisfactory anchorage to the
superstructure, each hanger is allocated two stay ropes
on either side.

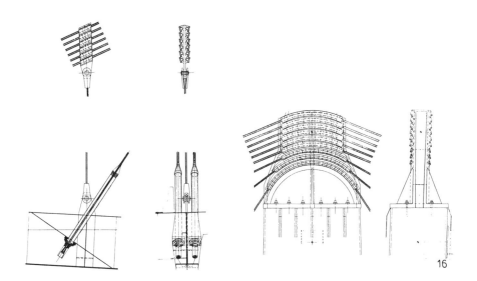

16

DEVELOPMENTS IN STEELS FOR CABLE-STAYED STRUCTURES

S.E. WEBSTER, B.M. ARMSTRONG
British Steel Technical, Swinden Laboratories, Rotherham,
England, UK
T.J. PIKE
British Steel, General Steels, Scunthorpe Works, England, UK

Abstract
The historical development of high tensile steel bridge wire is reviewed briefly as a
background to innovations in the metallurgical design of wire rod feedstock and in
steelmaking and casting technology. These developments are set in the context of service
requirements.

The factors influencing the properties of high tensile steel wire are discussed with
particular reference to the effects of composition, microstructure and process route. Recent
developments in the use of steels microalloyed with either chromium or vanadium to
provide higher strength whilst maintaining good ductility are described.

Major developments in steelmaking and casting technology have been introduced to
provide improved standards of quality and consistency. Massive capital investment at
British Steel's Scunthorpe Works underlines the industry's continuing commitment to
provide the bridge designer with the materials he needs.

Keywords: Wire Rope, Bridges, Steel Composition, Steel Properties, Fatigue,
Steelmaking, Casting

1   Introduction

The basic principles underlying the manufacture of bridge wire for suspension cables have
remained unchanged in the Century following construction of the Forth Rail Bridge. The
cables for the Brooklyn bridge built in 1883, were made from galvanised cold drawn
pearlitic high carbon steel wire with a composition and microstructure broadly similar to
those adopted in the 1980s. However, advances in metallurgical design and
manufacturing technology have brought about continuous improvements in steel quality,
consistency and economy over the intervening years and steel maintains its competitive
position over other materials.

The paper briefly reviews the historical trends in property requirements and
manufacturing technology before considering, in more depth, recent developments in both
product metallurgy and processing which provide the standards of integrity required by
modern bridge designers.

2   A century of progress

2.1   Wire technology
High tensile bridge wire has, for the last hundred years, been made from high carbon steel
by rolling to rod and cold drawing. Fig. 1 shows the trend in tensile strength over the
period. The wires for the Brooklyn Bridge were made from high carbon steel of somewhat

variable composition. Analysis of individual wires during refurbishment of the bridge reported by Carlson (1987) showed carbon to range from 0.55 to 0.91%. The mean strength of the wires was in the region of 1100 N/mm².

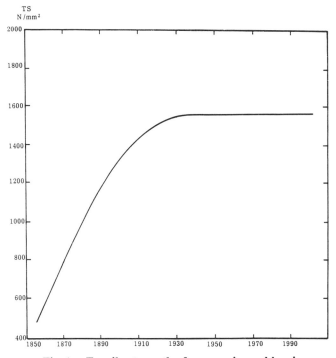

Fig. 1. Tensile strength of suspension cable wire

Strength levels increased significantly in the first half of the century but have remained broadly constant from 1930 to the present day. The strength increases achieved up to 1930 were a consequence of tighter control of steel composition, permitting a closer approach to the optimum carbon content, coupled with the introduction of the "patenting" heat treatment designed to provide the optimum pearlitic microstructure for cold drawing. More recently, the adoption of controlled post-rolling cooling on rod mills has permitted direct drawing of hot rolled rod, albeit with some shortfall in both strength and drawability compared to the best lead patenting practice. The implications of this are considered in detail later in the paper.

2.2 Cable technology

Individual wires with diameters of between 3 and 7 mm are typically used for bridge cables. Although the majority are round, others shaped to a Z or triangular profile are used.

There is a wide variety of rope types available and, in theory, they can be made to any length. However, in practice, the limitations are the capacity of manufacturing equipment, the weight of the product for transportation and handling and the ability of the rope to be coiled. Weight per unit length increases rapidly with diameter and the stiffness increase limits the minimum coil diameters for handling ropes.

The common types of wire rope which are used in bridge construction are:

-   Spiral Ropes

    These are round ropes manufactured by winding a number of individual wires around a central core. The successive layers are generally spun with the opposite direction of helix starting with a straight core in order to minimise the rotation characteristics of the rope. This rope type is divided into open, half locked and fully locked ropes but the former and latter variants are usually used for bridge construction.

    The nominal modulus of elasticity for these multiwire helical strands is usually 15-25% below that for single wires in the range of 1.5-1.7×10^5 N/mm^2. At a diameter of 100 mm a 50 t reel would provide almost 1000 m of rope with a breaking load in excess of 800 t.

-   Parallel Wire Ropes

    Parallel wire ropes generally use ungalvanised wires and have an elastic modulus of 2×10^5 N/mm^2. The load carrying capacity of such a system is greater than for spiral strand so that rope diameters can be reduced by approximately 10%. However, the parallel ropes must have an external sheathing to contain the wires. The sheathing also provides the protection from corrosive attack which is required for the ungalvanised steel wires.

    A variant of this approach is the use of parallel wire strand (PWS). This product consists of bundles of galvanised wires which are bound together with a plastic tape and are socketed at both ends. It has been used to build up main cables in suspension bridges or as an isolated component for stays. The former has generally used 5 mm diameter wires and the latter 7 mm diameter wires. The two most frequently used shapes are regular and modified hexagons.

-   Main Cables

    These cables can be made from spiral strand rope fastened together, either into an hexagonal shape or occasionally into a circle. The alternative is to use parallel wires and these can be assembled by the traditional air spinning method or by the use of PWS.

2.3 Steelmaking and casting

For much of this century, steel was made by the processes of Bessemer and Siemens who developed their converters and open-hearth processes in the 1860s/70s. Indeed, it was not until 1952 that commercial use of oxygen was exploited in experiments at Linz in Austria, which eventually led to the 'LD' process whereby oxygen is blown into the steel through a lance. Subsequent developments established the method (also known as BOS) to its dominant position today, accounting with its many variants for well over 50% of world steel output. The early quality range of mainly low carbon strip steels has been extended to high carbon, alloy and stainless steels.

Since the early 1950s, an increasingly complex and important range of secondary steelmaking techniques have been developed (Lange 1988). These proceed in the ladle after tapping from the BOS vessel and may involve vacuum degassing, injection, or heating processes separately or in combination. Many also provide some refining and alloying capability. Steel quality is thus essentially achieved in this secondary process. Today most modern plants employ secondary steelmaking in some form to improve quality and productivity.

Casting was exclusively into moulds to produce ingots until the first continuous casting of steel in 1948 - (Schrewe 1989). The continuous process has become increasingly important, with many developments to extend the product quality and shape ranges. Today steels for wire may be produced from direct continuous casting of billets, but frequently bloom casting is employed where the advantages of full shrouding are used to produce steel of high purity.

Thus a typical modern process route for high quality, high carbon steels for bridge wires involves primary BOS steelmaking, secondary steelmaking, and continuous casting. Cast blooms are subsequently rolled to billets which are then inspected and rectified, thus providing a high quality feedstock for subsequent rod rolling.

3 Metallurgical developments in high tensile steel wire

3.1 Structure property relationships
The two most important properties of wire rod for drawing into bridge wire are its tensile strength and ductility. The steel composition universally adopted has a carbon content designed to provide a fully pearlitic structure in the control cooled condition. This structure, illustrated in Fig. 2, comprises alternative laths of iron carbide (Fe_3C) and α-iron (ferrite).

Cementite   Ferrite

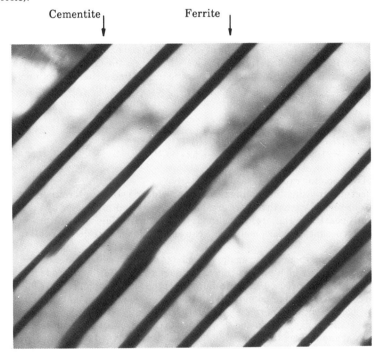

**Fig. 2. Thin foil electron micrograph of pearlitic structure in as-rolled rod.
The spacing of the cementite laths is 0.0001 mm**

In order to provide a basis for development, work was undertaken at British Steel's Swinden Laboratories by Brogan and McIvor (1980) in the late 1970s to establish the quantitative dependence of strength and ductility upon composition and microstructure. Relationships were established for a wide range of commercially available high carbon steel rod from both UK and foreign sources. The factors examined included steel composition and the pearlite morphology including both the interlamellar spacing and cementite lath thickness. The equations relating tensile strength and ductility to the composition and pearlite morphology were:

$$TS\,(N/mm^2) = 780 + 2.67\,S_0{}^{1/2} + 97(\%Si) - 12(\%\alpha)$$
$$+ 7663\,(\%N_f) \tag{1}$$

$$R\,of\,A\,\% = 57.5 + 18.8\,Log_{10}\,(\%\alpha + 1) - 267000t$$
$$- 52\,(\%P\text{-}0.001)^{1/2} \tag{2}$$

Where:-

S_0 is the pearlite interlamellar spacing
α is free ferrite
N_f is free nitrogen
t   is cementite lath thickness

The roles of carbon and manganese are accounted for in the microstructural features through their influence on the ferrite and pearlite morphology. These equations indicate that refining pearlite by reducing interlamellar spacing and lath thickness increases both strength and ductility. This feature has been appreciated in general terms for many years and has been the driving force behind the development of rod patenting and controlled cooling on rod mills. Both these treatments serve to depress the temperature at which the austenite transforms to pearlite, thereby producing the desired structural refinement.

Further work undertaken at Swinden Laboratories, using a laboratory simulator to examine the effects of rod mill cooling provided the following relationship defining the strength of plain carbon steels:-

$$TS\,(N/mm^2) = (267\,(\log CR)\,\text{-}293) + 1029(\%C) + 152(\%Si)$$
$$+ 210(\%Mn) + 442(\%P)^{1/2} + 5244(\%N_f) \tag{3}$$

Where CR is the cooling rate in °C sec^{-1} at 700°C.

The effect of increasing the cooling rate is to provide the structural refinement required for increased strength. There are, however, constraints imposed by the need to develop a fully pearlitic structure. If cooling is too fast, brittle low temperature transformation products such as martensite can form and these are detrimental to performance.

On the other hand there is also an upper bound to strength which is defined by the maximum cooling rate which can be achieved by forced air cooling on the rod mill, particularly for the thick rods, 10-13 mm diameter, used for drawing into wire for suspension cables. This upper bound is about 100 N/mm^2 lower than that which can be achieved by lead patenting. There was, therefore, an incentive in the 1980s to develop steels which would bridge the gap and this has led to the introduction of microalloyed grades (Jaiswal and McIvor 1985; Jaiswal, Kirckaldy and McIvor 1985; Jaiswal and McIvor 1989).

3.2 Development of microalloyed steels

Microalloying elements can be used to strengthen steels in three basic ways - by depressing the transformation temperature to provide a finer pearlite, by precipitation hardening and by solid solution strengthening. In principle all three mechanisms can be used, either singly or in combination. For high carbon steel rod, an initial study identified chromium (for depression of transformation temperature), vanadium (for precipitation strengthening) and silicon (for solid solution strengthening) as the preferred elements for conferring additional strength. The influence of these elements has been studied exhaustively in the laboratory, in full scale trials and in production, to provide a range of steels to meet specific product requirements within the constraints imposed by viable rod mill processing regimes. As both the base levels of strength and the increases associated with microalloying additions are critically dependent upon transformation temperature, and therefore upon the austenite grain size of the rod as it emerges from the mill and the subsequent controlled cooling rate, it was necessary to study these effects in some detail.

The strengthening coefficients for chromium, ΔCr, and vanadium, ΔV, were found to be non-linear and dependent on both cooling rate and composition. Typical values of strength increment obtained using these agents in 0.8% C steels are:

Cooling Rate °C	$\Delta Cr\ N/mm^2$		$\Delta V\ N/mm^2$	
	0.15Cr	0.30Cr	0.05V	0.10V
10	35	100	45	120
25	75	150	60	90

The magnitude of the effect for vanadium passes through a minimum at an intermediate value of cooling rate that is composition dependent.

The critical rate of cooling at which martensite first appears decreases, i.e. the risk of wire drawing problems increases, as the austenite grains become coarser. In order to rank the microalloying agents in terms of "martensite risk", it was necessary to select a standard austenite grain size for the controlled cooling treatments. For a grain size of ASTM 7-8, which characterises the relationship for tensile strength presented in equation 3 and is commercially relevant, the situation for critical cooling rate, CCR, measured at 700° C, is as follows - (Clarke and McIvor 1989).

$$CCR\ ^\circ C/s = 97 - 19(\%Si) - 70(\%Mn) - 50(\%Cr) - 224(\%P) \tag{4}$$

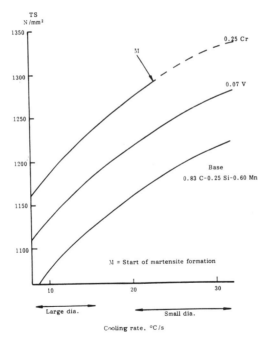

Fig. 3. Useful range of strength for microalloyed commercial rod

Vanadium additions at the levels used commercially do not have a significant effect on the tendency to form martensite. Consequently, of the microalloying agents, chromium exhibits the greatest potency for causing a brittle microstructure to form.

As a result, vanadium was chosen to enhance the properties of small diameter rod, chromium being too potent in forming martensite at relatively fast rates of cooling. The concentration of vanadium, of 0.07%, represented a compromise between strength increment and cost. For large diameter rod, which experiences relatively slow rates of controlled cooling, 0.2-0.3% Cr was preferred on the grounds of economy.

Figure 3 shows the ranges of useful strength that can be achieved in vanadium and chromium microalloyed steels compared with strength levels for plain carbon steel. Both the microalloyed steels provide a rod which matches the strength achieved by lead patenting. These steels are in regular production at British Steel Scunthorpe Works.

4   Design and performance criteria

4.1 Service conditions

There are several loading modes which must be considered for a cable stayed bridge. 'Dead' load is provided by the weight of the suspended structure and this, therefore, dictates the base load on the cables. 'Live' load arises primarily from traffic loading and the assumptions which are made regarding the type, frequency and distribution of the traffic. Vehicular movement, principally stops and starts, can also cause a cable stayed bridge to sway longitudinally. Obviously the precise value will vary with the design of a bridge but it must be remembered that in calculating fatigue damage, imposed stress ranges can be additive if they occur simultaneously.

Because cable stayed bridges are usually employed to provide long spans, wind effects cannot be ignored. Wind loading results in both a static and dynamic load and in a suspended structure the response is governed by the aerodynamic behaviour of the deck. The failure of the Tacoma Narrows Bridge was the result of poor aerodynamic properties and prediction of such properties is not an easy task.

In addition to wind effects on the cables as a result of the deck behaviour, the cables themselves can be influenced directly. The small mass per unit length in combination with the lack of bending stiffness make a single cable sensitive to vibration. The response of a cable to this 'vortex shedding' depends upon the tension in the cable, the mass/unit length, the cable length and the amount of sag in an inclined cable. Vibration induced in this manner can be of a higher frequency than that caused by traffic.

The various loading modes pertinent to cable stayed bridges, therefore, impose two basic types of load on the cables; axial and bending. The combination of these loading modes experienced by a particular cable will depend, to a large extent, upon the design and constructional details of the bridge. The main locations of bending forces are:-

-    Around the saddles, cable clamps and anchorages of the main cables in suspension bridges.
-    At the end terminations of the hanger and stay cables. The latter aspect is obviously accentuated if the end connection is unable to rotate and the bending is allowed to localise at the socket neck.

Load Histories

Fatigue damage occurs by the repeated application of stress cycles to a component. The two aspects which must be quantified are, therefore, the value of the applied stress range and the number of times it is applied. Over a wide range, the static load on the component does not influence fatigue crack growth rate; it is purely the applied stress range. The majority of fatigue tests are carried out using constant amplitude loading but load application is random. In order to compare the fatigue damage which results from a random load application with that which occurs as a result of constant amplitude loading the principle of incremental damage as defined by Miner's Rule is often used. This states that:-

$$\Sigma \frac{N_{pi}}{N_i} < 1$$

where N_{pi} is the actual repetitions of stress range i
         N_i is the number of repetitions that can be allowed for σ_i according to the design curve being used.

In order to apply even this simplified principle the two aspects required are; a reliable design line and a good evaluation of the stress ranges experienced by the component in question.

Projected loading cycles can be predicted from the relevant standards (e.g. BS 5400: Part 2) but actual measurements to relate projected or actual loading cycles to the response of cables are surprisingly very sparse. Some measurements have been made on the Severn Bridge (Flint and Neil Partnership 1983) and a more comprehensive set of measurements has been made on behalf of the Humber Bridge Board. In the latter evaluation two hangers were instrumented with strain gauges attached to the actual wires of the hangers. For an average axial strain range of 217 μm there is a bending component of 22 μm at the bottom connection of the shorter hanger studied. These correspond to relatively low stress ranges (45 and 5 N/mm^2 respectively) but during normal use occasional stress ranges of up to 202 N/mm^2 were recorded. Overall there was little evidence of any large number of cycles at other than the lowest stress levels recorded.

A large programme of work concerned with the monitoring of traffic conditions and related stress measurements has been carried out under ECSC funding (ECSC Contract No. 7210.KD). This work was primarily concerned with the evaluation of steel decks, but it did involve an extensive measurement of traffic types and frequency. Some measurements were made on the deck of the Forth and Wye cable supported bridges. It is noted that 2.9% of the axle loads on the Forth and 2.3% on the Wye Bridge exceed the UK legal load limit of 100 kN. Indeed on the Wye Bridge, 19 vehicles exceeded 150 kN.

The manner in which traffic loads are transferred to the bridge hangers differs for inclined and vertical hangers. In the former there is a slow build up of load as the vehicle approaches a hanger, then a rapid fall as it passes the hanger. The cycle is not uniform about the mean load with the 'compressive' part being less than the increase in tension. This loading pattern differs from that seen in vertical hangers which tend to show a steeper fall from the maximum load but little 'compressive' part of a cycle. The load range experienced by inclined hangers is therefore greater than for vertical hangers. The terminations used for wire ropes can introduce a number of features which are detrimental to the fatigue behaviour of the rope. Consequently, end terminations can have a major influence upon the rope fatigue properties measured and those of an actual installation.

4.2   The fatigue properties of wire rope

The majority of fatigue data which are available have been obtained in air although some tests have been carried out in seawater. In addition, the vast majority of tests have been completed using constant amplitude fatigue cycling and not the more structurally relevant variable amplitude cycling. The traditional test methods for wire ropes are axial loading and bending over sheave (BOS). In the former a suitably terminated rope is cycled under axial tension to failure, the result being plotted on an S-N graph. The usual variables are mean load and the fluctuating fatigue load. The ratio of load range to minimum guaranteed breaking load of the rope has usually been used to define the load axis. The definition of failure is more variable and can range from the detection of a small number of broken wires to complete collapse of the rope. It should be noted that because of the nature of stranded and spiral wire ropes an individual wire can carry a high load only a short distance from a complete break. One particular study by Smith et al (1978) showed that one rope cycled to 70% of its expected life still exceeded the guaranteed breaking load. The achievement of a high strength level after fatigue testing should not, therefore, be used to claim that the rope is 'undamaged' as the rate of wire failures can accelerate rapidly.

Bending over sheave testing requires the rope to be passed around a sheave, or sheaves, whilst sustaining a static load. The relevance of such tests to bridge cables is slight as they were designed to accommodate the requirements of haulage ropes. They could, however, provide an indication of rope behaviour for the situation of the rope being passed over a saddle, either on a tower or a cable clamp on a main cable.

The reliable fatigue data which are available for wire ropes suggest that, in common with the usual behaviour of metallic materials, the mean load does not have a major effect upon fatigue life but the load range applied does. The load range can be expressed as a function of the minimum breaking load of the rope or, more directly, in terms of stress range. The latter method has been used to compare the available data for large diameter ropes in Fig. 4 as it allows a more direct comparison with other fatigue data. The results shown are for axially loaded tests only and all except one were carried out in air. It is unlikely that the one series of tests in seawater can be considered to show any effect of that environment because of the test parameters used.

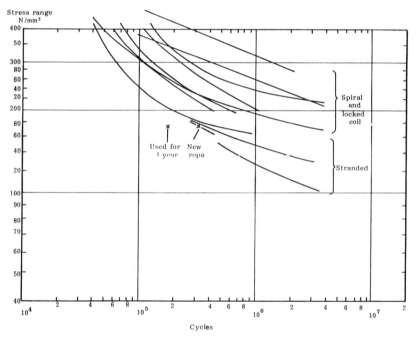

Fig. 4. Summary of fatigue data for wire ropes under axial loading

It can be seen from Fig. 4 that, in general, stranded ropes show the lowest fatigue lives. Locked coil, spiral and parallel wire ropes seem to have a similar behaviour but no reliable information is available to demonstrate a fatigue threshold value. It must also be remembered that the data presented are essentially for the mean fatigue behaviour. Design lines are usually for -2σ from the mean. Exposure to an aggressive environment can also reduce the available fatigue life. BS 6235 recommends that design lives be halved for unprotected steel used in sea-water. The data in Fig. 4 suggest threshold stress ranges of 77 N/mm² for stranded ropes and 102 N/mm² for the other rope types at 2 x 10⁷ cycles.

The only substantial set of published data - Hobbs (1977) and Hobbs and Ghavami (1978) - to indicate the effect of bending loads is given in Fig. 5. This data set suggests a limiting angular change of ±0.6°. Obviously in practice hanger ropes will be subjected to a combination of axial and bending loads. Wire failures will be governed by the local stress range. Hence the possibility of additive effects should be considered in the definition of a realistic design line.

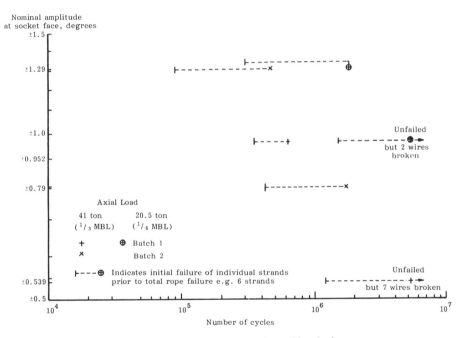

Fig. 5. Effect of bending upon fatigue life of wire rope

Under variable amplitude loading, it is also likely that fatigue crack growth will be started by the high loads imposed but once started a crack may be propagated by the lower stress range cycles. Consequently, it is dangerous to assume that a fatigue crack growth limit derived from constant amplitude tests is relevant to behaviour under variable amplitude loading. In addition the effective fatigue limit is related to the factor (1-R) where R is the ratio of the minimum:maximum stress level. Because of the high mean load on hanger cables the R ratio is usually high, 0.6 to 1, and therefore the effective fatigue limit is substantially lower than the apparent level determined from constant amplitude tests at a low R value.

Using the derived fatigue data to estimate lives based on the limited amount of actual stress ranges measured shows that the majority of damage is caused by the moderate stress levels, 35-75 N/mm², as these present the most severe combination of number of cycles and stress range. Damage is still caused by the low stress magnitude cycles because of the R ratio effect mentioned earlier. Indeed the greatest amount of damage is caused by the low stress ranges because of the large number of cycles imposed.

The above observations imply that the total volume of traffic, not just that of commercial vehicles, can affect the life observed from hanger ropes.

The observations made concerning those stress ranges which can be anticipated to cause fatigue damage show the necessity for more detailed information concerning the actual strains (stresses) imposed on bridge cables.

Conventional design methods tend to discount the lower stress ranges as being undamaging. These observations also emphasise the need to plan very carefully any fatigue programmes on wire ropes using variable amplitude loading. The use of block loading and the unstructured removal of low load ranges may have a marked effect on the results obtained.

For a given structural situation fatigue performance is critically dependent upon steel quality, which, in turn is related to manufacturing standards. The production routes which have been adopted by British Steel to achieve high levels of both internal and surface quality in wire rod feedstock are described in the following Section.

5   Steelmaking and casting

5.1   Product requirements
The specified requirements for bridge wire now generally relate to the chemical composition and the mechanical properties, tensile strength and ductility, Table 1.

Table 1   **Bridge wire specification**

Typical Chemical Composition Requirements	C	Si	Mn	P	S
	0.81/0.85	0.18/0.32	0.50/0.70	≤ 0.030	≤ 0.030

ASTM A586-86 ASTM A603-88 Wire*, Diameter $\oslash$	Stress at 0.7 % Extension N/mm^2	Tensile Strength N/mm^2	Total Elongation on 250 mm %	Wrap Test	
	≥ 1030	≥ 1520	≥ 2.0	≥ 2 turns on 3 $\oslash$	

* The properties depend on the class of zinc coating and the wire diameter.

However, to achieve these properties, and to produce the finished rope or cable, there are further metallurgical requirements which must be considered. These additional requirements become more important with increasingly demanding drawing and stranding operations and for higher strength, higher ductility wires. In fact, wiredrawers frequently specify limiting levels of tramp residual elements (Cu, Sn, As, etc.), sulphur and phosphorus, nitrogen, non-metallic inclusions, segregations and surface quality. These are in the domain of control of the steel producer.

5.2   Steelmaking
Steel for bridge wire is manufactured in the UK at British Steel's Scunthorpe Works. This is an integrated iron and steel plant employing modern processes to manufacture a wide range of carbon and low alloy steels.

The processes employ hot blast furnace iron of low residual tramp element content. Control of tramp element content is widely believed to be necessary in high carbon wire steels to promote good drawability, consistent heat-treatment response, and to ensure adequate final ductility (Leigh and Duckfield 1974). Table 2 shows the typical levels of tramp elements in Scunthorpe Works' steels compared to electric arc steels.

Table 2   Typical residual tramp element levels

	Cr	Mo	Ni	As	Co	Cu	Sn	Σ'
Scunthorpe Works BOS Steel	0.026	0.002	0.024	0.008	0.004	0.025	0.003	0.080
Typical Electric Arc Steel	0.07	0.02	0.06	0.016	0.015	0.08	0.009	0.239
Proposed General Rod Specification (Leigh and Duckfield 1974)	≤0.08	≤0.02	≤0.12			≤0.08	≤0.025	≤0.25

$$\Sigma' = Cr + Cu + Mo + Ni + Sn$$

All iron is desulphurised before being charged to the BOS vessel, by a two-stage process of soda ash additions and deep injection of magnesium-based compounds. The sulphur content can be reduced from about 0.06% in the hot iron to as low as 0.002% depending on the steel grade to be manufactured.

Primary steelmaking proceeds in 300 tonne capacity BOS vessels, employing an optimum refining path determined by computer models from audiometric and decarburisation rate measurements. The highly oxidising conditions promote good carbon and phosphorus removal. Main alloying and deoxidation proceeds at tap in the ladle prior to secondary processes. Several secondary processes are available, and are utilised according to final steel requirements (see Table 3).

Table 3   Scunthorpe works steelmaking facilities and effects

Process Stage	Effects	
Primary Steelmaking	● Carbon and Phosphorus removal ● Deoxidation at tap	● Nitrogen decrease ● Alloying: main additions at tap
Secondary Steelmaking	Specific Effects	Common Effects
Argon stir	● Composition control	
Powder injection	● Sulphur removal ● Inclusion shape control	● Deoxidation ● Inclusion composition control ● Inclusion removal ● Temperature control
Vacuum degassing	● Degassing (H, N) ● Vacuum decarburising ● Significant alloying ● Composition adjustment	
Ladle furnace	● S and P removal ● Inclusion shape control ● Heating ● Significant alloying ● Composition adjustment	

A key requirement for successful secondary steelmaking is separation of the phosphorus-rich primary steelmaking slag from the liquid steel (Pike 1989). Separation is effected by a combination of features: detailed attention and maintenance of taphole condition, detection of slag at the taphole for electromagnetic coils together with rapid backward tilting of the vessel, and by the use of slag dart stoppers which float at the slag-metal interface and block the taphole before issue of slag during ladle filling. Re-ladling is an extremely effective form of separation, and this is now one of the standard process routes employed at Scunthorpe Works since commissioning in 1989 of a modern ladle furnace unit.

Injection into the melt of argon gas or of powdered calcium-rich particles (carried by argon gas) are the simplest secondary steelmaking processes available prior to casting. Both cause intense stirring and mixing actions, promoting temperature and composition homogenisation and flotation of the non-metallic inclusions formed during steel deoxidation. The use of powdered calcium-rich particles allows desulphurisation to proceed to very low levels, e.g. 0.002% S.

Vacuum degassing is now an integral part of the secondary steelmaking processes at Scunthorpe Works, particularly for the production of high carbon steels. Melt circulation in the RH system encourages inclusion agglomeration, flotation, and removal to produce high quality clean steel. Moreover, during the degassing cycle, sampling and chemical analysis are conducted, followed by secondary alloying and composition adjustment for precise control to target chemical specifications. The foregoing secondary steelmaking processes generally require elevated tapping temperatures to accommodate the heat losses which arise during processing. This results in cost penalties, and encourages refractory wear and higher nitrogen and phosphorus contents. The ladle furnace installed recently at Scunthorpe Works provides a heating capability which overcomes these problems to a large degree by allowing reduced tapping temperatures. The unit is a multifunction vessel. Heat is provided by an arc struck between carbon electrodes and the steel; electromagnetic stirring causes metal circulation to promote homogenisation and inclusion removal; gas and powder injection facilities are included; and substantial alloying facilities are employed. A tight cover and argon atmosphere minimise absorption of oxygen and nitrogen during processing, and synthetic slags rich in lime are formed to promote deoxidation, desulphurisation and inclusion removal. Alloy steels up to more than 5% alloys can be produced employing the ladle furnace.

5.3 Continuous casting
Perhaps the most notable advantage of continuous casting is the inherent homogeneity of chemical composition conferred by the continuous and rapid mode of solidification.

An indication of the chemical uniformity of the bloom caster products is shown in Fig. 6.

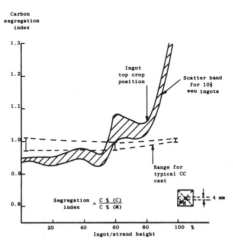

Fig. 6 Improvements in segregation levels in continuously cast feedstock

In the wire rod market sectors, this advantage has been progressively exploited as other developments have become available to improve quality, such that now most of the high carbon steel for wire rod is cast on a four-strand bloom caster.

The important features which promote excellent quality, above that obtainable in ingot steels, concern segregation control and steel cleanness. Both of these are affected partly by the preceding process stages, particularly those of secondary steelmaking.

Segregation control begins with very careful attention to casting temperature. Most of the secondary steelmaking processes include temperature control and homogenisation as part of their operations. This is particularly the case with the ladle furnace which allows relatively easy adjustment of temperatures. Once the target temperature has been achieved, casting under controlled conditions of speed and cooling can promote excellent segregation control characteristics, resulting in levels better than those which can be achieved by ingot casting (Fig. 7), particularly since the machine was converted to 4-strand casting in 1986.

Control of steel cleanness is a second key feature for high quality high carbon wire rod. The cleanness of molten steel in the ladle is promoted by the secondary steelmaking operations, by careful control of the deoxidation process and of the refractory regime, by flotation of inclusions by melt circulation during stirring and vacuum degassing, and by the use of refining slags in the ladle furnace.

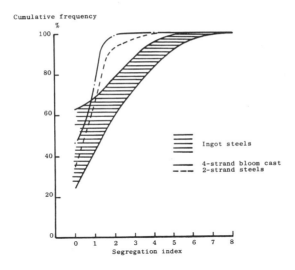

Fig. 7   Improvements in segregation levels in rod from continuously cast feedstock

At the continuous casting machine, the major concern is to prevent reoxidation of the steel, and this is accomplished to the highest degree in large section bloom casters such as the Scunthorpe Works machine, where full shrouding employing submerged nozzles is possible. The total shroud system is also vital to prevent nitrogen absorption during casting, enabling the production of low nitrogen, high carbon steels.

The surface quality of bloom cast steels is inherently superior to ingot steels. With good rolling practice, this superiority is carried through to the finished billets.

5.4 Billet production

After heating to rolling temperature, the large continuously cast blooms (750 x 355 mm) are rolled to billets in a modern bloom and billet mill, involving two reversing breakdown stands and a continuous finishing train of up to 10 stands in alternate vertical-horizontal configuration. The finished billet sizes for rod mills are typically 115 mm and 140 mm square, and up to 2 tonnes in mass.

The mill process routes are designed to build-in quality at all stages, but increasingly rod rollers demand final inspection and rectification of their billets. This demand is now being met by a major new inspection and rectification line, which includes shot blasting, surface inspection (Elkem 'Therm-o-matic'), internal inspection by an ultrasonic system ('Sonomatic'), and machine grinders.

The 'Therm-o-matic' device detects heat emission peaks from any defects when the billet is excited by an induced HF current. Four IR cameras scan the full billet surface, sending detection results to the machine computer which supervises paint sprays to mark any defects which exceed specified depth and length thresholds. The marks are used to direct grinding operations to rectify billets as necessary at one of four machine grinders.

The ultrasonic machine by 'Sonomatic' is an immersion system based on the pulse-echo technique. An array of up to 32 probes send ultrasonic pulses into the billet and subsequently detect the reflected signals which are sorted electronically into 'good' and 'defect' echoes. Again the results are passed to a machine computer which controls paint sprays to mark any defective parts of billets for removal.

A process control computer supervises the whole conditioning line, and records very detailed quality information for product development purposes.

An example of the inspection results, which demonstrates the superior and more consistent surface quality of the bloom derived material is given in Fig. 8.

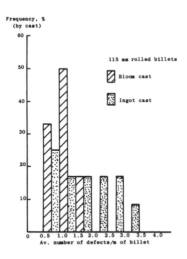

Fig. 8. Surface quality on high carbon billets rolled from ingot and bloom cast steel (Therm-o-matic results)

5.5 Bridge Wire Steels

The foregoing has indicated the major features of the process routes for steels for bridge wires. These are shown in summary form, together with the conferred product features, in Table 4.

These facilities integrate into a system of world-rank standards for the provision of carbon steels, carbon-chromium steels, and low alloy steels for use as bridge wire.

Table 4   Quality aspects of scunthorpe production route

Process	Process Feature	Product Feature
Ladle Furnace	• Lower tap temperatures • Lower refractory wear • Slag separation/Synthetic slag • Precise 'T' Control • Precise composition control • Major alloying	• Lower N, H, P • Cleaner steel • Cleaner steel • Improved segregation characteristics • Improved composition consistency • Extended alloy steel range
Bloom Caster (4 Strand)	• Large cast size • Temperature control (ladle pre-heat, lid) • Large section size (750 mm × 355 mm) • Submerged nozzles • Fully shrouded casting • Air mist cooling	• Chemical uniformity • Clean steel • Minimum superheat • Minimum segregation • Short casting time, slow speed • Excellent internal quality • Diverse products • High reductions • Aluminium-treated steels • Clean steel • Low nitrogen • Excellent surface quality
Automated Billet Conditioning Line	Surface Inspection (Therm-O-Matic) • Defect depth detection: 0.3 mm - 2.5 mm • High detection confidence Internal Ultrasonic Inspection (Sonomatic) • Defect size detection: 3.5 mm • Full length 2-D testing Automated Inspection Line • Continuous operation • High productivity • Computer QC information system Grinder • Additional capacity	 • High grade surface • Consistent surface quality • Freedom from smallest defects • Assured internal quality • Increased availability of high grade billets • Product developments • Improved presentation

6 Summary and conclusions

The factors influencing the properties of high tensile steel cables have been discussed with particular reference to the effects of steel composition, microstructure and production technology. Properties of the finished wire are shown to be critically dependent upon both composition and processing history. Modern rod mills have controlled cooling facilities designed to provide the wire drawer with the optimum microstructure in the rod and this has largely eliminated the need for lead patenting.

Recent developments in the use of steels micro-alloyed with either chromium or vanadium to provide higher strength levels whilst maintaining good drawability and ductility have been described and the potential for further improvements in this area discussed.

Major developments in steelmaking technology have been introduced in recent years. BOS steelmaking provides the benefits of high purity, low residual steel in large batches to give improved consistency. Secondary steelmaking permits further refinement and control of both composition and steel purity. Continuous casting with fully shrouded pouring to maintain high levels of purity ensures high levels of cleanness and both internal and surface quality. Massive capital investment in modern steelmaking and inspection equipment at British Steel's Scunthorpe Works underlines the industry's continuing commitment to provide the bridge designer with the materials he needs.

7 Acknowledgements
The authors wish to thank Dr. M.J. Pettifor, Chief Metallurgist and Mr. J.J. Gorman, Director, Scunthorpe Works and Dr. R. Baker, Director of Research and Development, British Steel plc for permission to publish this paper.

8 References

BS5400, Part 2: 1978, Specification for loads.

Brogan, J and McIvor, I.D., 1980, Quantitative relationships between the structure and properties of high carbon pearlitic rod. BISPA wire product group technical Conference, Stratford upon Avon pp 1-7.

Carlson, D.P., 1987, Roping the Brooklyn Bridge - The second centenary conference proceedings. Interwire '87, Atlanta pp 33-38.

Clarke, B.D. and McIvor, I.D., 1989, Effect of phosphorus on microstructure and strength of high carbon steel rod. Ironmaking and steelmaking 16(5), pp 335-344.

ECSC Contract No. 7210.KD. Measurement and Interpretation of Dynamic Wheel Loads and Bridges. Phases 1 and 2 European Commission, Brussels.

Flint and Neil Partnership, 1983, M4 Severn Crossing - Structural Feasibility Study.

Hobbs, R.E., 1977, Fatigue of Socketed Cables - Tests on 16 mm specimens, CESLIC Report 50.

Hobbs, R.E. and Ghavami, K., 1978, Fatigue of socketed cables - in-line and transverse tests on 38 mm specimens. CESLIC Report SC2.

Jaiswal, S. and McIvor, I.D., 1985, Metallurgy of vanadium microalloyed high-carbon steel rod, Material Sci. Technol. (4), pp 276-284.

Jaiswal, S., Kirckaldy, A. and McIvor, I.D., 1985, Microalloyed high carbon wire rod, Wire Industry, 52, pp 779-782.

Jaiswal, S. and McIvor, I.D., 1989, Microalloyed high carbon steel rod, Ironmaking and Steelmaking, 16 (1), pp 45-54.

Lange, K.W., 1988, Thermodynamic and kinetic aspects of secondary steelmaking processes. International Material Review, 33(2), pp 53-89.

Leigh, C.T. and Duckfield, B.J., 1974, proceedings BISPA Technical Conference 'Wire Rods', Harrogate, p 51.

Pike, T.J., 1989 Improved steels for the rod and wire industry. Ironmaking and Steelmaking 16(3), pp 168-173.

Schrewe, H.F., 1989, continuous casting of steel. Fundamental priciples and practice, English Edition. Stahl u Eisen.

Smith, H.L. et al., 1978, Increased fatigue life of wire ropes through periodic overloads OTC 3256, Houston.

DYNAMIC BEHAVIOR OF CABLE-STAYED BRIDGES UNDER MOVING LOADING

F. VENANCIO-FILHO
Dept Civil Engineering, Rutgers University, Piscataway,
New Jersey, USA
A. EL DEBS
Engineering School, University of Sao Paulo, Sao Carlos, Brazil

Abstract
In this paper the dynamic effects of a moving load on the structure of a cable-stayed bridge are analyzed. The bridge structure is mathematically modelled with two types of finite elements: bar and beam elements. Special bar elements are used to model the cables. Beam elements are used to model the bridge superstructure and the towers. The model for the vehicle is a mass in contact with the structure-spring and damper-suspended mass. The motion equations of the system formed by the mathematical model of the structure and the vehicle model are derived through standard methods of structural dynamics. These equations are nonlinear, due to cable behavior and the moving load, and are solved by an iterative step-by-step integration method based on a cubic interpolation of the inertia forces. Their solution provides time histories of displacements and internal forces in the structure. Impact coefficients are determined by comparing the maximum dynamic responses with the static ones. Example of a typical structure of a cable-stayed bridge under moving load is presented.
Keywords: Cable-stayed bridges, Moving Loading, Nonlinear Structures, Structural Dynamics, Impact Coefficients.

1 Introduction

The analysis of the dynamic behavior of bridge structures under moving loads has attracted the attention of structural engineers since the last decades of the last century. Early investigations were concerned with the behavior of railway bridges and, subsequently, with highway bridges. Dynamic effects were not important in face of the relatively low speed of the vehicles. High speed transportation vehicles have increased the importance of dynamic analysis however.

The first dynamic analysis of structures under moving loads involved a simply supported beam with the consideration of two limiting cases: 1) a load with mass

traversing on a massless beam; 2) a massless load
traversing on a beam with uniform mass. Limiting case 1
applies to railway bridges and has been analyzed by Stokes
(1883). Case 2 which applies to highway bridges was
studied by Krylov (1905), Timoshenko (1911), and Inglis
(1934). The second author solved the differential
equation of the forced vibrations of a simply supported
beam traversed by a load with constant velocity by modal
superposition. A maximum dynamic deflection of 1.5 times
the static deflection was found when the traversing time
is half the first natural period of the beam. Later on it
was recognized by Eichmann (1953) and Warburton (1964)
that the dynamic effect is less conservative. The maximum
dynamic deflection calculated by these authors was 1.743
the static deflection for a traversing time equal to 0.81
the first natural period. This conclusion was
subsequently supported with the use of matrix and finite
element methods
    Matrix and finite element methods were first employed
in the sixties. The bridge structures were modelled as an
assemblage of beam finite elements with lumped masses.
Fleming and Romualdi (1961) used a stiffness formulation
for the motion equations of a three-span continuous beam
traversed by a load with its mass split in a sprung and an
unsprung one. A numerical step-by-step integration method
was used to solve the equations. Veletsos and Huang
(1970) analyzed the problem of a cantilever beam under a
massless moving load through a flexibility formulation of
the motion equations which were integrated by the modal
superposition method. Single and three-span continuous
beams were analyzed through a flexibility approach by Wen
and Toridis (1962). Venancio-Filho (1966) analyzed single,
cantilever and continuous beams, and frames using a
stiffness formulation and the modal superposition method.
Munirudrappa (1969) used the same approach to tackle the
case of grids. Yoshida and Weaver (1971) analyzed beams
and plates traversed by a massless load and by a load with
mass. Fallabela (1975) considered the case of beams
traversed by an idealized vehicle consisting of a mass
suspended through a spring and damper upon a mass in
contact with the structure. In this case the motion
equations are nonlinear and were solved by a step-by-step
integration method. Olson (1987) considered beams and
frames traversed by the same idealized vehicle and solved
the motion equations by a modified modal superposition
method in order to take into account the problem
nonlinearity.
    Cable-stayed bridge structures traversed by the
idealized vehicle present two kinds of nonlinearities:
one due to the nonlinear structural behavior of the cables
and the other due to the idealized vehicle itself. In
this paper the problem of cable-stayed bridge structures
traversed by the idealized vehicle is addressed. A

mathematical model of the bridge structure is considered which is constituted by special bar finite elements which model the cables and by beam finite elements which model the tower and the superstructure. The nonlinear motion equations of the integrated system, idealized vehicle-mathematical model of the structure, are formulated using standard methods of structural dynamics and are solved by a high-order step-by-step integration method specially suitable for the nonlinear equations.

An example of an actual cable-stayed bridge is presented in which the dynamic effects are given in terms of impact coefficients for displacements and forces.

2   The Mathematical Model

2.1 Structure
The mathematical model of the structure consists of an assemblage of beam and special bar finite elements. The special bar element is used to model the cables taking into account its nonlinear behavior and the beam element is used to model the bridge superstructure and the towers. A typical mathematical model is depicted in Fig.1. The stiffness matrix of the beam element is composed by the conventional and the geometric stiffness matrix which takes into account the influence of normal forces.

The stiffness matrix of the special bar element is determined through an axial stiffness given, according to Fleming and Egeselli (1980), by fEA_o where A_o is the cable cross-sectional area, E is the modulus of elasticity and

$$f = \frac{12N^3}{12N^3 + EA_oG^2\cos^2\partial} \tag{1}$$

In Eq.1 f is a factor which takes into account the influence of the cable curvature in its effective axial stiffness, N is the cable force in the chord direction, ∂ is the angle between the chord and the horizontal, and G is the cable total weight.

2.2 Vehicle
The idealized vehicle is formed by a sprung mass m_1, a suspension constituted by a spring with stiffness k and a damper with coefficient of viscous damping c, and an unsprung mass m_2 in contact with the structure, Fig 2a.

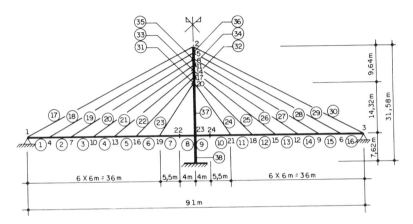

Fig.1   Mathematical model and example

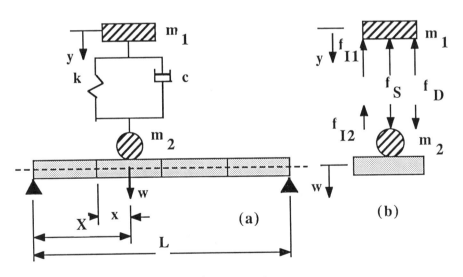

Fig.2   Idealized vehicle-structure

337

3 Motion Equations

Consider the idealized vehicle acting upon a generical beam finite element of the mathematical model, Fig.2a. This element is the loaded element. The following equations are formulated, Fig.2b:

- relative deflection of m_1 and m_2

$$\Delta = y - w \tag{2}$$

- spring force

$$f_S = k\Delta = k(y-w) \tag{3}$$

- damper force

$$f_D = c\dot{\Delta} = c(\dot{y}-\dot{w}) \tag{4}$$

- dynamic equilibrium of mass m_1

$$m_1\ddot{y} + c(\dot{y}-\dot{w}) + k(y-w) = 0 \tag{5}$$

- dynamic equilibrium of mass m_2 and structure

$$m\,\ddot{D} + c\,\dot{D} + k\,D = N^t f_o \tag{6}$$

In Eq.19 $\underset{\sim}{m}$ is the mass matrix of the structure, $\underset{\sim}{c}$, the damping matrix, $\underset{\sim}{k}$, the stiffness matrix, and $\underset{\sim}{D}$, $\underset{\sim}{\dot{D}}$, and $\underset{\sim}{D}$, the vectors of nodal displacements, velocities, and accelerations, respectively; $\underset{\sim}{N^t}$ is a vector with zero entries but those corresponding to the nodal displacements of the loaded element. The non-zero entries are the values of the beam interpolation functions computed in the section where the load is acting.

The force f_o acting in the loaded element is given by, Fig.2b,

$$f_o = (m_1 + m_2)g - f_{I_2} + f_S + f_D \tag{7}$$

338

where $f_{I_2} = m_2 \ddot{w}$ is the inertia force of mass m_2.
Introducing in Eq.7 this value of f_{I_2} and f_S and f_D from
Eqs.3 and 4 respectively, and the result in Eq.6 gives

$$\underset{\sim}{m} \ddot{\underset{\sim}{D}} + \underset{\sim\sim}{c} \dot{\underset{\sim}{D}} + \underset{\sim\sim}{k} \underset{\sim}{D} = \underset{\sim}{N}^t [(m_1 + m_2)g - m_2\ddot{w} + k(y-w) + c(\dot{y}-\dot{w})] \tag{8}$$

The time derivatives of $w(x,t)$ are given by

$$\dot{w}(x,t) = \frac{\partial w}{\partial x} \dot{x} + \frac{\partial w}{\partial t} \tag{9}$$

$$\ddot{w}(x,t) = \frac{\partial^2 w}{\partial x^2} \dot{x}^2 + 2 \frac{\partial^2 w}{\partial x \partial t} \dot{x} + \frac{\partial w}{\partial x} \ddot{x} + \frac{\partial^2 w}{\partial t^2} \tag{10}$$

The function $w(x,t)$ is interpolated from the nodal displacements of the loaded element through the interpolation functions of this element as

$$w(x,t) = \underset{\sim}{N}(x) . \underset{\sim}{D}(t) \tag{11}$$

The partial derivatives of $w(x,t)$ which appear in Eq.9 10 are obtained from Eq.11 as

$$\frac{\partial w}{\partial x} = \underset{\sim,x}{N} \underset{\sim}{D} \; ; \; \frac{\partial^2 w}{\partial x^2} = \underset{\sim,xx}{N} \underset{\sim}{D} \tag{12a,b}$$

$$\frac{\partial w}{\partial t} = \underset{\sim}{N} \dot{\underset{\sim}{D}}; \; \frac{\partial^2 w}{\partial x \partial t} = \underset{\sim,x}{N} \dot{\underset{\sim}{D}} \; ; \; \frac{\partial^2 w}{\partial t^2} = \underset{\sim}{N} \ddot{\underset{\sim}{D}} \tag{12a,b}$$

The position of the vehicle, moving with constant acceleration a_o, in the loaded element is given by

$$x = v_o t + \frac{1}{2} a_o t^2 \tag{13}$$

where v_o is the velocity when it enters in the element. The time derivatives of x from Eq.13 are

$$\dot{x} = v_o + a_o t \; ; \; \ddot{x} = a_o \tag{14a,b}$$

Substituting Eqs.12 and 14 into Eqs.9 and 10 the following equations are respectively obtained:

$$\dot{w} = (v_o + a_o t) \underset{\sim}{N}_{,x} \underset{\sim}{D} + \underset{\sim}{N} \dot{\underset{\sim}{D}} \tag{15}$$

$$\ddot{w} = (v_o + a_o t)^2 \underset{\sim}{N}_{,xx} \underset{\sim}{D} + 2(v_o + a_o t)\underset{\sim}{N}_{,x} \dot{\underset{\sim}{D}} +$$

$$a_o \underset{\sim}{N}_{,x} \underset{\sim}{D} + \underset{\sim}{N} \ddot{\underset{\sim}{D}} \tag{16}$$

The introduction of $\dot{w}, w$, and $\ddot{w}$ from Eqs.11,15, and 16 respectively, into the appropriate terms of Eqs.5 and 8 results finally in the matrix equation

$$\underset{\sim}{M} \ddot{\underset{\sim}{u}} + \underset{\sim}{C} \dot{\underset{\sim}{u}} + \underset{\sim\sim}{K}\underset{\sim}{u} = \underset{\sim}{F} \tag{17}$$

where

$$\underset{\sim}{u} = \begin{bmatrix} \underset{\sim}{D} \\ y \end{bmatrix}; \quad \dot{\underset{\sim}{u}} = \begin{bmatrix} \dot{\underset{\sim}{D}} \\ \dot{y} \end{bmatrix}; \quad \ddot{\underset{\sim}{u}} = \begin{bmatrix} \ddot{\underset{\sim}{D}} \\ \ddot{y} \end{bmatrix} \tag{18a,b,c}$$

$$\underset{\sim}{M} = \begin{bmatrix} (\underset{\sim}{m}+\underset{\sim}{m}^{*}) & \underset{\sim}{o} \\ \underset{\sim}{0} & m_1 \end{bmatrix} \tag{19a}$$

$$\underset{\sim}{C} = \begin{bmatrix} (\underset{\sim}{c}+\underset{\sim}{c}^{*}) & -c\underset{\sim}{N}^{t} \\ -c\underset{\sim}{N} & c \end{bmatrix} \tag{19b}$$

$$\underset{\sim}{K} = \begin{bmatrix} (\underset{\sim}{k}+\underset{\sim}{k}^{*}) & -k\underset{\sim}{N}^{t} \\ [-c(v_o + a_o t)\underset{\sim}{N}_{,x} -kN] & k \end{bmatrix} \tag{19c}$$

$$\underset{\sim}{F} = \begin{bmatrix} (m_1 + m_2)g \ \underset{\sim}{N}^{t} \\ 0 \end{bmatrix} \tag{19d}$$

$$\underset{\sim}{m}^{*} = m_2 \ \underset{\sim}{N}^{t} \underset{\sim}{N} \tag{19e}$$

340

$$c^* = 2m_2(v_o+a_ot)\underset{\sim}{N}^t \underset{\sim}{N}_{,x} + c\underset{\sim}{N}^t \underset{\sim}{N} \qquad (19f)$$

$$k^* = m_2(v_o+a_ot)^2 \underset{\sim}{N}^t \underset{\sim}{N}_{,xx} + m_2 a_o\underset{\sim}{N}^t \underset{\sim}{N}_{,x}$$
$$+ k\underset{\sim}{N}^t\underset{\sim}{N} + c(v_o+a_ot)\underset{\sim}{N}^t\underset{\sim}{N}_{,x} \qquad (19g)$$

Eq.17 is a nonlinear second order matrix differential equation. The nonlinearities stem from the cable nonlinear behavior and the moving mass which affect the mass, damping, and stiffness matrices.

4   Solution of the Motion Equations

The motion equations expressed by Eq.17 are solved by a step-by-step integration method. Consider initially this equation written as

$$\underset{\sim}{R} = \underset{\sim}{M} \ddot{\underset{\sim}{u}} = \underset{\sim}{F} - \underset{\sim}{C} \dot{\underset{\sim}{u}} - \underset{\sim}{K} \underset{\sim}{u} \qquad (20)$$

in which $\underset{\sim}{R}$ is the vector of inertia forces and $\underset{\sim}{R}_E$, the one of restoring forces

    The method of solution considered is based upon a cubic interpolation of inertia forces [Argyris, Dunne, and Angelopoulos (1973)] which is very suitable for nonlinear problems.
    The total time interval in which the structural response is to be calculated, is divided into intervals τ. The vector of inertia forces in each interval is interpolated through cubic Hermitian interpolation functions from the vectors of inertia forces and its derivatives at the beginning and the end of the interval, respectively, $\underset{\sim}{R}_o$ and $\dot{\underset{\sim}{R}}_o$, and $\underset{\sim}{R}_1$ and $\dot{\underset{\sim}{R}}_1$. In this way the velocities and displacements at the interval end are given by

$$\underset{\sim}{M} \dot{\underset{\sim}{u}}_1 = \underset{\sim}{M} \dot{\underset{\sim}{u}}_o + \frac{\tau}{12} (6\underset{\sim}{R}_o + \tau\dot{\underset{\sim}{R}}_o + 6\underset{\sim}{R}_1 - \tau\dot{\underset{\sim}{R}}_1) \qquad (21a)$$

$$\underset{\sim}{M} \underset{\sim}{u}_1 = \underset{\sim}{M} \underset{\sim}{u}_o + \tau\underset{\sim}{M} \dot{\underset{\sim}{u}}_o + \frac{\tau^2}{12} (21\underset{\sim}{R}_o + 3\tau\dot{\underset{\sim}{R}}_o + \underset{\sim}{R}_1 - 2\tau\dot{\underset{\sim}{R}}) \qquad (21b)$$

In Eqs.21 u_o and $\dot{u}_o$ and u_1 and $\dot{u}_1$ are the displacements and velocities at the beginning and the end of the interval, respectively.

The derivative with respect to time of $\underset{\sim}{R}$ from Eq.20 is

$$\dot{\underset{\sim}{R}} = \dot{\underset{\sim}{F}} - \underset{\sim}{C} \ddot{\underset{\sim}{u}} - \dot{\underset{\sim}{R}}_E \tag{22}$$

where

$$\dot{\underset{\sim}{R}}_E = \frac{\partial\, \underset{\sim}{R}_E}{\partial\, \underset{\sim}{u}} \dot{\underset{\sim}{u}} = \underset{\sim}{K}\,(\underset{\sim}{u})\, \dot{\underset{\sim}{u}} \tag{23}$$

If the time interval τ is sufficiently small Eq.23 can be linearized providing

$$\underset{\sim}{R}_{E_1} = \underset{\sim}{R}_{E_0} + \underset{\sim}{K}_o\,(\underset{\sim}{u}_1 - \underset{\sim}{u}_o) \tag{24a}$$

$$\dot{\underset{\sim}{R}}_{E_1} = \underset{\sim}{K}_o\, \dot{\underset{\sim}{u}}_1 \tag{24b}$$

where $\underset{\sim}{K}_o$ is the stiffness matrix at the beginning of the interval.

In order to accelerate the convergence of the iterative process expressed by Eqs.21 the stiffness matrix $\underset{\sim}{K}_o$, Eqs.24, is periodically updated as

$$\bar{\underset{\sim}{K}}_o = \frac{1}{2}\,(\underset{\sim}{K}_o + \underset{\sim}{K}_1) \tag{25}$$

where $\underset{\sim}{K}_1$ is a stiffness matrix calculated at the interval end.

5 Example

The cable-stayed bridge structure of Fig.1 is analyzed as an example of the proposed method of analysis. The structure overall dimensions, the divisions into structural elements, and the node and element designations

342

are shown in this figure. The superstructure elements are of reinforced concrete with $E = 45$ kN/mm^2, $\gamma = 22.8$ kN/m^3 and have cross-sectional area and moment of inertia $A = 2.10$m^2 and $I = 0.044$m^4, respectively. The tower elements are of steel with $E = 205$ kN/mm^2, $\gamma = 77$ kN/m^3 and have $A = 0.50$m^2 and $I = 0.177$m^4. The cables have cross-sectional areas varying between 0.09 and 0.14m^2. Structional damping is considered through a Rayleigh damping matrix $(c = \partial m + \beta k)$ where ∂ and β are

determined from typical values of the first and second natural frequencies of cable-stayed bridges as 0.0139 and 0.063, respectively. The vehicle weight is 250 kN with 80% in the sprung part and 20% in the unsprung one. The spring stiffness is $k = 87$kN/m and the damping coefficient is $c = 6$kNs/m. Some results of the analysis for a vehicle with constant velocity of 28m/s (100km/h) are displayed in Figs.3,4, and 5. In these figures time histories of selected node displacement and forces (dashed line) are presented jointly with the "statical" time-histories (solid line) which are calculated by neglecting the inertia and damping forces. Impact coefficients are calculated as the relation between maximum dynamic and static values. From Figs.3,4, and 5 the following coefficients are found: displacement of node 10:1.35; force in cable element 20:1.04; normal force in beam element 4:1.01. For a velocity of 60km/h the corresponding coefficients are 1.17, 1.03, and 1.01 respectively.

6 Conclusion

A structural dynamics method for the analysis of cable-stayed bridges under moving loading is developed. Nonlinear effects due to cable behavior and the moving loading mass are taken into account. An example of a typical cable-stayed bridge structure is presented in which impact coefficients for joint displacements and forces in the cables and in the superstructure are calculated. These coefficients are of the order of 1.17 to 1.35 for displacements and of the order of 1.01 to 1.04 for forces.

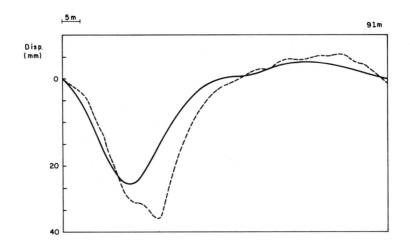

Fig.3   Displacement of node 10

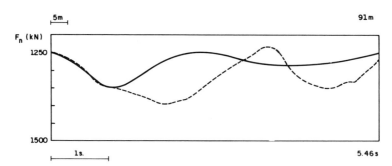

Fig.4   Force in cable element 20

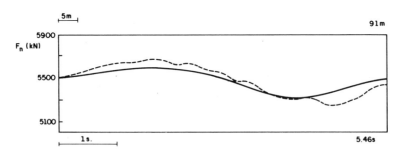

Fig.5   Normal force in beam element 4

7 References

Angyris, J.H., Dunne, P.C., and Angelopolous (1973) Nonlinear oscillations using the finite element technique, Computer Methods in Applied Mechanics and Engineering, 2.

Eichman, E.S. (1953) Note on the maximum effect of a moving force on a simple beam, J. Appl. Mech., Trans. ASME, 562.

Fallabela, J.E. (1975) Dynamics of framed structures under moving loads, M.Sc. Thesis, COPPE, Federal Univ. of Rio de Janeiro.

Fleming, J.F. and Egeselli, E. (1980) Dynamic behavior of a cable-stayed bridge, Earthquake Engineering and Structural Dynamics, 8, 1-16.

Fleming, J.F. and Romualdi, J.P. (1961) Dynamic response of highway bridges, ASCE J. Struc. Div., 87(ST7), 31-61.

Inglis, C.E. (1934) A mathematical treatise on vibration in railway bridges, Univ. Press, Cambridge.

Krylov, A.N. (1905) Uber die, erzwungenen schwingungen von gleichformigen elastischen staben, Mathematische Annalen, 61.

Munnirudrappa, N. (1969) Dynamic response of orthogonal bridge grid under moving force, M. Tech. Dissertation, I.I.T. Bombay.

Olson, M. (1987) Analysis of structures subjected to moving loads, Ph.D. Thesis, Lund Institute of Technology.

Stokes, G.G. (1883) Discussion of a differential equation relating to the breaking of railway bridges, Mathematical and Physical Papers, 2, 69-86.

Timoshenko, S.P. (1911) Erzwungene schwingungen prismatischer stabe, Z. Math. Phys., 59(2163), 203.

Veletsos, A.S. and Huang, T. (1970) Analysis of dynamic response of highway bridges, ASCE J. Engr. Mech. Div, 96(EM5), 593-620.

Venancio-Filho, F. (1966) Dynamic influence lines of beams and frames, ASCE J. Struc. Div., 92(ST2), 371-385.

Wen, R.K. and Toridis, T. (1962) Dynamic behavior of cantilever bridges, ASCE J. Engg. Mech. Div., 88(EM4), 27-43.

Warburton, G.B. (1964) The dynamical behavior of structures, Pergamon Press.

Yoshida, D.M. and Weaver, W. Finite-element analysis of beams and plates with moving loads, Intl. Assoc Bridge Struc. Engr. 31(1), 179-195.

PART SIX
RAILWAY BRIDGES

STEEL RAILWAY BRIDGES: RECENT DEVELOPMENTS

A.C.G. HAYWARD
Cass Hayward and Partners, Chepstow, UK
P.J.G. WIGLEY
British Railways Board, London, UK

Abstract
The paper describes steel rail bridges of short and medium
span. Developments arising from availability of better
steels, erection by large craneage and publication of new
codes are covered. Evolvement of steel rail bridges since
the 1950's is outlined. The legacy of nineteenth century
bridges having trough girders with track directly fixed is
shown to influence modern bridges using ballasted track,
necessitating floors of minimal depth. Development is
described of the trapezoidal box girder type with steel
floor. A 1989 redesign of the box girder range is described
for single or double tracks with skews up to 55° and maximum
span of 39m. Developments to improve maintenance include
better access inside permanently ventilated boxes and
movement bearings. Plans for future development allow for
factors such as the European loading, deformation criteria
under highspeed traffic and a wider structure gauge. The
paper includes some examples of recent bridges.
Keywords: Bridges, Railway, Steel, Box Girders, Fatigue.

1   Introduction

1.1   Early cast and wrought iron bridges
Building of the railways in the first half of last century
coincided with the introduction of cast and wrought iron in
bridges. Brilliant engineers such as Brunel and Robert
Stephenson exploited these materials to produce economic and
adventurous bridges, examples of which survive today.

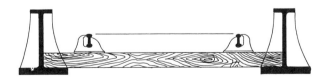

Figure 1.   Bridge over River Dee, Chester 1847

Failures also occurred, one being a new cast iron girder
over the River Dee near Chester by Stephenson in 1847
(ref.1) The three spans of 33m had girders each formed in
three lengths trussed together by inclined tension rods.
Failure probably resulted from lack of understanding of the
details. The collapse drew attention to shortcomings in the
common practice of merely resting transverse members on main
girder bottom flanges without positive connections (see
fig.1). In 1882 a cast iron bridge 25 years old collapsed
under a train in Inverythan killing four people. The cause
was a hidden blowhole near the bottom flange close to a
flange bolted splice. This led to cast iron being pro-
hibited for underline bridges on new lines. In 1891,
another cast iron girder of 8.15m span whose deck dated from
1860 failed at Norwood Junction with five passengers
injured. The rail chairs were spiked direct to 100mm timber
flooring with cross girders resting on the bottom flanges of
the main girders. A programme of cast-iron replacements
followed after 1891, although the events had been over-
shadowed by the major Tay Bridge disaster in 1879, heralding
the use of steel as the leading structural material in the
Forth Railway Bridge in 1890. Wrought iron, which had
gradually replaced cast iron remained in use having similar
qualities of high strength and ductility to steel.

     Many early U.K. railway bridges used cast or wrought iron
trough sections containing longitudinal timbers to which the
track was fixed directly without use of ballast (see fig.2).
This form of construction had maintenance difficulties and
did not permit track adjustment. Other bridges had half
through girders with shallow timber or trough floors which
had inadequate connections for buckling restraint of the top
flange by U-frame rigidity. Both types achieved very
shallow construction depth, seemingly an attribute, but
which has severely restricted later reconstructions and the
use of ballasted track. This contrasts with North America
where early bridges tended to use waybeam type decks having
girders located directly beneath the rails achieving more
generous construction depth. This meant that more
flexibility was possible in replacements using the more
straightforward and economic deck type of bridge.

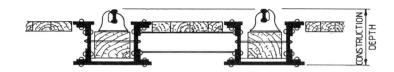

Figure 2. Trough girders and longitudinal timbers

The use of riveting as a reliable means of connecting
elements in wrought iron and steel had many attributes.
Potentially it provided connections rigid enough to achieve
U-frame stability of the main girder top flange yet
possessing some flexibility to absorb undesigned strains.
However decking to half through bridges often consisted of
transverse troughing without cross girders and bearing on
the girder bottom flange. Where they were provided then
such cross girders were often riveted to the main girder
flange, but not connected to the web such that eccentricity
occurred which caused the girders to tilt inwards and this
led to long term problems at bearing areas. Main girders
themselves were not designed to take account of inherent 'U'
frame action, stability being achieved by the use of low
working stresses and the limited rigidity of the connections. 'U' frame action as first introduced into BS153
(refs.2 & 3) in 1953 was not taken into account. At bridge
ends the floor often merely rested upon the abutment. On
skew bridges this resulted in quite large areas of deck
being supported independently of the main girders. Proper
expansion and articulation of the bridge was thereby prevented and deck ends corroded because maintenance was not
posssible. Various problems occurred with half through
bridges with inadequate connections between deck and main
girder where conditions of fixity inadvertently introduced
by the form of connection caused failures, especially prevalent on the centre girder on double track 3 girder
bridges. Rotation of centre girder connections caused
repeated flexure of the girder web when traffic alternately
used one track or the other. Web corrosion was exacerbated
by the inaccessibility of these areas where concealed by
ballast or boarding (ref.4). Such methods of construction
tended to be used up to the time that welding was introduced
in railway bridgework. (see fig.3).

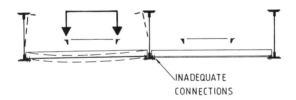

INADEQUATE
CONNECTIONS

Fig.3.   Early half through double track 3 girder bridges

1.2   Introduction of Welding
Welding had been accepted as a means of bridge repair before
1939, and the first road bridge using welded girders had
been built at Billingham in 1931. The first major U.K. use
for railway bridges was in 1948 when seven bridges were
washed away on the Scottish Border and were replaced by
welded girder structures. Six were of deck construction.

The seventh was of half through double track three girder
type having two 18m spans with floor of encased steel cross
girders (ref.5). It was the precursor of the standard half
through plate girder designs (types A to E) with insitu
concrete floors and cross girders. Proper design of
connections was now of increasing importance because of the
rigidity inherent with welded fabrication or in employing
high strength friction grip (h.s.f.g.) bolts for site joints
which had been pioneered in the U.K. by British Rail
engineers. In 1955 it was reported (ref.6) as being the
practice to design main girders assuming an effective length
for buckling instability of three times the distance between
stiffeners. Thus 'U' frame action was appreciated, but not
taken into account rationally. However, safe designs
resulted because the effects of end fixity were now being
taken into account when designing the floor to main girder
connections. Favoured practice was to use riveted girders
for spans exceeding 27.4m but welded for shorter spans.
This period represents the gradual changeover to welded
fabrication which was first used for truss girders in 1961
at Wheatley for a half through 32m span. This was followed
by the major Chepstow Bridge reconstruction where Brunel's
91m through span was replaced by deck construction with
underslung trusses and integral floor carrying ballasted
track (ref.1)

1.3 Standard bridges
As mentioned many early rail bridges had minimal construction depth often with track directly fastened so that
when replacement was required and ballasted track introduced
(desirable 300mm ballast under sleeper plus 368mm for track
using concrete sleepers) then half through decks were
usually dictated. Exceptionally a reduction in headroom
could be allowed. Thus the legacy of nineteenth century
bridges has significantly influenced modern forms. In U.K.
conditions this includes new bridges because most of these
have been to cross new highways where depths of approach
cutting must be minimised to limit land-take.
   A series of simply supported half through plate girder
bridges up to 34m span (the types 'A' to 'E') were developed
from the 1950's. These had concrete floors with rolled
section cross girders connected by shear plate connections
face bolted to the main girder webs. Type 'A' single track
decks for spans up to 15.2m had closely spaced shallow I-
section girders topped flush with rail level, being suitable
for multiple track situations involving piecemeal replacement of existing decks. A close girder spacing allowed use
of 200mm depth cross girders with concrete encasement flush
with the flanges (see fig.4). The inaccessible spaces
between girder flanges were filled with a single skin of
engineering brickwork. This design was later replaced by
the type 'Z' (see fig.5) and this minimises the inaccessible

areas which had started to cause maintenance problems.
Floor concrete in the type 'A' was also found to have poor
durability and details were improved. Types B and C used
deeper main girders (for single and double tracks
respectively) which terminated below the platform clearance
and have been superseded by the box girder type.

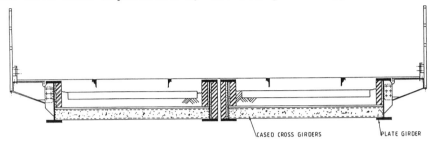

CASED CROSS GIRDERS   PLATE GIRDER

Fig.4   Former type 'A' bridge

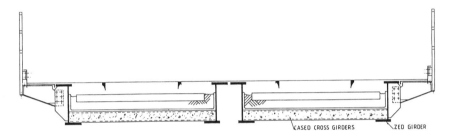

CASED CROSS GIRDERS   ZED GIRDER

Fig. 5   Type 'Z' bridge

Types D and E (see fig.6) (single track and double track 3
girder respectively) provide for longer spans by girders of
still greater depth positioned outside the structure gauge
width. All the standards originally used grade 43 mild
steels but grade 50 high tensile steel was later introduced
for the flanges and webs of the longer span type 'E' main
girders where fatigue did not govern the working stresses.
Later innovations included the use of g.r.p. permanent
formwork for floor soffits (ref.7). In 1973 the standards
were affected by the introduction (ref.8) of RU loading for
British Rail main lines as now published in B.S.5400
(ref.9), that in B.S.153 being appropriate to steam
traction. More emphasis was placed on individually designed
main girders based on a minimum weight of steel. This
philosophy is now changing back towards the use of a
standard design, first because the cost of steel is a much
smaller part of the whole, secondly there is much wider
range of craneage available and lastly design resources are
currently in short supply.

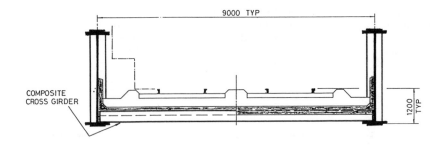

Fig.6   Type 'E' bridge

1.4   Erection constraints

A majority of bridges are necessarily constructed under
severe working conditions in that rail traffic must be
maintained during the work. (ref.10,11).   This particularly
affects the construction of foundations where temporary
works are often extensive. For example new abutments are
usually built in restricted conditions within steel sheet
piled cofferdams spanned by temporary waybeams.   Depending
upon site restraints then decks are usually erected by one
of the following methods.

(a)   'Rolling-in' transversely from pre-assembly area
      during track occupation.
(b)   Erection by road or rail mounted cranes during
      track occupation.
(c)   Construction insitu whilst tracks are temporarily
      diverted.

Availability of large mobile lorry mounted or tracked cranes
within the last 10 to 15 years has made method (b)
increasingly viable.   This encourages the complete prefab-
rication of decks which avoid use of insitu concrete.   The
trapezoidal box girder with steel floor fulfills these
requirements and this type was therefore developed as a
standard.

2 Standard steel box girders and transverse ribbed floors

2.1   Background

Steel half through trapezoidal box girder underbridges
originated in the 1950's.   Various floor types were used

including precast concrete and steel. In 1975 the Western
Region of British Railways had prepared standard designs to
B.S.153 for spans up to 25m using grade 43 steel throughout. The type is used principally where construction depth
is at a premium although it is also used on busy lines where
the time available for erection is very limited because main
girders and floors may be lifted as one. A more economic
solution where these restrictions do not apply is the half
through types 'Z' or 'E' plate girder bridges mentioned
previously.

2.2 1989 Standard steel box girders

The 1989 Standard covers double track 2 girder, double track
3 girder and single tracks design to B.S.5400 as shown in
figure 7. The bridges provide new or replacement decks from
12m to 39m span in 3 metre increments and are designed for
ease of erection, minimum maintenance and low dead weight.
They are especially suitable where minimum construction
depth is essential. Erection may be by placing of girders
and floor units separately, by lifting of complete spans or
by rolling-in. A set of Standard Drawings provides full
data on dimensions and sizes for a full range of spans,
skews, widths and traffic intensities. For each individual
bridge the designer will issue a general arrangement drawing
specifying the span and other salient dimensions and refer
to the Standard Drawings. From these the Contractor will
prepare his workshop drawings for submission to the
Engineer.

The concept consists of half through trapezoidal box
girders, the inner sloping face retaining the track ballast,
and proportioned to achieve the shortest possible floor span
whilst located within the platform clearance of the
structure gauge. Floors are of minimal depth comprising
steel plate protected by a waterproofing membrane and
stiffened with transverse ribs which span between the main
girders through virtually pin-jointed shear plate
connections. The floors are fabricated in transversely
jointed units bolted together using "Huck" fasteners.
Non-structural ballast plates protect the floor to main
girder connections. Parapets are cantilevered from the main
girder outer webs incorporating a walkway and service duct.

Access for inspection and maintenance is provided
throughout the length of the girders and ventilation is
provided through louvred openings in the end diaphragm. A
minimum number of intermediate diaphragms are provided
within the girders consistent with the need to control
distortional warping stresses and to facilitate fabrication.
Access man holes through the diaphragms are made flush with
the bottom flange to allow removal of injured personnel.
Other than this the design seeks to avoid transverse welds
or attachments welded to the main girders for fatigue
reasons. Flange thicknesses are limited to 70mm maximum in

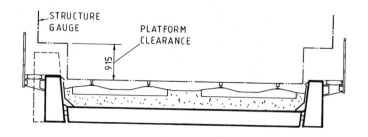

DOUBLE TRACK-2 GIRDER

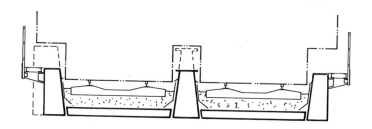

DOUBLE TRACK-3 GIRDER

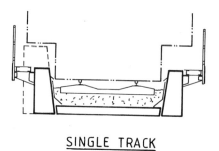

SINGLE TRACK

Fig.7 1989 Standard steel box girders
and transverse ribbed floors

compression and 60mm in tension. Doubler flanges are
necessary for the longer spans where opportunity is also
taken to employ grade 50 steels because fatigue tends not to
control the working stresses. Box corner welds are full
penetration butt welds. Two different section sizes are
possible for main girders, floor plates and other
fatigue-critical elements to cater for different traffic
intensities as specified. Skew angles up to 55° are catered
for. For the double track 2 girder type, two floor widths
"Narrow" (6750mm) and "Wide" (up to 8600mm) are available.

Main girder bearings are of knuckle type for spans up to
18m. For longer spans then roller bearings are provided at
the free end with knuckles at the fxed end. All bearings
are of fabricated steel rather than being of proprietary
form. For trimmer girders of longer span (i.e. for skews
exceeding 30° on double track 2 girder bridges) an
intermediate trimmer bearing is used to restrict live load
deflection within recently imposed limits for high speed
trains (ref.12). Main girders are suitably stiffened and
incorporate a jacking plate in front of each bearing to
permit removal for inspection or replacement.

Parapets incorporating walkway flush with main girder top
and service duct beneath, are provided except for the 30m,
33m, 36m and 39m spans where girders are located outside the
platform clearance. Parapets may be of either steel 3 rail,
2 rail/metal bar infill or solid GRC panels, depending on
site requirements. Steel cantilevers support the parapet
and are bolted to the main girder outer web using grade 4.6
bolts which are designed intentionally to fail first in case
of a derailed vehicle mounting the walkway and threatening
instability of the main girder. A steel plate fascia is
provided to give a clean external appearance and allows
removal of the walkway and duct if required during
maintenance. Generally floors using tee transverse ribs are
used. Exceptionally trough section ribs may be provided for
decks up to 20° skew only. Fatigue is a particular limiting
factor in the floor design and the results of recent British
Railway research have been used in addition to the
requirements of B.S.5400.

3   **Non-standard bridges**

3.1 Deck type
Where the construction depth allows, and for spans where
simple beam construction is not feasible (above about 20m),
fully welded deck type construction has been used. The
bridge in this case can be erected in a single lift, thus
eliminating any waterproofing or painting at site. Extra
care has to be taken with the weld detailing because fatigue
is likely to be a governing factor in many areas of the
design. At locations where extended track possession times
are available and where piece-meal erection is economic,

composite construction has been used successfully from 1964 (ref.13), the deck being formed from insitu concrete with permanent formwork or from precast units with insitu joints (see fig.8).

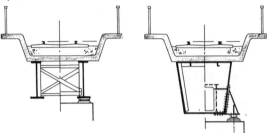

Fig.8   Composite deck construction

3.2   Through trusses

Recently a few larger bridges (up to 120m span) have been required in order to carry existing lines over new motorways and here warren truss type bridges have been used. (Ref.14) One of these is described in the paper. The main considerations are the inclination of the diagonals of the truss, the type of cross section of the members, the positioning of the site joints and transference of longitudinal traction and braking forces from the track into the substructures. Generally the diagonals of these bridges are inclined at between 50° and 65°; a balance being struck between the depth of the girder, the number of bays and aesthetics. The chords have generally been box sections (typically 1000 x 1000mm) with internal access throughout the length of the structure, whilst the diagonals and cross girders have been formed from built up 'H' sections. A gap of about 5mm is left between members such that all axial loads are transmitted by shear through the joints bolts. In the case of box sections the gap is sealed by a compressible waterproof material but is left open between 'H' type members because water cannot collect and natural ventilation will tend to keep the surfaces dry.

These structures are always trial erected before leaving the workshop and site jointing is kept to a minimum. Where construction depth allows it is preferable to have a continuous deck supported on longitudinal rail bearers which oversail the cross girders. This undoubtedly leads to more effective waterproofing because the deck joints can be reduced in number and can be positioned at points of contraflexure. In addition, by placing the rail bearers on sliding bearings (except at the fixed end of the structure) it allows the deck to move freely relative to the main girders so that built in stresses due to the temperature differences between the girders, which are exposed to the sun, and the deck which is not, are kept minimal. Disadvantages are that greater construction depth is required and the longitudinal forces are taken on a single

cross girder. The latter effect can however be over-come
without impairing long term maintenance.

3.3 Future developments

In terms of future developments, it must be always be borne
in mind that the main principles on which railway bridges
are based do not change. Thus ease of erection, maintenance
and durability are the most important factors because of the
high cost of carrying out remedial work. In addition the
consequence of any misjudgement can be potentially very
serious so that changes generally only take place gradually.

   With regard to materials, use is being made of higher
grade steels sometimes in combination with normal grades
where considerations of fatigue and brittle fracture allow.
There has also been a gradual increase in the maximum permitted thickness of plate (currently up to 70mm in compression flanges) as the quality of steel has improved.

   There have been enormous developments in craneage so that
it is now not uncommon for whole spans weighing 60 tons or
more to be lifted in as a single unit. Not only does this
reduce the erection time, but also minimises site jointing.
The latter can be a source of potential weakness especially
as many bridges are erected during cold and wet February
mornings in order to keep the disruption of British Railways' customers to a minimum. The requirements of the
Health and Safety at Work Act and more recently the Control
of Substances Hazardous to Health have also affected the
design of bridges. Access during construction and maintenance is now much more carefully regulated (the Forth Bridge
itself is a good example) and processes such as the
application of protective treatment now have to be
systematically assessed.

   The introduction of BS.5400 has led to designs which
have a more consistent level of reliability, but as the
speeds of train increase, the deformation of the structure
under live load is becoming as critical as its ability to
withstand the loads. Thus there are now specific requirements for limiting both the twist of the deck and the
deflection of the bridge due to passing traffic. The latter
limits depend on the bridge span, the train speed and the
level of passenger comfort required (ref.12)

   The assessment of fatigue damage is also being investigated. This is particularly important for railway bridges
where the live load often represents a large proportion (up
to 90% for cross girders) of the total load. The effect of
plate thickness and the type of variable amplitude loading
on fatigue life are both being studied at present. In the
future there will be a continuing emphasis to design out
welds with a low detail classification and to specify weld
quality consistent with "fitness for purpose".

   Much of the current work is being undertaken under the
aegis of the International Union of Railways (UIC) and it is
this body which has been the focus for setting European and

International Railway standards since the early 1960's. As
the introduction of Eurocodes approaches, together with the
greater European harmonisation planned for 1992 (aided by
the Channel Tunnel) the work of the UIC presently has added
significance, and British Railways is fully involved in this
work. There is a danger however that in trying to establish
standards that are acceptable to the International community, the envelope of safety will become so large as to
lead to unecomomic designs. This will need careful
watching!

4. Some examples of recent bridges

4.1 Box girder example. Atlantic Road Bridge

This bridge carries the South London lines over both the
Chatham Main Lines and Atlantic and Popes Road at Brixton
Station in South London. Construction depth was critical
and the design was further complicated by the track
geometry. The lines are on a horizontal curve of 408m and
are at the summit of a vertical curve.

A half through type bridge was chosen based on the
standard box girder type described above. However, the
girders are fully continuous and curved in plan not only to
minimise the construction depth of the deck but also to
minimise the overall width of the bridge which had to be
accommodated on existing abutments. Even so, a specially
designed deck was need to achieve the required construction
depth of 905mm between rail level and bridge soffit.
Fatigue was the critical factor particularly in the design
of the floor plates to resist local bending.

The bridge was fully trial erected at the fabricator's
works, and then erected during a three week blockade of the
South London lines and a weekend closure of the Chatham Main
Lines for the erection of the steelwork. Guidance was
sought from both the Local Authorities and English Heritage
to ensure that the new structure harmonised with its
Victorian surroundings.

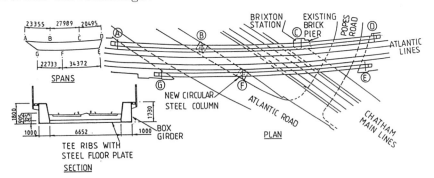

Fig.9   Atlantic Road Bridge

4.2 Warren truss example. Lingard Lane Bridge No.70A
This bridge carries the Romley-Redditch lines on a skew
alignment over the new Manchester outer ring road which is
constructed to motorway standards. The span required was
120m and this is one of the largest single spans constructed
on British Railways for some time. Its steel floor is
supported on cross girders at 12m centres. Continuous
longitudinal steel rail bearers carry the floor between
cross girders.

    Having been fabricated and trial erected at the steel
fabricator's yard, the bridge was erected piecemeal by the
side of the tracks and then slid in during a weekend closure
of the railway lines. The bridge contains 1900 tonnes of
steel and weighed 2,500 tonnes in total (including ballast
and track). Because of the large cost of renewal an
additional material factor of 1.05 was used in the design,
principally as a hedge against corrosion, nonetheless all
external surfaces were treated with a full protective
treatment consisting of Chlorinated Rubber paint and metal
spray to the external surfaces and Micaceous Iron Oxide
paint to the internal surfaces of the chord members

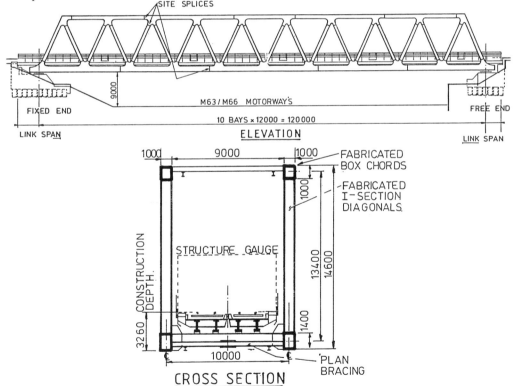

Fig 10. Lingard Lane Bridge No.70A

4.3 Composite example - Integral Rail & Road Bridge at Ebbw Vale. (ref.15)

This bridge has two 25m spans and carries a single track railway owned by British Steel and roadway across rail sidings in a reclaimed industrial area. Track and road are carried on an integral deck with running rails set flush with the roadway pavement. Composite steel and precast unit construction was used to achieve rapid implementation. Loading specification was RU railway loading. The structure was also checked for HA loading but this did not govern. The floor uses precast units of full width and depth with pockets left open for shear connectors welded to the top of the twin steel plate girders. Precast units use grade 50/20 concrete and incorporate holding-down bolts for parapets and rail fixings. Pockets to accommodate the shear studs are provided at 490mm intervals along the lines of the main girders, and these were filled with concrete after final adjustment of line and level of the units. All units are identical except for those at the ends which are incorporated into cast insitu transverse trimmer beams. Girders are simply supported with a 'half-joint' where supported on the middle pier. Plan bracing between the top flanges of the girders provides lateral rigidity while retaining the advantage of a torsionally flexible super-structure in accommodating relative rotation of the supports due to any future differential settlement.

The direct track fixing and flush infill meant that, at the ends of the bridge, a transition onto ballasted track had to be incorporated and run-on slabs were provided. Fabricated steel trackway channels are used to isolate the fixings for the 113A flat bottom rails from the roadway pavement to facilitate future rail replacement.

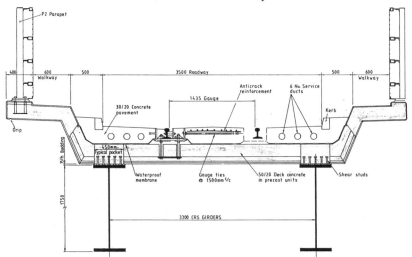

Fig.11   Elevation & Cross Section. Bridge at Ebbw Vale

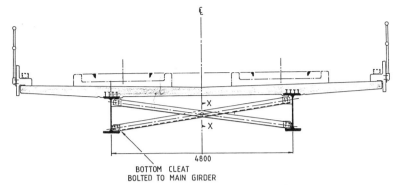

Fig.12a London Docklands Light Railway
Universal beams up to 26m spans

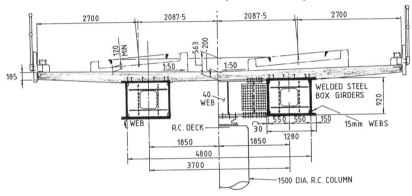

Fig.12b London Docklands Light Railway
Box girders, for curved skew spans on single columns

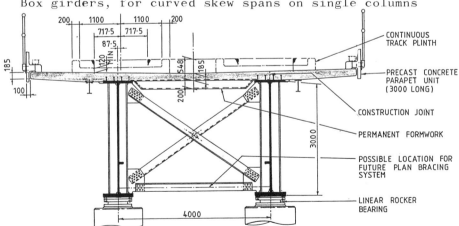

Fig.12c London Docklands Light Railway
Plate girders up to 65m span

4.4   Composite example - Viaducts for London Docklands Light
      Railway (ref.)
When opened in July 1987 the London Docklands Light Railway
was the first of its kind in Britain. The railway 12km long
has 3.8km of new viaducts. Even prior to the opening
ceremony planning was well under way towards upgrading and
extending the railway. The Canary Wharf complex and other
developments required immediate upgrading. Doubling of the
length of the trains and increasing their frequency has
meant that viaducts have needed strengthing especially for
fatigue effect, despite comparatively light loading.
     Viaducts are of continuous composite construction on
piled foundations using twin steel members of either;
     Universal beams (fig.12a) -   up to 26m spans and curved
                                     down to 60m radius
                                     curvature.
     Box girders (fig.12b)   -   up to 26m spans for curved
                                     skew crossing on single
                                     columns.
     Plate girders (fig.12c)   -   up to 65m spans with curved
                                     soffit, over water.
Strengthening, mainly in order to reduce fatigue stresses
has consisted of adding flange material at intermediate
supports, the placing of extra shear connectors using 20mm
force fitted spring steel fasteners through top flanges, and
the provision of shock transversing units (STU's) to allow
extra braking forces to be shared among more than one
section of viaduct. These works currently in hand, are
being performed whilst the railway is in operation.

Acknowledgements.

Clients: London Docklands Development Corporation
           London Underground Limited
Designer Contractor: GEC - Mowlem Railway Group
Consulting Engineers: W.S. Atkins Consultants Limited
Subconsultants for
Superstructures & Strengthening: Cass Hayward & Partners
of some Viaducts
Steel Work Subcontractors: Cleveland Bridge & Butterley
Engineering
Bridge at Ebbw Vale
Clients: Welsh Development Agency Blaeneau Gwent B.C.
Project lead consultants: Ryan Keltecs plc (Cardiff)
Bridge Designers: Cass Hayward & Partners, Chepstow, Gwent
Main contractor: DMD Ltd., Cardiff
Steelwork subcontractor: Fairfield-Mabey Ltd., Chepstow

5 References

1   Berridge, P.S.A., **The Girder Bridge,** Robert Maxwell,
    1969.
2   B.S.153, **Steel Girder Bridges,**1953.
3   Kerensky, O.A., Flint, A.R. and Brown W.C. **The Basis o
    Design of Beams and Plate Girders in the
    Standard 153,** I.C.E. Proceedings Paper No. 48,
    February 1956.
4   Berridge, P.S.A., **Factors Governing the Choice Between
    Repairing, Strengthening and Reconstructing Railway
    Girder Bridges,** I.C.E. Proceedings Paper No 6702,
    September 1963.
5   Mann, F.W., **Railway Bridge Construction. Some Recent
    Developments,** Hutchinson Educational 1972.
6   Berridge, P.S.A. and Easton, F.M., **Some Notes on the Half
    Through Type Plate Girder Railway Bridge,** I.C.E.
    Proceedings Paper No. 58, May 1955
7   Bastin, R.D., **Selby Diversion of the East Coast Main
    Line. 3: Bridges,** I.C.E. Proceedings Part 1, Nov. 1983
8   Technical Note, No. 27 **Design Loading and Load Factors
    for Railway Bridges,** British Railways Board, June 1973
9   B.S.5400, Steel Concrete and Composite Bridges, Part 3
    Code of Practice for Design of Steel Bridges, 1982
10 Atkins, F.E.and Wigley, P.J.G., **Railway Underline
    Bridges: Developments Within Constraints of Limited
    Possession,** I.C.E. Proceedings, Part 1, 1988, 84, Oct,
    989-1007 (Ordinary Meeting 22 November 1988).
11 Banasiak, M.K. and Lees F.I., **Some Recent Bridges for
    British Rail,** Proceedings of B.C.S.A. Conference on
    Steel Bridges, 1968, I.C.E.
12 **Draft UIC Leaflet on Deformation of Bridges,** British
    Railways Board, December 1987.
13 Bonnet, C.F., **Two Railway Bridges of Composite
    Construction,** The Structural Engineer, November 1964.
14 Clark, D.J., and Watermann, B.J., **London Regional
    Transport Railway Bridge D29 at Hanger Lane Junction,**
    I.C.E. Proceedings, Part 1, December 1986.
15 Sadler N.L. and Matthews S.J., **An Integral Rail and Road
    Bridge at the National Garden Festival Site, Ebbw
    Vale,**
    The Structural Engineer, April 1989.
16 Pritchard B.P. and Hayward A.C.G., **London Docklands Light
    Railway New Viaducts,** Second International Conference
    on Short and Medium Span Bridges, Ottawa, 1986.
17 Hayward A.C.G. and Pritchard B.P., **London Docklands
    Light Railway. Upgrading for Heavier Traffic,**
    Symposium on Strengthening and Repair of Bridges,
    Leamington Spa
    June 1988, Construction Marketing Limited.

ON FATIGUE EVALUATION OF RIVETED CONNNECTIONS IN RAILWAY TRUSS BRIDGES

A. EBRAHIMPOUR, E.A. MARAGAKIS, D.N. O'CONNOR
Dept Civil Engineering, University of Nevada, Reno, Nevado, USA

Abstract
A riveted steel railway bridge on the Union Pacific
Railroad in Nevada was considered for estimating
remaining fatigue life. This bridge is on a proposed
route for transportation of hazardous materials through
the state of Nevada; upon approval the normal train
traffic is expected to increase. A modified bridge
structural analysis program was used to save member load
histories caused by the passage of both steam and modern
diesel trains. A fatigue model based on Miner's rule
and stress category D for riveted connections was chosen
to estimate the remaining fatigue life. Stress values
in five members of the bridge exceeded the endurance
limit of category D under train loading. Fatigue life
of the critical members will be reduced due to the
proposed increase in the train traffic.
Keywords: Railway Bridges, Fatigue Life, Steel Bridges,
Train Loads.

1 Introduction

Railway bridges are part of the United States' aging
infrastructure. The older bridges, most of them steel
structures, were built around the turn of the century;
the safety of these bridges for transporting both
freight and passengers is crucial. Today there is a
renewed interest in the safety of railroad bridges,
Foutch (1988). Safety is of prime concern when
railroads are used to transport high level nuclear waste
such as spent reactor fuel. This is important for
health and environmental reasons, and is a major public
concern.

A research project was undertaken to study the effect
of the increase in rail traffic due to operation of the
proposed nuclear waste repository at Yucca Mountain,
Nevada. The project consisted of a survey of the
bridges along the Union Pacific Railroad from North of

Las Vegas to the border of Nevada and Utah, modification
of an existing computer program for rating railway
bridges under modern diesel loading, and evaluation of
remaining fatigue life in steel bridge members. This
paper focuses on the latter aspect of the research
program, namely the fatigue in riveted connections of a
typical steel bridge on the proposed railway route. A
description of the bridge survey and modification of the
existing truss bridge analysis computer program will be
presented briefly.

2 Survey of existing railway bridges

An inventory of the railway bridges was conducted on the
proposed railroad line. There are 135 railway bridges
on this segment of the railroad with a total of 259
spans. These bridges range from short trestles to
medium size steel plate girders and trusses. The bridge
data were saved in ASCII computer files for future data
manipulation. The data indicate mile post location of
the bridges, number of spans for each bridge, type of
the structure, year of construction, design Cooper E-
rating, and authorized track speed.

In addition to the data obtained from the Union
Pacific, the investigators travelled to southern Nevada
for site inspection of several bridges and a detailed
inspection of the bridge located at mile post 409.16,
which was chosen for in depth study. The bridge was
maintained properly and had only minor visible corrosion
on various parts, including the lower bracing system.
In addition the bridge was measured accurately to verify
the original dimensions as given in the Union Pacific
structural plans.

The bridge (see Fig. 1) is a 150-foot (46-meter)
span, single track, riveted, through truss made with
open hearth steel. The center to center spacing of the
two trusses is 17.5 feet (5.3 m). Data on the member
cross-sections and connection details are not given here
for brevity.

3 Modification of the existing bridge rating program

The Association of American Railroads (AAR) program 4
(1975) performs matrix structural analysis and uses the
current specifications to compute the Cooper rating of
individual members of a steel truss railway bridge.
Truss members can be riveted or have pinned connections
and can be constructed of different types of steel.

One of the shortcomings of the AAR program 4 is that
it was written to evaluate the bridge rating based on

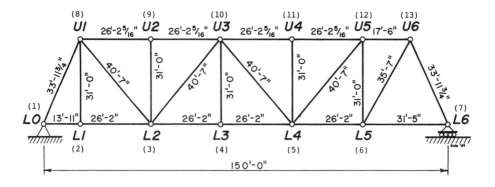

Fig. 1.   Union Pacific steel truss bridge at M.P.
           409.16, Caliente Subdivision (1 ft = 0.3048 m).

the Cooper E-80 train. The Cooper E-80 loading, which
consists of two steam locomotives and a uniform trailing
load of 8,000 pounds per foot (117 kN/m), is generally
accepted to be the most severe type of live load that a
bridge experiences. We modified this latter computer
program to analyze and rate trusses under special train
loading other than the conventional Cooper E-series.
The modified program uses a rating procedure based on
the percentage of live load capacity of the members.
Details of the computer modification is presented
elsewhere, Ebrahimpour, et al. (1989).
     The modified computer program can be used to
determine exact member forces under any arbitrary train
loading configuration defined by the user. As can be
seen in the following sections of this paper, this is of
special interest when predicting fatigue life under
repeated loading.

4   Bridge fatigue rating

One of the major concerns of railway bridge engineers
today is the safety of old riveted structures and the
potential fatigue damage that has accumulated, Fisher
(1984, 1987). Many of these bridges, including the one
selected for this study, were fabricated and placed into
service at the turn of the century. The uncertainties
associated with exact traffic loading in the past and
with the empirical nature of fatigue theory make
estimating fatigue life difficult.

4.1 Recommended practice for bridge fatigue rating

Both the American Association of State Highway and Transportation Officials (AASHTO, 1983) and the American Railway Engineering Association (AREA, 1989) use estimates for fatigue strength of different steel connection details based on experimental data. The major factors governing fatigue strength are the number of stress cycles, the magnitude of stress range, the type of stress range, and the type of construction details.

The stress range, S_R, is defined as the algebraic difference between the maximum and minimum calculated stress values. The AREA specification states that if both the live load and the dead load result in compressive stress in a member, fatigue need not be considered.

The type of stress is defined by the R-ratio which is the ratio of the minimum stress to the maximum stress in each loading cycle, Salmon and Johnson (1980). For example, an R-ratio of zero indicates stress variation from zero to a maximum tension value. On the other extreme, R=-1 indicates full stress reversal; that is, equal values for both compression and tension in each cycle. Typically the more negative R value results in lower fatigue life; thus the R value has a direct effect on the stress vs. number of cycles (S-N) fatigue curve as shown in Fig. 2. Although a range of R=0 to R=-1 can be assigned to the type of stress range, the AREA specifications uses only two kinds of stress ranges; tension (T) or stress reversal (Rev).

In a recent report, Fisher (1987) verified that for riveted bridges a simple check for category D provides a good estimate for the number of cycles required to develop fatigue cracks. Although category D constitutes a reasonable lower bound for fatigue crack development, one can use this category to conservatively estimate the remaining fatigue life (i.e., until failure) in such members. Fig. 3 shows a plot of S-N curve for category D.

4.2 Fatigue rating for Union Pacific bridge

Fatigue category D was chosen to estimate the remaining fatigue life of the Union Pacific bridge located at mile post 409.16. Only the primary load-carrying members and the corresponding riveted connections were considered for this part of the study, thus a separate analysis may be required in future to evaluate fatigue life for the floor beams and the secondary bracing system.

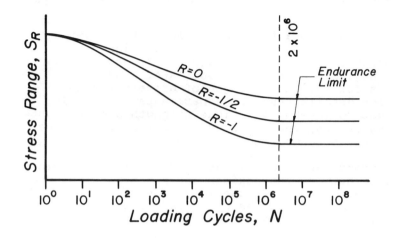

Fig. 2. Typical S-N fatigue curves.

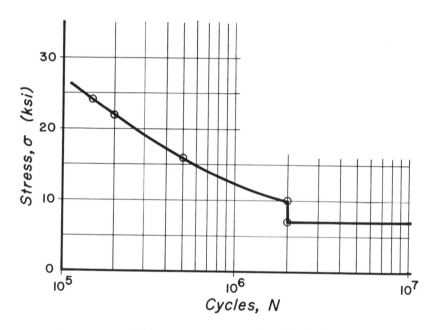

Fig. 3. Fatigue curve constructed from AREA
Specifications (1 ksi = 6.89 MPa).

4.2.1 Estimate of rail traffic

Data on the average number of trains per day on the
Caliente Subdivision for the period between 1978 and
1988 was obtained from the Union Pacific Railroad. As
shown in Table 1, the average traffic is 17.09 trains
per day or a total of 6238 trains per year. As shown in

the following sections, member stress histories produced
as a result of fully loaded freight cars are lower than
the corresponding stresses for the locomotives. The
freight car stresses are also below the endurance limit
for category D (7.0 ksi or 48.3 MPa), and therefore do
not contribute toward cumulative fatigue damage of
members. For the reasons mentioned, It is important to
properly estimate the number of locomotives, both steam
and diesel-electric, during the lifetime of the bridge.

Bridge 409.16 was built in 1911. It is assumed that
the repository will be open in 2003 and operate for 25
years, U.S. Department of Energy (1988). As estimated,
an increase in traffic of two trains per day was assumed
until the year 2028, when normal traffic will resume.
Table 2 shows the yearly number of diesel-electric and
steam locomotives used in the western United States, AAR
(1988). It is assumed that the average number of trains
on the Caliente line without the repository will stay
constant at 6238 trains per year for the remaining life
of the bridge. From Table 2, the yearly number of steam
and diesel locomotives can be calculated as shown in
Table 3.

Cooper E-55 was used to represent steam locomotives,
as this is the design rating of the bridge. For the
diesel loading, we assumed three SD-60 locomotives per
train. The SD-60 is among the heaviest diesel
locomotive commonly used today.

4.2.2 Computer generation of member load histories

In order to perform fatigue rating for individual
members of a truss bridge, we had to assemble the load
history experienced by each member due to passage of
each train. This is necessary to determine which
members experienced tensile forces as well as the number
of cycles within each stress level. The structural
analysis routine of the AAR bridge rating program was
modified to save force values for all members as the
train moves across the bridge. Previously the truss
rating program saved only the highest force values for
each member and discarded the remaining values.

For each joint on the lower chord the train is moved
from left to right, placing each axle in turn on that
joint. For each axle placement, loads are calculated
for all members. In many cases, the length of the
locomotive consist is longer than the panel lengths;
therefore, segments of the load history will overlap.
To account for this overlap, the x-coordinate of the
leading axle was also written to the output data file.
The data was then imported into LOTUS 1-2-3 and sorted
according to the position of the lead axle. Typical
sorted load histories are plotted in Figs. 4 and 5.
Figure 4 shows the load histories for the bridge members

Table 1.   Daily number of trains on the Union Pacific,
           Caliente Subdivision

Year	Average Number of Trains per Day
1978	17
79	19
80	20
81	21
82	17
83	15
84	15
85	14
86	15
87	16
88	19
Mean	17.09
Standard Deviation	2.34

Table 2.   Locomotives in Service (AAR, 1988)

Year	Diesel-Electric		Steam	
	Number	Percentage	Number	Percentage
1929	22	0	56,936	100
39	510	1	41,117	99
44	3,049	7	39,681	93
47	5,772	14	35,108	86
51	17,493	45	21,747	55
55	24,786	81	5,982	19
60	28,278	99	261	1
65	27,389	100	29	0
70	26,796	100	13	0
75	27,985	100	12	0
80	28,243	100	12	0
85	22,869	100	0	0
87	19,956	100	0	0

Table 3. Steam and diesel-electric trains on the Union
Pacific, Caliente Subdivision

| Period | No. of Years | Number of Trains | |
		Diesel-Electric	Steam
1911-1929	18	0	112,284
1929-1939	10	0	62,380
1939-1944	5	312	30,380
1944-1947	3	1,310	17,404
1947-1951	4	3,493	21,459
1951-1955	4	11,228	13,724
1955-1960	5	25,264	5,926
1960-1965	5	30,878	312
1965-2003	38	237,044	0
2003-2028	25	155,950 (174,200*)	0
Total	117	465,479 (639,679*)	264,367

* Numbers in parentheses indicate the projected number
of trains with the repository being in operation.

under Cooper E-55 loading; Fig. 5 presents load
histories resulting from three SD-60s and a train of
100-ton cars. Many members experience secondary cycles
under diesel loading because diesel locomotives have
only a few axles with wide axle spacing. Table 4 lists
the primary stress range values, S_{R1}, and secondary
stress range values, S_{R2}, for all members under tension
or stress reversal.

4.2.3 Evaluating fatigue using Miner's rule
Miner's rule assumes that fatigue occurs as a result of
cumulative damage due to load cycles at various stress
levels. This can be mathematically expressed as:

$$D = \sum_{i=1}^{p} (n_i/N_i) = n_1/N_1 + n_2/N_2 + \cdots + n_p/N_p \qquad (1)$$

where

D = fraction of fatigue life used;
n_i = number of actual cycles at stress range S_{Ri}; and
N_i = number of cycles to failure at stress range S_{Ri}.

373

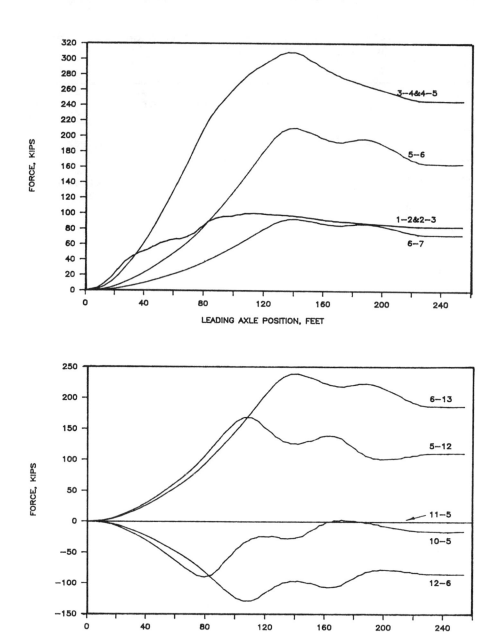

Fig. 4.   Member load histories for Cooper E-55 loading
          (1 kip = 4.45 kPa; 1 ft = 0.3048 m).

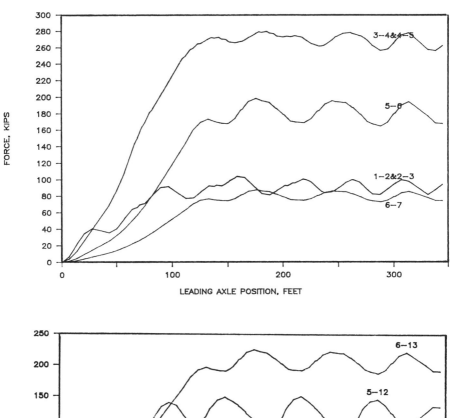

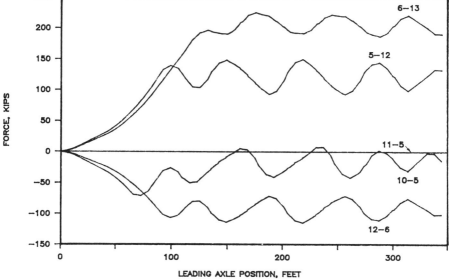

Fig. 5. Member load histories for SD-60 loading
(1 kip = 4.45 kPa; 1 ft = 0.3048 m).

Table 4. Stress range values for primary and secondary loading under Cooper E-55 and SD-60 loads (1 ksi = 6.89 MPa).

| Member | Cooper E-55 | | SD-60 | |
	S_{R1} (ksi)	S_{R2} (ksi)	S_{R1} (ksi)	S_{R2} (ksi)
1-2	6.57	--	6.84	1.19 (4)
2-3	6.57	--	6.84	1.19 (4)
3-4	8.69	--	7.81	0.41 (3)
4-5	8.69	--	7.81	0.41 (3)
5-6	8.25	--	7.79	1.21 (3)
6-7	6.11	--	5.71	0.89 (3)
1-8	--	--	--	--
8-9	--	--	--	--
9-10	--	--	--	--
10-11	--	--	--	--
11-12	--	--	--	--
12-13	--	--	--	--
13-7	--	--	--	--
8-2	4.18	1.64 (2)	4.30	3.01 (4)
8-3	7.36	--	6.92	1.38 (4)
9-3	--	--	--	--
3-10	5.71	--	1.09	--
10-4	5.11	1.64 (2)	5.16	3.15 (4)
10-5	8.67	--	4.34	3.10 (4)
11-5	--	--	--	--
5-12	8.67	2.08 (2)	7.69	2.57 (4)
12-6	--	--	--	--
6-13	7.58	--	7.10	1.07 (3)

Numbers in the parentheses denote the number of secondary cycles for the corresponding member.

By comparing values of Table 4 with the 7.0 ksi (48.3 MPa) endurance limit for category D, the stresses in five members of this bridge exceed the endurance limit under both Cooper E-55 and SD-60 loading if more than 2 million cycles were to occur. These members are 3-4, 4-5, 5-6, 5-12, and 6-13. Failure in any one of these members could result in collapse of the structure. Note that stresses in members 8-3 and 10-5 exceed the endurance limit value only under steam locomotives and have accumulated a total of 264,367 cycles (i.e., less than 2 million cycles) and are thus not prone to fatigue problems.

4.2.4 Fatigue life without the repository
Using Miner's rule the fatigue life of the five members mentioned were found using the following formula (see Table 3 for the number of cycles used).

$$\frac{465,479 + 264,369 + n_1}{2,000,000} = 1$$

$$n_1 = 1,270,154 \text{ cycles}$$

Where n_1 is the number of remaining cycles in the critical member after the year 2028. Knowing that the bridge will be 117 years old in 2028, the total fatigue life of the critical members becomes

$$\text{Fatigue life} = 117 + \frac{1,270,154}{6238} = 321 \text{ years}$$

4.2.5 Fatigue life with repository in operation
Assuming a repository life of 25 years, the additional number of cycles for diesel trains will be 639,679 (see Table 3). The new fatigue life was calculated as follows:

$$\frac{639,679 + 264,369 + n_2}{2,000,000} = 1$$

$$n_2 = 1,095,954 \text{ cycles}$$

Where n_2 is the number of remaining cycles after the year 2028. The total fatigue life of the critical members becomes

$$\text{Fatigue life} = 117 + \frac{1,095,954}{6238} = 293 \text{ years}$$

Therefore the difference in fatigue life for the critical members is 28 years.

5 Summary and conclusions

This paper dealt with the analytical aspect of railway bridge safety associated with the proposed transportation of nuclear waste to the Yucca Mountain repository in Nevada. In particular the paper addressed the reduction in fatigue life of the critical members of a typical steel railway bridge due to proposed increase in traffic. The AAR rating computer program number 4 was modified to rate steel truss bridges under any type of special loading directly, without computing the corresponding Cooper E-rating.

A simplified fatigue rating was performed using Miner's rule and fatigue strength category D of the AREA specification. The stresses in five members of the Union Pacific steel bridge M.P. 409.16 exceeded the endurance limit stress value of category D, thus are limited to only 2 million load cycles. The fatigue life for these critical members will be reduced by 28 years as a result of the operation of the nuclear waste repository.

6 References

American Association of State Highway and Transportation Officials (1983) **Standard Specifications For Highway Bridges.** 13th Edition, Washington, D.C.

American Railway Engineering Association (1989) Specifications for Steel Railway Bridges. **Manual for Railway Engineering.** Chapter 15, Washington, D.C.

Association of American Railroads (1988) **Railroad Facts.** Information Public Affairs Department, Washington, D.C.

Associations of American Railroads Technical Center (1975) **Program No. 4: Computer Program for Rating of Railway Truss Bridges**, Chicago, Illinois.

Association of American Railroads (1987) **Bridge Research** Research Report 1986-1987, pp. 54-55.

Ebrahimpour, A. Maragakis, E. A. and O'Connor, D. N. (1989) Survey and Evaluation of Nevada's Transportation Infrastructure, Task 7.4: Railway Bridges, **Report for the State of Nevada**, Nuclear Waste Project Office, Carson City, Nevada.

Fisher, J. W. (1984) **Fatigue and Fracture in Steel Bridges.** John Wiley and Sons.

Fisher, J. W. Yen, B. T. and Wang D. (1987) **Fatigue and Fracture Evaluation For Rating Riveted Bridges.** NCHRP Report 302, Transportation Research Board, National Research Council, Washington D.C.

Foutch, D. A. (1988) A Summary of Railway Bridge Research Needs, **Bridge Research in Progress, Proceedings**, Des Moines, Iowa, pp. 45-52.

Salmon C. G., and Johnson, J. E. (1980) **Steel Structures, Design and Behavior.** Second Edition, Harper and Row, New York, N.Y.

U.S. Department of Energy (1988) **Section 175 Report: Secretary of Energy's Report to the Congress pursuant to Section 175 of the Nuclear Waste Policy Act, As Amended.** Office of Civilian Radioactive Waste Management, Washington, D.C.

PART SEVEN

MASONRY ARCH BRIDGES

ANALYSIS OF MASONRY ARCH BRIDGES BY A FINITE ELEMENT METHOD

B.S. CHOO, M.G. COUTIE, N.G. GONG
Dept Civil Engineering, University of Nottingham, Nottingham, UK

Abstract
The behaviour of a masonry arch bridge under loading is analysed in this paper using a non-linear straight tapered beam element computer program. Collapse solutions are obtained resulting from the 'effective arch ring' that excludes cracking and yielding portions in the arch ring. The load distribution on the arch ring and the lateral passive fill pressure are discussed. Comparisons between the numerical results and the results of load tests show reasonable agreement. The effects of different arch shapes and the properties of the materials are assessed.
Keywords: Masonry Arch Bridges, Finite Element Method, Tapered Beam Element.

1 Introduction

The arch bridge is one of the oldest forms of bridge in the world. It has proved to be durable and reliable with limited maintenance. There are presently over 30,000 masonry arches in service with British Railways alone. Most of those were built in the last century. However, over the years, these bridges deteriorate, resulting in some structural damage (for example ring separation, loss of some bricks and cracking). It is difficult and unnecessary to replace all the masonry arch bridges on the railway system with modern bridges. But the question is - can these old bridges still carry heavier and heavier modern traffic? To answer this, it would be necessary to assess the actual strength of the arch bridge in its present deteriorated condition. The mechanism and 'MEXE' (Military Engineering Experimental Establishment) methods are in common use in the UK, but their usefulness is limited. With the development of computer based numerical methods, the finite element method has been used to simulate the complex behaviour of masonry arch bridges[1,2,3]. The arch ring has been represented by a string of short beam elements.The fill material above the arch is treated simply as dead load on the arch ring.

The one dimensional finite element program developed by the authors is fully described in Ref. 3, this gives details of the method used and characteristics of the tapered beam element employed. In this paper, the effects of varying the material properties of arch and fill material, and load distribution assumptions are discussed. Their influence on collapse loads is examined. However, the method is described briefly as follows.

2   Analysis method

2.1   Tapered beam element and effective arch ring

In this paper, a finite element method based on the straight tapered beam element, of unit thickness, is used. The arch ring is simulated by a string of tapered beams. As nodal cracks develop, the tension zone of each element is assumed to be ineffective and is neglected. In addition, when the compressive stress reaches the crushing (ie yielding) value of the material, it also results in ineffective zones. The rest of the structure, the so called 'effective arch ring', is the uncracked and uncrushed portion of the arch ring. The stiffness of an element is therefore dependent on only the effective region in compression as shown in Fig. 1.

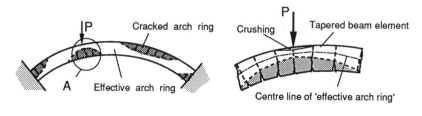

Fig. 1. Tapered beam elements and the effective arch ring.

2.2   Solution procedure

The arch ring is divided into a series of elements, each corresponds to several bricks, or stone blocks, typically 50 elements in total. When the arch ring is under the jth load increment $\{\Delta P\}_j$, the stiffness of the arch ring, $[K(d)]$, which is a function of the effective depth d, is reduced because of the development of cross section cracks. The depths can be calculated at each nodal cross section by satisfying equilibrium for the axial forces and the moments respectively. Therefore, the solution procedure at each load increment is to seek a set of global displacements $\{\delta\}_i$ which satisfies both equilibrium conditions at all nodal points. At each load increment, an iteration is performed to take account of the changing depth d and the changing ring geometry.

The iterative procedure used in this program may be summarized as,

$$[K(d)]_i \{\delta\}_i = \{\Delta P\}_j \qquad (1)$$

$$\{d\}_{i+1} = S \{\delta\}_i \qquad (2)$$

where S is an operator which calculates the depths of the effective arch ring.

The convergence criterion is based on the changes of the effective depth. This iterative procedure continues until the convergence criterion

$$\|\{d\}_{i+1} - \{d\}_i\| \leq \varepsilon \qquad (3)$$

is satisfied, where ε is a small positive number, say 0.1 (mm).

The iteration procedure adopted converges quickly. For example, for an arch ring of depth 711 mm, the convergence tolerance ε was specified as 0.1 mm and convergence was attained with only 12 iterations for the first load increment. Usually, the number of iterations is inversely proportional to the number of load increments.

2.3 Load on the arch ring

The load on the bridge is increased incrementally. Initially, the magnitudes of the load increments are kept fairly constant. However, as the total load P on the bridge approaches its collapse value, the magnitude of the load increment ΔP is reduced automatically using a procedure which has been included in the program. This is based on the desirability of limiting the consequent increases of displacement. Only the first load increment need be specified.

The total loading on an arch bridge comprises its self weight and imposed loading. The self weight includes the weight of the arch ring and the weight of the fill above the arch. The magnitude of the self weight can be calculated from the geometry of the bridge. The imposed loading is applied above the fill and it is not easy to determine its exact distribution on to the arch ring through the fill. In this paper the spread of the imposed loading is limited by the angle θ which may vary from 0° to 45°, that is, the imposed loading on the arch ring is confined to the arc AB in Fig. 2. The distribution of load on the arc AB is assumed to be the vertical pressure σy due to the patch load on the edge of a semi-infinite elastic plane of unit thickness, representing the fill and the arch. For a uniform patch load p per unit area applied on the fill as shown in Fig. 2, the vertical pressure on the arc is [5]

$$\sigma_y = -\frac{p}{\pi}[\beta_2 - \beta_1 + \frac{1}{2}(\sin 2\beta_2 - \sin 2\beta_1)] \tag{4}$$

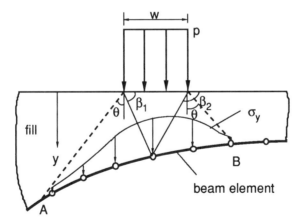

Fig. 2. Distribution of vertical fill pressure on the arch ring.

As the load width w in Fig. 2 approaches zero the angle β_1 tends towards the angle β_2. In the limit of a point load P, the vertical pressure on the arc AB is given by [4]

$$\sigma_y = -\frac{2P}{\pi y} \sin^4 \beta_1 \qquad (5)$$

In addition, the load distribution on the arch ring may be specified directly rather than using the above method. It has been shown[9] that the collapse load is not sensitive to the way load is distributed on to the arc AB, provided the loaded arc length is constant.

However, higher collapse loads may be obtained as the angle θ increases, especially in arch bridges of small span/rise ratio. In such cases, the arc length AB should be reduced to allow for sliding which may occur near the springing along the interface of the arch ring and fill material. Thus all the load is applied on to the arch ring and is not dispersed into the abutment through the fill. For example, in a semi-circular arch bridge, the loaded arc length obtained by this method is similar to that obtained experimentally by Smith[6] as shown in Fig. 3.

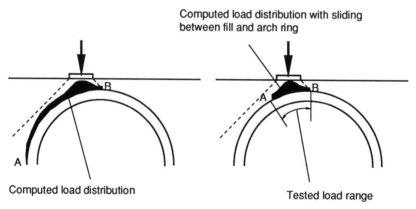

Computed load distribution with sliding between fill and arch ring

Computed load distribution

Tested load range

Fig. 3. Range of the load distribution on the arch ring.

2.4 Fill element

The effect of lateral soil pressures on the collapse load is significant in masonry arch bridges, especially for small span/rise ratios. Due to the deformation of the arch ring, passive lateral fill pressure provides some resistance to the arch ring against collapse. The ratio of lateral to vertical pressure in the passive state is termed the coefficient of lateral passive fill pressure K_p. To simulate the lateral resistance of the fill, one dimensional fill elements as shown in Fig. 4(a) are used. These elements come into effect only when the arch ring moves horizontally into the fill. The stress/strain relationship of the fill element is bi-linear as shown in Fig. 4(b). An elastic linear relationship is maintained until the

stress in the fill reaches the failure value $K_p\sigma_y$. However, the contribution of the fill element is neglected in the region of the crown due to sliding failure along the interface of the fill and the arch ring resulting from the effects of passive pressure.

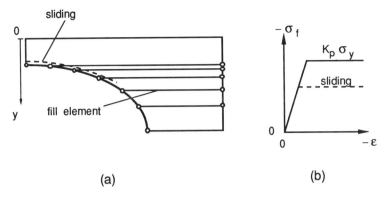

Fig. 4. Fill element and its stress/strain relationship.

2.5 Movement of arch ring centre line

Load tests on masonry bridges [6,7,8] have shown that both the maximum vertical and horizontal displacements are unlikely to exceed the half of the arch ring depth. At each load increment, the deformation of the arch ring is so small compared with the dimensions of the arch ring that a linear calculation could be carried out. Nevertheless, the co-ordinate of the arch centre line is modified at each load increment as shown in Fig. 5(a). This change in the arch ring centre line is in addition to that resulting from cracked zones and yielded portions of the arch as shown in Fig. 5(b). The modified position based on the previous load increment is used as the datum for the computations of the next load increment.

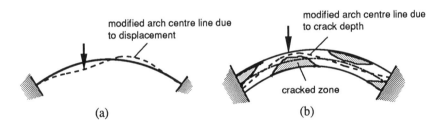

Fig. 5. Changing of arch ring geometry.

3 Numerical results

The program was used to assess the behaviour of a model masonry arch and redundant bridges which were loaded to collapse, namely, a) Towler's model, b) Bridgemill bridge, a shallow parabolic arch bridge, and c) Bargower bridge, which was semi-circular in shape. Their dimensions and loadings are shown in Fig. 6

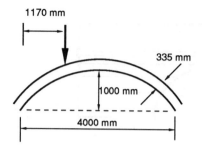

(a) Towler's test model

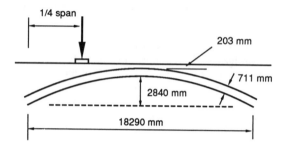

(b) Bridgemill bridge

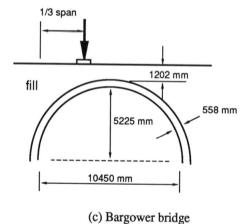

(c) Bargower bridge

Fig. 6. Dimensions and loadings.

Initially, comparison is made with the test results obtained by Towler as this avoids the effects of fill material. The model arch was a 3-course brick arch ring of 4 metre span and 1 metre rise. A line load across the bridge width was applied at approximately one-third point of the span. The load due to fill was simulated by weights applied to small concrete steps which were cast in-situ on to the arch ring. The experimental E value of the brickwork is 9,400 N/mm^2 and the compressive strength is 16 N/mm^2. It should be noted that an artificially lower E value of 3,000 N/mm^2 was adopted by Towler and Crisfield for their finite element analyses because of the softer experimental load deformation characteristics. A comparison of these results is shown in Fig. 7. The test collapse load was 117 kN and the computed collapse loads by the present method are 120.3 kN and 119.5 kN using E values of 9,400 N/mm^2 and 3,000 N/mm^2 respectively.

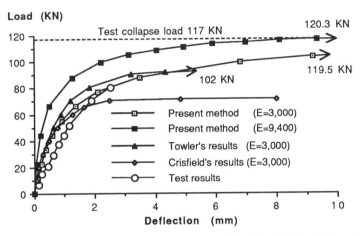

Fig. 7. Load/deflection relationship under loading for Towler's model arch,
E values: (N/mm^2).

Bridgemill bridge was a parabolic stone voussoir arch bridge. It was loaded to failure (3,000kN), using a line load at a quarter point of the span. The bridge was built one hundred years ago in red sandstone. Specimens from the arch ring tested in the laboratory indicated that the elastic modulus and compressive strength of red sandstone were 15,000 N/mm^2 and 43.8 N/mm^2. However, the elastic modulus and compressive strength for the combined stone and mortar material were estimated as 5,000 N/mm^2 and 5 to 8 N/mm^2 respectively.

Results by the present method for E values ranging between 5,000 N/mm^2 and 15,000 N/mm^2, and for compressive strength σ_c values of 5 N/mm^2 and upwards are shown in Fig. 8 and 9 respectively. For the fill material, the coefficient of the lateral pressure K_p and E value were assumed to be 3 and 100 N/mm^2 respectively . It is shown in Fig. 8 that the magnitude of elastic modulus of the arch ring significantly affects the predicted initial stiffness but not the collapse load. In Fig. 9, it can be seen that the load-deflection curve is sensitive to the compressive strength of the arch material. Clearly, using the compressive strength of the masonry 43.8 N/mm^2 provides an upper bound solution whilst the estimated value of 7 N/mm² for the composite arch material (mortar and stone) provides a more realistic collapse load.

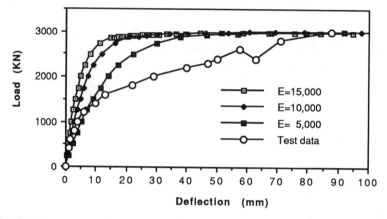

Fig. 8. Effect of the E-value (N/mm²) for Bridgemill Bridge (σ_c = 7 N/mm²).

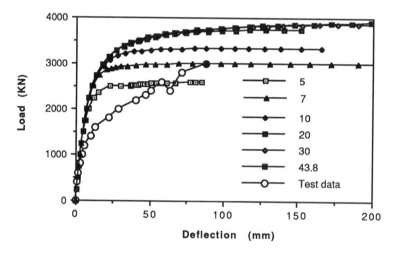

Fig. 9. Effect of σ_c(N/mm²) for Bridgemill bridge (E = 10,000 N/mm²).

Bargower bridge was a semi-circular stone voussoir arch bridge with a 1.2m thick fill at the crown. The collapse load for the structure was 5,600kN. Specimens from the arch ring tested in the laboratory indicated that the elastic modulus and compressive strength were 14,100 N/mm² and 33.3 N/mm² respectively. The analytical load/deflection curves obtained for Bargower bridge are shown in Figs. 10 and 11.It can be seen that they are not sensitive to variation in E values and are generally closer to the test curve than those obtained for Bridgemill Bridge. However, the initial predicted stiffness is softer than that obtained experimentally. This may be because the initial deformation due to the imposed load takes place in the relatively softer fill material rather than the arch ring.

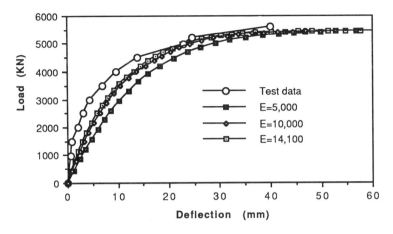

Fig. 10.   Effect of E (N/mm^2) for Bargower bridge (σ_c=5 N/mm^2).

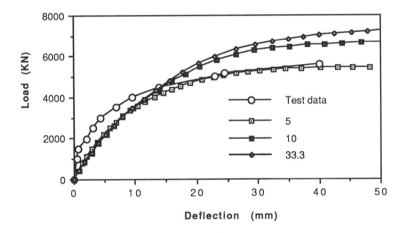

Fig. 11. Effect of σ_c (N/mm^2) for Bargower bridge (E=10,000 N/mm^2).

The effect of the coefficient of lateral fill pressure K_p for both Bridgemill (E = 10,000 N/mm^2, σ_c = 7 N/mm^2) and Bargower bridges (E = 10,000 N/mm^2, σ_c = 5 N/mm^2) can be seen in Fig. 12. For Bridgemill bridge, a shallow arch bridge, the influence of K_p on the ratio of predicted to test collapse load is insignificant for values of K_p beyond 2. The load ratio varies by only 10% for a wide range of K_p. However, for the semi-circular arch Bargower bridge, the effect of K_p is very significant.

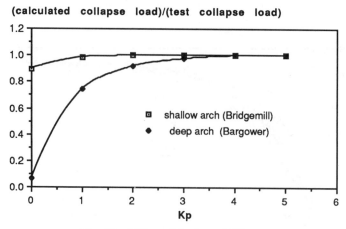

Fig. 12. Effect of the lateral fill pressure.

The effect of the position of imposed load as a ratio of the span is presented in Fig. 13. For a shallow arch (Bridgemill), the lowest collapse load is obtained when the load is applied around 0.2 and 0.3 of the span. For a semi-circular arch (Bargower), the worst position for applying a line load is at about 0.3 of the span. Results for load positions less than 0.3 of the span are considered to be unreliable because it is difficult to estimate the proportion of the imposed load which is dispersed into the abutment. For Towler's model test, the lowest collapse loads are obtained for load positions which are less than the third point of the span.

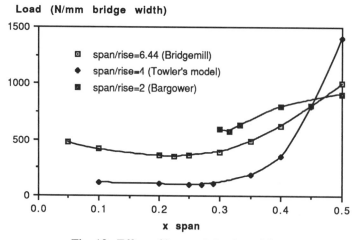

Fig. 13. Effect of imposed load positions.

The development of four cracks in the critical zones of Bridgemill arch during analysis are shown in Fig. 14. Due to only the dead load, cracks 1 and 4 of depth of 328 mm, developed at the springings. As the imposed load is applied, crack 4 which is at the extrados of the arch ring closed gradually until the load reached approximately half the magnitude of the collapse load. The crack started to open up again at the intrados when the imposed load exceeded about 1,300 kN . It is clear that a four hinge mechanism was formed when the arch approached its load carrying capacity. Note that cracks 1 and 2 closed a little at higher load levels. This was because of load redistribution due to yielding of the arch material in the uncracked compression region.

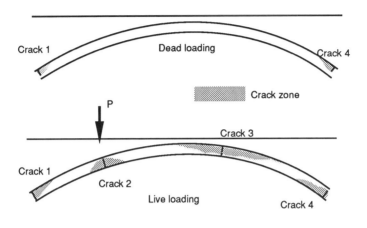

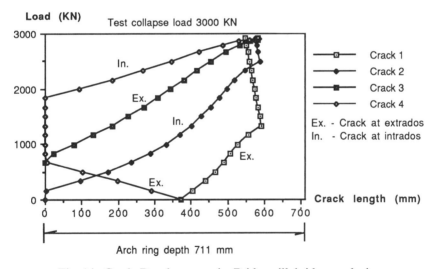

Fig. 14. Crack Development in Bridgemill bridge analysis.

4 Conclusions

The finite element program based on the tapered beam element can be used to obtain reasonably accurate load-deformation and cracking characteristics of brick and stone arch bridges. It has been shown that the elastic modulus value of the arch material does not affect the ultimate collapse load predicted by the program even though it can significantly affect the elastic stiffness of the structure. The compressive strength of the arch material is an important parameter in the computation. It has been shown that taking the actual compressive strength of the bricks or stones and neglecting the mortar strength results in the highest predicted collapse load. However, more realistic collapse loads may be obtained if a lower level of compressive strength which more closely represents the brick/mortar composite is used.

The effect of the coefficient of lateral fill pressure K_p is much more significant for deep arch bridges than for shallow arch bridges, particularly for low values of K_p. This is only to be expected as the mobilisation of the lateral fill pressures is dependent on the horizontal deformations of the arch ring. Using values of K_p greater than 2 will result in predicted collapse loads which are within 10% of the upper bound value for this parameter.

Load tests on arch bridges are frequently conducted by applying line loads at either the quarter or one third points of the span. This analysis shows that the load position for achieving the lowest collapse load for the arch bridge is between the 0.2 to 0.35 points of span, depending on its shape. Applying the line load at the crown position results in a relatively higher collapse load.

A four hinge mechanism is required for collapse of an arch. A distribution of cracks in the four positions corresponding to the four hinge mechanism can be obtained using this program and the development of any of these cracks in the arch ring with increases in applied load can also be traced.

5 Acknowledgment

The work described in this paper is supported by the British Railways Board as part of a larger study into the behaviour and deterioration of masonry arches. The authors are grateful to Mr C Lemmon of British Rail Research for his help.

6 References

[1]   Towler, K. and Sawko, F. Limit state behaviour of brickwork arches, 6th Int. brick masonry conference, Rome, May 1982

[2]   Crisfield, M.A., A finite element computer program for the analysis of masonry arches, TRRL laboratory report 1115

[3]   Choo, B.S., Coutie, M.G. and Gong, N.G., Finite element analysis of masonry arch bridges using tapered elements, to be published

[4]   Timoshenko, S. and Goodier, J.N., Theory of elasticity, Second edition, McGraw-Hill Book Company, Inc., New York, 1951

[5]   Flugge, W., Handbook of engineering mechanics, McGraw-Hill Book Company, Inc., New York, 1962

[6]   Smith, F.W. and Harvey, W.J., Full-scale test of a masonry arch, Proc. of SERC-R.M.O. conference, June, 1989.

[7]   Hendry, A.W., Davies, S.R. and Royles, R., Test on stone masonry arch at Bridgemill-Girvan, TRRL contractor report 7, 1985

[8]   Hendry, A.W., Davies, S.R. and Royles R., Load test to collapse on a masonry arch bridge at Bargower, Strathclyde, TRRL contractor report 26, 1986

[9]   Coutie, M.G., Choo, B.S. and Gong, N.G., One dimensional finite element analysis of masonry arch bridges, Contract Report 1, NUCE/ST/22, University of Nottingham, 1989

THE BEHAVIOUR OF MASONRY ARCH BRIDGES – THE EFFECTS OF DEFECTS

C. MELBOURNE
Dept Civil Engineering and Building, Bolton Institute of
Technology, Bolton, England, UK

Abstract
There are in excess of 40 000 arch bridges in the UK, many of which were
constructed using brickwork barrels. All carry loads well in excess of those
envisaged by their designers. There is currently a nationwide research
programme aimed at enhancing our understanding of the behaviour of these
complex structures with a view to improving our assessment methods.

The paper discusses the general areas of uncertainty with regard to
establishing a theoretical physical model which can be used to assess the load
carrying capacity of masonry arch bridges.

An integral part of any assessment is an appraisal of the significance of
defects which exist within the structure. Two such problems are spandrel wall
separation and ring separation. The former occurs when the spandrel walls
separate from and cease to give direct support to the arch barrel. The latter
occurs in multi-ring brickwork arches when separation between successive
brick rings disrupts structural homogeneity of the arch barrel.

The paper describes tests that have been conducted in the Institute's Large
Scale Testing Facility on both model and large scale segmental brickwork
arches. The results are discussed with regard to understanding the behaviour
of arch bridges in general and the significance of defects in particular.
Comparison with a proposed theoretical physical model is made and
conclusions are drawn regarding the model's validity and limitations.
Keywords: Masonry, Arch Bridges, Defects.

1 Introduction

The principles of arch construction were understood by the Mesopotamian and
Egyptian builders some five thousand years ago[1]; but the arch was probably
only used in buildings. It was likely that the Chinese first employed the arch
in bridge construction; the earliest being constructed about 2900BC.

The Romans were the first to make wide use of the arch in Europe. The
earliest examples date back to about 600BC and were in sewer construction.
By 100BC, bridges of 15m to 21m span were being built using stone.

Some arch bridges were built in Britain during the Middle Ages; but the great era of arch bridge building began with the construction of the canals in the second half of the eighteenth century and ended when the railway network was substantially complete at the beginning of the twentieth century. Very few masonry arch bridges have been built since the First World War. There are in excess of 40 000 masonry arch bridges in the UK, many are over 120 years old and all carry loads well in excess of those envisaged by their designers.

Design of the ancient arches was probably based upon empirical rules relating proportion and shape. It was not until the work of La Hire[2] and Couplet[3] in the eighteenth century that theories to explain arch behaviour were propounded. The work of Couplet was particularly significant and is essentially the same as the mechanism method which has re-emerged from the twentieth century work of Pippard[4] and Heyman[5]. This followed a century of preoccupation with elastic methods of analysis. Recently, both approaches have been refined using computers[6-9] but they all disregard the three dimensional nature of the problem.

A physical theory for the behaviour of masonry arch bridges must describe adequately all the relevant aspects of the mechanical behaviour and it must be arranged in such a way that real engineering problems may be solved. The two main materials found in such bridges are masonry and soil; since both materials do not behave in a simple way, it is usually fruitless to attempt to obtain a physical theory that is, at the same time, simple and exact. An attainable goal, therefore, is the development of a physical theory which enables engineering calculations which are reasonably straightforward and the results sufficiently accurate, for the particular purpose of a bridge assessment.

An integral part of any assessment is an appraisal of the significance of defects which exist within the structure. Common problems associated with the arch barrel relate to deterioration of the masonry and mortar, longitudinal cracking and spandrel wall separation, transverse and diagonal cracking, ring separation, transverse and diagonal cracking, ring separation in multi-ring brick arches and arch distortion. The causes of these defects primarily relate to environmental and changing loading conditions. The purpose of this paper is to consider two types of defect : ring separation in multi-ring brick arches and spandrel wall separation. Both types of defect have been identified by the industry as being commonplace and significant, although to date, engineers have only been able to take account of them in the form of a subjective condition factor.

Ring separation occurs in multi-ring brick arches and is associated with the loss of bond between successive rings caused by weathering and/or stress cycling of the mortar.

2 Model Tests

The models were of a parabolic profile with a span of 1000mm, span/rise ratio of 3:1 and a width of 500mm. The arch ring comprised two rings or courses of brickwork either bonded or unbonded around the full or part arc of the arch. In total the ring thickness was approximately 100mm. The models were constructed using half scale bricks (Fletton or Raewell). The arches were constructed without spandrel walls.

Because there were no spandrel walls, the fill was contained by the perspex and wooden sides of the test rig. Prior to laying the bed face, each brick was coated with oil in order to minimise the effects of the bond strength on the model behaviour. A 1:1:6 (cement, lime, sand) mortar was used; it was mixed by volume and achieved an average 28 day compressive strength of 4 N/mm^2. Sand was used as backfill and was compacted by vibration to a depth 100mm over the crown.

A knife edge load (KEL) was applied incrementally and monotonically upto failure at either the ¼ or crown point.

Table 1 presents the results of the tests. Arches 1 - 4 were built such that the two rings of brickwork were fully bonded together using a mortar (by bonding it is meant adhesion rather than brick bonding using "headers").

Arches 5 - 8 were built with the mortar between the rings being replaced by damp sand to simulate loss of adhesion.

Arch 9 was built with ring separation over the central half of the span.

All of the models which were loaded at the ¼ span failed due to the formation of a four hinge mechanism.

For those models loaded at the crown, failure was due to the development of a "classical" five hinge mechanism. At failure in the "bonded" arches no or little ring separation occurred and formation of the hinges was as a monolithic "single" ring. In the "unbonded" arches separation increased with load as measured by embedment strain gauges. Development of the hinges in the models with initial ring separation was such that two thrust lines developed, one in each of the inner and outer rings. Hinges in the two rings were coincident and developed simultaneously. Collapse was sudden and caused by almost total physical separation of the two rings and full development of the hinges.

Table 1   Model Test Results

Arch No	Inter-ring Bedjoint Material	Brick Type	Loading position	Experimental Ultimate load (kN)
1	mortar	Fletton	¼	3.2
2	mortar	Fletton	¼	18.6
3	mortar	Fletton	Crown	31.0
4	mortar	Raewell	¼	16.5
5	sand	Fletton	¼	6.9
6	sand	Fletton	¼	8.1
7	sand	Fletton	Crown	8.8
8	sand	Raewell	¼	6.8
9	mortar/sand	Raewell	¼	9.8

3 3 Metre Span Bridge Tests

Two 3 metre span bridges were built and loaded to failure in the Bolton Institute's Large Scale Testing facility. The segmental arch barrel (radius 1875mm) had a span to rise ratio of 4:1 and consisted of two rings of brickwork using Class A solid engineering bricks. The brickwork was built in a "stretcher" bond with no bonding between the rings other than through the mortar in the "bonded" case and the damp sand in the "ring separation" case. The spandrel, wing and retaining walls were built in English bond using concrete commons. The spandrel walls were not attached to the arch ring. An average gap of 10mm was provided between the spandrel walls and the arch ring. This ensured that the effects of both ring separation and spandrel wall separation could be studied.

Compaction of the "graded" 50mm limestone backfill was achieved using 100mm layers and a vibrating compacting "wacker" plate. The bridges were filled to 300mm above the crown.

Both bridges were subjected to three loading conditions. Firstly, a 25.7kN KEL was applied at 250mm centres across the span to simulate a rolling load. Secondly, a 50kN KEL was incrementally applied at the north quarter point, crown and south third point. Finally, a KEL was applied incrementally at the quarter point through to collapse. The elastic tests confirmed that the structure responded to the loading as a local effect, with soil pressure and brickwork strain changes being confined to the vicinity of the loading. This has been observed in field studies[10] and other fullscale tests[11].

Both arches failed by the formation of four hinge mechanisms. In each case the spandrel walls cracked and rotated about the abutment remote from the KEL, figure 1. The sequence of hinge formation is given in Table 2.

Fig. 1 3m span arch barrel - ring separation

Table 2 Sequence of hinge formation

		Load, kN	
Hinge No.	Position	Bonded Arch	Unbonded Arch
1	Under load point, north quarter point	240	220
2	South quarter point	300	240
3	South abutment	400	320
4	North abutment	480 (540 Max.)	320 (360 Max.)

In the bonded arch some ring separation occurred at the crown but the hinges formed at intrados and extrados. On the other hand, the unbonded arch produced extensive ring separation shortly after the formation of the second hinge. It is significant to note that the first hinge formed at approximately the same load in each test. As no ring separation cracking had occurred at this stage, it confirmed that the two arches were comparable. The unbonded arch deteriorated more rapidly after the formation of the first hinge and carried an ultimate load of 360kN - a 33% reduction in carrying capacity compared with the bonded arch.

As with the model tests, once ring separation occurred each ring formed its own pattern of hinges which interacted with each other.

An assessment of the bridge using the presently accepted 'MEXE' method[12,13] gave an equivalent KEL of 200kN. This represents a load which is 85% of that required to cause the formation of the first hinge.

The soil/structure interaction was monitored using pressure cells not only in the extrados but also in the spandrel walls and the backfill. The overall picture which emerged was one of a compacted backfill exerting pressure greater than earth pressures at rest and which restrained the arch and dispersed the applied KEL. Additionally, there was a frictional/cohesive resistance between the backfill and the spandrel walls and the extrados which restrained the arch initially and which increased as the arch swayed into the backfill as hinges formed. This movement into the backfill not only increased the longitudinal horizontal soil pressure but also the stress on the back of the spandrel walls and hence the resistance to movement.

4   A Theoretical Physical Model

To date the theoretical modelling of arch bridges has only considered a 2-dimensional idealisation of the arch barrel interacting with the backfill[6-9]. Over sophistication of this model is misplaced considering the 3-dimensional nature of the problem. It would be better to develop a cruder but more realistic 3-dimensional theoretical physical model which allowed an engineering assessment of the relative significance of each of the parameters. Figure 2 shows a suggested theoretical physical model. The pre-requisite for any discussion is to recognise that the arch will always respond in the most efficient way to resist the applied loading. Generally, compaction of the backfill induces soil pressures which are in excess of those calculated using earth pressure at rest coefficients[14]. With time there will be some relaxation of these pressures at depth; however, this level of "pre-compression" can be expected in a bridge in good condition at shallow depths due to the constant compaction by the traffic. The resultant longitudinal soil stress will act upon the extrados.

Additionally, similar pressures will exist on the back of the spandrel walls and will provide a restraint to longitudinal movement and thus enhance the potential restraint offered by the spandrel wall and soil mass. As the load is applied and the arch sways, it will try to lift the spandrel wall. The lower bound restraint offered by the wall will be due to its welf weight; appropriate allowance may be made for the effects of pilasters and wing walls. It is important to consider the magnitude of each of these effects to determine which of them is critical. The width of the bridge will influence this.

Under load the bridge will respond if the pressure is greater than the "pre-compressed" stress levels in the backfill. Above these levels the fill is further compressed resulting in an increase in the longitudinal and transverse soil stresses. The first hinge forms under the load. The load at which this occurs is dependent upon the "effective" arch which supports it. This in turn will depend upon the extent of the local restraint the spandrel wall and backfill give the arch. Formation of further hinges and the onset of mechanism behaviour depends upon the general resistance of the spandrel wall and backfill. If spandrel wall separation is present then only the friction on the back of the walls or the self weight effects whichever is the lesser should be used in addition to the main self weight and longitudinal soil pressure restraints.

Using the above theoretical physical model an ultimate load of 470kN was calculated for the 3m span arches which compared well with the bonded arch test, Table 2.

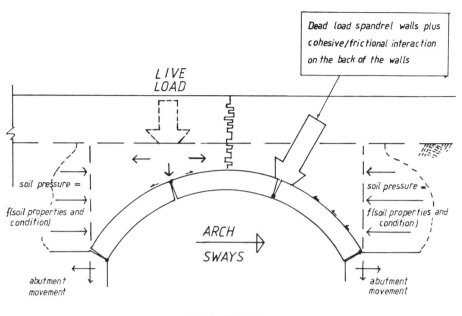

LIVE
LOAD

Dead load spandrel walls plus cohesive/frictional interaction on the back of the walls

soil pressure =

f(soil properties and condition)

ARCH

SWAYS

soil pressure =

f(soil properties and condition)

abutment
movement

abutment
movement

ELEVATION

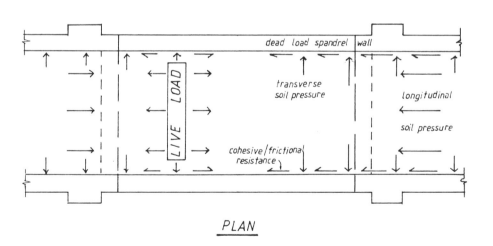

dead load spandrel wall

LIVE LOAD

transverse soil pressure

longitudinal

soil pressure

cohesive/frictional resistance

PLAN

Figure 2 Theoretical Physical Model

5 Conclusions

1) If free to do so, an arch bridge will fail due to the formation of a 4 hinge mechanism.

2) Using a modified mechanism analysis, incorporating the longitudinal backfill pressures, spandrel wall stiffening and the backfill cohesion/friction structural interaction with the spandrel wall and arch, the onset of mechanism behaviour and the collapse load can be predicted. The prediction of the onset of mechanism behaviour could be used to set a serviceability limit state.

3) At all stages of loading, consideration must be given to the possibility of local failure (eg punching shear, ring separation, snap-through, spandrel wall instability) and adequate factors of safety applied.

4) The presence of total or partial ring separation caused a reduction in the ultimate load carry capacity of up to 56% for the models and 33% for the 3m span bridges

5) Passive soil pressures were not observed in any of the tests even at gross deformation

6) Where spandrel wall separation existed the cohesion/friction resistance on the back of the wall made a significant contribution to the stiffening of the arch.

6 References

1 Van Beck, G.W. "arches and Vaults in the Ancient Near East" Scientific American July 1987
2 La Hire, P. de "Sur la construction des voutes dans le edifices" Memoires de L'Academie Royale des Sciences, 1712.
3 Couplet "De la poussee des voutes" Histoire de L'Academie Royale des Science, 1730.
4 Pippard, A.J.S. and Baker, J.F. "The Analysis of Engineering Structures, London, 1962, Arnold.
5 Heyman, J. "The Masonry Arch" First Edition London 1982 Ellis Horwood.
6 Crisfield, M.A. and Packham, A.J. "A Mechanism Program for Computing the Strength of Masonry Arches" Transport and Road Research laboratory Research Report 124, 1987.
7 Harvey, W.J. "Application of the Mechanism Analysis to Masonry Arches "The Structural Engineer Vol.66 No.5/1 March 1988.

8 Hughes, T.G. and Vilnay, O. "The Analysis of Masonry Arches" Proc. of the 8th IBMAC, Dublin 1988.

9 Coutie, M. and Choo "Assessment of the load carrying capacity of Masonry Arch Bridges using Tapered Finite Elements" To be published Proc. ICE 1900.

10 Melbourne, C. "The Construction of Mass Concrete Arch Bridges" Structural Faults and Repairs Conf., London June 1989.

11 Melbourne, C. and Walker, P.J. "Load Test to Collapse of a Full Scale Brickwork Masonry Arch Bridge" TRRL contractors Report to be published 1990.

12 "The Assessment of Highway Bridges and Structures" Department of Transport Roads and Local Transport Directorate, Departmental Standard BD 21/84, Department of Transport March 1984.

13 "The Assessment of Highway Bridges and Structures" Department of Transport Roads and Local Transport Directorate, Advice Note BA 16/84, Department of Transport, March '84.

14 Clayton, C.R.I. and Mililitsky, J. "Earth Pressure and Earth Retaining Structures" Surrey Univ. Press 1987.

7 Acknowledgements

The Author wishes to acknowledge the financial support of SERC, British Rail, TRRL and NAB and also the encouragement and support from the staff of the School of Civil Engineering and Building at Bolton Institute of Higher Education.

PART EIGHT
STRUCTURAL ANALYSIS OF BRIDGES

THE TAY BRIDGE DISASTER – A STUDY IN STRUCTURAL PATHOLOGY

T.J. MARTIN
British Steel Corporation, Ravenscraig, Scotland, UK
I.A. MACLEOD
Dept Civil Engineering, University of Strathclyde, Glasgow, Scotland, UK

Abstract
This paper presents new information about the 1879 collapse of the Tay Rail Bridge based on modern wind loading and analysis techniques. The likelihood of near simultaneous toppling and bracing failure is demonstrated and the effect of the presence of the train on the bridge is discussed.
Keywords: Bridge, Tay Bridge Disaster, Structural Engineering, Collapse, Wind Loading, Cast Iron, Wrought Iron, Structural Analysis, Holding Down Bolts, Bracing.

1 Background

The first Tay rail bridge was completed in February 1878 to the design of Thomas Bouch. At that time it was the longest bridge in the world consisting of 85 spans over a distance of almost two miles.

Bouch made his reputation as a railway engineer by building bridges quickly and on a tight budget. Because his structures were built economically the railway companies were always willing to engage his services. His bridges tended to be lattice girders supported on slender cast iron columns which were braced together with wrought iron ties and struts - Figure 1. The contractor who built his biggest bridges described them as 'flesh without muscle', Thomas (1970).

The disaster occurred on the stormy night of 28 December 1879. A train of one engine and six carriages crossing the bridge from south to north was lost when the structure of the navigation span section collapsed. There were no survivors of the 75 people aboard the train.

2 The Bridge

Of the thirteen navigation spans eleven were of 74.7 metres (245 feet) and two were 69.2 metres (227 feet). These were through girders (the 'High Girders') 8.2 metres (27 feet) high with 26.8 metres (88 feet) clearance above high water level. All the navigation spans collapsed.

405

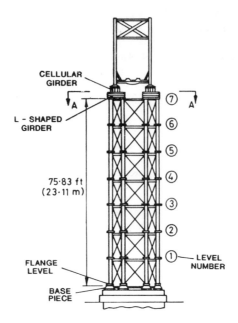

CELLULAR
GIRDER

L - SHAPED
GIRDER

75·83 ft
(23·11 m)

FLANGE
LEVEL

BASE
PIECE

⑦

⑥

⑤

④

③

②

①——LEVEL
NUMBER

Fig.1. Elevation of pier.

The spans were supported on piers consisting of six
crossed braced cast iron columns - Figure 2. Each column
was fabricated in seven sections (corresponding to the
seven tiers of bracing) connected via flanges each with
eight 28.5 mm (1-1/8 inch) bolts. The columns each have a
rake of 305 mm (1.0 feet) over their height of 25.165 m
(76.0 feet). The outer columns (Type A - see Fig 2) were
457 mm (18.0 inches) outside diameter, the other columns
(Type B) being 381 mm (15.0 inches) diameter. The wall
thickness in each case was 25.4 mm (1.0 inches). The
columns were filled with concrete.

The bracing members were connected to pairs of integ-
rally cast lugs on the columns.

The 'vertical' bracing consisted of:

(a) horizontal members of two back to back wrought iron
     channels 6x2-1/2x1/2 inches connected to the lugs
     by two 28.5 mm bolts.
(b) diagonal crossbracing members of 114 x 12 mm
     (4-1/2x1/2 inches) cross section. These were
     connected at one end by a single 28.5 mm bolt and
     at the other to pairs of 114 x 9.5 mm (4-1/2x3/8
     inches) sling plates via a cotter arrangement. The

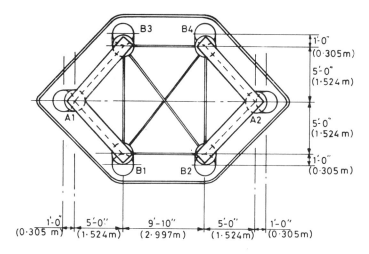

Fig.2. Section AA plan of pier

cotters were driven in on erection to tighten the
braces. Tne sling plates were connected to lugs by
a single 28.5 mm dia bolt.

At the level of each column connection there was a horizontal cross bracing arrangement as shown in Figure 2. The
diagonals at these levels were 38.1 mm (1-1/2 inches) dia
rods, single bolted to lugs on the columns. The ties were
back to back channels.

The columns were fixed at their bases to base pieces
which were in turn fixed to the foundation by four 44.5 mm
(1-3/4 inch) dia bolts which passed through two courses of
stone of the foundation.

A wrought iron girder L-shaped (in plan) transferred the
deck loads to the piers - Figure 2. A wrought iron cellular girder running parallel to the track was placed above
each of the L-shaped girders. Immediately above the cellular girders were the longitudinal lattice girders forming
the sides of the bridge. The cellular girders were placed
equidistant between the outer and two inner columns. This
arrangement led to a quarter of the total girder weight
being borne by the outer columns and an eighth of the
weight being borne by each of the four inner columns. The
main girders were either bolted to the tops of the cellular
girders or supported on rollers. Main girder spans were
connected to each other by cover plates top and bottom
except at expansion joints which were provided every 4th or
5th span.

The foundations for the piers supporting the navigation
spans consisted of wrought iron caissons 9.449 m (31.0

feet) diameter lined with 457 mm (18.0 inches) of brick-work and filled with concrete. Upon the caissons were constructed hexagonal shaped concrete foundation blocks. The upper part was faced with four courses of stonework, the top being 1.524 m (5.0 feet) above spring high tide level.

3 The Analytical Model

Since the main thrust of evidence in the past points to the collapse being caused by inadequate strength of the piers of the bridge, detailed attention was focussed on their behaviour.

A space frame model of a single pier structure was crea-ted. The members were treated as beam elements with ben-ding and axial deformation. No second order effects were considered except that the compression diagonals were removed. The columns were treated as being continuous and all other members were pin-connected. The finite sizes of the columns were neglected. The column bases were treated as being fixed except that a 5 mm uplift was imposed on the windward column in some load cases.

Table 1. Section properties

	Torsion Const. J (Ft^4)	Iyy (Ft^4)	Izz (Ft^4)	Area of section (Ft^2)
18" Column	0.363	0.178	0.178	1.132
15" Column	0.181	0.089	0.089	0.808
Horizontal Struts	4.02E-5	2.33E-3	1.94E-3	0.069
Wind Bracing	9.04E-6	1.83E-4	2.27E-6	0.0156
Horizontal Bracing	2.4E-5	1.2E-5	1.2E-5	0.0123

Table 2. Material properties

	Modulus of Elasticity (Kips/Ft^2)	U.T.S. (Tons/in^2)	U.C.S (Tons/in^2)	Poisson's Ratio
Wrought Iron	4.03E6	20	16.71	.28
Cast Iron	1.15E6	9.1	32.5	.25
Concrete	6.27E5	-	-	.2

SOURCE: Court of Enquiry (1880)

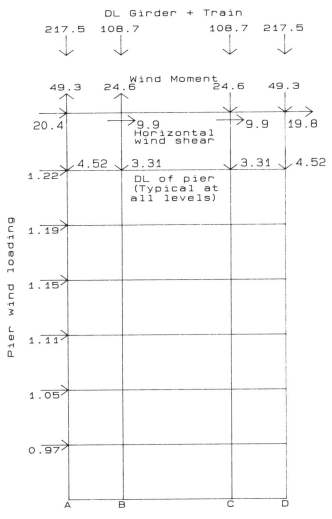

Fig.3. Typical loading (Kips) on a pier with train and wind gust
velocity V = 60.9 Miles/hr.

The member properties used were as shown in Table 1.
Material properties are given in Table 2. Figure 3 shows
the wind, dead and live loads used in a typical run (train
on bridge, wind velocity at deck level 60.9 mph). The
effect of the wind loading from the superstructure above
the top of the model was simulated by a shear and moment at
the top of the model. The moment was treated using pairs
of equal and opposite vertical loads.
The loading was increased in steps noting the condition
of the members.

4 Loading

Wind loading. The calculation of the forces on the structure due to wind was based on CP3 Chapter 5, British Standards Institution (1977). For structures of horizontal
dimension greater than 50 m, this Code specifies the use of
a 15 sec gust period to compensate for the incoherence
effects caused by the turbulence in the air. As the high
girders were 75 m in length, wind loading on them was based
on a 15 sec gust while loading on the pier was based on a 3
sec gust.

For the calculations the train was positioned centrally
over pier 32. The wind loading code does not allow for a
train inside the girder. To allow for this, the drag
coefficient over the presented area of the train was taken
as 2.2 in comparison with 3.1 over the presented area of
the girder members.

The drag coefficient for the pier was estimated using
the CP3 Code for towers. To allow for increase in wind
velocity with height over the pier, the wind loading was
calculated for each tier of the pier.

Train load. Values were obtained from data used by the
Court of Enquiry, (1880).

5 Results

Figure 4 shows a plot of maximum load in a wind bracing
member against gust wind velocity at deck level. Four
basic variations are considered:

    (a)   with the train,
    (b)   without the train,
    (c)   with the windward column fully fixed,
    (d)   with a fixed upward (arbitrary) deformation of 5 mm
           at the windward column.

Also shown on the four curves of Figure 4 are the loads at
which the first column bolt would reach yield and ultimate
load. With fixed windward column the failure is in the
bolts of that column. With 5 mm uplift failure is in the
bolts of the 15 inch columns.

Two potential modes of failure are:

    (a)   Tensile failure of the wind bracing members.
           Assessment of this is based on tests carried out
           for the original enquiry. The test specimens were
           held in lugs taken from the original columns and in
           all but one case the failure was in the cast iron
           lugs. Figure 4 shows three levels of wind bracing
           strength - minimum, average and maximum.
    (b)   Tensile failure of the bolts connecting the column
           sections. These were also tested for the enquiry.

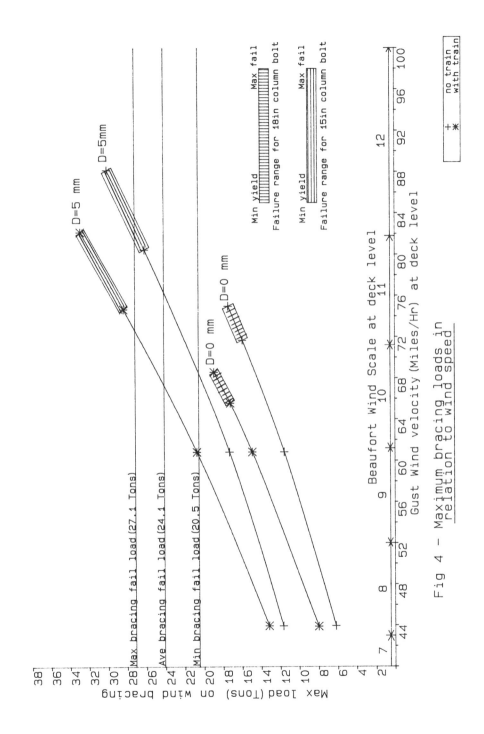

Fig 4 – Maximum bracing loads in relation to wind speed

From the results of these tests a minimum yield
load of 15.2 tons and a maximum failure load of
17.7 tons were used for Figure 4.

From Figure 4 the following conclusions are drawn:

(a) The presence of the train significantly increases
the bracing load for a given wind load. However
the drag coefficients used to take account of the
train are speculative and the difference due to the
train in Figure 4 may not be quite as great as
shown.

(b) The uplift of the windward column significantly
increases the bracing load and also reduces the
effective bolt load. The reason for allowing uplift is that the anchor bolts for the base piece
only penetrated two courses of masonry of the top
foundation. The report of the Enquiry, Court of
Enquiry (1880), section 35, notes:
'... but the joints of the masonry of the hexagonal piers had in almost every case been
severely shaken, and in two instances the two
upper courses of stone on the west side had been
wrenched off and tilted up on end.'

With the train on the bridge and 5 mm uplift, the maximum bracing load is reached at a wind speed of about 73
mph. This is the top end of Force 11 on the Beaufort
scale. Such a situation is realistic.
It seems likely that in a severe gust there would have
been some uplift of the windward column base. How much, we
cannot tell but the model results show that such an effect
increases the probability of bracing failure.
A very interesting feature of the analysis is that the
maximum wind bracing load is not at the base of the piers.
Figure 5 shows the distribution of bracing load in the
bracing planes between columns B1 and B2 (Figure 2) at gust
velocity of 60.9 mph (at deck level). With uplift the load
at the second level is significantly the highest. Conventional understanding would predict the highest load at the
bottom level. In fact, the moments at the bases of the
columns cause a redistribution of bracing load.
There has been speculation that bracing failure did
occur above the base level. For two of the piers (nos 29
and 30, the first two navigation span piers on the south
side), the base level of bracing remained in place.
The Court of Enquiry (1880), paragraph XII, says
'The distance at which the girders were found from the
piers, and the position of the wreckage on the piers,
is such as would result from a fracture and separation
taking place in the piers somewhere above the base of
the columns;'.

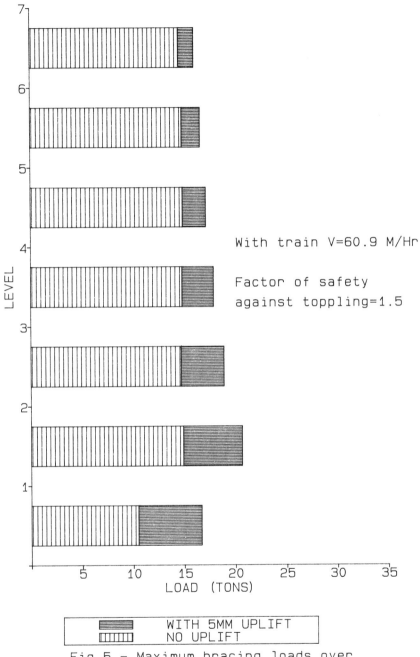

With train V=60.9 M/Hr

Factor of safety
against toppling=1.5

WITH 5MM UPLIFT
NO UPLIFT

Fig 5 - Maximum bracing loads over
height of pier

6 Conclusion

While theories of failure involving the initial derailment
of the train have been proposed, the results of this study
show that the structure of a pier of the bridge could have
failed due to the effect of the wind. The presence of the
train would exacerbate the situation and uplift of the
windward columns would increase the load in the weak
bracing members towards failure.

7 References & Bibliography

Blackburn J. (1980) The Tay Bridge Disaster. **Foundry Trade
J.**, Feb 1980.
British Standards Institution. (1972) **CP3**, Chapter 5, Part
2, Wind Loading.
Buchan A. (1880) The Tay Bridge storm of 28th December
1879. J of Scottish Met Soc, Vol 5, 355-360.
Court of Enquiry (1880) Report upon the circumstances
attending the fall of a portion of the Tay Bridge.
Francis A J. (1980) Introducing Structures. Pergamon
Press, p 126.
Hopkins H J. (1970) **A span of bridges**. David & Charles,
Newton Abbot.
Houghton E L & Carruthers B C. (1976) **Wind forces on buil-
dings and structures**. Edward Arnold.
Nock O S. (1981) **Railway Archaeology**. Patrick Stevens,
Cambridge.
Prebble J. (1979) **The High Girders**. Penguin.
Rolt L T C. (1970) **Victorian Engineering**. Penguin Books.
Sibly P G & Walker A C. (1977) Structural accidents and
their causes. **Proc Instn Civ Engrs**, 62, 191-208.
Simpson C. (1979) The Tay Bridge Disaster. **Engineering**,
Dec.
Smith C S. (1881) Wind pressure on bridges. **Min Proc Instn
Civ Engrs**, 69, 145.
Smith D W. (1976) Bridge Failures. **Proc Instn Civ Engrs**,
60, 367.
Thomas J. (1970) **The Tay Bridge Disaster**. David & Charles,
Newton Abbot.

SENSITIVITY ANALYSIS FOR SHORT-SPAN STEEL BRIDGES

M. MAFI
Civil Engineering Dept, Union College, Schenectady, New York, USA

Abstract
The effects of several variables on the optimum design
configurations of composite and noncomposite steel bridges are
studied. In other words, the sensitivity of design of such bridges
is investigated with respect to the following variables: spacing
between beams, different loading types, ultimate strength of
concrete, yield strength of steel, degree of interaction between
steel and concrete, variations in unit costs and provision of
lateral bracing.
Keywords: Short-span Bridges, Variable Study, Sensitivity Analysis,
Variables in Bridges.

1 Introduction

1.1 Background
The results of a research effort to study the design of composite
and noncomposite steel bridges and to provide engineers with a
simple tool to obtain optimal design of such bridges were presented
by this author in a separate paper, Mafi (1990). The computer
program developed in that study enables an engineer to obtain the
optimum design of composite and noncomposite steel bridges for
different values of preassigned variables. These include, but are
not limited to:

(a) Span of bridge
(b) Width of bridge
(c) The yield stress of steel to be used, usually 36 or 50 ksi
(d) Minimum thickness of slab mandated by certain agencies
(e) The degree of interaction between steel and concrete
(f) Ultimate strength of concrete used
(g) Any limitation on depth of members
(h) Different permit loads

1.2 Scope of this paper
Even though the computer program developed in the aforementioned
study provides the optimum design for any combination of preassigned
variables in five to six seconds, it would still take a substantial
amount of time if one is interested in obtaining the optimum results
for all possible permutations of different values of several
preassigned variables. Upon consideration of the need of some

415

engineers to reduce time spend on the study of design, it was
decided to use this powerful tool to study systematically the
interaction of several important variables in the hope of developing
some rules that could be applied by people who:

(a) Are interested in optimum design, but want to cut down on the
    number of computer runs
(b) Don't have the computing facilities or don't have the time
    to use them
(c) Are content with near optimum designs

In this paper the relationships among some of the more important
variables and their effects on the cost of selected bridges will be
examined. Several graphs and tables will be provided to guide the
designers in their efforts to minimize cost. A VAX 11/785 at Union
College was used to obtain data and a LOTUS 123 program was used to
create the bar graphs with an HP plotter. The influence of the
following variables on the optimum design of composite and
noncomposite short-span steel bridges will be studied here:

(a) Spacing between the beams
(b) Different loading types
(c) Ultimate strength of concrete
(d) Yield strength of steel
(e) The degree of interaction between steel and concrete
(f) Variation in unit costs
(g) Provision of lateral bracing

Spans of up to 100 feet long and different bridge widths were
considered.

2 Influence of spacing between stringers on optimum cost

Given any width, the aforementioned program determines the best
spacing between stringers so that the length of the overhang
portions on both sides of the cross-section remains within the
allowable ranges as specified by AASHTO. Several bridge widths were
considered. For the sake of brevity, only the information for a
bridge width of 47.5 feet will be presented here. The conclusions
drawn here are equally applicable to other widths. The influence of
different spacings between stringers on the optimum cost of bridges
is shown in Fig. 1 and Fig. 2 for noncomposite and composite decks,
respectively. It should be noted that each value of spacing
corresponds to a certain number of stringers. In the legends of
Figs. 1 and 2, each value for spacing between stringers is followed
by the corresponding number of beams in parenthesis. The
information in Fig. 1, which is for noncomposite decks, shows that
the largest spacing among the six shown, i.e., the spacing of 9.75
feet, gives the minimum cost for spans of up to 50 feet. In spans
of up to 50 feet, the spacing of 9.75 feet and 7.92 feet seems to be
equally good. For spans of 60 feet and 70 feet, the spacing of
7.92 feet provides a more economical design.

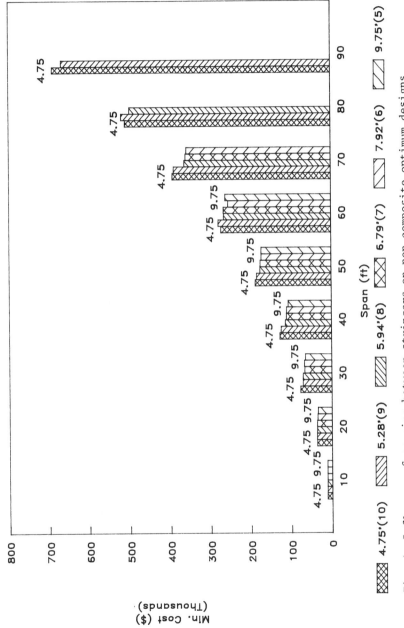

Fig. 1. Influence of spacing between stringers on non-composite optimum designs.

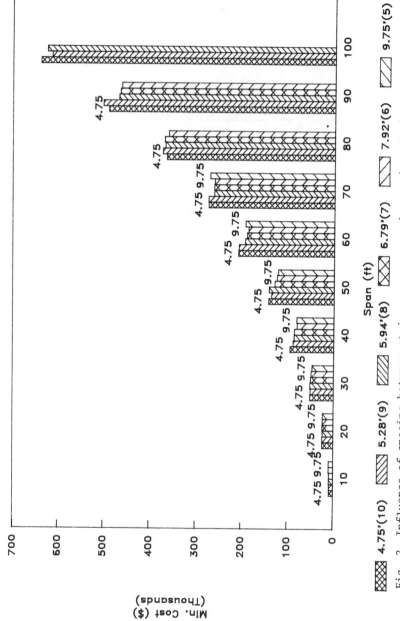

Fig. 2. Influence of spacing between stringers on composite optimum designs.

Generally speaking, one can say that as span lengths of noncomposite decks increases, one should decrease the spacing between beams or increase the number of stringers. But, for any span length, the larger spacing between the beams seems to provide the optimum design.

For composite decks, however, one cannot make such a statement. Referring to Fig. 2, one can see that for spans of up to 90 feet a spacing of 7.92 feet is capable of providing the least expensive design. In fact, for spans of 70 feet it is clear that increasing spacing between stringers increases the cost.

3 Influence of different loadings on optimum cost

Ordinarily, the bridges on main highways are designed for HS20 load as specified in the AASHTO manual. However, some highway departments use higher loads for their designs. For example, 125% of HS20 is sometimes used and may be designated as HS25. Special loads are also used in issuing permits for extra-heavy loads. One such load is shown in Fig. 3, and will be referred to as "permit load" in this paper. In order to observe the effect of different loadings, the optimum designs for spans of up to 100 feet were obtained for all three loads, i.e., HS20, the so-called HS25 and the permit load. Fig. 4 shows that using 25% higher load than HS20 does change the configuration and results in a more expensive optimum design. However, for all the spans considered, there was no difference in the optimum design for HS25 and the permit load. This clearly shows that the examination of the permit load is not necessary for this span range.

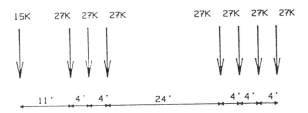

Fig. 3. Permit load.

4 Influence of concrete strength on optimum cost

The price one must pay for concrete increases as the strength required increases. Based on information obtained from several suppliers, an increase in purchase price for an increase in strength required was incorporated in the program, and different configurations were studied. Several different span lengths and widths were considered and designed for several different beam spacings. Fig. 5 shows the diagram for the noncomposite bridge with a span of 30 feet and a width of 37.5 feet. The conclusions drawn for this particular bridge are applicable to all spans of less

419

than 100 feet and of other widths. This bridge was designed with 3, 4, 5 and 6 beams for concrete strengths of 4000, 4500, 5000, 5500 and 6000 psi. As shown in Fig. 5, regardless of strength of concrete used, the optimum configuration remains that of 4 beams. As the concrete strength goes up, the total cost goes up slightly, but the optimum configuration does not change. Fig. 6 contains the same information as Fig. 5, but for a composite bridge. For concrete strength of 4000 psi, the optimum configuration has 4 beams, but for concrete strength of 6000 psi, the optimum configuration should have 5 beams. One should note, however, that for this particular bridge the configurations with 4 and 5 beams are really very comparable, particularly when one notes that the difference between the two was minimal to begin with. From these results, it seems reasonable to conclude that concrete strength within the specified range does not play an important role when considering optimum cost. Therefore, satisfaction of the minimum requirements of the controlling agency should be sufficient.

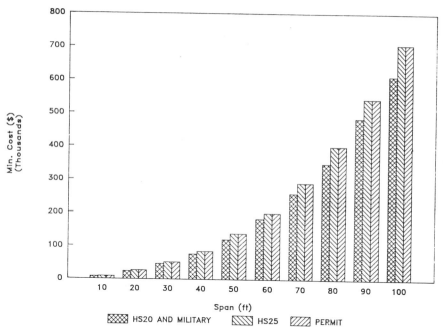

Fig. 4. Influence of different loadings

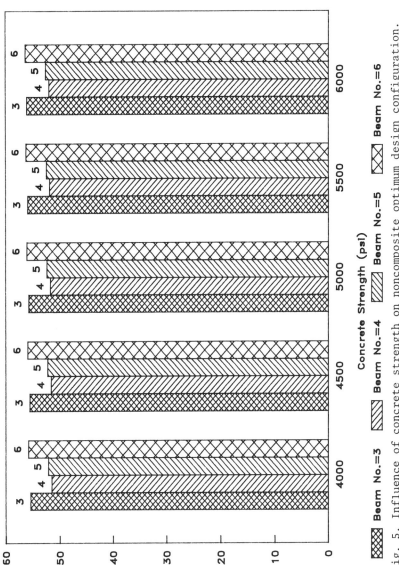

Fig. 5. Influence of concrete strength on noncomposite optimum design configuration.

Beam No.=3   Beam No.=4   Beam No.=5   Beam No.=6

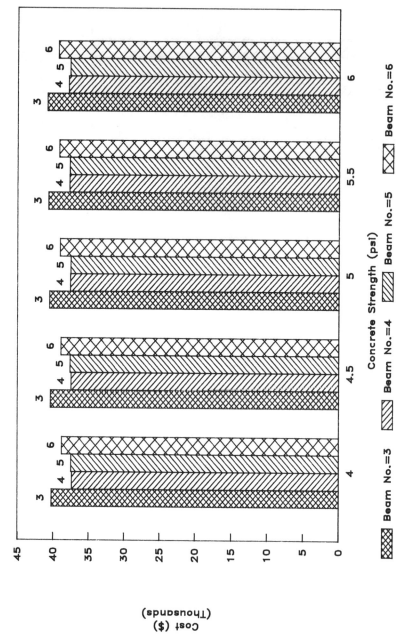

Fig. 6. Influence of concrete strength on composite optimum design configuration.

5 Influence of using different types of steel on optimum configuration

Two different yield strengths of steel were considered: 36 ksi and 50 ksi. The price of the higher-trength steel is obviously more than the lower-strength steel. The values of 1.0, 1.3, 1.6 and 1.9 were used as the ratio of cost of steel with Fy=50 ksi to the cost of steel with Fy=36 ksi. The optimum costs for five different designs of noncomposite bridges are shown in Fig. 7 for spans of 10 feet to 100 feet. The first design is for steel with yield strength of 36 ksi. The other four designs are for steel with yield strength of 50 ksi that can be purchased at prices that are equal to, 30% higher, 60% higher and 90% higher than F36 steel. As can be seen from Fig. 7 for spans of up to 60 feet, using a higher-strength steel at prices that are up to 30% higher than regular steel does not change the total cost. For spans of more than 60 feet, using the 30% more expensive steel will, in fact, provide a less costly design. Similar conclusions can be drawn with regard to composite bridges from Fig. 8.

6 Influence of interaction between concrete and steel on optimum cost

Bridges of different widths and different spans were considered in order to study the effect of degree of interaction between concrete and steel. As Fig. 9 shows, the composite design is less costly than the noncomposite counterpart for every span. This conclusion was also true when considering differently priced studs. Even after tripling the price of in-place studs, the composite bridge was determined to be less costly than the noncomposite in any span. This conclusion is in disagreement with the widely held belief that composite bridges are more economical than noncomposite bridges for spans greater than certain lengths.

7 Influence of variation in costs on optimum design

The cost data and unit prices which were used to study the sensitivity of the optimum design of bridges with respect to different factors were average values from data that was obtained from several highway agencies and construction cost data catalogues. However, costs for different material do vary in different areas and at different times. In order to ensure that the conclusions drawn in this study are not true only for the specific cost values used, it was necessary to study the sensitivity of the optimum design to price fluctuation.

Three sets of unit prices were selected from the spectrum of available price data: low, average and high. A set of optimum designs for spans of 10 feet to 100 feet and of different widths was obtained for the average unit prices to be used as a base set for comparison. For the same geometric configurations, another set of optimum designs was obtained for the high unit prices. The two

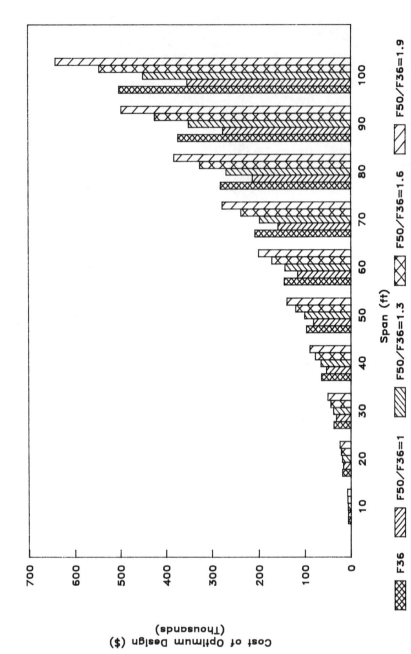

Fig. 7. Influence of using different types of steel on optimum cost of noncomposite systems.

Cost of Optimum Design ($) (Thousands)

Span (ft)

F36   F50/F36=1   F50/F36=1.3   F50/F36=1.6   F50/F36=1.9

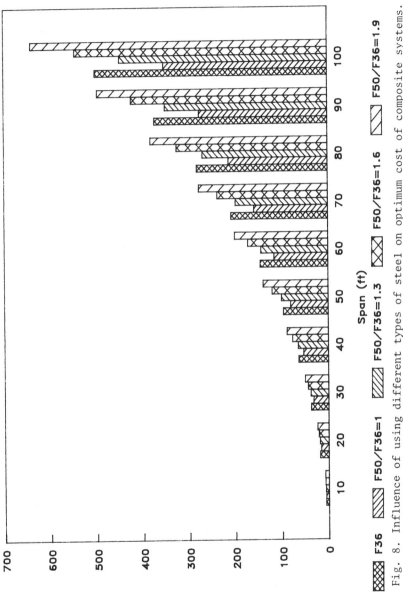

Fig. 8. Influence of using different types of steel on optimum cost of composite systems.

F36   F50/F36=1   F50/F36=1.3   F50/F36=1.6   F50/F36=1.9

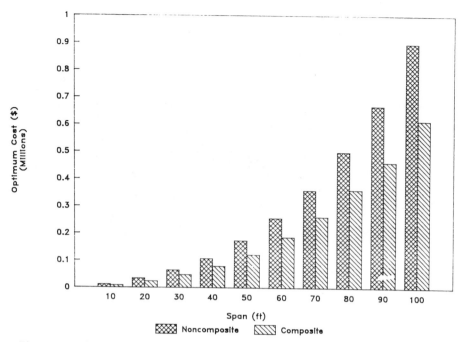

Fig. 9. Influence of interaction between concrete and steel on
optimum cost.

sets were compared. The results for a bridge width of 37.5 feet are
summarized in Table 1. As shown in Table 1, although the total
prices for higher-priced materials are increased for all the bridge
spans, the optimum design configurations don't change. In other
words, the 19% increase in the unit price of concrete and 92%
increase in the unit price of steel didn't change the number of
beams, the steel sections and the thickness of concrete slabs for
optimum design of any of the spans, except for the composite design
with the span of 20 feet. The last column in Table 1 shows that,
for this particular bridge, if the base configurations were used
with the high prices, there would be a difference in price of only
0.35%.

Since similar conclusions were drawn for other widths, one can
say that the optimum designs generated by the program and, hence,
all the preceding sensitivity analyses are valid for different unit
prices and are not sensitive to reasonable fluctuations in unit
prices.

8 Influence of provision of lateral bracing on the optimum cost

In order to evaluate the advantage of providing deeper and stronger
bracing in simply supported bridges, a study was made with regard to

Table 1. Influence of variation in unit costs on optimum designs

SPAN	TYPE*	Base Set: Average Unit Prices$					2nd Set: High Concrete and Steel Unit Prices$					% INCREASE IN TOTAL COST	% INCREASE IF BASE CONFIGURATIONS WERE USED WITH HIGH PRICES
		NUMBM	SPACE ft	SECTION	TSLAB In.	COST $	NUMBM	SPACE ft	SECTION	TSLAB In.	COST $		
10	NC	5	7.50	W18×40	8.0	9517.	5	7.50	W18×40	8.0	14889.	56	0
10	C	9	4.17	W12×14	8.0	7379.	9	4.17	W12×14	8.0	10984.	49	0
20	NC	5	7.50	W24×76	8.0	28430.	5	7.50	W24×76	8.0	47778.	68	0
20	C	5	7.50	W21×44	9.5	20996.	4	9.67	W24×55	8.5	32806.	56	0.35
30	NC	4	9.67	W33×130	8.5	54725.	4	9.67	W33×130	8.5	93896.	72	0
30	C	4	9.67	W27×84	8.5	40639.	4	9.67	W27×84	8.5	66619.	64	0
40	NC	5	7.50	W36×135	8.0	87659.	5	7.50	W36×135	8.0	154560.	76	0
40	C	4	9.67	W33×118	8.5	68355.	4	9.67	W33×118	8.5	115995.	70	0
50	NC	4	9.67	W36×230	8.5	143412.	4	9.67	W36×230	8.5	256502.	79	0
50	C	4	9.67	W36×150	8.5	102138.	4	9.67	W36×150	8.5	176986.	73	0
60	NC	5	7.50	W36×230	8.0	205877.	5	7.50	W36×230	8.0	374346.	82	0
60	C	5	7.50	W36×160	8.0	151651.	5	7.50	W36×160	8.0	269924.	78	0
70	NC	6	6.25	W36×245	8.0	297490.	6	6.25	W36×245	8.0	547575.	84	0
70	C	6	6.25	W36×170	8.0	215958.	6	6.25	W36×170	8.0	390756.	81	0
80	NC	6	6.25	W36×300	8.0	408887.	6	6.25	W36×300	8.0	757788.	85	0
80	C	5	7.50	W36×245	8.0	290943.	5	7.50	W36×245	8.0	529908.	82	0
90	NC	9	4.17	W36×245	8.0	552824.	9	4.17	W36×245	8.0	1032460.	87	0
90	C	6	6.25	W36×245	8.0	383315.	6	6.25	W36×245	8.0	704850.	84	0
100	NC	10	3.75	W36×260	8.0	716857.	10	3.75	W36×260	8.0	1344191.	88	0
100	C	6	6.25	W36×300	8.0	511985.	6	6.25	W36×300	8.0	948111.	85	0

Increase in conc. price ≈ 19%
Increase in steel price ≈ 92%

* NC = noncomposite systems
  C = composite systems

bridges with widths of 37.5 feet and with spans ranging from 10 feet to 100 feet. Both composite and noncomposite systems were considered. In one case, the beams were allowed to deflect independently, and in another case, they were assumed to have equal deflection.

The provision of sufficient bracing results in less costly designs for certain non-optimum configurations. However, for the optimum designs the configurations were the same for both cases. The one exception was for a noncomposite bridge with a span of 40 feet.

For this single case, in which the configuration did change, the decrease in cost was only 2%. Consideration of the cost of providing sufficient bracing, which is not included in the listed price, will reduce this apparent saving to a negligible amount, if not eliminate it. Thus, when considering diaphragms and bracings in the design of simply supported short-span bridges, it should be sufficient to satisfy minimum code requirements, due to the fact that virtually no cost saving would be realized by providing deeper and stronger diaphragms and bracings.

9 Conversion table

1 inch = 0.0254 m
1 foot = 0.3048 m
1 lb/ft^3 = 157.086 Newton/m^3
1 ksi = 6.89 x 10^6 Newton/m^2

10 References

American Association of State Highway & Transportation Officials (AASHTO), Standard Specification for Highway Bridges, (1989).

American Institute of Steel Construction. Manual of Steel Construction, 8th ed., 1980.

Better Road, (1986), Nov., p. 42.

Fox, R.L. (1971) Optimization Methods for Engineering Design, Addison-Wesley, Reading, Ma.

Hegarty, M.J. (1986), Design variables and systems scenario of low-volume bridges, Problem Report, West Virginia Univ., Morgantown, WV.

Mafi, M. (1985) Cost-effective, Short-span Bridge System. Ph.D. Thesis, Pennsylvania State University, State College, Pa.

Mafi, M. (1988) A basic expert system for type selection of bridges (BEST Bridge), in Proceedings of ASEE Conference, Portland, Oregon,

National Cooperative Highway Research Program, Bridges of Secondary Highways and Local Roads: Rehabilitation and Replacement. NCHRP Report 222, May 1980.

LIVE LOAD ANALYSIS OF BRIDGES MODELLED AS PLANE FRAMEWORKS

A. FAFITIS
Dept Civil Engineering, Arizona State University, Tempe,
Arizona, USA
J.R. McKELLAR
Tudor Engineering Company, Phoenix, Arizona, USA

Abstract
For the purpose of identification in this paper, an in-plane structure modeled only by its horizontal members is called a continuous beam; and an in-plane structure modeled by its horizontal and vertical members is called a plane frame.
Truck and lane live load analysis of bridges is most often accomplished by modeling the bridge superstructure as a continuous beam. The stiffnesses of the piers and abutments are not taken into account as a simplifying assumption in such analysis.
A plane frame influence line generation computer program and a plane frame box girder analysis computer program were developed to study the accuracy of the continuous beam model and for use in design. The programs were written in FORTRAN 77 for IBM PC and compatible microcomputers.
Several design parameters of bridges were analysed as continuous beams and as plane frames by use of the programs. It was found that for shorter span bridges, the difference in bending moments of continuous beams versus plane frames, is greater than for longer span bridges, and that this difference is significant for shorter spans.
Keywords: Bridge, Box , Prestressed Concrete, Influence Lines

Introduction

Bridge superstructures have traditionally been analysed as continuous beams; such procedure neglects the stiffness of the piers. Live load analysis of bridges has long been performed with the aid of influence lines. Books of tables, containing influence lines for various configurations of continuous beams, have been widely used to aid the bridge designer in live load analysis. Computer programs are also available that perform live load analysis of bridges modeled as continuous beams. Depending upon the superstructure and pier type, and the pier-to-superstructure connection details, the continuous beam model may be adequate. However, where moment resisting connections are to be modeled between superstructure and pier, significant differences may occur in comparing bending moment influence line values of a bridge modeled as continuous beams versus plane frame modeling of the same bridge.

This paper presents the results of a parametric study which compares the bending moment influence lines at pier supports for a 3-span, concrete box girder superstructure, supported by concrete wall piers. The study was accomplished by the use of two FORTRAN 77 microcomputer programs compiled on a IBM PC compatable computer [7]. Program INFLU generates bending moment influence lines at specified joints for either continuous beam or plane frame structures. The influence line value is computed at tenth points for each structural member. Program BOX generates analysis results for continuous beam or plane frame prestressed concrete box girder bridges, or degenerate

429

cases thereof. Analysis results include axial forces, shearing forces, bending moments, deflections and concrete stresses at the extreme top and bottom fibers for each structural member at twentieth points. In tandem, the programs constitute a design aid or a research tool for the analysis of bridge structures modeled either as continuous beams or plane frames.

Program Influ

The placement of moving loads poses a question as to where the moving loads shall be place to achieve the maximum stresses in the bridge. The answer to that question lies in the investigation of the structure's influence lines.

The stiffness method of matrix structural analysis was used to solve the problem of statically indeterminate framed structures which is encountered in influence line generation [2,3,5,9,11].

Bending moment influence lines at supports are of prime interest for the analysis of frame structures. Shearing force, reaction force and intermediate span bending moment influence lines may be derived from the bending moment influence lines at supports by a matter of simple statics. Progam INFLU calculates the bending moment influence lines at supports of continuous beams and plane frames. The results are computed at tenth points for all the members of the structure. Bending moment influence lines can be computed for several joints by a single run of the program. The program reads a user defined, ASCII input data file and writes to a user specified, ASCII output file. The output file is so configured that it can be printed with margins to any standard printer and be immediatly inserted into a 8 1/2" x 11" sized report.

To calculate joint moment influence lives, program INFLU takes advantage of Castigliano's theorem, which says that the partial derivative of strain energy δU of a system with respect to an applied force δPy or δM is equal to that component of displacement δ or θ at the point of application of the force which is in the direction of the force [10].

$$\delta U/\delta Py = \delta \tag{1}$$

$$\delta U/\delta M = \theta \tag{2}$$

For a unit force (cause) equation 1 & 2 give

$$\delta M = \delta/\theta \tag{3}$$

and for $\theta=1$

$$\mathbf{M} = \delta \tag{4}$$

Eq. (4) states that, for a structure, the moment influence line at a joint is the deflected shape of the structure where a unit rotation is forced upon the joint.

From Maxwell-Betti's reciprocal theorem, the deflection (translation) of point 1 due to a unit couple applied at point 2 δ_{12} is numerically equal to the rotation at point 2 due to a unit load applied at point θ_{21} [4], as shown in Fig. [1].

$$\delta_{12} = \theta_{21} \tag{5}$$

Therefore, a unit couple must be exerted on the hinged joint of a structure to develop the deflected shape, to some scale, which is required by Eq. (4). Program INFLU imposes a scaling factor to the joint rotations caused by a pair of opposite-hand unit couples on the hinged joint to develop a unit rotation of the hinged joint. The result is the moment influence line at the joint.

An equation may be written for each member to express the influence line ordinates as a function of the member joint rotations δ_i, δ_j , member length L, and variable position along the member x. The equation is:

$$\text{INFLU}(x) = Ax^3 + Bx^2 + Cx + D \tag{6}$$

Where: $A = (\theta_j + \theta_i)/(L^2)$

$B = -(\theta_j + 2\theta_i)/(L)$

$C = \theta_i$

$D = 0$

By differentiating Eq. (6), and solving for the positive root of the resulting quadratic equation between the limits of 0 and L, the maximum influence value and its position for each member may be established.

Program Box

Program BOX is a plane frame, stiffness method, structural analysis program which has been enhanced to treat the analysis of prestressed concrete box girder bridges. The program box may be used to investigate other superstructure cross-sections such as single and double-tee girders and rectangular sections. Variable cross-sections for the superstructure are not possible. Understructure cross sections are not calculated by program BOX and shall be calculated by the user and input to the data input file. Variable understructure cross-sections may be approximated through use of incremental sized member models of the piers. The cable profile designated in program BOX is parabolic, but other type cable profiles may be investigated by setting the prestressing force to zero and by imposing equivalent prestressing loadings on the bridge computer model [8]. Figure 2 is a least squares fit of prestressed concrete box girder bridge maximum span to superstructure depth data from 173 international bridges [6]. The figure shows the data scattering as well as the the least squares fit described by:

$$H = (Lm/(22.58888 * Exp (-0.000664799*Lm))$$

where $H =$ superstructure depth (Ft.), $\qquad$ (7)

$Lm =$ bridge maximum span length (Ft.).

Fig. [2] , or Eq. (7), is useful in obtaining a first approximation of the depth of a bridge of a known maximum span.

Comparison of Continuous Beam vs. Plane Frame

The example bridge is a three-span structure with spans of 160 ft., 190 ft., and 170 ft. from left to center to right, and piers 55 ft. in height as shown in Fig. [3]. The superstructure is a prestressed concrete box girder of nine cells, top slab 10″, bottom slab

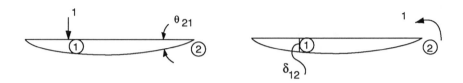

Fig.1.   Betti-Maxwell Reciprocal Theorem

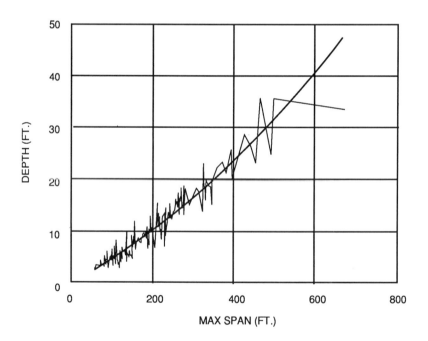

Fig. 2. Span Depth Approximation

of 6", and an overall depth H = 7'-6". The parabolic centroid of the prestressing cable profile and the points of cable inflection are shown in Fig. [4 A.]. The magnitude of the prestressing force is chosen at 28,600 Kips and the equivalent prestressing loads are shown in Fig. [4 B.].

The three loadings cases investigated for a matter of comparison in this example are:

1. Self Weight Dead Load (Dead Load or DL)
2. Dead + Prestressing Load (DL+PS)
3. Dead + Prestressing + Live Load (DL+PS+LL)

The live loading is AASHTO HS 20-44 lane loading for maximum moment [1]. The live lane loading is a combination of an uniform load and two concentrated loads which are used depending upon whether the analysis is for shear or moment. The uniform load is 0.640^k per foot of load lane. The concentrated load is either 18^k for moment or 26^k for shear. An additional concentrated load is allowed for continuous spans involving land loading. AASHTO stipulates 12 ft. lane widths across the superstructure. In this case a 76 ft. wide bridge yeilds 6 lanes. Therefore values of 6 times the live lane loads or 3.84 ^{k}/ft. uniform and 108^k concentrated respectively are herein used for maximum moment analysis at the supports. AASHTO uses scaling factors to reduce the stresses caused by loading three or more lanes with live load simultaniously. This scaling is not important here since, for comparison sake, both the continuous beam and the plane frame will be loaded with the same loads.

Continuous Beam Solution

The continuous beam model of this example is the same as Fig [3], neglecting the columns. Program BOX first must be run to determine the section properties of the superstructure and analyze the bridge for the first loading case.
Loading cases 2 and 3 may be analized by preparing a second data input file and running program BOX after obtaining the continuous beam influence lines for moment at the interior supports. Program UNFLU is employed for obtaining the influence lines. Fig. [5] shows the moment influence lines for both interior supports of the continuous beam. The influence line values for moment at the interior supports indicate maximum influence when spans 2 and 3 are loaded with the live lane loading. The maximum support moment for this structure is induced at joint 3. An examination of influence line shows that the concentrated loads should be placed at distances of 117.8 ft. and 71.9 ft. right of joints 2 and 3 to respectively to achieve the maximum moment influence at joint 3.

Plane Frame Solution

A significant advantage in using programs BOX and INFLU in a plane frame analysis of bridges over a continuous beam analysis is that the piers may be analyzed at the same time as the superstructure.
Fig. [6] shows the frame influence lines for moment at the interior supports. By inspection the placement of the live lane loadings is different for the plane frame compared to the continous beam.

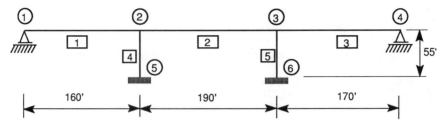

Fig. 3.   Example Bridge Profile

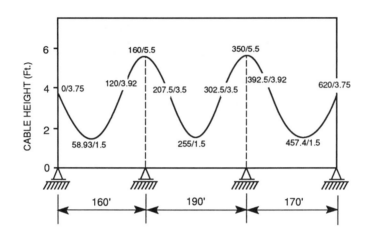

Fig. 4. A. Prestressing Cable Profile

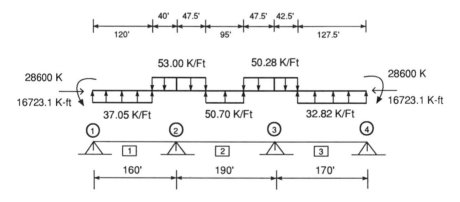

Fig. 4. B. Prestressing Equivalent Loads

434

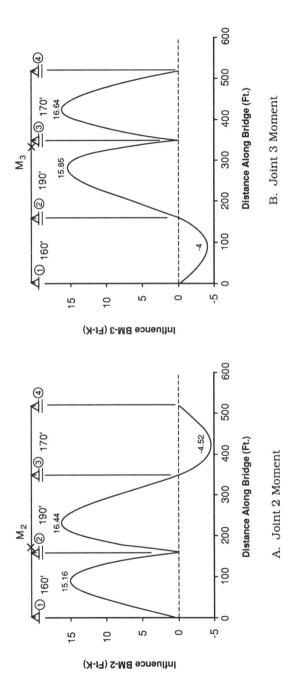

A. Joint 2 Moment

B. Joint 3 Moment

Fig. 5.   Beam Example Influence Lines

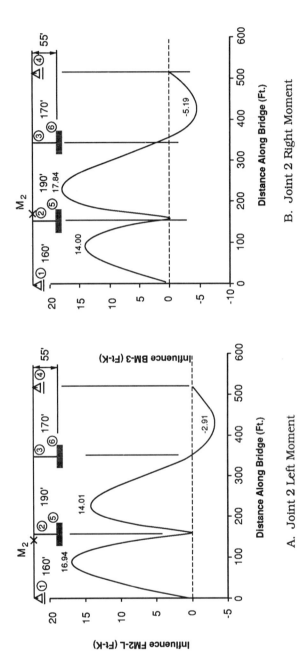

Fig. 6. Frame Example Influence Lines

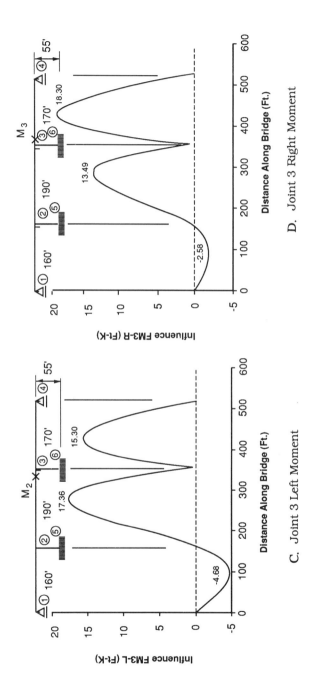

Fig. 6. Frame Example Influence Lines

Comparison of Results

Table [1] shows a comparison of the influence values computed for plane frame and continuous beam analysis discussed above. Differences as great 17.5% exist in this example betyween influence line magnitudes of the frame and beam analysis. Table [2] shows a comparison between the program results of the frame and beam analysis. Differences as great as 2.9% are found in this comparison. The 17.5% difference in the influence line values indicates that the two methods vary significantly in live load sensitivity.

Table 1 Comparison: Fram Vs. Beam Max. Influ. Line Values

MAX INFLU VALUE	FRAME				BEAM				COMPARISON			
	Joint 2		Joint 3		Joint 2		Joint 3		Joint 2		Joint 3	
	MEM 1	MEM2	MEM 2	MEM 3	MEM 2	MEM 3	MEM 1	MEM2	MEM 1	MEM 2	MEM 2	MEM 3
HINGE LEFT	16.94	14.01	17.36	15.30	15.16	16.44	15.85	16.64	10.51%	-17.34%	8.70%	-8.76%
HINGE RIGHT	14.00	17.84	13.49	18.51	15.16	16.44	15.85	16.64	-8.29%	7.85%	-17.49%	10.10%

Table 2 Comparison: Fram Vs. Beam Maximum Analysis Values

MAX VALUE	FRAME RESULTS			BEAM RESULTS			COMPARISON		
	DL	DL+PS	DL+PS+LL	DL	DL+PS	DL+PS+LL	DL	DL+PS	DL+PS+LL
Shear	2361	2473	2963	2625	2467	2980	.23%	.24%	-.57%
Moment	83720	26875	39805	82783	26624	39210	1.12%	.93%	1.49%
Defl.	-.1646	.0588	.06933	-.1649	.05836	.07016	-.18%	.75%	-1.20%
Str. Max	1896	1793	2086	1875	1787	2072	1.11%	.33%	.67%
Str. Min	-1385	628	526	-1369	646	536	1.16%	-2.87%	-1.90%

For a basis of study, the before mentioned three-span bridge is used with variable span lengths based upon the constant span lengths of 160 ft., 190 ft., and 170 ft. Proportional structures are analyzed with the constant of proportionality of the spans being α. The piers remain constant for the comparison. The section has a variable height **H** and variable section properties based upon the span to depth relationship shown in Fig. [2]. Table [3] shows the results of running programs INFLU and BOX for five values of α. Fig. [7] shows a plot of the results of Table [3]. Part A shows a comparison between a continuous beam and a plane frame analysis of the maximum moment influence line values at joint 3. Part B shows a plot of the percent difference between the analyses vs. the length variable α. It can be seen that by holding the span properties proportional, and keeping the column properties constant, the percent difference between the two analyses increases as the relative span lengths decrease. Fig. [8] shows maximum dead load bending moment at joint 3 plotted versus the length variable α Even though the differences are not so apparent as those from Fig. [7], the trend is the same in that the

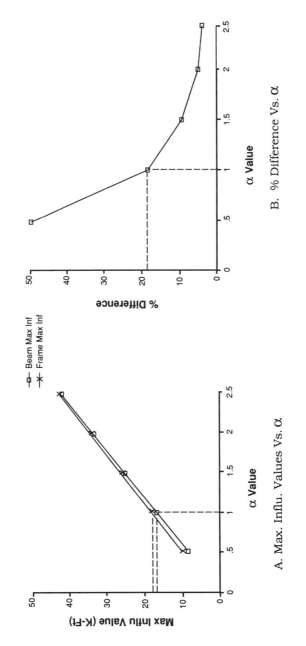

A. Max. Influ. Values Vs. α

B. % Difference Vs. α

Fig. 7. Variable Span: Joint 3 Moment Influ. Line Results

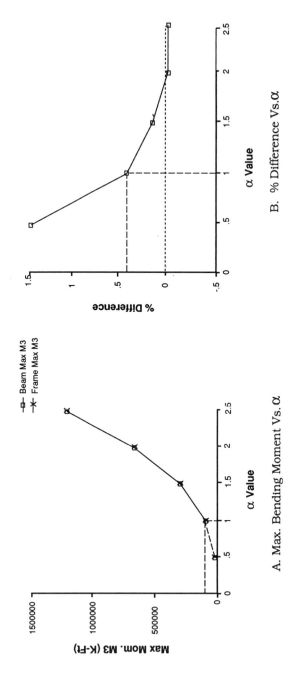

A. Max. Bending Moment Vs. α

B. % Difference Vs. α

Fig. 8 . Variable Span: Joint 3 Moment Analysis Results

differences between the values of the two analyses increases with decreasing length variable α. These two figures show that the differences between a continuous beam analysis a plane frame analysis increase with decreased span length.

Table 3. Variable Span and Depth Analysis

| α | H | A | I | INFUL | INFLU | M3 |
	FT	FT**2	FT**4	LT K-FT	RT K-FT	K-FT
0.5	4.5	137.4	379.4	7.92	8.32	16959
1.0	10	193.3	2678.7	15.85	16.64	95394
1.5	16	254.9	8508.9	23.77	24.96	283135
2.0	23	3227.1	21162	31.69	33.28	645843
2.5	28	378.7	344992.7	39.62	41.6	1168318

A. Continuous Beam Results

| α | H | A | I | INFUL | INFLU | M3 |
	FT	FT**2	FT**4	LT K-FT	RT K-FT	K-FT
0.5	4.5	137.4	379.4	6.09	9.78	17203
1.0	10	193.3	2678.7	14.58	17.63	95774
1.5	16	254.9	8508.9	22.84	25.68	283466
2.0	23	3227.1	21162	31.02	33.79	645694
2.5	28	378.7	344992.7	38.98	42.08	1168113

B. Frame Results

Research Summary

A bridge which is more sensitive to live load influence would be subject to larger analysis result differences between the two methods as is evident in the large percentage differences between the influence line values. It follows logically that shorter span bridges, bridges where the dead loads and prestressing loads comprise a lesser percentage of the total bridge loading than longer span bridges, have higher differences between joint bending moment influence line values, and hence, between joint bending moment results.
A plane frame analysis, such as the stiffness method analysis used in programs BOX and INFLU, has the following major advantages over a continuous beam analysis:

- Simultaneous superstructure-understructure analysis;
- Axial deformation capability which is essential in the development of prestressing forces;
- Greater sensitivity to live loading influence, especially for short span bridges.

REFERENCES

1. American Association of State Highway and Transportation Officials (AASHTO), 1977, "Loads", **Standard Specifications for Highway Bridges**, 2nd ed., American Association of State Highway and Transportation Officials, Washington D.C., pp.18-42.

2. Fafitis, A. "Non-Linear Analysis of Framed Structures by Computer-1." **Indian Concrete Journal**, pp.98-104, April 1982.

3. Forsythe, G.E., M.A. Malcolm, and C.B. Moler, 1977, "Linear Systems of Equations", **Computer Methods for Mathematical Computations**, Prentice Hall, New Jersey, pp.48-55.

4. Gaylord, E.H., and C.N. Gaylord, 1968, "Structural Analysis", **Structural Engineering Handbook**, McGraw-Hill, New York, pp.1-8.

5. Gere, J.M., and W. Weaver, Jr. **Analysis of Framed Structures**, Princeton, New Jersey, Van Nostrand, 1965, 475pp.

6. Maisel, B.I., and R.A. Swann, **The Design of Concrete Box Spine-Beam Bridges**, CIRIA Report 52, Construction Industry Research and Information Association, London, 1974, pp.52.

7. McKellar, John R. **Computer Analysis of Prestressed Concrete Box Girder Bridges**, Master of Science Project, Arizona State University, 1987, 277pp.

8. Naaman, A.E., 1982, "Continuous Beams and Interminate Structures", **Prestressed Concrete Analysis and Design**, McGraw Hill, New York, pp.349-364.

9. Tuma, J.J., and R.K. Munshi. **Schaum's Outline of Theory and Problems of Advanced Structural Analysis**, New York, McGraw-Hill, 1971, 280pp.

10. Ugural, A.C., and S.K. Fenster, 1981, "Energy Methods", **Advanced Strength and Applied Elasticity, The S.I. Version**, Elsevier Science Publishing Co., Inc., New York, pp.288-292.

11. Weaver, Jr., W. **Computer Programs for Structural Analysis**, Princeton, New Jersey, Van Nostrand, 1967, 385pp.

AN IMPROVED NUMERICAL APPROACH FOR DYNAMIC ANALYSIS OF BRIDGES TO MOVING VEHICLES

M.R. TAHERI, M.M. ZAMAN, A. ALVAPPILLAI
School of Civil Engineering and Environmental Science, University
of Oklahoma, Norman, Oklahoma, USA

Abstract
A general algorithm is developed to study the transient
response of highway bridge structures due to traversing
heavy vehicles. The approach is based on a combined
structural impedance and finite element methods and
accounts for the complete vehicle-bridge dynamic
interaction. Thin plate theory is assumed for the bridge
deck. The predicted response shows good agreement with the
available experimental results. A parametric study over a
wide spectrum of velocities and mass ratios indicates that
there are three critical frequency ranges: a sub-critical,
a critical, and a super-critical region. The vehicle mass
inertia is more pronounced in the super-critical region
where the transient deflection of the bridge propagates in
a wave-like manner.
Keywords: Vehicle-Bridge Interaction, Bridge Dynamics,
Structural Impedance Method, Moving Mass, Moving Force.

1 Introduction

The importance of dynamic interactions between bridges and
moving heavy vehicles, such as locomotives, has been
recognized by bridge engineers since the nineteenth
century. Although the necessity of modeling bridges more
accurately as plates has been known, analytical solutions
for plate models are very cumbersome due to mathematical
complexities involved. Simplifications have commonly been
introduced by assuming a simple beam theory and neglecting
part of vehicle-bridge dynamic interactions.

In modeling a highway bridge as a plate, the
mathematical difficulties are compounded by complicated
vehicle-bridge dynamic interaction and due to difficulties
in representing the behavior of the bridge to traversing
multi vehicles. With the advent of high speed computers
and efficient numerical techniques, more accurate modeling
and analysis have become feasible. By using numerical
techniques, bridges can be modeled as two-dimensional
plates, Wilson and Tsirk (1967). In addition, the dynamic

interactions due to moving loads on the bridge can be treated as components of an integrated structural system, Ting et al. (1974) and Taheri (1987).

Many analytical studies in the area of bridge dynamics, Wilson (1973), Chung (1977), Chung and Genin (1978) and Genin and Chung (1979), have used a beam-mass model and the solutions have been obtained by series expansion method or modal expansion techniques along with series truncation procedures and an iterative algorithm to include the coupling terms in the dynamic equilibrium equation. Within the similar mathematical framework, Fourier analysis has been used to study the problem, Chiu (1971), Stanisic et al. (1974) and Genin et al. (1974). The modal expansion technique appears to yield satisfactory results if the moving mass solution is small perturbation of the corresponding moving force solution, as is the case when the mass ratio is small and the vehicle travels at a low speed. Numerical results, Chiu et al. (1971), demonstrated that for such cases only a few modes are needed for the analysis. However, for a large mass ratio and for vehicles traveling at higher speeds, a large number of modes are required and the convergence of the iterative process cannot be guaranteed. To remove the iterative process, finite difference integration techniques have been suggested, Genin et al. (1974). It usually requires excessive computations, and thereby some of the advantages of obtaining approximate solutions are lost. A different discretization which employs the finite element method was introduced by Yoshida and Weaver (1971) and by Daily et al. (1973). The method is particularly advantageous when two- or three-dimensional structures are considered.

The structural impedance method based on the principle of superposition for vehicle and structural system has been developed by Ting et al. (1974) and extended by Taheri (1987). This approach is capable of considering the vehicle-structure interactions and can be used for any prescribed boundary conditions because the effects of boundaries are embedded in the influence functions. The algorithm is based on two fundamental concepts: (1) introduction of an influence function, and (2) use of mass position as the pseudo time variable. This method is capable of solving multi dimensional problems and is simpler than most of the above numerical methods. In essence, the structural impedance method allows one to separate the modeling of the moving vehicle, the bridge deck and their interactions. Thus, a change in one model does not affect the basic modeling of the other components.

In this paper an algorithm is presented for analysis of dynamic interactions between a moving vehicle and the bridge. The formulation is based on the structural impedance approach and utilizes influence functions, which mathematically are Green's functions. The time variable is evaluated using Newmark's β-method, Newmark (1959). The

algorithm presented herein can be applied to the general
moving mass problems as well as simplified moving force and
static problems.

2   Mathematical model

To account for the complete vehicle-bridge interaction, the
moving vehicle and the bridge deck are considered as a
single system and the transverse inertia effect of the
moving vehicle is taken into account. The bridge deck is
assumed to be a thin rectangular elastic orthotropic or
isotropic plate with arbitrary boundary conditions. The
vehicle is modeled as a set of independent discrete
suspension units moving at the same velocity. Such an
idealization eliminates the inertia effects due to roll,
pitch and yaw motions of the vehicle. The mass of the
vehicle is lumped on the suspension systems. The
suspension units consists of linear springs and dampers.
All movements of the suspension units, except the vertical
motions, are constrained. The contact between the bridge
deck and the moving vehicle is assumed to be a point
contact. The vehicle may travel at an arbitrary speed $v(t)$
and arbitrary acceleration $\dot{v}(t)$. The vehicle-bridge model
is depicted in Fig. 1. In this figure, a is the length of
the bridge, b is the width of the bridge, t is the
thickness of the bridge deck, i represents the number of
suspension units, M_i, k_i, c_i and u_i are the vehicle lumped
mass, elastic spring, damping coefficient and displacement
of the ith suspension unit, respectively, F_i is interaction
force between the bridge and the ith suspension system, ξ is
position of the first suspension unit, and s_i is distance
between the first mass and the ith suspension unit.

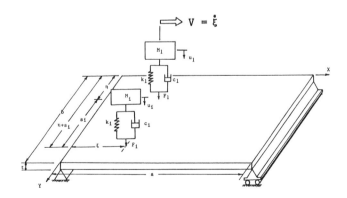

Fig.1. Vehicle-bridge model.

3   Mathematical formulation

The governing equation for the bridge is obtained
by considering that the dynamic deflection of the bridge is
caused by the weight and inertia of the vehicle, and by the
inertia of the bridge deck. Using the influence functions,
which mathematically represent Green's functions, and
expanding the basic impedance equation proposed by Ting, et
al. (1974), to account for a two-dimensional bridge deck,
the bridge dynamic deflection can be expressed in an
integral form as

$$w(x,y,t) = \sum_{i=1}^{I} G(x,y,\xi,\eta+s_i) \; F_i - \int_0^a \int_0^b G(x,y,\alpha,\epsilon) \; m(\alpha,\epsilon) \; (\partial^2/\partial t^2) \; w(\alpha,\epsilon,t) \; d\alpha \; d\epsilon \tag{1}$$

where $w(\xi,\epsilon,t)$ is the dynamic deflection of a point (x,y) at
time t, $G(x,y,\alpha,\epsilon)$ is the flexibility function, α and ϵ are
the integration parameters, $m(\alpha,\epsilon)$ is the mass per unit area
of the bridge deck, and I is the total number of suspension
units. According to D'alembert's principle, equation (1)
indicates that the total dynamic deflection of a point on
the bridge at time t is equal to displacement induced by
the interaction force F_i and the inertia forces of the
bridge deck. Note that the effect of boundaries is
included in the influence functions. A comparison of this
equation with the commonly used partial differential form,
Ting et al. (1974), shows that the integral formulation
eliminates the higher-order spatial derivatives, the delta
function and explicit boundary conditions. Therefore, the
numerical advantages of this equation are apparent.
    Considering the equilibrium of the spring-damper, the
interaction force, F_i, can be expressed as

$$F_i = [M_i \; g - M_i \; u_i(t)] \; \Delta[x-\xi, \; y-(\eta+s_i)] \tag{2}$$

where M_i is the mass of the ith suspension, g is the
gravitational constant, $u_i(t)$ is the acceleration of ith
suspension units, Δ is the Dirac delta function which is
equal to unity at $x = \xi$ and $y = \eta+s_i$ and zero elsewhere.
Assuming a linear spring and damper and considering the
relationship between the displacement, u_i, of the mass and
the deflection, w, of the bridge deck, the equation of
motion for vehicle mass can be obtained as

$$M_i \; d^2u_i(t)/dt^2 + c_i \; du_i(t)/dt + k_i \; u_i(t) = c_i \; \partial w(\xi, \eta+s_i, \; t)/\partial t + k_i \; w(\xi, \eta+s_i, \; t) + M_i \; g \tag{3}$$

Substituting equation (2) into (1) results in the equation

of motion for the bridge as

$$w(x,y,t) + \int_0^a \int_0^b G(x,y,\alpha,\epsilon) \, m(\alpha,\epsilon) \, (\partial^2/\partial t^2) \, w(\alpha,\epsilon,t) \, d\alpha \, d\epsilon =$$

$$\sum_{i=1}^{I} G(x,y,\xi,\eta+s_i) \, M_i \, (g - d^2u_i(t)/dt^2) \, \Delta[x-\xi, \, y-(\eta+s_i)] \quad (4)$$

For convenience the bridge deflection at any instant of time, t, is defined in terms of the mass position, ξ, as a pseudo time variable as

$$w(\alpha,\epsilon,t) = w(\alpha,\epsilon,\xi) \quad (5)$$

The time derivatives of equation (5) can be evaluated by applying the chain rule, as

$$\partial w(\alpha,\epsilon,t)/\partial t = \dot{\xi} \, \partial w(\alpha,\epsilon,\xi)/\partial \xi \quad (6)$$

$$\partial^2 w(\alpha,\epsilon,t)/\partial t^2 = \ddot{\xi} \, \partial w(\alpha,\epsilon,\xi)/\partial \xi + \dot{\xi}^2 \, \partial^2 w(\alpha,\epsilon,\xi)/\partial \xi^2 \quad (7)$$

Similarly, the following equations are obtained for the vehicle displacement in terms of ξ

$$u_i(t) = u_i(\xi) \quad (8)$$

$$du_i(t)/dt = \dot{\xi} \, \partial u_i(\xi)/\partial \xi \quad (9)$$

$$d^2u_i(t)/dt^2 = \ddot{\xi} \, \partial u_i(\xi)/\partial \xi + \dot{\xi}^2 \, \partial^2 u_i(\xi)/\partial \xi^2 \quad (10)$$

where $\dot{\xi} = d\xi/dt$, $\ddot{\xi} = d^2\xi/dt^2$ are the speed and acceleration of the first suspension unit, respectively.
To obtain the governing dynamic equations (GDE) in terms of the unknowns, bridge and suspension units deflections, the following operation are performed: (i) Substitute equations (5) through (10) into equations (3) and (4); (ii) obtain an approximate solution of the GDE by expressing the integral expressions in a finite difference form using the pseudo time variable ξ; (iii) use Newmark's β-method for time integration. From the operations performed in steps (i) through (iii), two equations are resulted that can be arranged into a matrix form in terms of the unknowns, vehicle deflection, $u(\xi_j)$, and bridge deflection, $w(\xi_j)$, as

$$[A] \, \{X\} = \{E\} \quad (11)$$

$$[A] = \begin{bmatrix} [p_1] & [p_5] & \langle\Delta_j\rangle \\ \hline [z_5] & [z_1] + [z_9] \end{bmatrix} \quad (12)$$

$$\{X\} = \left[\frac{\{u(\xi_j)\}}{\{w(\xi_j)\}}\right] \qquad (13)$$

$$\{E\} = \left[\frac{\begin{array}{l} [p_2]\ \{u_i\ (\xi_{j-1})\} - [p_3]\ \{u_i'\ (\xi_{j-1})\} - \\ [p_4]\ \{u_i''(\xi_{j-1})\} - [p_6]\ <\Delta(\xi_{j-1})>\ \{w(\xi_{j-1})\}+ \\ [p_7]\ <\Delta(\xi_{j-1})>\ \{w'(\xi_{j-1})\} + \\ [p_8]\ <\Delta(\xi_{j-1})>\ \{w''(\xi_{j-1})\} + \{p_9\} \end{array}}{\begin{array}{l} [z_6]\ \{u_i(\xi_{j-1})\} - [z_7]\ \{u_i'(\xi_{j-1})\} - \\ [z_8]\ \{u_i''(\xi_{j-1})\}+ [z_2]\ \{w(\xi_{j-1})\} - \\ [z_3]\ \{w'(\xi_{j-1})\} - [z_4]\ \{w''(\xi_{j-1})\} + \{z_{10}\} \end{array}}\right] \qquad (14)$$

in which

$$[p_1] = [e_i\ (\ddot{\xi}_j\ \delta/h\ \beta + (\dot{\xi}_j)^2/h^2\ \beta) + ((f_i\ \dot{\xi}_j\ \delta/h\ \beta) +1)] \qquad (15)$$

$$[p_2] = [e_i\ (\ddot{\xi}_j\ \delta/h\ \beta + (\dot{\xi}_j)^2/h^2\ \beta) + (f_i\ \dot{\xi}_j\ \delta/h\ \beta)] \qquad (16)$$

$$[p_3] = [e_i\ (\ddot{\xi}_j\ (1 - \delta/\beta) - (\dot{\xi}_j)^2/h\ \beta) + f_i\ \dot{\xi}_j\ (1 - \delta/\beta)] \qquad (17)$$

$$[p_4] = [e_i\ (\ddot{\xi}_j\ h(1- \delta/2\beta) - ((\dot{\xi}_j)^2/\beta)(1/2 - \beta)) +$$
$$f_i\ \dot{\xi}_j\ h(1 - \delta/2\beta)] \qquad (18)$$

$$[p_5] = [(f_i\ \dot{\xi}_j\ \delta/h\ \beta) + 1] \qquad (19)$$

$$[p_6] = [f_i\ \dot{\xi}_j\ \delta/h\ \beta] \qquad (20)$$

$$[p_7] = [f_i\ \dot{\xi}_j\ (1 - \delta/\beta)] \qquad (21)$$

$$[p_8] = [f_i\ \dot{\xi}_j\ h\ (1 - \delta/2\beta)] \qquad (22)$$

$$[p_9] = \{e_i\ g\} \qquad (23)$$

$$[z_1] = \sum_{k=1}^{N+1} \sum_{q=1}^{M+1} [G]\ ([m]\ (\ddot{\xi}_j\ \delta/h\ \beta + (\dot{\xi}_j)^2/h^2\ \beta))\ f_q\ f_k \qquad (24)$$

$$[z_2] = \sum_{k=1}^{N+1} \sum_{q=1}^{M+1} [G]\ ([m]\ (\ddot{\xi}_j\ \delta/h\ \beta + (\dot{\xi}_j)^2/h^2\ \beta))\ f_q\ f_k \qquad (25)$$

$$[z_3] = \sum_{k=1}^{N+1} \sum_{q=1}^{M+1} [G]\ ([m]\ (\ddot{\xi}_j\ (1 - (\delta/\beta)) - (\dot{\xi}_j)^2/\beta))\ f_q\ f_k \qquad (26)$$

$$[z_4] = \sum_{k=1}^{N+1} \sum_{q=1}^{M+1} [G]\ ([m]\ (\ddot{\xi}_j\ h\ (1 - \delta/2\beta) -$$
$$((\dot{\xi}_j)^2/\beta)(1/2 - \beta)))\ f_q\ f_k \qquad (27)$$

$$[z_5] = [G] \{\Delta\}_j [M_i] (\ddot{\xi}_j \, \delta/h \, \beta + (\dot{\xi}_j)^2/h^2 \, \beta) \qquad (28)$$

$$[z_6] = [G] \{\Delta\}_j [M_i] (\ddot{\xi}_j \, \delta/h \, \beta + (\dot{\xi}_j)^2/h^2 \, \beta) \qquad (29)$$

$$[z_7] = [G] \{\Delta\}_j [M_i] ((\ddot{\xi}_j (1 - (\delta/\beta)) - (\dot{\xi}_j)^2/h \, \beta)) \qquad (30)$$

$$[z_8] = [G] \{\Delta\}_j [M_i] (\ddot{\xi}_j \, h (1-\delta/2\beta) - ((\dot{\xi}_j)^2/\beta)(1/2 - \beta)) \qquad (31)$$

$$[z_9] = [I] \qquad (32)$$

$$[z_{10}] = [G] \{\Delta\}_j [M_i] \, g \qquad (33)$$

In equations (11) through (33), f_q and f_k are the weighting coefficients whose values depend on the integration scheme chosen, d and b are the parameters used in Newmark integration, N+1 and M+1 are discrete station points along the length and the with of the bridge, respectively, and $e_i = M_i/k_i$, $f_i = c_i/k_i$. The other variables are defined below:

$\langle \Delta_j \rangle$ = a row matrix indicating the vehicle position,
$\{\Delta\}_j$ = a column matrix indicating the vehicle position,
$[I]$ = an identity matrix,
$\{w(\xi_j)\}$ = deflection of bridge, vehicle at j^{th} location,
$\{u(\xi_j)\}$ = displacement of the suspensions at j^{th} location,
$[G]$ = influence coefficient matrix, and
$\{m\}$ = bridge mass matrix.

4 Numerical results

4.1 Verification
The versatility and accuracy of the algorithm are verified by comparing the numerical results with the exact solutions

Table 1. Dynamic magnification factors for moving force on beam

T_1/T	Velocity (in./sec)	w_f/w_s		
		SIM	Yoshida & Wear	Eichmann (Exact)
1/8	614	1.042	1.055	1.045
1/4	1228	1.082	1.112	1.108
1/2	2456	1.266	1.252	1.250
1	4912	1.662	1.700	1.707
2	9824	1.518	1.540	1.550

449

Table 2. Dynamic magnification factors for moving force on plate

T_1/T	Velocity (in./sec)	w_f/w_s		
		SIM	Yoshida & Wear	Wilson & Tsirk
1/8	515	1.042	1.042	-----
1/4	1030	1.032	1.088	1.111
1/2	2060	1.234	1.200	1.216
1	4120	1.525	1.568	1.510
2	8240	1.359	1.390	-----

and with other available numerical and experimental results.

The dynamic amplification factors for a 1-lb moving force on a simply supported beam are shown in Table 1. The beam given by Yoshida and Weaver (1971) is used for verification. In this table T_1 is the fundamental period, T is the time required for the moving vehicle to traverse the beam span and w_f/w_s is the ratio of the moving force deflection and the static deflection at the center of the beam. The dynamic amplification factors for a 2-lb moving force on a simply supported square plate are also shown in Table 2. The results obtained from the structural impedance method (SIM) and the other numerical results, Yoshida and Weaver (1971), are in excellent agreement. Since the exact solution for moving mass is not available, the results are compared with experimental studies, Ayer

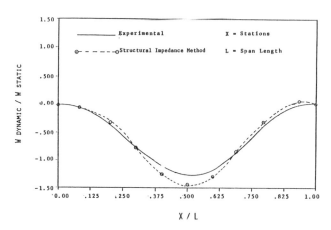

Fig 2. Comparison of SIM and experimental transverse deflection of single-span beam; $M/M_b = 1/2$, $\pi v/L\omega_1 = 1/4$.

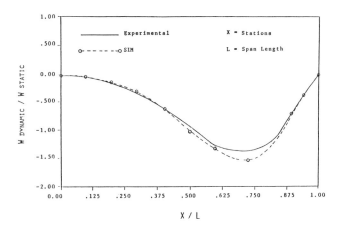

Fig 3.   Comparison of SIM and experimental transverse
deflection of single-span beam; $M/M_b = 1/2$, $\pi v/L\omega_1 = 1/2$.

et al. (1951). Three velocity ratios with a constant mass
ratio is considered for the comparison. The results of
comparison are presented in Figs. 2 through 4. In these
figures, L is the span length, v is the vehicle velocity, M
is the vehicle mass, M_b is the mass of the beam and ω_1 is
the fundamental frequency of the beam. A very good
agreement is observed between the results obtained from the
structural impedance method (SIM) and those observed
experimentally.

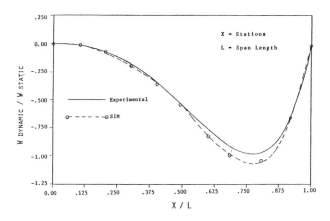

Fig 4.   Comparison of SIM and experimental transverse
deflection of single-span beam; $M/M_b = 1/2$, $\pi v/L\omega_1 = 3/4$.

4.2 Parametric Study

The algorithm presented here facilitated the study of vehicle-bridge interaction problems for vehicles moving at high speeds, and equally important, the algorithm allowed for a parametric study over a wide spectrum of velocities and mass ratios. Figs 5 through 7 show the time history of the displacement of a simply supported bridge for different velocities and mass ratios. In these figures M_p represents the bridge mass. The results for a low, medium

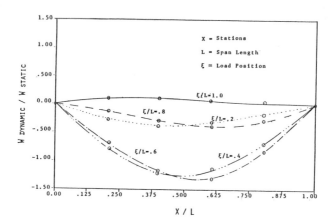

Fig. 5. Time history of deflection of a simply supported bridge; $M/M_p = 1/2$, $\pi v/L\omega_2 = 1/4$.

and relatively high longitudinal velocities, are shown in Figs. 5, 6 and 7, respectively. It is observed that as the

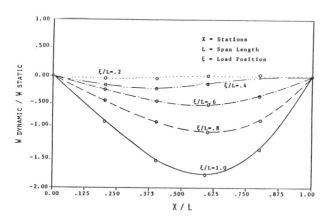

Fig. 6. Time history of deflection of a simply supported bridge; $M/M_p = 1/2$, $\pi v/L\omega_2 = 1.0$.

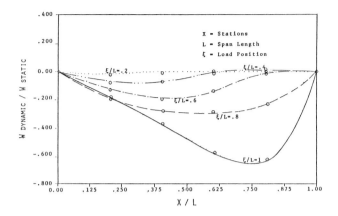

Fig. 7. Time history of deflection of a simply supported
bridge; $M/M_p = 1/2$, $\pi v/L\omega_2 = 2.0$.

velocity increases, the point of maximum displacement
shifts more to the right and it progresses in a wave-like
manner.

5   Concluding   Remarks

An algorithm is developed for dynamic response of bridges
to moving vehicles. The method considers complete vehicle-
bridge interaction (moving mass), with the moving force
approximation and the static solution as a special case.
The algorithm utilizes the influence functions which are
calculated for static loadings. It is observed that when
the vehicle-bridge mass ratio is small or the vehicle
travels at a low speed, the moving force solution yields
good approximations. A parametric study shows that there
are three frequency ranges: a sub-critical, a critical and
a super critical. The mass inertia of the vehicle is more
pronounced in the super-critical region where the transient
displacement of bridge propagates in a wave-like manner.
Such behavior can only be accurately predicted by including
the complete vehicle-bridge interactions, i.e. the moving
mass algorithm.

6   References

Ayre, R.S., Jacobson, L.S. and Hsu, C.S. (1951a **Transverse
    Vibration of One and of Two-Span Beam under the
    Action of a Moving Mass Load,** Proceedings of the
    First U.S. National Congress of Applied Mechanics, pp.
    81-90.

Chiu, W., Smith, R. and Wormley, W.N. (1971) Influence of Vehicle and Distributed Guideway Parameters on High-Speed, Vehicle-Guideway dynamic Interactions. **J. Dynam. Syst. Meas. Control, Trans. ASME**, 93(1), 25.

Chung, Y.I. and Genin, J. (1978) Stability of a Vehicle on a Multi span Simply Supported Guideway. **J. Dynam. Syst. Meas. Control, Trans. ASME**, 100 (4), 326.

Daily, G., Caywood, W.C. and O'Connor, J.S. (1973) A General Purpose Computer Program for the Dynamic Simulation of Vehicle-Guideway Interaction. **AIAA J.** 11 (3), 278.

Eichmann, E.S. (1953) Note on the Auxiliary Effect of a Moving Force on a Simple Beam. **J. of Appl. Mech.** 562.

Genin, J. and Chung, Y.I. (1979) Response of a Continuous Guideway on Equally Spaced Supports Traversed by a Moving Vehicle. **J.Sound and Vibration** 67 (1).

Genin, J., Ginsberg, J.H. and Ting, E.C. (1974) Longitudinal Track-Train Dynamics: A New Approach. **J. Dynam. Syst. Meas. Control, Trans. ASME**, 96 (4), 466.

Newmark, N.M. (1959) A Method of Computation for Structural Dynamics. **ASCE, J. of Eng. Mech. Div.**(85), 67-94.

Stanisic, M.M., Euler, J.A. and Montgomery, S.T. (1974) on a Theory Concerning the Dynamic Behavior of Structures Carrying Moving Masses. **Ing. Arch.** 43 (5), 295.

Taheri, M.R.(1987) Dynamic Response of Plates to Moving Loads: Structural Impedance & Finite Element Methods. Ph.D. Dissertation, **Purdue University.**

Ting, E.C., Genin, J. and Ginsberg, J.H. (1974) A General Algorithm for the Moving Mass Problem. **Journal of Sound and Vibration**, 33 (1), 49.

Wilson, E.N. and Tsirk, A. (1967) Dynamic behavior of Rectangular Plates and Cylindrical Shell. **Civil Engineering Department, New York University** Report No. S-67-7

Wilson, J.F. (1973) Response of Simple Spans to Moving Mass Loads. **AIAA J.**, 11, 4.

Yoshida, D.M. and Weaver, W. (1971) Finite-Element Analysis of Beams and Plates with Moving Loads. **Publ. Intl. Assoc. Bridge Struc. Engr.**, 31 (1), 179-195.

PART NINE
BRIDGE DYNAMICS

THE DYNAMIC BEHAVIOUR OF BRIDGES: PREDICTION AND PERFORMANCE

J.W. SMITH
Dept Civil Engineering, University of Bristol, Bristol, England, UK

Abstract
The analysis of long span bridges should take into account
geometric effects which result in gravity stiffening of the
cables in suspension bridges and stress softening of
towers and girders of cable-stayed bridges. The linearised
theory for small deflections is available in current
versions of commercial computer programs. Full scale tests
have confirmed the existence of many modes with closely
spaced frequencies well within the range of excitation by
the dynamic component of the wind. The significance of
multiple-support excitation of bridges in earthquakes is
considered. Experience in real earthquakes has shown
that short span bridges are the most vulnerable.
Keywords: Dynamic, Bridges, Non-linear, Vibrations,
Suspension, Aerodynamic, Earthquake, Performance.

1 Introduction

The non-linear behaviour of suspension bridges due to
large deflections is known to be sufficiently important
to warrant inclusion in design calculations for static
loads. An early study of the dynamic behaviour of suspension chains by Pugsley (1949) assumed that the chain
was inextensible and was limited to small deflections.
The stiffening effect of the tension in the main cables
of a suspension bridge was appreciated as being of
fundamental importance. In contrast, the compressive
forces in the towers and deck girders of cable-stayed
bridges result in a loss of stiffness which was taken
into account in dynamic analysis by Fleming and Egeseli
(1980).
 The dynamic response of long span bridges under wind
loading has been studied intensively since the collapse
of the Tacoma Narrows Bridge in 1940. Vincent (1962)
measured the oscillation of the Golden Gate Bridge in the
wind. Prediction of the performance of suspension bridges
in the wind has largely been based on wind tunnel tests

on sectional models (Scanlan, 1983). This form of testing
has proved useful for designing the profile of deck
structures to minimise aerodynamic instability but is
not reliable for observing the effects of buffeting by
turbulent wind which is three dimensional by nature.
Davenport (1982) has carried out tests on complete models
of bridges in a boundary layer wind tunnel and has
developed a method for the analysis of response due to
turbulence buffeting.

A number of highway bridges in California collapsed
during the 1971 San Fernando earthquake (Lew et al, 1971).
This was mainly due to failure of the piers or supports.
However, this alarming experience prompted a great deal
of research into the behaviour of all types of bridges,
including suspension bridges, under the effects of earthquakes. In particular, consideration has been given
to the possibility of differential support excitation
of long structures due to travelling earthquake wave
(Abdel-Ghaffar and Rubin, 1982).

Human response to vibration is an important serviceability criterion in the design of bridges. There have
been reported cases of car drivers abandoning their
cars on long span bridges after becoming alarmed by
large oscillations, possibly amplified by the vehicle
suspension, with consequent disruption to traffic. Also
it is known that many pedestrians will not use certain
bridges in high winds because of their fear of the large
oscillations. Much research has been done on human
response to motion, including the low frequency range,
and data is now available to establish limits to ensure
satisfactory performance.

2 Non-linear behaviour of long span bridges

The non-linear stiffness of a cable may be appreciated
by considering the differential element shown in Fig.1.

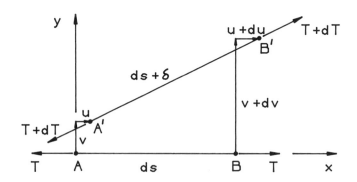

Fig.1 Large displacements of a cable

The element, length ds, has an initial tension T due to the gravity loading of the bridge structure. When live loading is superimposed the element deflects to a new position A'B'. The displacements of the ends A and B are denoted by u,v and u+du,v+dv respectively. By considering the geometrical changes it can be shown that the strain due to live load is given by

$$\epsilon = \frac{\delta}{ds} = \frac{du}{ds} + \frac{1}{2}\left\{\left(\frac{du}{ds}\right)^2 + \left(\frac{dv}{ds}\right)^2\right\} \tag{1}$$

where δ is the elongation of the element. Higher order terms have been ignored. The first term is the same as in small deflection theory while the second term, which is non-linear, is due to the finite rotation of the element. Note also that even if the dT increase in cable tension is ignored the rotation of the force T will result in changes in horizontal and vertical components. The existence of the initial tension T therefore gives rise to a geometric stiffness. This effect is often known as "stress stiffening".

Przemieniecki (1968) showed that the matrix equation for a pre-loaded rod element of length L is given by

$$\begin{bmatrix} F_1 \\ S_1 \\ F_2 \\ S_2 \end{bmatrix} = \frac{AE}{L}\begin{bmatrix} 1 & 0 & -1 & 0 \\ 0 & 0 & 0 & 0 \\ -1 & 0 & 1 & 0 \\ 0 & 0 & 0 & 0 \end{bmatrix}\begin{bmatrix} u_1 \\ v_1 \\ u_2 \\ v_2 \end{bmatrix} + \frac{T}{L}\begin{bmatrix} 0 & 0 & 0 & 0 \\ 0 & 1 & 0 & -1 \\ 0 & 0 & 0 & 0 \\ 0 & -1 & 0 & 1 \end{bmatrix}\begin{bmatrix} u_1 \\ v_1 \\ u_2 \\ v_2 \end{bmatrix} \tag{2}$$

where F_1, S_1, F_2, S_2 are the x and y components of force at nodes 1 and 2; u_1, v_1, u_2, v_2 are the corresponding displacements; AE is the unit stiffness of the rod; T is the preload. This may be written as

$$F = (K_e + K_g)u \tag{3}$$

where K_e is the usual elastic stiffness matrix and K_g is the geometric component of the stiffness matrix associated with rotation of the element as it is displaced. A further component of the stiffness matrix K_c can be derived to take account of second order non-linear terms. This has been ignored by most analysts in the linearised theory. However, Abdel-Ghaffar and Rubin (1983) included second order terms and used perturbation analysis to deal with the resulting non-linear equations.

Although special computer programs have been written by

researchers when studying the problem, commercially
available computer programs now exist which possess the
capability of including geometric or stress stiffening
(De Salvo and Swanson, 1988). First, a static analysis
is required under dead load. Then the forces in the
elements are used to calculate the geometric stiffness
matrices which are added to the elastic stiffness matrices.
Finally, eigenvalues are extracted from the resulting
system to obtain the modes and frequencies. Lumped or
consistent mass matrices may be used.

3 Suspension Bridges

Early studies of suspension bridge vibration adopted
a linearised version of the deflection theory extended
to include dynamic equilibrium. Continuum methods have
been useful in the past but are inconvenient when applied
to real bridges with non-simple boundary conditions.
Therefore it is natural that discrete methods of analysis,
and finite elements in particular, should be attractive
(Abdel-Ghaffar, 1980).
    Three dimensional analysis of the free vibrations
of suspension bridges was achieved by Dumanoglu and Severn
(1985). Their analysis of the Humber Bridge, in which
they employed over 900 elements and 3500 degrees of
freedom, took 100 secs of CPU time on a CRAY-1 computer
to obtain 20 modes of vibration. The model took into
account gravity stress stiffening but was restricted
to small deflection linearised theory. Some of the modes
are illustrated in Fig.2. Furthermore, by treating
vertical and horizontal modes separately they were able
to obtain a total of 40 modes. The northern sidespan
of the Humber Bridge is shorter than the Barton sidespan
to the south and the lack of symmetry has a profound
effect on the vibration modes as can be seen in the case
of modes 3 and 17 shown in the figure. The fundamental
mode is lateral motion of the deck. Mode 10 is interesting
in that it consists of lateral motion of the cables with
the deck remaining relatively motionless. Mode 14 is
the first torsional mode.
    Brownjohn et al (1987) verified the theoretical analysis
by measuring the ambient vibration of the Humber Bridge
under wind and traffic. The outputs from servo accelerometers located at suitable points on the bridge were
processed by a spectrum analyser. For example, the
auto-power spectrum obtained from the vertical motion
of a reference accelerometer located on the main span
of the bridge is shown in Fig.3. The large number of
resonant peaks is very striking. However, by careful
analysis of the outputs of a pair of mobile accelerometers
it was possible to identify the natural frequencies and
mode shapes of a total of 26 vertical modes, 16 lateral
modes, 11 torsional modes and some additional tower modes

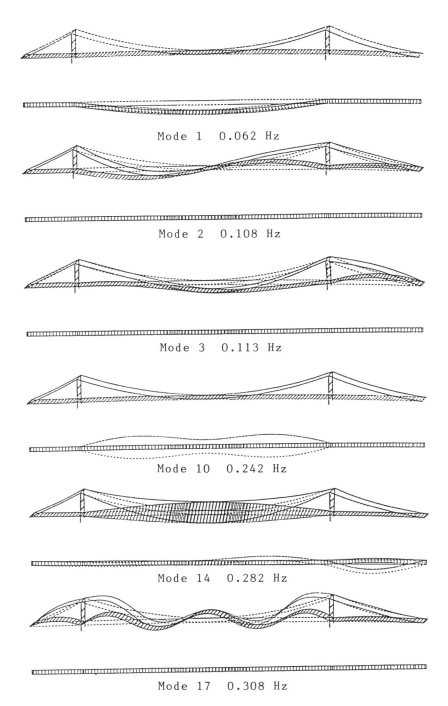

Mode 1   0.062 Hz

Mode 2   0.108 Hz

Mode 3   0.113 Hz

Mode 10   0.242 Hz

Mode 14   0.282 Hz

Mode 17   0.308 Hz

Fig.2   Mode shapes of the Humber Bridge

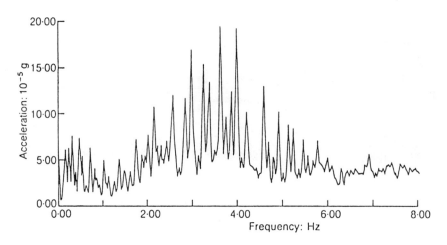

Fig.3 Main span vertical response

with frequencies up to 2.0Hz.
    Very good correlation was obtained with the theoretical
results. As an illustration of this the behaviour of the
bearing connections between the deck and the towers
was studied. They were designed to accommodate free
longitudinal movement, being fabricated in the form
of A-frame rockers. However, the experimental observations
indicated that free movement at the Barton end was being
restricted to some extent and therefore an alternative
pinned condition was assumed in the analysis. The
analysis was also carried out for the design condition
and the results are compared with experiment in Table 1.

Table 1. Theoretical vertical modes for different end
           conditions

Design condition		Hinged condition	Experiment
Mode	Frequency (Hz)	Frequency (Hz)	Frequency (Hz)
V1	0.108	0.115	0.116
V2	0.115	0.155	0.154
V3	0.168	0.179	0.177
V4	0.207	0.220	0.218
V5	0.240	0.239	0.240
V6	0.307	0.307	0.310
V7	0.313	0.314	0.317

A further important result of the experimental study of the
Humber Bridge was the effect of wind speed on modal
frequency. Non-linear behaviour would be exhibited by
a change of natural frequency with amplitude and hence
with wind speed. During the course of the tests a wide
range of wind speeds was experienced with the maximum
being gale force 8 on the Beaufort scale. The variation
of frequency with windspeed for the first and fourth
vertical modes are shown in Fig.4. Nonlinear behaviour
of the cables should exhibit itself as displacement
stiffening. However, the overall change in frequency
was very small in both cases, and well within the scatter
band. It may therefore be concluded that linear geometric
stiffening is sufficiently accurate for suspension bridge
analysis.

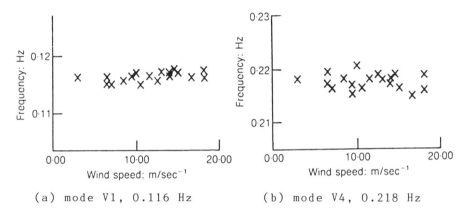

(a) mode V1, 0.116 Hz   (b) mode V4, 0.218 Hz

Fig.4   Modal frequency against wind speed.

4 Cable-stayed bridges

So far we have only considered the effect of stress stiffening, which occurs when the element force shown in Fig.1
is in tension. However, if the element force is in
compression the geometric behaviour of the element leads
to a reduction in stiffness. This may be noted in Equation
(2) when the element force is changed to -T. A well
known example of geometric stress softening is the reduction in frequency of transverse vibration of a beam-
column (Przemieniecki, 1968). This is illustrated in
Fig.5 where it may be seen that frequency (non-dimensional)
reduces linearly with compressive force and reaches the
limit at the Euler buckling load.
     The essence of cable-stayed bridges is that the cable
tensions are anchored by the deck girder instead of at
the bridge abutments. Consequently, the girder is in

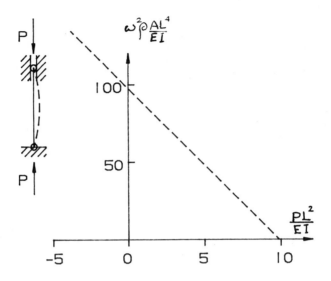

Fig.5   Variation in natural frequency of a beam-column

compression over most of its length and this will have
an effect on the natural frequencies.   Also the towers
will be in compression, as they are in suspension bridges,
and will tend to behave like beam-columns.   These stress
softening effects should be taken into account in the
analysis.   Furthermore, the inclined cables themselves,
although appearing to be straight to the naked eye,
may sag significantly, especially in long span bridges.
Therefore, geometric stress stiffening applies to the
cables.
    These effects were taken into account by Fleming
and Egeseli (1980) who used an equivalent modulus of
elasticity for the cables given by

$$E_{eq} = E/\{1 + [(WH)^2AE/12T^3]\} \qquad (4)$$

where E is the stiffness modulus of the straight cable,
W is its weight per unit length, H is the horizontal
projection of the cable, A is the cross-sectional area
and T is the tension in the cable due to external loading.
This procedure is useful for finite element analysis
of cable-stayed bridges since it permits the use of
a single straight element for each cable stay (Dumanoglu
et al, 1988).

5 Dynamic response to turbulent wind

Natural frequencies and mode shapes are interesting but
for design purposes it is the dynamic response to certain
forms of excitation that is of critical importance.
There are several different phenomena that give rise
to dynamic response of structures in the wind. These
include buffeting, vortex shedding, galloping and flutter.
   Buffeting by turbulent wind is probably the most
significant form of excitation for large structures such
as suspension bridges. A critical speed generally exists
at which large displacements may occur but the motion
tends to be modified by turbulence. Tanaka and Davenport
(1983) noticed that the sub-critical response of the
Golden Gate Bridge was predominantly due to buffeting.
Clear evidence of the importance of turbulence buffeting
was provided by Brownjohn et al (1987) who showed that
the response of the Humber Bridge was proportional to
the second power of the wind velocity.
   Some ineraction with vortex shedding evidently occurs
with some cross-sections but this may be minimised by
aerodynamic design of the cross-section or by the use
of turning vanes (Tappin and Clarke, 1985). Flutter
is a coupled motion, often being a combination of bending
and torsion, and it may be reduced by using a torsionally
stiff box girder deck as pioneered in the Severn Bridge.
   Turbulence of the wind may be treated as a 'stationary
random process'. It is random in the sense that the
velocity cannot be predicted at any instant of time.
It is stationary in the sense that certain statistical
properties, such as the mean and variance of the wind
speed, do not depend very much on the part of the wind
velocity record from which they were calculated. The
dynamic response of a structure subjected to random
loading cannot be expressed in a deterministic form but
instead has to be expressed in terms of stationary statistical properties. This approach was adopted by Davenport
(1962) who developed the spectral response method of
analysis. A detailed description of this method and
its application to long span bridges was given by Wyatt
(1980) who derived a series of design charts for analysing
line-like structures.
   Davenport took advantage of the fact that the dynamic
response of a structure can be treated as the combination
of uncoupled modal responses. Each mode of vibration
behaves as a single degree of freedom and may be treated
independently. The essence of the method is contained
in the expression for the spectral density of the modal
response of the mth mode. This is given by

$$S_{ym}(n) = \rho^2 \overline{V}^2 C_d^2 b^2 \frac{H_m^2(n)}{K_m^2} \chi^2(n) S_x(n) \int_0^L \int_0^L \phi_m(x) \phi_m(x') \gamma(\Delta x; n) dx dx'$$

(5)

where ρ is the density of air, $\overline{V}$ is the mean wind velocity
at the bridge deck level, C_d is the drag coefficient, b is
the transverse dimension of the bridge. $H_m^2(n)$ is the
mechanical admittance and K_m^2 is the generalised stiffness
of the mth mode. $\chi^2(n)$ is the aerodynamic admittance,
being a correction to take account of the distorting
effect of the presence of a structure on the air flow.
$S_x(n)$ is the spectral density of horizontal gustiness of
the wind and varies with frequency of turbulence, n.
$\phi_m(x)$ is the shape function of the mth mode of vibration
and is normalised such that it has a maximum value of 1.0.
$\gamma(\Delta x; n)$ is the square root of the "coherence" function
and is a measure of the correlation between wind velocities
at two points x and x'. Hence it takes into account the
spatial nature of turbulence which is related to the size
of eddies. It is the spatial variation of turbulent wind
speed that results in the need for the double integration.
    The general form of Equation (5) is shown by the solid
line in Fig.6. The broad hump is governed by the shape
of the spectrum for horizontal wind velocity which is only
modified slightly by the aerodynamic admittance. However,
the mechanical admittance has the effect of introducing
a sharp spike in the vicinity of the natural frequency, f_m,
in that mode. The variance of the dynamic component of

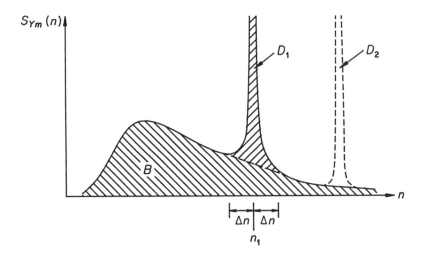

Fig.6   Spectral density for modal response

modal response is obtained by integrating Equation (5):

$$\sigma_d^2(Y_m) = \int_o^\infty S_{ym}(n)dn \simeq \pi \xi f_m S_{ym}(f_m) \tag{6}$$

where Y_m is the generalised displacement of the mth mode and ξ is the critical damping ratio. The approximation is justified by separating the non-resonant or broadband response (B) from the resonant dynamic response (D). The integral corresponding to the fundamental mode D_1 is shown hatched. The variance of the modal response for the second mode may be obtained in the same way by integrating the area D_2. Finally, the modes may be combined using

$$\sigma_d^2(x) = \phi_1^2(x)\sigma_d^2(Y_1) + \phi_2^2(x)\sigma_d^2(Y_2) + \cdots \tag{7}$$

The square root of this value is the root mean square (r.m.s.) response, which is a useful measure of the dynamic performance of a large structure.

The spectral response method has been applied to tall buildings with great success. The reason for this is that the natural frequencies are well separated and only the lowest one or two modes of even the largest buildings are within the range of wind excitation. However, in the case of long span bridges the situation is very different.

First of all, let us consider the spectral density of horizontal gustiness of the wind. This has been published by ESDU(1985) as the non-dimensional parameter $nS_x(n)/v_*^2$ where v_* is the surface friction velocity and is determined geographically. The variation of this spectrum with frequency at a height of 50m is shown in Fig.7 for the case of a reference wind speed of 25m/s over open level country. It should be noted that the

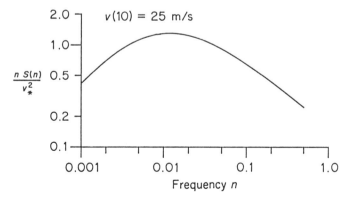

Fig.7 Spectrum of horizontal gustiness

spectrum reduces rapidly as the frequency approaches 1.0Hz. However, if we return to the results of the Humber Bridge study we should be impressed by the fact that no fewer than 40 modes were found to have frequencies ranging between 0.063 Hz and 0.816 Hz. These all lie well within the range of excitation by the wind.

Of course, the Humber Bridge has the longest single span in the world but longer bridges are planned or under construction. Therefore, it is very clear that there is an urgent need for an extension of the available spectral response method in order to include a large number of modes, many of them being very closely spaced in terms of frequency. There are obvious difficulties especially if it is desired to perform the integrals in Equation (5) over three-dimensional mode shapes. A discretized version of the expression would be a logical first step in this direction.

An interesting problem concerned with wind excitation of large bridges occurs in the case of strengthening work. Two examples illustrate this point. The first concerns the strengthening of the Wye Bridge connecting England and Wales. The original design comprised two towers each with a single cable passing over the top and radiating out to its anchorages at deck level on both sides of each tower. Thus the deck was divided into seven spans. The strengthening involved extending the towers and replacing the single cables with twin cables. More anchorage points on the deck were required resulting in a subdivision into eleven spans. Consequently, the bridge is now much stiffer and has a higher fundamental frequency. The second example concerns the re-surfacing of the Rama IX Bridge in Bangkok. The original asphalt surfacing, which broke up prematurely under traffic, was 80mm in thickness and weighed approximately 3500 tons, a substantial proportion of the dead load. A much thinner 40mm surfacing is being considered to replace the original. The lighter weight will increase significantly the fundamental frequency.

The question that arises with both of these problems is: what will the change of frequency do to the aero-dynamic stability of these bridges? It will probably improve the response to turbulence buffeting because as the fundamental frequency rises, so the spectral response will be calculated further along the downward slope of the wind spectrum shown in Fig.7. However, the critical wind speed would probably change also.

6 Earthquake loading of bridges

Earthquake damage to bridges is not likely to be a major problem in the UK. However, many UK consulting engineers are involved in design of bridges for overseas clients, often in countries with high seismic risk.

The performance of bridges in earthquakes was well illustrated in the 17th October 1989 earthquake affecting the San Francisco Bay area. The Golden Gate Bridge and the suspension spans of the Bay Bridge were undamaged. On the other hand, a 1.3 km length of the concrete double deck Cypress viaduct collapsed and a short span of the steel frame section of the Bay Bridge fell off its bearings. This behaviour was consistent with many years previous experience of damage to bridges in earthquakes.

Experience gained from the study of a crop of highway bridge failures during the 1971 San Fernando earthquake (Lew et al, 1971) revealed the following:

(a) There was a need for multiple column piers. Bridges with single column supports proved to be extremely vulnerable because of the lack of an alternative load path in the event of that column failing.

(b) Continuity of reinforcement between piers and foundations and across joints is important. There were examples of columns being uprooted from the foundation piles.

(c) The bearing areas of girders were in many cases unable to cope with excessive lateral ground movements and bearings were literally pulled out from under the end supports of bridges precipitating some catastrophic collapses.

These experiences were repeated in 1989. Over much of its length the upper deck of the Cypress viaduct in Oakland was supported by twin column piers with the columns tapering down to the lower deck level. The joint at lower deck level was relatively lightly reinforced and tended to behave like a pin during horizontal swaying. Failure appeared to have initiated at this joint leading eventually to collapse of the upper deck. The short span of the Bay Bridge which failed was a classic example of lack of bearing area. The bearings were only 125mm wide. Longitudinal movement of the steel girder structure caused the locating bolts at the end of the span to shear and the span fell off its bearings at that end.

Long span suspension bridges have proved to be very robust in earthquakes because of their inherent flexibility which makes them capable of accomodating large ground motions. The towers are generally thought to be the most critical components when subjected to earthquake motion in the longitudinal direction (Dumanoglu and Severn, 1987).

Much thought has been given to the problem of multiple-support excitation or asynchronous support motion which could affect long span structures. It is usually assumed that all parts of the base support of a structure

experience the same ground motion. This may be reliable
for tall buildings whose base dimensions are small.
However, the wavelength of a travelling earthquake wave
could be typically 500m and therefore long bridges could
experience considerable discrepancies in ground motions
at their supports. Dumanoglu and Severn (1987) have
studied this problem and shown that, although dynamic
stresses are generally less than with synchronous excitation, the pseudo-static stresses due to relative motion
were significant.

Analysis of the problem is straightforward. All that
is necessary is to apply ground motion time histories
to the supports as prescribed boundary conditions in
a finite element model, with solution by direct integration.
This procedure is within the capabilities of most standard
finite element programs. Unfortunately, the precise
details of independent support motions are not known in
advance. One approach is to impose a known earthquake
motion to each support with an appropriate phase difference
introduced to simulate the effect of a travelling wave.
This is called asynchronous support motion. In practice
earthquake motions are very complex and it has been found
that large differential movements can be very localised
and result in opening of wide cracks in the ground.
Therefore, it is likely that long multispan viaducts
are in fact more vulnerable than long single span bridges
and hence the need for large bearing areas at supports.

7 Performance criteria

The predicted dynamic response of bridges should be
compared with certain criteria for judging their performance. The three most important criteria are

    -resistance to overall collapse
    -satisfactory human response to motion
    -adequate fatigue resistance

7.1 Resistance to collapse
The maximum response of a structure subjected to random
loading such as turbulent wind cannot be evaluated in a
deterministic sense. However, the probability of the
response exceeding a certain magnitude may be determined
from the root mean square value of the response. A
practical procedure was suggested by Davenport (1964)
to derive a peak factor by which the r.m.s. component
during a storm wind would be exceeded with a 50% probability. First the load effects such as bending moment
or shear have to be obtained by integrating the vibratory
inertia forces in each mode over the span of the bridge.
The total resonant load effect is then obtained by summing
modal contributions. The maximum total load effect may

then be obtained from

$$E_{max} = \overline{E} + \sqrt{\{[g_b\sigma_b(E)]^2 + [g_d\sigma_d(E)]^2\}} \qquad (8)$$

where $\overline{E}$ is the load effect (bending moment, shear etc.)
due to the mean wind, $\sigma_b(E)$ is the r.m.s. broad band
or non-resonant load effect due to gusts, $\sigma_d(E)$ is the
r.m.s. resonant load effect combined from all the modes,
g_b is the broad band peak factor, and g_d is the resonant
peak factor. Details of how to calculate the peak factors
are given by Davenport (1964). Their values are generally
in the region of 3.0 to 4.0.

Specification of a design earthquake is a difficult
problem because earthquakes are, by their very nature,
notoriously unpredictable. A rational procedure would
be to determine the probability of exceeding a specified
input ground motion and design the structure so that
the probability of damage or collapse was acceptable
taking into account the penalty of failure in terms of
loss of life or economic cost. Much effort has been
put into developing such a procedure for the design of
buildings in seismically active regions (ATC, 1978).
However, the design of large bridges to resist earthquakes
is still carried out by relatively unrefined procedures.
For example, the Bosporus Bridge in the seismically active
region of Turkey was designed to resist equivalent quasi-
static forces resulting from a ground acceleration of
0.125g, in combination with dead load, temperature and
reduced live load (Davis,1988). This procedure was
justified by the fact that the effects of live load and
wind were more severe. It is certainly true that long
span bridges have a good record of performance in earthquakes. The performance of short-span and multi-span
highway bridges has already been discussed.

7.2 Human response

Human beings are surprisingly sensitive to vibration and
are often disturbed by intensities that are well below
those required to overstress the structures they use.
Psychological factors are very important. The sensation
of motion of a long-span bridge tends to undermine confidence in the security of the structure.

Irwin (1978) suggested a base curve for acceptable
human response to vibration of bridges under frequent
forms of excitation such as traffic and wind. The curve
is for vertical motion and is shown in Fig.8. Bridges
are predominantly susceptible to vertical vibration,
being wide but generally shallow in depth. The curve
compares well with the lower limit for standing persons
obtained by Leonard (1966). He showed that body motion
and muscular coordination during walking tends to increase
tolerance to the vibration. There is obviously an upper

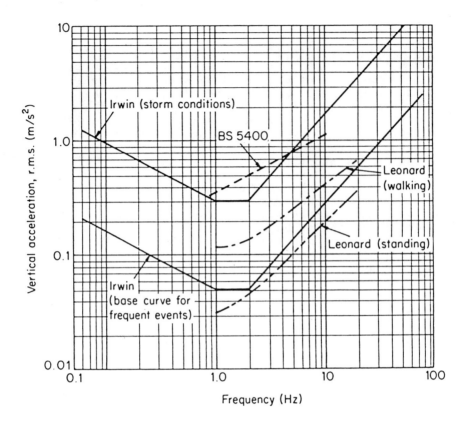

Fig.8 Human response to vibration of bridges

limit above which there is difficulty in walking. Irwin
proposed an upper limit curve for storm conditions which
approximates closely to the recommendation given for
footbridges in BS5400(1978).

7.3 Fatigue endurance
Bridges on major trunk roads may be loaded by as many as a
million heavy lorries per year which would amount to
about 5×10^8 axle loads during a life of 120 years. However,
the maximum stress under axle load is mainly a static
or pseudo-static effect. The dynamic components due
to vibration under traffic are usually small enough to
be neglected.
    Wind loading gives rise to many oscillations of stress
during the lifetime of a structure. Smith (1980) calculated the effect of the dynamic component of wind on
the fatigue endurance of the Wye Bridge in the UK. The
numerical data suggested that fatigue was not a significant
problem in that context.

8 Conclusions

Commercial finite element programs are now available which are capable of taking into account geometric stiffening in the analysis of long span bridges. The linearised theory has been shown, by comparison with tests on full-scale bridges, to be satisfactory for purposes of engineering design. Theoretical methods have been developed to include secondary non-linear effects due to large displacements. These effects are generally small but can result in coupling between vertical and torsional modes when the natural frequencies are very close.

It has been found that the longest span suspension bridges have many modes of vibration within the range of excitation by turbulent wind. The spectral analysis method needs to be developed into a discrete format so that more modes, many of which have complicated shapes, can be included.

Earthquake loading has proved to be a severe test of the design of short span bridges and multi-span viaducts in seismically active regions. Suspension bridges and other large bridges have a good record of performance in earthquakes. Wind loading, combined with dead and live load, is still probably the critical load case.

Human sensitivity to vertical motion is an important consideration in the design of bridges. Data exists to establish limits to ensure satisfactory performance over a wide range of frequencies.

9 References

Abdel-Ghaffar,A.M. (1980) Vertical vibration analysis of suspension bridges. J.Struct.Div.,ASCE,v106,ST10,p2053.

Abdel-Ghaffar,A.M. and Rubin,L.I. (1982) Suspension bridge response to multiple-support excitations. J.Engng.Mech., ASCE,v108,EM2,419-435.

Abdel-Ghaffar,A.M. and Rubin,L.I. (1983) Non-linear free vibrations of suspension bridges: theory. J.Engng.Mech., ASCE,v109,EM1,313-329

ATC (1978) Tentative Provisions for the Development of Seismic Regulations for Buildings. Applied Technology Council,ATC 3-06, National Bureau of Standards, USA.

Brownjohn,J.M.W., Dumanoglu,A.A., Severn,R.T. and Taylor,C. (1987) Ambient vibration measurments of the Humber Suspension Bridge and comparison with calculated characteristics. Proc.ICE,v83,part 2,561-600.

BS 5400 (1978) Steel, Concrete and Composite Bridges, Part 2: Specification for Loads. British Standards Institution,London.

Davenport,A.G. (1962) The response of slender, line-like structures to a gusty wind. Proc.ICE,v23,389-408.

Davenport,A.G. (1964) Note of the distribution of the largest value of a random function with application to gust loading. Proc.ICE,v28,187-196.

Davenport,A.G. (1982) Comparison of model and full-scale tests on bridges. Wind tunnel modelling for civil engineering applications (Reinhold,TA ed.) Cambridge Press

Davis,D.C.C.(1988) Discussion on paper 9172. Proc.ICE,Part2 v85,p731.

De Salvo,G.J. and Swanson,J.A.(1988) ANSYS: User Manual. Swanson Analysis Systems, Houston, PA.

Dumanoglu,A.A., Garevski,M. and Severn,R.T. (1988). Dynamic characteristics and seismic behaviour of Jindo bridge, South Korea. Struct.Eng.Review,v1,n3,141-150.

Dumanoglu,A.A. and Severn,R.T. (1985) Asynchronous seismic analysis of modern suspension bridges: Part 1: Free vibration. Research Report, University of Bristol.

Dumanoglu,A.A. and Severn,R.T.(1987) Seismic response of modern suspension bridges to asynchronous vertical ground motion. Proc.ICE, v83,part 2,701-730.

ESDU (1985) Characteristics of atmospheric turbulence near the ground. Item 85020, Engng.Sci.Data Unit, London.

Fleming,J.F and Egeseli,E.A.(1980) Dynamic behaviour of cable-stayed bridges. Earth.Engng.Struct.Dyn,v8,n1,1-16.

Irwin,A.W.(1978) Human response to dynamic motion of structures. The Structural Engineer,v56A,237-244.

Leonard,D.R.(1966) Human tolerance for bridge vibrations. TRRL Report LR34, Trans.and Road Res.Lab., UK.

Lew,H.S., Leyendecker,E.V. and Dikkers,R.D.(1971) Engineering aspects of the 1971 San Fernando Earthquake. Nat.Bur.Standards, Ser.no.40, USA.

Przemieniecki,J.S.(1968) Theory of matrix structural analysis. McGraw-Hill, New York.

Pugsley,A.G.(1949) On the natural frequencies of suspension chains. Quarterly J.Mech.& App.Mechs,II(4),412-418.

Scanlan,R.H.(1983) Aeroelastic simulation of bridges. Proc.ASCE,v109,ST12,p2829.

Smith,I.J.(1980) Wind induced dynamic response of the Wye Bridge. Engng.Structs.,v2,p202.

Tanaka,H. and Davenport,A.G.(1983) Wind induced response of the Golden Gate Bridge. J.Eng.Mech.,ASCE,v109,EM1,p296.

Tappin,R.G.R. and Clark,P.J.(1985) Jindo and Dolsan bridges: design. Proc.ICE,v78,part 1,1281-1300.

Vincent,G.S.(1962) Golden Gate Bridge vibration studies. Trans.ASCE,v127,part 2,667-707.

Wyatt,T.A.(1980) Evaluation of gust response in practice. Paper 7, Conf. on Wind Engineering in the Eighties, CIRIA, London.

DYNAMIC BEHAVIOUR OF CABLE-STAYED BRIDGES

J.C. WILSON, T. LIU, W. GRAVELLE
Dept Civil Engineering, McMaster University, Hamilton, Ontario,
Canada

Abstract
An extensive program of full-scale ambient vibration measurements has
been conducted to evaluate the dynamic response of a 542 meter cable-
stayed bridge. A total of 25 modal frequencies and associated mode
shapes were identified for the deck structure within the frequency
range of 0-2 Hz. Of these 25 modes, 10 were vertical modes, 10 were
coupled torsion-transverse modes, 3 were torsional modes, and 2 were
transverse modes. There was considerable dynamic interaction between
the deck and tower structures. Estimations were also made for damping
ratios for the first two modes. It is clearly evident from the
results of this test program that the dynamic properties of the cable-
stayed bridge are characterized by the occurrence of many modes having
closely-spaced frequencies and dimensional complexity in their mode
shapes.
Keywords: Cable-stayed Bridge, Ambient vibrations, Dynamics,
Experimental, Modes

1 Introduction

Cable-stayed bridges possess complicated dynamic behaviour when
excited by forces such as wind, traffic or earthquakes. This paper
investigates their dynamic behaviour with results of a full-scale
vibration measurement program conducted on a newly constructed cable-
stayed bridge in the United States. The test program was conducted to
determine dynamic properties of the 542 meter Quincy Bayview Bridge
which crosses the Mississippi River in Illinois (Liu, 1989). The test
program consisted of measurement of the vibration response of the
bridge to ambient levels of wind and traffic excitations. Detailed
information on experimentally evaluated dynamic characteristics of a
full-scale cable-stayed bridge is of current value in providing a
reference for verification of numerical modelling techniques. It is
of future value in leading towards a more complete understanding of
the dynamic response of cable-stayed bridges to environmental forces.
In this paper, emphasis is placed on experimental aspects and results
of the research program. Comparisons of the experimental measurements
with results of a three-dimensional finite element model are also
included.

2 The bridge

The Quincy Bayview Bridge, shown in Figure 1, (opened in September 1987), carries two lanes of traffic. The bridge deck consists of a concrete road surface, two main steel girders on the outer edges of the deck, and five steel stringers between the girders. The deck is continuous over supports at the two towers. The towers are supported on end-bearing piles driven through soft river bottom soil to rock.

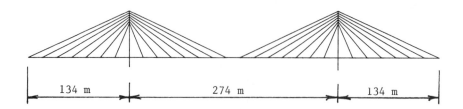

134 m   274 m   134 m

Fig. 1. Elevation view of the Quincy Bayview bridge

3 Ambient vibration measurement program

3.1 Test procedures

The schedule of the test program was developed after examination of previous experimental studies of suspension bridges (Abdel-Ghaffar and Scanlan, 1985; Brownjohn et al., 1987). However, the dynamic behaviour of a cable-stayed bridge is significantly different than that of a classical suspension bridge. A unique feature of the dynamic characteristics of cable-stayed bridges is that many of the modes possess multi-dimensional mode shapes and many modal frequencies are spaced at close intervals in the low frequency end of the spectrum. Thus, special attention must be paid to the design of a vibration measurement test program in order to obtain sufficient measurements at proper locations on the bridge in order to identify these complex dynamic properties.

In this test program all vibration measurements were conducted only on the eastern (Illinois) side as the bridge is very nearly symmetric about the mid-span. To identify vertical and torsional motions of the deck, a reference station (R) and a moving station (E) were established, as shown in Figure 2. Each station consisted of two vertically oriented transducers placed at the outer edges of the deck. The reference station was located at position R11 (11th cable anchor point from the east end) on the main span and the moving station was positioned successively at each cable anchor point along the deck. Information on vertical and torsional components can be obtained by adding and subtracting time domain signals from the two transducers located on the outer edges of the deck. The mode shapes can be

476

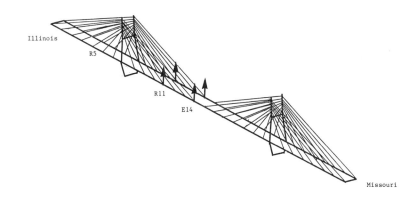

Fig. 2. Measurement of vertical and torsional motions

determined by comparing the spectral amplitudes at the moving station with the amplitudes at the reference station.

Figure 3 shows a typical test configuration for measuring the transverse vibration motions using two horizontally oriented trans-ducers. The three transducers at the reference station measure simul-taneous motions in the vertical, torsional, and transverse directions to provide three-dimensional amplitude and phase information. Dynamic interaction between the deck and the tower was examined by simultaneously measuring horizontal motions of the top of the tower and vertical and transverse motions of the deck. A total of 55 different test setups were used on the deck, tower and pier.

3.2 Instrumentation and data analysis
A 4-channel microcomputer-based structural vibration measurement system was used to measure, record and analyze the dynamic response of the bridge. Data were sampled at 50 points per second per channel with a total sampling time of 10 minutes per test configuration. This duration was based on the total number of planned test configurations, on the available data storage capacity of the microcomputer, and on some preliminary on-site spectral analyses. A few sets of records were also taken for longer durations to assist in determination of damping characteristics. After some preliminary investigations it was decided that post-processing spectral analyses would be limited to the frequency range of 0-2 Hz using a spectral frequency resolution of 0.01 Hz.

4 Experimental modal properties

4.1 Frequencies and mode shapes
A total of 25 vibration modes of the deck were identified within the frequency range of 0-2 Hz. These included vertical, torsional, trans-

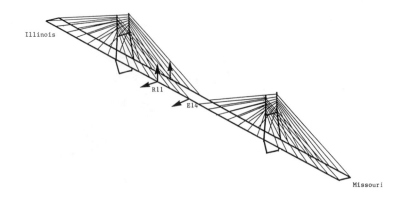

Fig. 3 Measurement of transverse motions and phase

verse, and coupled torsion-transverse modes. Longitudinal modes were
not identified in the tests. Table 1 lists the experimentally
identified deck modes and their classification by mode shape. Of
these 25 modes, 10 are pure vertical bending modes, 3 are pure
torsional modes, 2 are pure transverse bending modes, and 10 are
coupled torsion-transverse modes. The fundamental mode of the bridge
deck is a pure vertical bending mode with a natural frequency of
0.37 Hz. The first coupled torsion-transverse mode has a frequency of
0.56 Hz. Data points for the first four experimentally determined
vertical mode shapes are indicated by the "+" marks in Figure 4, and
the first two torsion-transverse modes are shown similarly in Fig. 5.
Experimental modal amplitudes are not shown for the centre span of the
third vertical mode because the reference station (R11) was at a node
of this mode. The smooth curves shown in these figures are from a
finite element solution, to be discussed later in the paper.

In the context of this experiment, a parameter called the **modal
coupling ratio** has been defined. This ratio is intended to provide a
measure of the degree to which different directional components of a
mode contribute to the complete mode shape. Here, the coupling ratio
for a coupled torsion-transverse mode can be expressed by dividing the
maximum vertical amplitude of the torsional component (at the edge of
the deck) by the maximum transverse modal amplitude. These ratios are
shown in parentheses in Table 1. The maximum amplitudes are the maximum absolute amplitudes occurring along the span for each component of
the mode (i.e., torsion, transverse), regardless of whether they
occur at the same or at different locations. The modal coupling
ratios for all coupled torsion-transverse modes, except the first mode
(Mode 3, f = 0.56 Hz), are greater than 1.0, which means that the
modal responses of these coupled torsion-transverse modes are
dominated by torsional components. In many cases, the vertical
amplitude of the torsional component greatly exceeds the transverse
amplitude.

TABLE 1.   Experimentally identified modes of the Quincy Bayview
           Bridge (deck)

Mode	Modal Frequency (Hz)	Mode Classification and Modal Coupling Ratio
1	0.37	Vertical
2	0.50	Vertical
3	0.56	Coupled Torsion-Transverse (0.71)
4	0.63	Transverse
5	0.70	Coupled Torsion-Transverse (2.45)
6	0.74	Coupled Torsion-Transverse (2.13)
7	0.80	Vertical
8	0.80	Torsion
9	0.89	Vertical
10	0.89	Coupled Torsion-Transverse (6.82)
11	1.06	Vertical
12	1.11	Coupled Torsion-Transverse (---)
13	1.18	Coupled Torsion-Transverse (2.78)
14	1.37	Vertical
15	1.40	Coupled Torsion-Transverse (4.55)
16	1.43	Vertical
17	1.44	Coupled Torsion-Transverse (5.35)
18	1.46	Vertical
19	1.47	Torsion
20	1.48	Vertical
21	1.68	Transverse
22	1.71	Coupled Torsion-Transverse (1.38)
23	1.75	Vertical
24	1.80	Coupled Torsion-Transverse (4.42)
25	1.92	Torsion

(---) R11 is at a node of the 1.11 Hz mode

   Although the frequency resolution used in the data analysis was
reasonably high (0.01 Hz), some spectral resolution problems were
encountered. Two pairs of very closely spaced modes were found: mode 7
(vertical) and mode 8 (torsion) at 0.80 Hz, and mode 9 (vertical) and
mode 10 (coupled torsion-transverse) at 0.89 Hz. Even with an
increased frequency resolution of 0.005 Hz, it was not possible to
obtain a clear separation of the frequencies of these modes. However,
other evidence in the form of phase and coherence values made it
possible to determine that separate shapes were involved in these
modes.

FIRST VERTICAL MODE

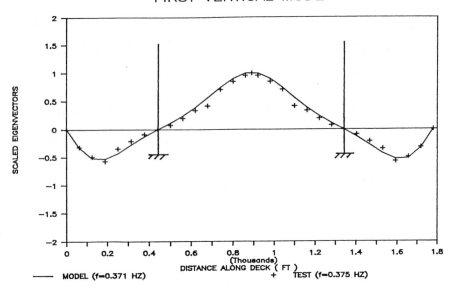

SECOND VERTICAL MODE

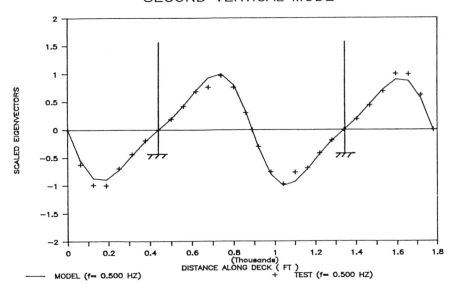

Fig. 4. Vertical mode shapes

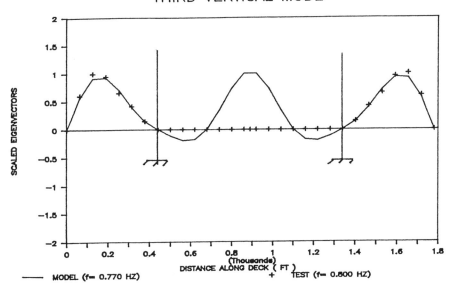

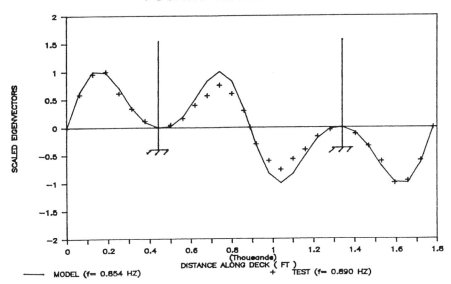

Fig. 4(cont'd) Vertical mode shapes

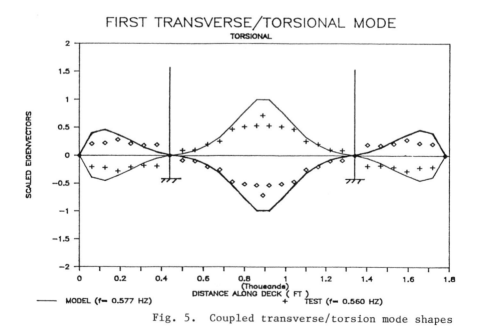

Fig. 5. Coupled transverse/torsion mode shapes

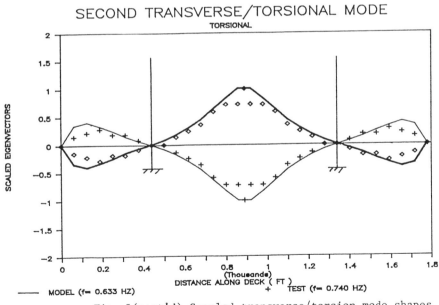

Fig. 5(cont'd) Coupled transverse/torsion mode shapes

4.2 Damping

A limited number of records of 20 minutes time duration were processed to examine modal damping ratios. The damping ratios for the first deck bending mode and first coupled torsion-transverse mode were estimated by employing the half power bandwidth method. The estimated ranges of the damping values for the first vertical bending mode and first coupled torsion-transverse mode of the deck are approximately 2.0-2.6% and 0.9-1.8%, respectively.

5 Comparisons with a finite element model

The mode shapes and frequencies evaluated in the experimental test program were compared to predictions of a linear elastic finite element model (Gravelle, 1990). Figure 6 shows a comparison of the experimental and finite element frequencies for the vertical and transverse/torsion modes. In general the finite element model predicts slightly lower frequencies than measured. Within the bandwidth of 0 to 2 Hz, there is very good agreement for the vertical frequencies, and reasonably good agreement in the more complicated coupled transverse/torsion frequencies. A comparison of experimental and analytical shapes of the first four vertical and first two transverse/torsion modes are shown by the solid lines in Figs. 4 and 5. The agreement in the computed and measured mode shapes is judged to be very good. More complete details are provided by Gravelle (1990).

6 Conclusions

(1) This test program has provided conclusive evidence of the complex dynamic behaviour of the Quincy Bayview Bridge, a design which is typical of many modern cable-stayed bridges. The dynamic response is characterized by many closely spaced coupled modes (25 modes with fre-quencies less than 2 Hz). The fundamental mode of the bridge (symmetric vertical) is 0.37 Hz and the first torsion mode which is coupled with a transverse response is 0.56 Hz. In terms of the total number of modes within the frequency range of 0-2 Hz, the vibration of the Quincy Bayview Bridge is dominated by vertical modes and coupled torsion-transverse modes. No experimental evidence was found to suggest that vertical vibration of the bridge deck is coupled with either torsional vibration or transverse vibration within the studied frequency range. For each coupled torsion-transverse mode except the first one (Mode 3, f = 0.56 Hz), the contribution of the torsional component predominates in the mode.

(2) The range of damping values found for the first vertical mode was 2.0-2.6% and for the first coupled torsion-transverse mode was 0.9-1.8%. These damping values provide a general indication of the level of damping which existed for low levels of ambient vibration, however these values should be used with the same caution that is appropriate for all experimental estimates of full-scale structural damping.

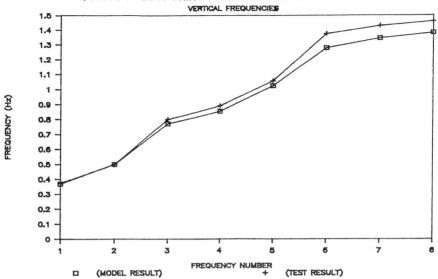

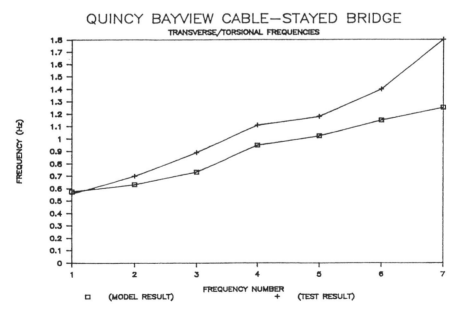

Fig. 6. Comparisons of experimental and analytical
        modal frequencies

(3) There is considerable interaction between the deck and the towers, and many tower modes were associated with deck motions within the frequency range of 0-2 Hz. However, the measurements were not sufficiently detailed to determine complete mode shapes of the tower.

(4) No evidence was found to suggest nonlinear dynamic behaviour of the Quincy Bayview Bridge at low levels of ambient excitation. The consistency of occurrence of modal peaks for different data records, and consistent phase and coherence information, indicate that the ambient dynamic response of this bridge structure is linear.

(5) A three-dimensional linear elastic finite model appears to be capable of representing accurately much of the complex dynamic behaviour (modal frequencies and mode shapes) of the bridge. It confirms the experimental observation that transverse and torsional motions of the deck are coupled, but that vertical vibration of the deck is not coupled with either torsion or lateral motions of the deck.

7 Acknowledgements

The authors would like to thank Don Kennedy and Ming Wong for their valuable assistance in the field test. The cooperation of Floyd Jacobsen and his staff from the Illinois Department of Transportation was a great asset. Appreciation is also extended to the Natural Sciences and Engineering Research Council of Canada for their support.

8 References

Abdel-Ghaffar, A. M., and Scanlan, R. H. (1985). "Ambient Vibration Studies of Golden Gate Bridge: I. Suspended Structure", **Journal of Engineering Mechanics, ASCE**, Vol. 111, No.EM4, pp. 463-482.

Brownjohn, J. M., Dumanoglu, A. A., Severn, R. T., and Taylor, C. A. (1987), "Ambient Vibration Measurements of The Humber Suspension Bridge and Comparison With Calculated Characteristics", **Proc. Instn. Civ. Engrs.** pp. 561-600.

Gravelle, W. (1990), Master's Thesis, submitted to McMaster University, Hamilton, Ontario, Canada.

Liu, T. (1989), "Full-Scale Ambient Vibration Measurements on a Cable-Stayed Bridge", Master's Thesis, submitted to McMaster University, Hamilton, Ontario, Canada.

NONSTATIONARY ANALYSIS OF SUSPENSION BRIDGES FOR MULTIPLE SUPPORT EXCITATIONS

C-H. HYUN
Nuclear Safety Center/KAERI, Daejon, Korea
C-B. YUN
Dept Civil Engineering, Korean Advanced Institute of Science and
Technology, Seoul, Korea
M. SHINOZUKA
Dept Civil Engineering, Princeton University, Princeton, USA

Abstract
A method for the nonstationary response analysis of suspension bridges subjected to earthquake ground excitations at multiple supports is developed. The equation of motion is formulated for the vertical motion of the bridge girder using a continuous coordinate system. The nonstationarity of the earthquake excitation is modelled by utilizing a time varying envelope function. The correlation effects between ground excitations at different supports are investigated, for various cases of travelling seismic waves with different wave propagation velocities. The nonstationary responses are obtained in terms of time dependent variance functions. Expected peak responses are also evaluated, thereafter. Numerical results from the example analysis indicate that the correlation effects between different support excitations are very significant on the response of the suspension bridge.
Keywords: Nonstationarity, Suspension Bridge, Earthquake, Multiple Support Excitation, Variances.

1 Introduction

For the seismic response analysis of suspension bridges, it is important to consider the nonstationary characteristics and the correlation effect of the earthquake excitations at multiple supports. The nonstationarity of the bridge response is caused by the nature of earthquake motion as well as the flexible nature of the structure accompanied with relatively long natural periods.

Recently, many studies have been reported on the seismic analysis of suspension bridges (Abdel-Ghaffar and Rubin 1982; Abdel-Ghaffar and Stringfellow 1984; Baron et al. 1976; Rubin et al. 1983). However, in most of the cases, the randomness of the earthquake excitation and structural response, particularly their nonstationary characteristics, are not properly analyzed. That is, the response analysis is carried out either by the ordinary spectral analysis which is based on the stationary assumption or by the time history analysis using a limited number of measured or artificially generated earthquake records. More rigorous approaches for the nonstationary response analysis have been reported for several cases of simpler structures (Lin 1965; Shinozuka et al. 1967 and 1968). By treating the nonstationary excitation as a sequence of random pulses, Lin developed a method to compute the response in terms of the log-characteristic functional. By modelling the earthquake ground motion as a filtered Poisson process, Shinozuka et al. developed a procedure to obtain the time dependent variance function of the response.

In this paper, a method for the nonstationary response analysis of suspension bridges subjected to multiple support seismic excitations is developed. The equation of motion is constructed for the vertical motion of the bridge girder using a continuous coordinate system. Effects of the horizontal and vertical ground accelerations at each support are included in the formulation. The nonstationary ground motion is represented by a product of a stationary process and a deterministic envelope function of time. The bridge responses are obtained as time dependent variance functions by using a similar approach to the one proposed by Shinozuka et al (1968). For the computational efficiency, the envelope function of the earthquake acceleration and the correlation function of its stationary component are expressed in terms of exponential and sinusoidal functions, so that the time dependent variance functions of the responses can be computed analytically. Expected peak responses are also evaluated. The correlation effects between ground motions at different supports are investigated for various cases of travelling earthquake waves without energy dispersion. Parametric studies are carried out for various shear wave velocities of the soil foundation.

2 Equation of motion

2.1 General equation of motion
Considering the vertical motion only, the linearized equation of motion of a typical three-span suspension bridge subjected to earthquake excitations at supports can be obtained as (Timoshenko and Young 1965) (Fig. 1)

$$\frac{w_{ti}}{g} \cdot \frac{\partial^2 \eta_i}{\partial t^2} + C_\eta \frac{\partial \eta_i}{\partial t} + E_i I_i \frac{\partial^4 \eta_i}{\partial x_i^4} - H_w \frac{\partial^2 \eta_i}{\partial x_i^2} + \frac{w_{ti}}{H_w} H_\eta(t) = 0 \tag{1}$$

$$H_\eta(t) = \frac{E_c A_c}{L_c} \sum_{i=1}^{3} \left\{ \frac{dy_i(x_i)}{dx_i} \eta_i(x_i,t) \Big|_0^{l_i} + \frac{w_{ti}}{H_w} \int_0^{l_i} \eta_i(x_i,t)dx_i + \xi_i(x_i,t) \Big|_0^{l_i} \right\} \tag{2}$$

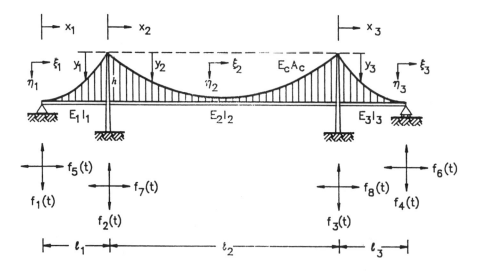

Fig. 1.   Schematic diagram of suspension bridge

where $\eta_i = \eta_i(x_i, t)$ is the total (not relative to the supports) vertical motion of the girder of the i-th span from a fixed datum; w_{ti} is the dead weight of the bridge per unit span length; E_i and I_i are the Young's modulus and moment of inertia of the bridge cross section; g is the gravitational acceleration constant; C_η is the structural damping coefficient; H_w is the sum of the horizontal components of the initial tension in two main cables; $H_\eta(t)$ is the fluctuating component of the horizontal cable tension due to the girder and support motions; x_i is the longitudinal (i.e., horizontal) coordinate of the i-th span; E_c, A_c and L_c are the modulus of elasticity, cross-sectional area and virtual length of the main cables; ξ_i is the horizontal displacement of the main cables at the tower tops or anchorages; l_i is the length of the i-th span; and $y_i(x_i)$ is the initial cable profile of the i-th span from the horizontal line due to the dead load and is expressed as

$$y_i(x_i) = \begin{cases} y_0(x_1) + \left[h - \dfrac{h}{l_1}x_1 \right] & , \text{ for } i = 1 \\[2mm] y_0(x_2) & , \text{ for } i = 2 \\[2mm] y_0(x_3) + \dfrac{h}{l_3}x_3 & , \text{ for } i = 3 \end{cases} \tag{3}$$

in which $y_0(x_i)$ is the cable profile of the i-th span measured from its chord as

$$y_0(x_i) = \frac{w_{ti}l_i^2}{2H_w}\left[\frac{x_i}{l_i} - \left(\frac{x_i}{l_i} \right)^2 \right], \text{ for } i = 1,2,3 \tag{4}$$

and h is the tower height from the deck level of the bridge. It is to be noted that, in their previous studies (Abdel-Ghaffar and Rubin 1982; Abdel-Ghaffar and Stringfellow 1984; Rubin et al. 1983), Abdel-Ghaffar et al. employed a formulation which is based on $y_0(x_i)$ instead of $y_i(x_i)$ for evaluation of $H_\eta(t)$ in Eq. 2, which is inappropriate to investigate the seismic response due to nonuniform support movements. The boundary conditions for the vertical motions of the girder at the supports are

$$\eta_i(0,t) = f_i(t) , \quad \eta_i(l_i,t) = f_{i+1}(t) , \quad \eta_i''(0,t) = \eta_i''(l_i,t) = 0 , \text{ for } i = 1,2,3 \tag{5}$$

where $f_i(t)'s$, $(i=1,2,3,4)$ are the vertical ground displacements at the supports as in Fig. 1 and the double prime($''$) denotes differentiation with respect to x.

Using Eqs. 3-5, Eq. 2 can be rearranged as (Hyun 1989)

$$H_\eta(t) = \frac{E_c A_c}{L_c}\left\{ \sum_{i=1}^{3}\left[\frac{w_{ti}}{H_w}\int_0^{l_i}\eta_i(x_i,t)dx_i - \frac{w_{ti}l_i}{2H_w}[f_{i+1}(t) + f_i(t)] \right] \right.$$

$$\left. + \frac{h}{l_1}[f_1(t) - f_2(t)] + \frac{h}{l_3}[f_4(t) - f_3(t)] + [f_6(t) - f_5(t)] \right\} \tag{6}$$

where $f_i(t)'s$, $(i=5,6)$ are the horizontal ground displacements along the longitudinal axis of the bridge at two anchorage points. Eq. 6 implies that the tension variation and the vertical response of the bridge are not only affected by the vertical ground motions, but also by two longitudinal ground motions at the end anchorages.

Now, substituting Eq. 6 into Eq. 1, equation of motion of a suspension bridge in the vertical direction can be rewritten as

$$\frac{w_{ti}}{g}\cdot\frac{\partial^2\eta_i}{\partial t^2} + C_\eta\frac{\partial\eta_i}{\partial t} + E_iI_i\frac{\partial^4\eta_i}{\partial x_i^4} - H_w\frac{\partial^2\eta_i}{\partial x_i^2}$$

$$+\frac{w_{ti}}{H_w}\cdot\frac{E_cA_c}{L_c}\left[\sum_{k=1}^{3}\frac{w_{tk}}{H_w}\int_0^{l_k}\eta_k\,dx_k + \sum_{j=1}^{6}A_jf_j(t)\right] = 0 \qquad (7)$$

where

$$A_1 = \frac{h}{l_1} - \frac{w_{t1}l_1}{2H_w} \qquad\qquad (8.a)$$

$$A_2 = -\frac{h}{l_1} - \frac{w_{t1}l_1}{2H_w} - \frac{w_{t2}l_2}{2H_w} \qquad\qquad (8.b)$$

$$A_3 = -\frac{h}{l_3} - \frac{w_{t2}l_2}{2H_w} - \frac{w_{t3}l_3}{2H_w} \qquad\qquad (8.c)$$

$$A_4 = \frac{h}{l_3} - \frac{w_{t3}l_3}{2H_w} \qquad\qquad (8.d)$$

$$A_5 = -1 \qquad\qquad (8.e)$$

$$A_6 = 1 \qquad\qquad (8.f)$$

2.2 Equation of motion for relative vertical displacement

The response of a bridge subjected to ground excitations may be evaluated as a sum of the quasi-static response due to the support movements and the additional (vibrational) displacement relative to the quasi-static response as,

$$\eta_i(x_i,t) = \sum_{j=1}^{6}g_{ji}(x_i)f_j(t) + v_i(x_i,t) \quad , i=1,2,3 \qquad (9)$$

where $g_{ji}(x_i)$ is the static displacement (influence function) of the girder at x_i due to a unit displacement of the j-th component of the support movements while the other supports are held fixed; and $v_i(x_i,t)$ is the relative displacement of the girder to the quasi-static response.

Substituting Eq. 9 into Eq. 7 and eliminating the quasi-static responses of Eq. 7, the equation of motion for the relative vertical displacement of the girder can be obtained as

$$\frac{w_{ti}}{g}\ddot{v}_i + C_\eta\dot{v}_i + E_iI_iv_i^{IV} - H_wv_i'' + \frac{w_{ti}}{H_w}\frac{E_cA_c}{L_c}\sum_{k=1}^{3}\left[\frac{w_{tk}}{H_w}\int_0^{l_k}v_k\,dx_k\right]$$

$$= -C_\eta\sum_{j=1}^{6}g_{ji}\dot{f}_j - \frac{w_{ti}}{g}\sum_{j=1}^{6}g_{ji}\ddot{f}_j \qquad (10)$$

where the boundary conditions are

$$v_i(0,t) = v_i(l_i,t) = v_i''(0,t) = v_i''(l_i,t) = 0 \quad , \quad i=1,2,3 \qquad (11)$$

In this study, only the relative vibrational responses are investigated, because it has been reported that they are much larger than the quasi-static response components (Baron et al. 1976; Rubin et al. 1983). The numerical results in the present study also

490

indicated that the quasi-static response component accounts for at most 10% of the total response.

Utilizing the mode superposition as

$$v_i = v_i(x_i,t) = \sum_{n=1}^{\infty} \phi_n(x_i) q_n(t) \tag{12}$$

where $\phi_n(x_i)$ is the n-th mode shape of the bridge and $q_n(t)$ is the n-th modal coordinate, and neglecting the terms associated with the ground velocities in the right hand side of Eq. 10 whose influence is insignificant (Baron et al. 1976; Rubin et al. 1983), Eq. 10 can be rewritten as

$$\ddot{q}_n(t) + 2\zeta_n \omega_n \dot{q}_n(t) + \omega_n^2 q_n(t) = -\sum_{j=1}^{6} \Gamma_{jn} \ddot{f}_j(t) \tag{13}$$

in which Γ_{jn} is the n-th modal participation factor associated with the j-th support excitation and given by

$$\Gamma_{jn} = [\sum_{i=1}^{3} w_{li} \int_0^{l_i} g_{ji}(x_i) \phi_n(x_i) dx_i] / [\sum_{i=1}^{3} w_{li} \int_0^{l_i} \phi_n^2(x_i) dx_i] \quad , \quad j=1,2,...,6 \tag{14}$$

3 Formulation of nonstationary response

The duration of the strong earthquake ground motion is usually limited to a period in the order of 10-30 seconds. Hence, for a structure with a relative long fundamental natural period as in the present case of a suspension bridge, the duration of the earthquake excitation may not be long enough to make the structure vibrate upto the steady state level. Therefore, for the seismic response analysis of such structures, the nonstationary characteristics of the ground excitation and structural response must be properly incorporated.

In this study, the nonstationary ground excitation(acceleration) is modelled as (Shinozuka and Sato 1967)

$$\ddot{f}_k(t) = a_k(t) r_k(t) \quad , \quad k=1,2,...,6 \tag{15}$$

where $a_k(t)$ is the deterministic envelope function that describes the nonstationarity of the ground acceleration at the k-th support; and $r_k(t)$ is the stationary component.

Then, the nonstationary seismic response can be evaluate in terms of time dependent variance functions. The variance functions of the relative vertical displacements, $v_i(x_i,t)$ and bending moment of the bridge girder, $s_i(x_i,t)$ can be obtained as

$$E[v_i^2(x_i,t)] = \sum_{n=1}^{\infty} \sum_{m=1}^{\infty} \phi_n(x_i) \phi_m(x_i) \cdot E[q_n(t) q_m(t)] \tag{16}$$

$$E[s_i^2(x_i,t)] = \left[E_i I_i\right]^2 \sum_{n=1}^{\infty} \sum_{m=1}^{\infty} \phi_n''(x_i) \phi_m''(x_i) \cdot E[q_n(t) q_m(t)] \tag{17}$$

where $E[\cdot]$ denotes the expectation. The covariance functions, which are to be used for estimating maximum responses, can be calculated as

$$E[v_i(x_i,t) \dot{v}_i(x_i,t)] = \sum_{n=1}^{\infty} \sum_{m=1}^{\infty} \phi_n(x_i) \phi_m(x_i) \cdot E[q_n(t) \dot{q}_m(t)] \tag{18}$$

$$E[\dot{v}_i(x_i,t) \dot{v}_i(x_i,t)] = \sum_{n=1}^{\infty} \sum_{m=1}^{\infty} \phi_n(x_i) \phi_m(x_i) \cdot E[\dot{q}_n(t) \dot{q}_m(t)] \tag{19}$$

$$E[\dot{s}_i(x_i,t)\dot{s}_i(x_i,t)] = \left[E_i I_i\right]^2 \sum_{n=1}^{\infty}\sum_{m=1}^{\infty} \phi_n''(x_i)\phi_m''(x_i) \cdot E[q_n(t)\dot{q}_m(t)] \qquad (20)$$

$$E[\dot{s}_i(x_i,t)\dot{s}_i(x_i,t)] = \left[E_i I_i\right]^2 \sum_{n=1}^{\infty}\sum_{m=1}^{\infty} \phi_n''(x_i)\phi_m''(x_i) \cdot E[\dot{q}_n(t)\dot{q}_m(t)] \qquad (21)$$

The time dependent covariance functions of the modal coordinates and its time-derivatives in Eqs. 16-21 can be computed using convolution integral as follows:

$$E[q_n(t)q_m(t)] = \sum_{i=1}^{6}\sum_{j=1}^{6}\Gamma_{in}\Gamma_{jm}\int_0^t\int_0^t h_n(t-\tau_1)h_m(t-\tau_2)R_{\ddot{f}_i\ddot{f}_j}(\tau_1,\tau_2)d\tau_1 d\tau_2 \qquad (22)$$

$$E[q_n(t)\dot{q}_m(t)] = \sum_{i=1}^{6}\sum_{j=1}^{6}\Gamma_{in}\Gamma_{jm}\int_0^t\int_0^t h_n(t-\tau_1)\dot{h}_m(t-\tau_2)R_{\ddot{f}_i\ddot{f}_j}(\tau_1,\tau_2)d\tau_1 d\tau_2 \qquad (23)$$

$$E[\dot{q}_n(t)\dot{q}_m(t)] = \sum_{i=1}^{6}\sum_{j=1}^{6}\Gamma_{in}\Gamma_{jm}\int_0^t\int_0^t \dot{h}_n(t-\tau_1)\dot{h}_m(t-\tau_2)R_{\ddot{f}_i\ddot{f}_j}(\tau_1,\tau_2)d\tau_1 d\tau_2 \qquad (24)$$

where

$$R_{\ddot{f}_i\ddot{f}_j}(\tau_1,\tau_2) = E[\ddot{f}_i(\tau_1)\ddot{f}_j(\tau_2)] = a_i(\tau_1)a_j(\tau_2)R_{r_i r_j}(\tau_1-\tau_2) \qquad (25)$$

$$h_n(t) = \frac{1}{\omega_{nd}}e^{-\zeta_n\omega_n t}\sin\omega_{nd}t \qquad (26)$$

$$\omega_{nd} = \omega_n(1-\zeta_n^2)^{1/2} \qquad (27)$$

in which $R_{r_i r_j}(\tau_1-\tau_2)$ is the cross-correlation function of the stationary components of the support accelerations.

4 Support ground motions for travelling seismic waves

In general, the correlation of the ground motions at different supports is extremely complicated, particularly when the foundation condition varies along the path of the seismic wave propagation. Therefore, in this paper, special cases for propagating earthquake motions without energy dispersion are considered. Then, the cross-correlation function, $R_{\ddot{f}_i\ddot{f}_j}(\tau_1,\tau_2)$, of the ground accelerations at two supports i and j can be obtained as

$$R_{\ddot{f}_i\ddot{f}_j}(\tau_1,\tau_2) = \begin{cases} a_1(\tau_1-T_i)a_1(\tau_2-T_j)R_{r_1 r_1}(\tau_1-\tau_2-T_i+T_j)U(\tau_1-T_i)U(\tau_2-T_j), & i,j=1,2,3,4 \\ a_5(\tau_1-T_i)a_5(\tau_2-T_j)R_{r_5 r_5}(\tau_1-\tau_2-T_i+T_j)U(\tau_1-T_i)U(\tau_2-T_j), & i,j=5,6 \end{cases}$$

$$(28)$$

where $R_{r_1 r_1}(t)$ is the autocorrelation function of the stationary component of the vertical ground acceleration at the left anchorage; $R_{r_5 r_5}(t)$ is the one for the horizontal ground acceleration; and $T_1=0$ and T_i's (i=2,3,4) are time lags between the left anchorage and the i-th support($T_5=T_1$ and $T_6=T_4$); and $U(t-T_i)$ is a unit step function.

For evaluating $E[q_n(t)q_m(t)]$, $E[q_n(t)\dot{q}_m(t)]$ and $E[\dot{q}_n(t)\dot{q}_m(t)]$ in Eqs. 22-24, it is required to perform double integrations. If the envelope function $a_k(t)$ and the correlation function $R_{r_k r_k}(t)$ of the k-th support motion can be determined in proper

492

analytical expressions, the double integration may be carried out analytically. Considering the fact that the impulse response function $h_n(t)$ is given as a product of an exponential and a sinusoidal (i.e., complex exponential) functions, in this study, $a_k(t)$ is expressed in terms of exponential functions (Shinozuka et al. 1968) and $R_{r_k r_k}(t)$ is in terms of products of exponential and sinusoidal functions as

$$a_k(t) = \frac{e^{-\alpha'_k t} - e^{-\beta'_k t}}{e^{-\alpha'_k t_k^*} - e^{-\beta'_k t_k^*}} \quad , \quad (t_k^* = \frac{1}{\alpha'_k - \beta'_k} \ln \left[\frac{\alpha'_k}{\beta'_k} \right] , \ \beta'_k > \alpha'_k > 0) \tag{29}$$

$$R_{r_k r_k}(\tau) = \sum_{l=1}^{n_l} \gamma_{lk} e^{-a_{lk}|\tau|} [\cos \omega_{lk}\tau + \frac{a_{lk}}{\omega_{lk}} \sin \omega_{lk}|\tau|] \ , \ k=1,5 \tag{30}$$

where α'_k, β'_k, a_{lk}, ω_{lk}, and γ_{lk} are constants which may be determined based on the measured earthquake records; n_l is the number of functions required to express the dynamic characteristics of earthquake excitation appropriately; and index k indicates the vertical($k=1$) or horizontal($k=5$) components of propagating earthquake motions. Then, the double integrations can be efficiently carried out and analytical expressions of the covariance functions can be obtained. The power spectral density(PSD) function corresponding to the correlation function in Eq. 30 can be obtained as

$$S_{r_k r_k}(\omega) = \sum_{l=1}^{n_l} \frac{2 a_{lk} \gamma_{lk}}{\pi} \frac{a_{lk}^2 + \omega_{lk}^2}{[a_{lk}^2 + (\omega + \omega_{lk})^2][a_{lk}^2 + (\omega - \omega_{lk})^2]} \ , \ k=1,5 \tag{31}$$

In this paper the constant values in Eq. 31 are evaluated by curve-fitting to the PSD functions obtained from measured earthquake data.

5 Evaluation of nonstationary response

Substituting Eqs. 26, 29 and 30 into Eqs. 22-24, the time dependent covariance functions of the modal coordinates can be calculated analytically as (Hyun 1989)

$$E[q_n(t)q_m(t)] = I_{nm}^1(t) \tag{32}$$

$$E[q_n(t)\dot{q}_m(t)] = -\zeta_m \omega_m I_{nm}^1(t) + \omega_{md} I_{nm}^2(t) \tag{33}$$

$$E[\dot{q}_n(t)\dot{q}_m(t)] = \zeta_n \omega_n [\zeta_m \omega_m I_{nm}^1(t) - \omega_{md} I_{nm}^2(t)]$$
$$+ \omega_{nd}[-\zeta_m \omega_m I_{nm}^3(t) + \omega_{md} I_{nm}^4(t)] \tag{34}$$

where

$$I_{nm}^\delta(t) = \sum_{i=1}^{6} \sum_{j=1}^{6} \sum_{l=1}^{n_l} \sum_{p=1}^{4} \frac{\gamma_{lk} H_{nmp}}{8\omega_{nd}\omega_{md}} \sum_{i_1=1}^{2} \sum_{j_1=1}^{4} \left[e^{a_{lk} z_{ij}} V_{nmijlpi_1 j_1}^\delta(t) + e^{-a_{lk} z_{ij}} W_{nmijlpi_1 j_1}^\delta(t) \right]$$

$$, \text{ for } \delta=1,2,3,4 \tag{35}$$

and $z_{ij} = T_i - T_j$ is the time-lag of the travelling earthquake wave at the i-th support relative to that at the j-th supports; $k=1$ (for $1 \le i, j \le 4$) or 5(for $5 \le i, j \le 6$); and $V_{nmijlpi_1 j_1}^\delta(t)$, $W_{nmijlpi_1 j_1}^\delta(t)$ and H_{nmp}, are constants which can be obtained explicitly as functions of structural properties (ζ_n, ω_n, Γ_{in}) and parameters of earthquake excitations (α'_k, β'_k, a_{lk}, ω_{lk}, T_i) (Hyun 1989).

Finally, the expected maximum of the nonstationary response is evaluated using the approach by Shinozuka and Yang (1971) as

$$E[v_m] = \sigma[K + 0.5772K^{1-\alpha}] \tag{36}$$

where

$$K = (\alpha \ln \overline{N})^{1/\alpha} \ , \quad (\overline{N} = 2 \int_0^T v^+(0,t)dt \ , \quad v^+(0,t) = \frac{1}{2\pi} \frac{\sigma_{\dot{v}}(t)}{\sigma_v(t)} \left[1 - \rho^2(t)\right]^{1/2}) \tag{37}$$

and $E[v_m]$ is the asymptotic expected value of the maximum response v_m as the time duration(T) increases; $v^+(0,t)$ is the expected rate of upcrossing of level zero of the response at time t; $\rho(t)$ is the correlation coefficient between $v(t)$ and $\dot{v}(t)$; and α and σ are the parameters which can be determined by fitting the peak distribution of the response obtained in this study to the Weibull distribution.

6 Numerical examples and discussions

Example analyses have been carried out for the Golden Gate Bridge in the United States. Structural properties of the bridge are summarized in Table 1. Dynamic analysis has been performed using the first five symmetric modes. The corresponding natural periods were obtained as 8.15, 6.45, 3.92, 3.48 and 2.40 seconds. The asymmetric modes of the bridge are not used, since the response associated with them are found to be insignificant, as reported in Rubin et al. (1983). Modal damping ratio is assumed to be 2% for each mode. Dynamic responses are shown only at the midpoint of the center span for brevity.

Two cases of support earthquake excitations with different dynamic characteristics are considered. The parameters for the envelope and power spectral density functions are approximately evaluated based on two sets of earthquake records; i.e., the 1940 El Centro earthquake(NS and vertical components in Fig. 2) for Case I and the 1971 San Fernando earthquake(S16E and vertical components) for Case II. The power spectral density functions are estimated fairly conservatively during the curve fitting. The estimated parameters for Case I are summarized in Table 2. Fig. 3 shows the corresponding power spectral density functions of the stationary components. In this paper, dynamic responses have been calculated considering the relative vibrational responses only, since the numerical investigation indicated that the quasi-static components account for at most 10 % of total responses.

Fig. 4 shows the time-varying variance functions of the nonstationary responses due to travelling earthquake excitations(Case I) for various shear wave velocities of

Table 1. Parameters of Golden Gate Bridge

w_{ii} (kips/ft)	$i=1,3$	23.10
	$i=2$	22.70
E_i (ksi)		29000
I_i (ft^{2}in^2)	$i=1,3$	56000
	$i=2$	86400
H_w (kips)		106934
A_c (in^2)		1663.8
E_c (ksi)		29000
l_i (ft)	$i=1,3$	1125
	$i=2$	4200
L_c (ft)		7534
h (ft)		500

Table 2. Parameters of earthquake excitations

| Parameters | Case I | |
	$k=1$	$k=5$
α_k' (1/sec)	0.1167	0.0834
β_k (1/sec)	4.651	1.166
a_{1k} (1/sec)	1.980	2.973
ω_{1k} (rad/sec)	5.975	5.890
γ_{1k} (cm^2/sec^3)	193.8	4897.
a_{2k} (1/sec)	9.265	3.591
ω_{2k} (rad/sec)	28.94	11.05
γ_{2k} (cm^2/sec^3)	4089.	2462.

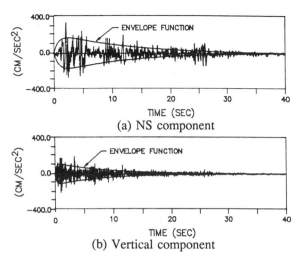

Fig. 2.   1940 El Centro earthquake acceleration records and assumed envelopes

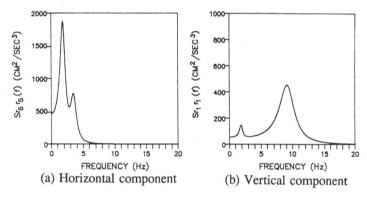

(a) Horizontal component   (b) Vertical component

Fig. 3.   PSD's of stationary parts of ground accelerations assumed in Case I

the soil foundation; i.e., V_k =200, 800, 5000 and 100000 ft/sec. The shear wave velocity of 5000 ft/sec is considered to be appropriate for the rock formation which is the
case for the Golden Gate Bridge (Baron et al. 1976). Results for two extreme cases
with fully-correlated and uncorrelated support excitations are also shown in the
figures for the purpose of comparison. The results indicate that the girder responses
show quite different nonstationary characteristics depending on the shear wave velocity of the soil foundation(i.e., the correlation between support excitations). It can be
seen that bridge responses reach their maximum values at about 10-20 seconds after
the earthquake starts, and then the responses die down very gradually. The responses
last significantly longer than the earthquake excitations. It is also interesting to
notice that, in the case of bending moment, variance functions particularly for
V_k =200 ft/sec show two apparent peaks, which are caused by the time lag of the pro-

pagating earthquake wave from one anchorage point to the other. From the results, it can be observed that the correlation effects between support excitations are very significant to the bridge response. In general, the uncorrelated assumption gives overly conservative response, while the fully correlated assumption predicts significantly unconservative response. The unconservative results for the fully correlated support excitations, are due to the fact that the vertical bridge motion is not affected by the uniform horizontal movements of the supports. On the other hand, for the partially correlated cases(i.e., those for travelling earthquake waves), the horizontal support motions contribute very significantly to the bridge response in the vertical direction as discussed later.

Table 3 shows the expected peak responses obtained in the present study along with those by the time history analysis method and the Vanmarke's method (Gasparini and Vanmarke 1976). In the time history analysis, twenty different nonstationary time histories are simulated based on the power spectral density function (Eq. 31) and the envelope function (Eq. 29). Then, the maximum responses are calculated for each case, and the averages of the results are obtained, thereafter. The results in the

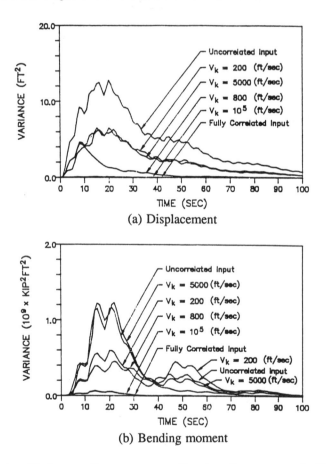

(a) Displacement

(b) Bending moment

Fig. 4.   Variances of responses at mid-point of center span in Case I

Table 3. Expected maximum displacement at mid-point of center span

(a) Case I   (unit : feet)

Input		Time history analysis[a]	Present method	Vanmarke's method[a] Confidence level(%)	
				50	95
Propagating	200	3.953(18.0)[b]	5.033	5.106	8.190
wave	800	3.135(20.0)	3.228	3.006	4.436
velocity	5000	4.794(17.8)	5.118	6.664	11.90
(ft/sec)	10^5	0.775(7.70)	0.703	1.011	1.515
Fully correlated input		0.688(10.0)	0.657	0.880	1.337
Uncorrelated input		---	7.187	7.499	16.76

(b) Case II   (unit : feet)

Input		Time history analysis[a]	Present method	Vanmarke's method[a] Confidence level(%)	
				50	95
Propagating	200	9.564(23.5)[b]	12.66	13.44	22.35
wave	800	6.800(25.5)	7.664	7.303	10.90
velocity	5000	9.790(20.4)	11.57	13.64	24.43
(ft/sec)	10^5	2.538(14.6)	2.874	3.402	5.164
Fully correlated input		2.372(15.7)	2.802	3.204	4.923
Uncorrelated input		---	17.09	16.10	36.32

[a] Average of maximum responses for 20 different simulated time histories.
[b] Coefficients of variation(%) of maximum responses.

Table 4. Expected maximum responses at mid-point of center span due to each component of ground motion (V_k=5000 ft/sec in Case I)

Support acceleration	Displacement(ft)	Bending moment(kip-ft)
Horizontal component only	5.076	66110.
Vertical component only	0.642	11980.
All component	5.118	66960.

tables show that the coefficients of variation of the maximum responses obtained by using different simulated earthquake motions are fairly large. However, the average values of the peak responses for the twenty simulated earthquakes are found to be reasonably close to the results by the present method. Unlike to the results of the present study, the peak values by Vanmarke's approach have been evaluated utilizing the variances computed by the (stationary) spectral analysis. In this approach, the power spectral density functions of the earthquake accelerations are obtained approximately, by utilizing the Fourier transforms of the simulated nonstationary earthquake records which were used in the time history analysis. It is also found that the the the results by the Vanmarke's method with 50% confidence level are reasonably compared with those by the present approach.

Finally, in Table 4, the contributions of the horizontal and vertical components of support accelerations are compared for the earthquake load Case I with V_k =5000 ft/sec. From the table, it can be seen that the contributions of the horizontal components of ground motions are dominant to the total response of the bridge girder. In other word, the responses of suspension bridges are more significantly affected by the horizontal (longitudinal) earthquake components than by the vertical components.

7 Conclusions

In this paper, an efficient method for the nonstationary response analysis of suspension bridges subjected to multiple support earthquake excitations is developed. In this approach, time dependent variance functions of the bridge responses are evaluated. The correlation effects of the earthquake excitations at different supports are investigated. It has been found that the correlation between support excitations plays a very important role on response characteristics of suspension bridges. In general, uncorrelated support motions give conservative responses. On the other hand, fully correlated support excitations yield very unconservative responses. It is also found that the effect of the horizontal (longitudinal) earthquake component on the responses of suspension bridges is more significant than that of the vertical component.

8 References

Abdel-Ghaffar, A. M. and Rubin, L. I. (1982). "Suspension bridge response to multi-support excitations." **J. of the Eng. Mech. Div.**, ASCE, 108(2), 419-435.

Abdel-Ghaffar, A. M., and Stringfellow, R. G. (1984). "Response of suspension bridges to travelling earthquake excitations: PART I. vertical response." **Soil Dynamics and Earthquake Engineering**, 3(2), 62-72.

Baron, F., Arikan, M., and Hamati, E. (1976). "The effects of seismic disturbances on the Golden Gate Bridge." **Report No. EERC 76-31**, Earthquake Engineering Research Center, University of California, Berkeley, C.A.

Gasparini, D. A., and Vanmarke, E. H. (1976). "Simulated earthquake motions compatible with prescribed response spectra." **MIT, Publication No. R76-4**.

Hyun, C. H. (1989). "Stochastic response analysis of suspension bridges for wind and earthquake loads." thesis presented to KAIST, Seoul, Korea, in partial fulfillment of the requirements for the degree of Doctor of Philosophy.

Lin, Y. K. (1965). "Nonstationary excitation and response in linear systems treated as sequences of random pulses." **J. of the Acoust. Soc. of Am.**, 38(3), 453-460.

Rubin, L. I., Abdel-Ghaffar, A. M., and Scanlan, R. H. (1983). "Earthquake response of long-span suspension bridges." **Report No. 83-SM-13**, Dept. of Civil Eng., Princeton University, Princeton, N.J.

Shinozuka, M., Itagaki, H., and Hakuno, M. (1968). "Dynamic safety analysis of multistory buildings." **J. of the Struc. Div.**, ASCE, 94(1), 309-330.

Shinozuka, M., and Sato, Y. (1967). "Simulation of nonstationary random process." **J. of the Eng. Mech. Div.**, ASCE, 93(1), 11-40.

Shinozuka, M., and Yang, J.-N. (1971). "Peak structural response to nonstationary random excitations." **J. of Sound and Vibration**, 16(4), 505-517.

Timoshenko, S. P., and Young, D. H. (1965). **Theory of structures**. 2nd Ed., Mc-Graw Hill, Inc., New York, N.Y.

9 Appendix A. Conversion to SI units

1 ft = 0.305 m, 1 in. = 0.0254 m, 1 kip = 4.45 KN, 1 ksi = 6.89 MPa

MICROCOMPUTER-BASED EXTRACTION OF CABLE-STAYED BRIDGE NATURAL MODES

A. NAMINI
Dept Civil and Architectural Engineering, University of Miami, Coral
Gables, Florida, USA

Abstract
In any design of cable-stayed bridges, dynamic natural modes must be computed
in order to analyze the proposed bridge against dynamic loadings. This paper
presents a computational formulation for the extraction of bending, torsional, and
sway natural modes using the subspace iteration method with starting iteration
vectors generated by the Lanczos method. The formulation is developed for the
IBM-compatible microcomputer environment implemented via an out-of-core
storage scheme, necessitated by the Disk Operating System's memory addressing
limitation of 640 kilobytes for program size. The resulting algorithm is then
applied to the Luling Bridge, a double-plan fan-type cable-stayed bridge located in
New Orleans, USA. Though runtime is slower than commercially available
mainframe programs, the algorithm's accuracy compares extremely well to
previous analyses.
Keywords: Cable-Stayed Bridges, Natural Modes, Microcomputer, Frontal Method.

1 Introduction

Cable-stayed bridge dynamic behavior involves local and global vibration
tendencies. Local behavior pertains to the dynamic properties associated with one
component of the overall bridge. For instance, cables may be excited into violent
oscillations due to resonance or galloping. This paper will not concern itself with
local vibration.

Global behavior pertains to the vibration tendencies associated with the entire
structure, whereby the deck, tower, and cable components unite to offer resistance
to dynamic loads. For instance, the original Tacoma Narrows suspension bridge
was destroyed by divergent oscillations due to aerodynamic flutter during a mild
gale of 68 km/hr. Other wind related dynamic phenomenon are: vortex-shedding,
which occurs when von Karman vortices are shed alternately from the upper and
lower edges of the deck equal to one of the bridge's natural frequencies; and
buffeting, which is the forced response due to random components of the wind,
either produced by natural turbulence or by upwind obstructions such as mountains
or buildings. Though not destructive, vortex-shedding produces a large steady-
state amplitude response which is discomforting to vehicle passengers and prone to
accelerate fatigue damage, while buffeting may induce alarming vibrations.

Other dynamic responses which must be accounted for are earthquakes and
moving vehicles. Earthquakes impart a sharp lateral acceleration at the base of
the bridge, which if strong enough may excite the lower natural modes.
Theoretically moving vehicles, such as many large trucks, if driven at high speeds

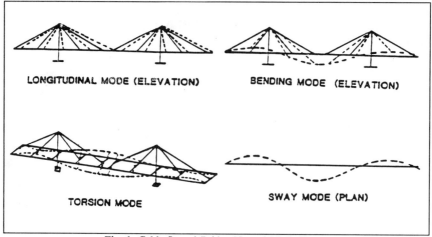

Fig. 1. Cable-Stayed Bridge Natural Mode Types

or with poor shock absorption mechanisms, many induce large vibrations though its probability is rather minimal.

All of the stated global dynamic tendencies have one common attribute. Namely, the excitation of a bridge natural mode, either bending, torsional, or sway. Therefore, any cable-stayed bridge design must at some point, computationally extract natural modes. In global dynamic behavior, response motion is dominated by only the lowest few natural modes. As shown in Figure 1, cable-stayed bridges have four predominant types of vibration: longitudinal, oscillation along the plane of the span; bending, oscillation vertical to the span; sway, oscillation transverse to the span; and torsional, oscillation pivoting about the deck's elastic axis. In general, longitudinal vibration is not important for cable-stayed bridges, because inclined cables provide horizontal stabilizing forces in the deck. Typically, natural modes are coupled between the different types, with straight bridges exhibiting a low degree, while horizontally or vertically curved spans displaying a high degree of coupling.

To extract cable-stayed bridge modes, previous investigators have developed methods suitable for finite element discretization. Tang (1971), Cheung and Kajita (1973), and Morris (1974) all determined natural modes using a linear-elastic lumped mass formulation, though they did not reveal their solution method. Izyumov, Tschanz, and Davenport (1977) also determined natural modes using a lumped mass formulation but considered the coupling between torsional and sway modes caused by the vertical eccentricity between the deck's center of gravity and elastic axis. Again, their solution method was not explicitly stated. Fleming and Egeseli (1976) stated that linear elastic lumped-mass formulation offers the best discretization model with the solution method being any acceptable eigensystem solution technique. More recently, Kumarasena, Scanlan, and Morris (1989) have analyzed the Deer-Isle suspension bridge, with a lumped-mass formulation while using a commercially available software package for the extraction of the lowest few natural modes.

This paper details a microcomputer-based algorithm for the determination of uncoupled natural modes using the subspace-iteration method (Bathe and Ramaswamy, 1980). In order to accelerate convergence, starting iteration vectors are generated by the Lanczos method for matrix tridiagonalization. The three-dimensional cable-stayed bridge's inertial forces are modeled with either a consistent or lumped mass formulation, while the elastic forces are modeled with space frame elements for the deck and tower, and the well-known Ernst rod element for the cables. Computationally, the entire algorithm is implemented via an out-of-core storage scheme, necessitated by the DOS's memory addressing limitation of 640 kilobytes. Finally, the resulting algorithm is applied to the Luling Bridge, a double-plan fan-type cable-stayed bridge which opened to traffic in 1983.

2 Equations of Motion

For a discretized bridge model with space frame elements used for deck and tower members, and Ernst (1965) rod elements used for cable members, the assembled equations of motion for undamped response can be expressed in global matrix form as,

$$[M] \{q_{tt}\} + [K] \{q\} = \{Q\} \tag{1}$$

where, [M] is the symmetrical mass matrix of order N, either lumped or consistent; [K] is the symmetrical stiffness matrix of order N; {Q} is the load vector of order N; {q} is response deformation vector of order N; and the subscript, tt, used to represent in tensor form, the second partial derivative of the deformation vector with respect to time, t, ie. $\partial^2 q/\partial t^2$, the response acceleration vector. The order of matrices, N, is the total number of distinct degrees of freedom used to discretize the bridge. Natural modes are defined to be response amplitudes for a null load vector. Therefore, with {Q} having all entries being zero, the response motion is assumed to be harmonic at all degrees of freedom and written along the complex plane as,

$$\{q\} = \{\Phi\} \exp(i\omega t) \tag{2}$$

where, {Φ} is response amplitude vector; and ω is response circular frequency. Upon substitution of Eq. 2 into Eq. 1, with {Q} being a null vector, valid solutions exist only if the following set equations are met,

$$[K] \{\Phi\} = \omega^2 [M] \{\Phi\} \tag{3}$$

Eq. 3 represents an eigenvalue problem, in which N unique solution exist, one for each degree of freedom of the modeled bridge. The eigenvalues are ordered as follows,

$$0 < \lambda_1 \leq \lambda_2 \leq \lambda_3 \leq ... \leq \lambda_N \tag{4}$$

where, $\lambda_i = \omega_i^2$, eigenvalue corresponding to the i-th lowest natural mode, i=1,2,3,...,N. For each eigenvalue, there is an associated eigenvector, $\{\Phi\}_i$. All eigenvalues are positive since both the mass and stiffness matrices are positive definite, physically requiring that no rigid body modes exist.

3 Subspace Iteration Method

For a desired set of P lowest eigensolutions, $(\lambda_i, \{\Phi\}_i)$, i=1,2,3,...,P, the subspace iteration method extracts the P eigensolutions with the following algorithm.

Establish starting iteration vectors $[X_1]$, arranged in a matrix of N rows by S columns, where S is the subspace size and $S > P$.

For $k = 1,2,3,...$ **until** Convergence **Do**

Perform an inverse iteration, solving for $[X^S_{k+1}]$,

$$[K] [X^S_{k+1}] = [M] [X_k] \tag{5}$$

Compute the stiffness and mass matrices projected onto the S-dimensional subspace,

$$[K^S_{k+1}] = [X^S_{k+1}]^T [K] [X^S_{k+1}] \tag{6}$$

$$[M^S_{k+1}] = [X^S_{k+1}]^T [M] [X^S_{k+1}] \tag{7}$$

Solve for the eigensolutions of the S-dimensional subspace,

$$[K^S_{k+1}] [Q^S_{k+1}] = [M^S_{k+1}] [Q^S_{k+1}] [\Omega_{k+1}] \tag{8}$$

Calculate improved approximations of the eigenvectors,

$$[X_{k+1}] = [X^S_{k+1}] [Q^S_{k+1}] \tag{9}$$

Check for convergence for all eigenvalues, i=1,2,3,...,P,

$$| (\lambda_i)_{k+1} - (\lambda_i)_k | / (\lambda_i)_{k+1} < \varepsilon \tag{10}$$

EndFor

Starting iteration vectors, $[X_1]$, must be linearly independent of each other, so that convergence will occur to different eigensolutions. Then, providing that the starting iteration vectors are not orthogonal to any of the desired P eigenvectors, the i-th diagonal entry of $[\Omega_{k+1}]$ converges to the eigenvalue, λ_i, and the i-th column vector of $[X_{k+1}]$ converges to the eigenvector, $\{\Phi\}_i$, as k approaches ∞.

Bathe (1977) states the ultimate rate of convergence, γ_i, of the i-th iteration vector to the desired i-th eigenvector will be

$$\gamma_i = \lambda_i / \lambda_{S+1} \tag{11}$$

and the ultimate rate of convergence of the i-th eigenvalue is

$$\gamma_i = (\lambda_i / \lambda_{S+1})^2 \tag{12}$$

According to Eqs. 11 and 12, convergence is best attained when the subspace is large in comparison to the number of eigensolutions sought, i.e. $S >> P$, while noting that when $S = P$, the P-th eigensolution converges extremely slowly. The computer effort to perform one subspace iteration is considerable for large three-dimensional idealizations. Bathe (1982) has stated that the subspace size for optimal runtime convergence is

$$S = \min \{ 2P, P+8 \} \tag{13}$$

which is based on a bandwidth stored stiffness and mass matrix, with the inverse iteration being solved in-core by the Cholesky decomposition method.

4 Lanczos Starting Iteration Vectors

The greatest difficulty in using the subspace iteration method is to choose starting iteration vectors, $[X_1]$, that are linearly independent to each other and are not orthogonal to any of the lowest desired eigenvectors. The better the starting iteration vectors approximate the true eigenvectors, the quicker the method converges.

There are currently three acceptable methods of generating starting iteration vectors. The first method consists of choosing random vectors, without regard to the known attributes of the bridge model discretization. In the second or standard method, a full unit vector is chosen for the first iteration vector. The elements of the next (S-2) vectors are set equal to zero, except for an unit entry corresponding to the next (S-2) degrees of freedom having the largest ratios of mass to stiffness. The last iteration vector is a random vector. The standard method implies that the degrees of freedom with the largest mass to stiffness ratios will be excited to the lowest eigensolutions. This is rarely the case for large three-dimensional idealizations.

The third, and by far the most powerful, was developed by Lanczos (1950) for the tridiagonalization of matrices. The Lanczos generated vectors will be used as starting iteration vectors for the subspace iteration method. The method computes vectors with the following procedure.

With $\beta_1 = 0$, mass normalize a full unit vector $\{X_0\}$ to obtain the first iteration vector, $\{X_1\}$, as

$$\{X_1\} = \{X_0\} / (\{X_0\}^T [M] \{X_0\})^{1/2} \tag{14}$$

For $i = 2,3,4,...$ **until S Do**

Perform an inverse iteration, solving for $\{X^L_i\}$,

$$[K] \{X^L_i\} = [M] \{X_{i-1}\} \tag{15}$$

Solve for major diagonal terms of tridiagonal,

$$\alpha_{i-1} = \{X^L_i\}^T [M] \{X_{i-1}\} \tag{16}$$

Compute the non-Mass-normalized iteration vector,

$$\{X^L_i\} = \{X^L_i\} - \alpha_{i-1} \{X_{i-1}\} - \beta_{i-1} \{X_{i-2}\} \tag{17}$$

Determine the Mass-normalized scalar, which is tridiagonal off-diagonal term,

$$\beta_i = (\{X^L_i\}^T [M] \{X^L_{i-1}\})^{1/2} \tag{18}$$

Mass-normalize the iteration vector,

$$\{X_i\} = \{X^L_i\} / \beta_i \tag{19}$$

EndFor

The set of Lanczos generated vectors are linearly independent of each other, and more importantly, provide a good first approximation of the desired eigenvectors.

5 Computational Solution Techniques

In any natural mode extraction formulation, one of the most difficult tasks is the execution of the inverse iteration, i.e. Eq. 5 of subspace iteration, and Eq. 15 of Lanczos method. The difficulty lies in storing the assembled stiffness, [K], and mass, [M], matrices in memory, and then solving for the equations in a reasonable amount of time.

The storage difficulty arises from the large number of elements contained within [K] and [M] which must be retained. For example, a cable-stayed bridge modeled with 100 distinct joints with six degrees of freedom per joint, yields [K] and [M] of order 600. The full matrices each contain 360,000 entries, and if double precision accuracy is used (8 bytes/entry), total storage requirement for [K] and [M] alone would be 5.76 Megabytes. On an IBM microcomputer environment, DOS only permits individual programs to be no greater than 640 kilobytes. This limit pertains to all portions of a program, i.e. code, data, heap, and stack. Obviously, storing [K] and [M] fully in memory is impossible.

Certain attributes of [K] and [M] may be used to reduce storage requirements. Since linear-elastic assumptions are adhered to in their development, [K] and [M] are both symmetrical, and in the case of [M] may be lumped. A lumped mass matrix contains nonzero elements only along the major diagonal. Thus, only the upper or lower triangular portion of the full matrix need be retained, which effectively reduces the storage requirement to 2.88 Megabytes. This is still too large. More efficient storage techniques retain only the banded portion of the matrices. A banded matrix is stored by rows (or columns) from the major diagonal to the maximum bandwidth (largest number of columns traversing nonzero elements). The efficiency of the banded matrix depends on the joint numbering scheme, which must minimize the difference in a member's near and far joint number. A further refinement of the banded matrix is the skyline matrix. The skyline matrix is stored by rows (or columns) from the major diagonal to the last nonzero element on that row (column).

The other difficulty discussed is the solution method, which goes hand-in-hand with the storage techniques just mentioned. A full matrix is extremely sparse, and therefore standard inverse matrix operations are unstable. The large number of computations needed to invert a matrix by standard inverse procedures propagates roundoff errors, thus producing an ill-conditioned solution. Other methods such as Gaussian Elimination (Crandall, 1956) can be easily used but, its effectiveness is reduced since properties of symmetry, positive definiteness, and bandedness are not utilized. This paper adopts the frontal method (Irons, 1970) since it affords the only solution technique for solving the inverse-iteration on a microcomputer.

6 Frontal Method

The frontal method is in reality a subset of Gaussian Elimination but with the strategy of reducing storage by assembling and eliminating degrees of freedom at the same time. As soon as the stiffness coefficients of a particular degree of freedom are completely assembled from the contributions of all members, this particular degree of freedom is eliminated by static condensation. The reduced equation is then stored on peripheral disk. After all degrees of freedom have been traversed, deformations are obtained in the reverse order of static condensation by backward substitution.

An any given moment in the solution algorithm, the computer stores only the upper triangular portion of a square, symmetric matrix containing the equations of the degrees of freedom currently being assembled. Therefore, memory requirements are bound by the constraint of the maximum number of degrees of freedom permitted to be assembled at any one time, so-called maximum frontwidth. During the assembly and condensation process, the degrees of freedom are assembled in a prescribed member-by-member order. Therefore, the most efficient front numbering scheme is to number members sequentially from the extreme left to the extreme right of the modeled bridge. Maximum frontwidth is not affected by the sequence to which joints are numbered. The frontal method's greatest strength is its stability and reduction of storage. The entire stiffness matrix, [K], of order N is never completely assembled, only small portions of it. Its weakness is its increased runtime. Disk input/output (I/O) during static condensation and backward substitution is notoriously slow in comparison to in-core data I/O.

7 Extraction of Uncoupled Natural Modes

As previously stated, DOS's memory limitation of 640 kilobytes retards any in-core solution algorithm. Therefore, in this study, the frontal method with a predefined constraint of 150 degrees of freedom as the maximum frontwidth has been adopted. With the memory barrier, it was predetermined that the maximum subspace size was nine. Theoretically, the first nine natural modes could thereby be determined via the subspace iteration method. However, the rate of convergence would be extremely slow for the higher modes (See Eqs. 11 and 12). Therefore, a predetermined limit of four natural modes and a minimum subspace size of six were adopted. Realistically, with only four modes determined, it is rather unlikely to compute the desired bending, sway, and torsional natural modes. In order to extract desired modes, a physical approach to force convergence to a particular mode type is formulated. For instance, when extracting bending modes, the bridge is constrained to only deform in a bending manner. Similarly for sway and torsion modes, the bridge is permitted to only deform in sway and torsional manners, respectively. Computationally, there are two techniques which forces convergence to a particular mode type.

The first technique, so-called Guyan reduction (1965) condenses all undesired degrees of freedom from the stiffness and mass matrices. It is not utilized here, because the condensation process adds considerable computer effort to the formulation. Instead, the second method, so-called forced-restraint (Namini, 1990), restrains unwanted degrees of freedom by applying a support condition at a particular degree of freedom. With the six degrees of freedom per joint labeled as follows: 1st dof = axial displacement, 2nd dof = flexural displacement in the bending direction; 3rd dof = flexural displacement in the sway direction; 4th dof = torsional rotation; 5th dof = flexural rotation in the sway direction; and 6th dof = flexural rotation in the bending direction, one can physically force convergence to a particular mode type. For example, to obtain bending natural modes, degrees of freedom associated with nonbending (3, 4, and 5) are restrained at all joints, as can be seen in Figure 2. Similarly, as illustrated in Figure 3, sway natural modes are found by restraining all nonsway (2, 4, and 6) degrees of freedom. Finally, in order to extract torsional modes, all nontorsional (3 and 5) degrees of freedom are restrained, but with the added restraint of bending displacement (dof 2) at the

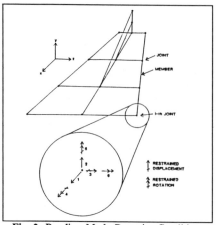

Fig. 2. Bending Mode Restraint Conditions

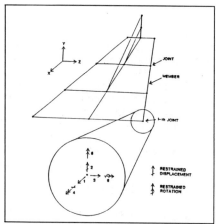

Fig. 3. Sway Mode Restraint Conditions

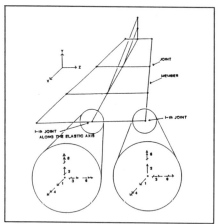

Fig. 4. Torsion Mode Restraint Conditions

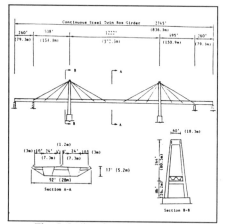

Fig. 5. Luling Cable-Stayed Bridge

elastic axis, which forces the torsional rotation about the elastic axis, as shown in Figure 4.

8 Example: Luling Cable-Stayed Bridge

In order to demonstrate the effectiveness of the computational methods previously described, an example cable-stayed bridge is chosen. The example bridge chosen is the Luling Cable-Stayed Bridge. The 837.6 m long, four-lane bridge on Interstate 310 spans the Mississippi River between the towns of Luling and Destrehan in St. Charles Parish, about 20 km west of New Orleans, USA, which opened to traffic in October, 1983. As shown in Figure 5, the bridge length is subdivided into five spans consisting of a 372.5 m center span between the towers, two 154.8 and 150.9 m anchor spans, and two 79.2 m approach spans.

Table 1. Luling Bridge Natural Modes

Mode Desig	Freq with Lumped Formulation (Hz)	Freq with Consistent Formulation (Hz)	Natural Mode Shapes	Previous Analysis Freq (Hz)
Bending				
1	0.4281	0.4309	Figure 6	0.3690
2	0.6827	0.6804	Figure 7	
3	0.9999	0.9998	Figure 8	
4	1.2480	1.2460	Figure 9	
Torsion				
1	1.2490	1.2640	Figure 10	1.2380
Sway				
1	0.6966	0.7011	Figure 11	0.5540

The cables are arranged in a double-plane fan configuration with 12 cables in each plane.

To investigate the algorithm, the first four bending modes are determined, while only the fundamental mode for torsion and sway types are extracted. For each extraction, a lumped and consistent mass formulation is used. Referring to Table 1, natural frequencies from these analyses are shown. As illustrated in Figure 6, the fundamental bending mode corresponds to the first symmetric case. The second bending mode corresponds to the first antisymmetric case, as shown in Figure 7. The third and fourth bending modes correspond to the second and third symmetric cases as illuminated in Figures 8 and 9, respectively. Finally, the fundamental torsion and sway modes correspond to their respective first symmetric case and are illustrated in Figures 10 and 11.

Also shown in Table 1 are the actual fundamental modes determined by the design engineers, Frankland and Leinhard - Modjeski and Masters (1974). To compare this paper's results to the consulting engineers' analysis, certain differences in modeling exist. They disregarded the modeling of the approach spans, since originally the bridge was to be only three spans, but eventually was changed to five spans. They did not revise their dynamic analysis since the addition of the approach spans would only increase the overall bridge rigidity, thus their original natural frequencies were conservative. For completeness, the Luling Bridge was modeled without the approach spans and the fundamental bending, torsion, and sway natural frequencies were computed to be 0.3641, 1.226, and 0.5390 Hz, respectively, which correlate closely to that of the consulting engineers' analysis.

For each of the natural modes found, the lumped mass formulation yielded lower natural frequencies than did the consistent mass formulation. This is easily explained since a lumped mass formulation tends to concentrate inertial forces, while consistent mass formulation tends to distribute inertial forces. Tong, Pian, and Bucciarelli (1971) offer an analytical explanation by recognizing that a lumped mass formulation is computationally derived using uncoupled inertial forces with no rotary inertia. These assumptions are essentially a Guyan reduction

Table 2. Execution Properties for Luling Bridge Natural Modes

Mode Desig	Subspace Size	Number of Iterations to Convergence	Execution Time (min)
Bending			
1	9	1	15
2	9	1	15
3	9	1	15
4	9	3	45
Torsion			
1	6	3	45
Sway			
1	6	1	15

of those rotary degrees of freedom. An analogy to monotonic convergence is inferred in that a discretized bridge model approaches its true solution from the stiffer side, i.e., an increase in degrees of freedom yields a more flexible structure.

The runtime expended to extract natural modes was rather considerable, as is shown in Table 2 with other execution properties. All extractions were performed on a Compaq 386/20, which is an IBM-compatible system, and was also configured with a math coprocessor and a virtual disk for enhancing runtime.

As is consistent with the convergence behavior of the subspace iteration method, the lower modes converge faster than due the higher modes. Also, the torsion mode, since being the highest frequency encountered exhibits the poorest convergence behavior. In microcomputer applications, a strategy of small subspace size which decreases runtime within each iteration but increases the number of iterations to convergence is adopted as opposed to a large subspace size with a small number of iterations to convergence. The author believes that

Fig. 6. First Bending Mode Corresponding to First Symmetric Case

Fig. 7. Second Bending Mode Corresponding to First Antisymmetric Case

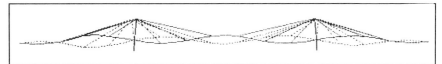
Fig. 8. Third Bending Mode Corresponding to Second Symmetric Case

Fig. 9. Fourth Bending Mode Corresponding to Third Symmetric Case

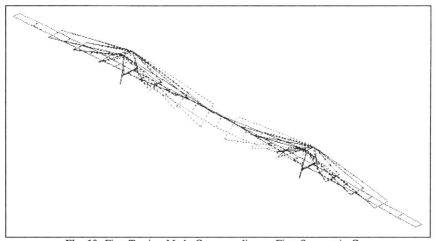
Fig. 10. First Torsion Mode Corresponding to First Symmetric Case

Fig. 11. First Sway Mode Corresponding to First Symmetric Case

the range of subspace size being $6 \leq S \leq 9$, while only permitting up to four modes of each type is a compromise in order to extract enough modes to represent response motion, while computationally performing execution within a reasonable amount of time.

9 Conclusions

The algorithm presented has proved reliable in accuracy, but has a high runtime for execution. The use of the subspace iteration method with starting iteration vectors generated by the Lanczos methods provides the most advanced computational technique for natural mode extraction of cable-stayed bridges. As implemented with the frontal method to compensate for DOS's program size limitation of 640 kilobytes, it is somewhat remarkable that natural modes are indeed found on a microcomputer environment.

10 References

Bathe, K.J. (1977) Convergence of Subspace Iteration in **Formulations and Numerical Algorithms in Finite Element Analysis** (eds K.J. Bathe, J.T. Oden, and W. Wunderlich), M.I.T. Press, Cambridge, Mass.

Bathe, K.J. (1982) **Finite Element Procedures in Engineering Analysis.** Prentice-Hall, Inc., Englewood Cliffs, New Jersey.

Bathe, K.J. and Ramaswamy, S. (1980) An Accelerated Subspace Iteration Method. **Comput Methods Appl Mech Eng, 23,** 313-331.

Cheung, Y.K and Kajita, T. (1973) Finite Element Analysis of Cable-Stayed Bridges. **IABSE, 33-II,** 101-112.

Crandall, S.H. (1956) **Engineering Analysis.** McGraw-Hill Book Company, New York.

Ernst, J.H. (1965) Der E-Modul von Seilin unter Berucksichtigung des Durchanges. **Der Bauingenieur, 40**(2).

Fleming, J.F. and Egeseli, E.A. (1976) Dynamic Behavior of Cable-Stayed Bridges, in *Proc Int Symposium Earthquake Structural Engng*, St. Louis, Missouri.

Frankland and Leinhard - Modjeski and Masters, Consulting Engineers. (1974) Section Model Testing - Vibration Analysis - Mississippi River Bridge (Luling).

Guyan, R.J. (1965) Reduction of Stiffness and Mass Matrices. **AIAA J, 3**(2), 380.

Irons, B.M. (1970) A Frontal Solution Program for Finite Element Analysis. **Int J Numer Methods Eng, 2,** 5-32.

Izyumov, N., Tschanz, T. and Davenport, A.G. (1977) A Study of Wind Action for the Weirton-Steubenville Cable-Stayed Bridge, Boundary Layer Wind Tunnel Laboratory, The University of Western Ontario.

Kumarasena, T., Scanlan, R.H. and Morris, G.R. (1989) Deer-Isle Bridge: Field and Computed Vibrations. **J Struct Eng ASCE, 115**(9), 2313-2328.

Lanczos, C. (1950) An Iteration Method for the Solution of the Eigenvalue Problem of Linear Differential and Integral Operators. **J Res Natl Bur Stand, 45,** 255-282.

Morris, N.F. (1974) Dynamic Analysis of Cable-Stiffened Structures. **J Struct Div ASCE, 100,** 971-981.

Namini, A.H., (1990), Cable-Stayed Bridge Analysis for Static and Dynamic Response Behaviors. in preparation to be submitted to **Comput Struct**.

Tang, M.C. (1971) Analysis of Cable-Stayed Girder Bridges. **J Struct Div ASCE, 97**(5), 1481-1496.

Tong, P., Pian, T.H.H. and Bucciarelli, L.L. (1971) Mode Shapes and Frequencies by Finite Element Method Using Consistent and Lumped Masses. **Comput Struct, 1,** 623-628.

PART TEN

BRIDGE ASSESSMENT, MAINTENANCE AND REPAIR

LOAD TESTING AS A MEANS OF ASSESSING THE STRUCTURAL BEHAVIOUR OF BRIDGES

A.E. LONG
Professor of Civil Engineering, Queen's University, Belfast, Northern
Ireland, UK
A. THOMPSON
Senior Research Officer, Queen's University, Belfast, Northern
Ireland, UK
J. KIRKPATRICK
Dept Environment (NI) Roads Service, Northern Ireland, UK

Abstract
Experience gained in the use of load testing to assess the in-situ
performance of three different types of bridge decks is presented in
this paper. As these tests have been carried out on a medium span
steel-concrete composite system, a number of short span precast
concrete composite decks and a long span steel box girder bridge
widely differing measurement techniques have been utilised. The
techniques range from the relatively simple, some 20 years ago, to
extremely sophisticated methods which take advantage of recent
developments in computer technology. Brief details of the monitoring
systems are presented along with typical test results which have been
found to give a valuable insight into the behaviour of these
structures. It is concluded that in-situ testing will be much more
widely used in the future.
Keywords: Bridge Decks, Composite, Concrete, Displacements, In-situ,
Loading, Monitoring, Steel, Strains, Testing.

1 Introduction

Proof testing has been carried out on over 200 bridges in Ontario,
Bakht (1987), where the main objective has been to establish the safe
load carrying capacity of existing bridges. In these tests it has
been found that there are nearly always aspects of the behaviour of
bridges which are unexpected and these often have an influence on the
load carrying capacity. Thus apart from being a most effective means
of determining the safe load carrying capacity of bridges they give a
useful insight into structural behaviour. When combined with modern
methods of monitoring strains and displacements they can be utilised
to validate analytical techniques. In such instances the findings
have much greater credibility with practising engineers than
validations based on model tests.

   For over 20 years the author has been involved in the full scale
testing of bridges and during this period the rapid developments in
electronics have had a profound influence. In the late 1960's in-situ
monitoring was difficult, time consuming and results were often
inaccurate. However with the developments in instrumentation,
especially designed for site use, in the 1970's and 1980's it is now
possible to monitor the behaviour of most bridge structures under load
with the expectation that useful results will be obtained. In order
to illustrate the impact of these developments three contrasting test
programmes will be considered:

i)   A medium span composite plate girder bridge in Canada (late
     1960's).
ii) Tests on M-beam bridge decks (late 1970's/early 1980's).
iii) Remote monitoring of a long span steel box girder bridge in
     N. Ireland (late 1980's).

2 Bay of Quinte tests

In the late 1960's transverse cracking of the concrete deck slab of
the multi span bridge over the Bay of Quinte in Ontario, Canada was a
source of concern to the then Department of Highways, Ontario. A
subsequent study of the bridge by staff at the Queen's University,
Kingston, Ontario revealed that the main cause of the cracking in the
composite slab was the sequence of concrete placements utilised during
the construction. This in turn was accentuated by shrinkage and live
load tensile stresses. As the bridge was considered by workmen to be
rather lively, possibly as a result of the cracking, it was decided to
carry out a simple dynamic test to determine whether the displacements
lay within acceptable limits.

As a significant number of the approach spans of the bridge were
not over water this allowed an extremely simple measuring system to be
used. At mid-span a pointer was fixed to the bridge deck so that
readings could be taken relative to a standard surveying staff which
was firmly placed on the ground. The relative displacement, which
occurred when a fully loaded truck moved across the bridge at around
50 mph, was monitored visually and recorded by using an 8 mm cine
camera. The maximum amplitude monitored was of the order of 10 mm and
was well within the limitations considered safe by the designers.
However it should be noted that prior to the use of the monitoring
system various engineers, including the author, had estimated that the
relative displacement would be of the order of 50 mm. Striking
evidence that the human body is very sensitive to movement and that
perceptions of displacements in structures are often excessive.

3 Tests on precast M-beam bridge decks

3.1 Background
A high proportion of highway bridges lie within the range of spans of
15 m to 29 m. Precast prestressed concrete beams are widely used as
they offer an economic solution to most railway/river crossings and
busy road intersections where interference must be kept to a minimum.
In the UK these standard beams have a bottom flange approximately 1 m
wide and are generally referred to as M-beams. They act compositely
with a cast in-situ slab and a grillage analysis is often used for
analysis. The accuracy of this analysis depends on the idealisation
of the grid mesh and the method utilised for the calculation of the
section properties of the grid members. These have been largely based
on the results of model tests and few, if any, full scale tests have
been carried out to check their validity.

In the 1970's as this type of beam was being used extensively in N.
Ireland a number of bridges was tested to establish their load

distribution characteristics with the objectives of improving current
design procedures and reducing construction costs.

A further improvement in the performance of this type of deck can
be achieved by increasing the spacing of the beams and in addition to
the obvious economic advantages this allows the range of spans to be
increased from 15 to 29 m to 8 to 29 m. This does however introduce
problems regarding the strength of the slab spanning between beams and
a one-third scale model bridge was tested to establish the strength of
the slab with the beams spaced at up to 2 m. These strength tests
were supplemented by in-situ tests on a bridge deck which were carried
out to ensure that serviceability requirements were not violated.

3.2 Transverse load distribution characteristics

The two T-beam (Fig. 1(a)) and two pseudo box (Fig. 1(b)) bridge decks
tested had previously been designed by the Department of Environment
for N. Ireland. The strain distributions across the decks were
monitored using Tyler surface mounting recoverable vibrating wire
gauges of 140 mm gauge length which were attached to the beam soffits.
In order to monitor the variation in temperature during the tests –
thermocouples were mounted on the beams adjacent to the strain gauges.
It should be noted that all tests were carried out on dull, overcast
days with little wind so as to minimise strains caused by temperature
variations.

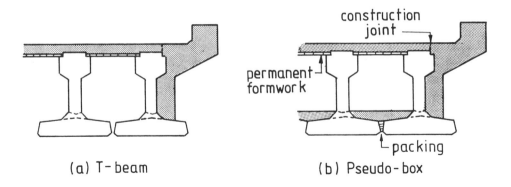

(a) T-beam   (b) Pseudo-box

Fig. 1. Precast concrete composite systems

Test procedure
Static loading was applied to each deck by two 6-wheel trucks, loaded
with crushed stone, the total weight of which has been accurately
measured prior to arrival on site. However as accurate estimates of
individual wheel loads were required, for use in a theoretical
grillage analysis, these were measured using a Digital Portable
Weighbridge, Snaith (1978). The loading applied to all decks
consisted of two trucks each weighing about 30 tonnes and three
loading positions at mid span were considered. Prior to the tests the
various positions for the trucks were carefully marked out on the deck
to ensure the locations of the wheel loads were known precisely. The
trucks were then driven onto the deck, positioned as required and a

complete set of readings taken. The resulting strain readings from
the vibrating wire gauges were then converted into concrete stresses
and corrected by taking account of the modulus of elasticity values of
the beams and overall equilibrium to give the moment absorbed by each
individual beam.

In order to establish the influence of the concrete edge feature on
the performance of the deck this procedure was carried out before and
after the upstand was cast.

Theoretical predictions
Theoretical analyses of all the test bridges were carried out using a
grillage analysis which allowed both series of tests with and without
the parapet upstand to be considered. The bridge decks were idealised
and section properties calculated in accordance with the recommendations by West (1973) which suggest that the longitudinal grillage
beams should be formed from about two physical beams and the overall
grid mesh should correspond approximately to the aspect ratio of the
deck. Section properties were proportioned in a similar manner.

The mid span distribution of beam moments for one of the bridges
which utilised T-beam construction is given in Fig. 2. The results of
tests with and without the parapet upstand are shown and these are
compared with the theoretical grillage analysis. Good overall
correlation was found to result for this type of deck indicating that
the method of idealisation is realistic.

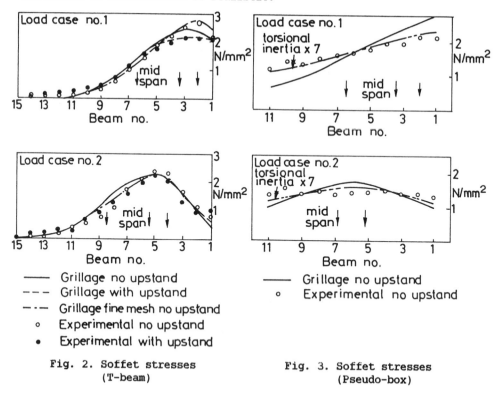

—— Grillage no upstand
– – – Grillage with upstand
—·— Grillage fine mesh no upstand
  ○   Experimental no upstand
  ●   Experimental with upstand

—— Grillage no upstand
  ○   Experimental no upstand

Fig. 2. Soffet stresses
(T-beam)

Fig. 3. Soffet stresses
(Pseudo-box)

However much poorer correlation was found when the theoretical predictions were compared with the test results for the pseudo box beam decks. From Fig. 3 it can be seen that it was necessary to adjust the torsional stiffness, from that recommended by West (1973), by a factor of seven to attain the same degree of accuracy as had been achieved for the Tee-beam decks. The reasons why this significant change is necessary, to accurately model the performance of this type of bridge deck, have since been explained and full details of this study are given in Kirkpatrick, Long and Thompson (1982). Spaced M-beam bridge decks are also considered in Kirkpatrick, Long and Thompson (1984).

3.3 Ultimate strength of deck slabs

An important aspect of the development of this type of bridge deck was the use of the beams at increased spacing and in order to establish the strength of the slabs spanning between beams a one-third scale model bridge deck was constructed in the laboratory and tested. The model was fully representative of a prototype and in particular the end diaphragms and parapet upstand were included to ensure the slab could develop its full potential of in-plane forces. Details of the model deck and the notation for the test panels are given in Fig. 4.

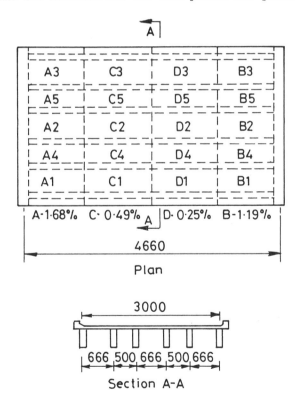

Fig. 4. Details of model bridge deck

The designs of the prototype slab for the two No. 112.5 kN wheel
loads using the equations of Westergaard indicated that steel
reinforcement of the order of 1.7% was required under the highway
pavement. For test purposes additional areas of reinforcement
equivalent to approximately 1.2%, 0.5% and 0.25% were provided in the
model and are represented by four areas of slab approximately 1200 mm
wide in the span-wise direction and the full width of the deck in the
transverse direction. Therefore with this banding of reinforcement
three panels equivalent to 2 m spacing and two panels equivalent to
1.5 m beam spacing are available providing a total of 20 test panels.

Hydraulic jacks were used to load the model with the loading for
the heavily reinforced panels A and B simulating a single 112.5 kN
wheel load of an HB vehicle. The lightly reinforced panels C and D
were subjected to either single or double wheel loading to check the
influence of the latter on the ultimate capacity.

Model test results
The ultimate load capacity of each test panel was the load which
caused the loading shoe simulating the wheel load to punch through the
slab in the characteristic manner. It was found that there was very
little variation in the ultimate load capacity of all the panels even
though the slab reinforcement varied from approximately 0.25% to
1.68%. In comparison with United Kingdom, BS5400 (1978), the Ontario
Highway Bridge Design Code (1979) predictions (Fig. 5) the results of
the tests on the one-third scale model showed considerable enhancement
to the design capacity of the standard slab with the M-beams spaced at
up to 2 m apart. This enhancement has been attributed to the
considerable in-plane restraint that is inherent in bridge slabs.
Fig. 5 clearly shows that the codes do not give a satisfactory
prediction of the punching shear capacity of typical bridge slabs and
a more appropriate method which allows for in-plane restraint was
developed, Kirkpatrick, Rankin and Long (1984).

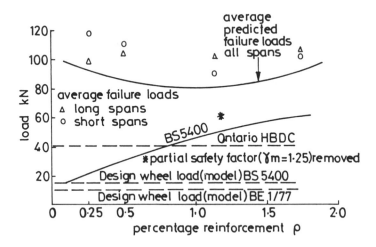

Fig. 5. Comparison with bridge codes

Full scale tests for serviceability

The widths of the cracks induced in the slabs were monitored during the model test and it was found that under service load conditions no cracks resulted. However as scale effects can affect the accuracy of these measurements full scale tests, Kirkpatrick, Rankin and Long (1986), were subsequently carried out on a bridge built by the DOE(NI) Roads Service. This bridge incorporated beams at 1.5 m and 2 m spacing and the reinforcement varied from 0.25% to 1.7% in the standard 160 mm thick deck slab.

The tests showed that current crack control formulae are not applicable because of the enhanced performance which results from the development of compressive membrane action. Initial cracking occurred at loads well in excess of the design service loads and even after cracks had been induced by severe overloading it was found that the slabs still satisfied the serviceability limit state requirements.

3.4 Commentary

As a result of this extensive series of full scale and model tests the DOE(NI) Roads Service have incorporated many of the findings in bridges built subsequently. Reinforcement levels have been reduced to 0.6% in the standard 160 mm deck slab and considerable further economies achieved by using slightly larger M-beams at 1.5 m and 2.0 m spacings. Indeed it has been estimated that the cost savings achieved, in the first bridge incorporating these features, were greater than the total cost of this study carried out in collaboration with Queen's University. A further potential benefit is that epoxy coated bars could, at these low reinforcing levels, be cost-effectively incorporated in the deck slabs and this could greatly enhance the long term durability of these structures.

It is of interest to note that prior to the commencement of the load distribution tests the advice of leading experts in the field was sought. Grave doubts were expressed as to whether any useful results would be obtained. However as can be seen from this brief description of these loading tests many valuable findings have been obtained. The success of the tests can be attributed to a test programme which was implemented with great care and the availability of equipment which allowed relatively small concrete strains to be measured accurately on site.

4 Tests on a major steel box girder bridge

4.1 Background

In 1985 the Department of the Environment (NI) asked the Civil Engineering Department of the Queen's University of Belfast for assistance in monitoring the recently completed Box Girder Bridge over the river Foyle, near Londonderry.

The Foyle Bridge is a high level crossing at Madam's Bank, some 3kM downstream from the Craigavon Bridge in the city centre. The bridge, designed by Freeman Fox and Partners, London and built by RDL-Graham Joint Venture, is 866 m long and comprises three main spans totalling 522 m constructed in steel together with a 344 m long approach viaduct which is of prestressed concrete box construction. Each carriageway is supported independently, so that the bridge consists, in effect, of two parallel structures supported on common concrete piers and abutments. The general arrangement is shown in Fig. 6.

The main steel sections were fabricated by Harland and Wolff in Belfast. Each 522 m long girder was built in three sections with the two site splices being located at the sixth points of the centre span. Each side span unit is 180 m long and varies in depth from 3 m to 9 m and weights 950 tonnes. The centre span units are 161.5 m long and weigh 700 tonnes. The units were transported to the site by barge and joined on site after being lifted into their final position, Hunter and McKeown (1984).

With a deck at 30 m above water level, the structure is very exposed, being completely open to both North and South and with almost no shelter from any other direction. The basic design wind speed given in CP3: chapter 5 for the area is 53 m/s, compared with 50 m/s for the Forth Bridge and 43 m/s for the Severn. Occupying, as it does, the most exposed site of any major bridge in the British Isles, the Foyle Bridge offers an excellent opportunity to monitor the in-service performance of a box-girder structure.

The project is concerned only with the steel section of the bridge. The objectives are:

* To investigate methods of continuously monitoring long-span bridges with the aim of developing a cost effective solution which could have application in similar situations elsewhere.
* To establish a "footprint" of the structural response to a range of test loads when the structure is in the new condition so that any future deterioration can be deleted.
* To monitor the behaviour of the structure under a variety of wind and traffic loadings.
* To compare the measured and predicted behaviour of the structure.

4.2 Monitoring system

Preliminary studies indicated that, as an initial requirement, data would be needed on:

Deflection at the mid-span point of the centre span.
Strains at mid-span and at the intermediate supports.
Temperature changes in the structures.
Wind speed and direction.
Movement of the expansion joint.
Several other requirements of a more general nature soon became apparent.

The bridge site is remote both from the University (120 km) and the DOE base in Coleraine (50 km). For security and safety reasons, access to the interior of the structure must be restricted. Since, in addition, studies of response to wind, temperature and traffic have to be made on a continuous basis, the system had to be capable of operating unattended for long periods and of storing the resulting data for subsequent analysis. Since automatic data collection generates very large volumes of data it quickly became apparent that it would be essential to have significant processing capacity on site.

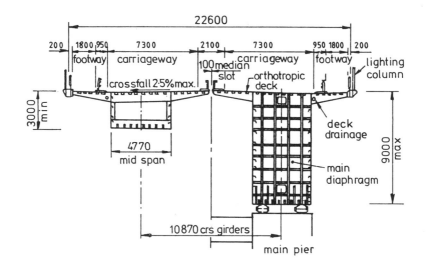

a) Cross – sections of steel girders

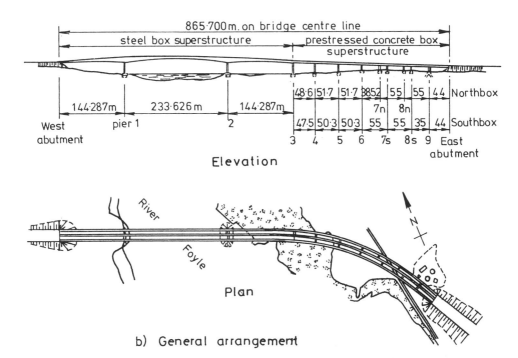

Elevation

Plan

b) General arrangement

Fig. 6. Details of the Foyle Bridge

This would allow a preliminary analysis of the raw data to be done immediately so that only the significant parts would be retained for transmission to the University for further analysis.

A schematic diagram of the monitoring system is shown in Fig. 7. The core of the system is a DEC PDP-11/53 minicomputer which runs under a multi-tasking operation system. This system can be accessed either locally from a terminal to the bridge, or remotely, via a modem and dial-up telephone line from either the University or the DOE offices.

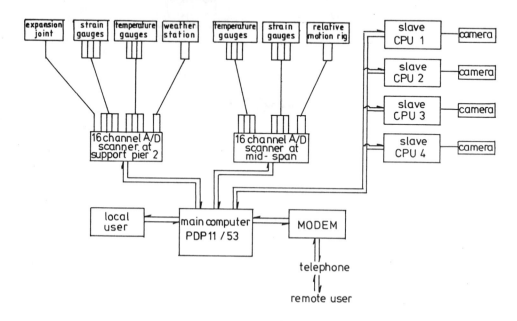

Fig. 7. Schematic of Foyle Bridge monitoring system

Measuring the deflections of such a large and inaccessible structure presents considerable difficulty; the fundamental problem lies in establishing a frame of reference from which to make measurements. Conventional methods, such as levelling or the use of displacement transducers were clearly impractical due both to the distance to the nearest fixed point and to the need to record the response to dynamic loadings.

Initially, consideration was given to the use of accelerometers but these were rejected as they could not measure the very small accelerations experienced during the slow moving deflections caused by heavy vehicles. The solution adopted is shown diagrammatically in Fig. 8. Lasers, fitted with beam expander and focusing optics, are mounted inside the box girders at mid-span so as to project a spot of light of approximately 3 mm diameter on to targets which are fixed over the main piers. Any movement (either linear or rotational) of the lasers will thus cause the position of the light spot on the target to vary.

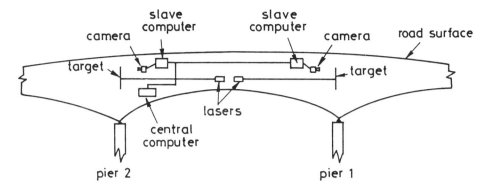

Fig. 8. Laser deflection system

The movement of each spot is tracked using a solid state camera linked to a BBC microcomputer which stores the resulting co-ordinates of the light spot in memory for later transmission to the central computer.

All the other data is collected using transducers which generate electrical analogue signals. These are connected to a pair of free-standing 16-channel analogue to digital converters, which are located as close as possible to the transducers so as to minimise the length of the cables carrying the analogue signals. The alternative, which was to incorporate the A/D converter in the main computer, would have involved cable runs of up to 150 m with all the attendant problems of signal attenuation and noise pickup. More detailed information on the monitoring system is given in Sloan, Kirkpatrick and Thompson (1990).

4.3 Typical test results
Since the system was installed in 1989 an enormous amount of data has been collected and a representative sample is shown in Figs. 9-12.

Fig. 9 shows the influence line for static mid-span deflections induced when a 75 tonne load was moved across the bridge. This test was carried out on a calm day when the bridge was closed to traffic. Each reading is the mean of 1024 camera scans, to remove the effect of vibrations induced in the structure when the load was moved from one position to the next. The position of the load is measured from the East end of the steel structure.

Fig. 10 shows the dynamic deflections recorded as a test load of 100 tonnes was driven over the bridge from East to West at approximately 43 km/hour. Initially, the structure was almost motionless, what small movement there was being caused, probably, by a light breeze. As the load passed over the bridge, the mid-span point first rose and then, as the load reached the centre span, deflected downwards by 220 mm before rising again as the load moved to the second side span. At the right hand side of the trace, the structure can be seen vibrating at its natural frequency of 0.4 Hz; the amplitude reducing after the source of the disturbance was removed.

Fig. 11 shows the movement recorded during storm conditions. It can be seen that these were significant wind induced displacements with peak to peak amplitudes of the order of 110 mm.

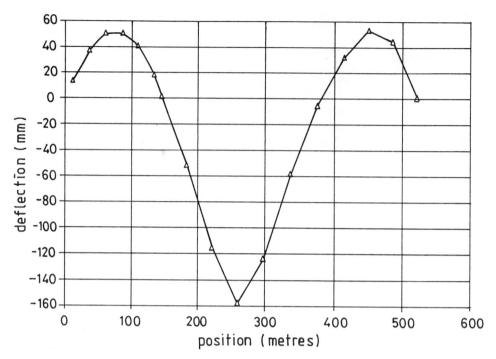

Fig. 9. Deflections at mid-span (75 tonne load)

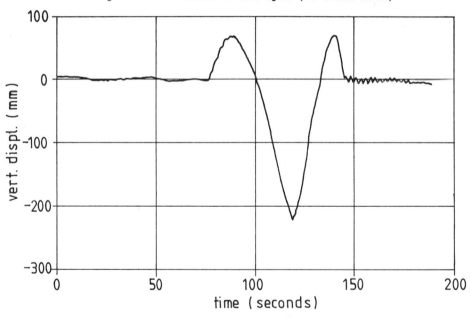

Fig. 10. Deflections due to 100t moving load

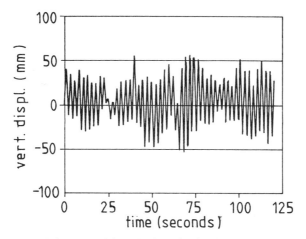

Fig. 11. Wind induced displacements

Fig. 12 is a plot produced from data collected on the movement of the expansion joint caused by temperature changes in the structure. A clear trend can be seen, with the "best fit" straight line having a gradient of 7.9 mm per degree change in steel temperature. This figures is about 2.0 mm larger than that to be expected from the expansion of the steel alone; the difference can be accounted for by the expansion of the concrete section of the bridge.

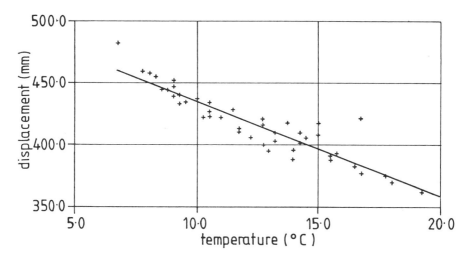

Fig. 12. Thermal movement at expansion joint

4.4 Commentary
The high quality of the results presented clearly demonstrates that an extremely versatile and accurate monitoring system has been developed

for a major structure. It is fully automatic and can be run equally well under local or remote control. In addition as the system has been designed to run unattended over long periods it is particularly suitable for continuous monitoring.

Over the next few years much useful data will be collected which will allow the DOE(NI) to check the in service behaviour of this major bridge. Furthermore the impact of any deterioration or damage can be quickly assessed. It is also anticipated that the data will be of value to designers of bridges of this type in that comparisons can readily be made with theoretical predictions. Any discrepancies can then be investigated so that modelling inaccuracies can be identified and future designs improved.

5 Concluding remarks

The loading tests reported in this paper have been utilised very effectively to assess the structural behaviour of different bridge types. Depending on the type of bridge in question and the information required it is now possible to state with confidence that most aspects of structural behaviour can now be monitored successfully using loading tests. This state of affairs has been greatly assisted by improvements in electronic instrumentation and the widespread availability of microcomputers. Now that sophisticated monitoring systems are available or can be developed it is likely than the 1990's will be a period when full scale testing will be used much more widely than at present. Thus as we look back in awe at the achievements of the engineers who created the Forth Rail Bridge we can also look forward to an exciting new era when existing structure can readily be assessed using loading tests.

6 Acknowledgements

The Department of the Environment for N. Ireland (Roads Service) and the Science and Engineering Research Council have jointly funded much of the work described in this paper and this support is gratefully acknowledged. The interest and support of Dr Corden Stevenson, Chief Engineer DOE(NI) Roads Service as well as the major contributions made by Barry Rankin and David Sloan is also acknowledged with gratitude.

7 References

Bakht, B. (1987) Bridge evaluation by proof testing. **Proceedings of IStructE/BRE Seminar on Structural Assessments**, Watford, Paper No. 25, 7 pages.

BS5400 (1978) Steel and concrete composite bridges. **British Standards Institution**, London.

Hunter, I.E. and McKeown, M.E. (1984) Foyle Bridge fabrication and construction of the main spans. **Proceedings of Institution of Civil Engineers**, Part 1, May, Vol. 76, pp 411-448.

Kirkpatrick, J., Long, A.E. and Thompson A. (1982) Load distribution characteristics of M-beam bridge decks. **The Structural Engineer**, 60B, No. 2, June, pp 34-43.

Kirkpatrick, J., Long, A.E. and Thompson, A. (1984) Load distribution characteristic of spaced M-beam bridge deck. **The Structural Engineer**, 62B, No. 4, December, pp 86-88.

Kirkpatrick, J., Rankin, G.I.B. and Long, A.E. (1984) Strength evaluation of M-beam bridge deck slabs. **The Structural Engineer**, Vol. 62B, No. 2, September, pp 60-68.

Kirkpatrick, J., Rankin, G.I.B. and Long, A.E. (1986) The influence of compressive membrane action on the serviceability of beam and slab bridge decks. **The Structural Engineer**, Vol 64B, No. 1, March, pp 6-12.

Ontario Highway Bridge Design Code (1979) Ministry of Transportation and Communications, Toronto, Ontario, Canada.

Sloan, T.D., Kirkpatrick, J. and Thompson, A. (1990) Remote computer aided bridge performance monitoring. **International Conference on Bridge Management**, Surrey University, March, 10 pages.

Snaith, M.S. (1978) A digital portable weighbridge. **The Journal of the Institution of Highway Engineers**, London, Vol. XXV, No. 7, pp 9-12.

West, R. (1973) Recommendations on the use of grillage analysis for slab and pseudo slab bridge decks. **C&CA/CIRIA Technical Report**, N.46.017.

THE MANAGEMENT OF INFORMATION IN BRIDGEWORKS MAINTENANCE

T. OBAIDE, N.J. SMITH
Project Management Group, Dept Civil and Structural Engineering,
UMIST, Manchester, England, UK

Abstract
Bridge owners have a statutory duty to ensure the safety of these structures in operation. When defects in a number of bridges are discovered, usually as a result of routine or special inspections or occasionally due to accidental damage, the owner has to decide the extent and the priority of the works to be undertaken. Effective decisions cannot be made without detailed bridge data and an appropriate management maintenance strategy. This paper reviews the problems in maintaining the bridge stock in the UK and considers the role of computer based management maintenance systems in manipulating data to improve the effectiveness of bridgeworks maintenance.
Keywords: Bridgeworks Maintenance, Bridges, Maintenance Strategy, Management Information System.

1 UK Bridgeworks Stock

The UK has one of the most varied bridgestocks in the world, ranging from masonry arch bridges, many of which were constructed over a hundred years ago, listed structures in wrought and cast iron, to present day steel box, prestressed concrete and cable stayed structures. The vast majority of these bridges are of small span, probably less than 15m. The total bridgeworks stock is estimated to be about 150,000 of which about 8,900 are the responsibility of the Department of Transport, (DTp), [1], and 129,000 the responsibility of the Local Authorities. In general terms the stock consists of about 70,000 Masonry and Brick arch and culvert structures mostly constructed prior to 1922, about 60,000 Concrete and 25,000 Metal bridges largely constructed post 1922.

Bridges have traditionally been designed to operate for long periods of time without requiring major maintenance. The current design life for road bridges is 120 years, however bridges which were constructed before 1922 and several which were constructed shortly afterwards, were not designed to standards which would be acceptable today. These older structures are now subjected to much greater loadings than could have been envisaged at the time of their design, which accelerates the need for maintenance work.

The public sector are the owners of most of the UK bridges; although there are a small number of private bridge owners. The owner has a duty to ensure the safety of these structures in operation. Recent Local Authority surveys, [2], have shown that on average 5% of the bridge stock is substandard. Nationally this would indicate a maintenance cost of about £24bn. This problem has been aggravated by the recent increases in the intensity and the axle weight of road traffic. Consequently in addition to routine maintenance and the repair of accidental damage, many bridges require strengthening works to accommodate these loadings. The modification of existing structures requires detailed and specific information so that the basic choice of either refurbishment of existing elements or complete replacement can be made.

2 Bridgeworks Maintenance

Bridgeworks maintenance can be defined as ensuring the safe, unrestricted passage of people, animals and vehicles as specified in the Construction and Use regulations, without limitations, [3]. The work is a combination of planned maintenance, refurbishment, replacement and the repair of accidental damage. The first three items are regarded as traditional bridgeworks maintenance and would usually be funded from the annual budget. However, the last item is often regarded as a special case, sometimes requiring immediate action and therefore special funding arrangements usually exist for these cases.

Maintenance work could be regarded as the action taken to prolong the useful life of a bridge at a minimum cost with least interference to its operational function. This approach contains contradictory aims as the minimum cost of maintenance is often only achievable if the bridge is closed. Frequently, the cost of diversions of pedestrians and vehicles exceeds the cost of the physical bridge repair. The expertise in the management of the maintenance process is in deciding upon cost effective minimum works which permit the bridge to be used without restrictions.

Although the determination of the bridge condition is the prime objective of bridgeworks inspections, it appears that the rate of deterioration of bridges is not standard or uniform. Additionally, it is suspected that the relatively old and relatively new bridges account for a higher proportion of the maintenance budget than would be expected. The reasons for expenditure on pre-1922 bridges have been outlined above but the more recent bridges, post 1960, were constructed at a time when the material specification had been revised, tenders concentrated on minimum capital cost and more sophisticated designs were being produced.

The implication of these factors is that more attention should be given to the collection and analysis of a wealth of information regarding the age, the design type, the materials, the site constraints, the condition and other relevant factors to facilitate maintenance decision making. Traditionally, maintenance has been regarded as an "after-the-fact" activity and was something done in response to an unacceptable condition arising. The trade-off between initial capital cost and maintenance expenditure over the design life raises the question of whether the life cycle cost of the bridge should be the basis for bridgeworks maintenance.

3 Maintenance Strategy

A bridgeworks maintenance strategy has to meet certain objectives, one of the most important being the statutory requirements for public safety. This has to be achieved by making the most effective use of limited resources, by efficient planning and selection of priorities, by using the most appropriate contract strategy and all within strict budgetary constraints.

The fundamental requirement for the management of the maintenance in the operation stage is relevant base data. This data is largely obtained from visual General and Principal Inspections, and it is the frequency of these inspections and hence the cost of collecting the data, and the reliability and accessibility of the data which influences the effectiveness of the management process. However it should be noted that a database stores historical data made available at the time of inspections or repair works and cannot be completely up to date.

Once a defect has been identified one of the key management decisions is to determine the level of intervention, which can be described as the appropriate time for particular types of maintenance work; repair, refurbishment or replacement. This aspect is a central problem for management and illustrates the difficulties of decision making under conflicting constraints. Detailed planning is difficult as, particularly on the older bridges, the full extent of the maintenance works required may not become known until the work has started on site. Many faults in different bridges may be identified at the same time but the budget will not allow all the works to be undertaken and hence the expenditure has to be made in such a way as to produce the 'outcome of least regret', [4].

The interrelationship between the available budget and priority spending, the age and type of the bridgeworks stock, statutory requirements and the vehicular flows using the bridges is not well defined. These types of uncertainty mean that management has to be flexible in the scheduling of routine maintenance work and will need additional funding from time to time. It is clear that without detailed bridgeworks information effective decisions cannot be made. There is no standardisation in precisely what data is recorded, how these records are stored and in what form they can be retrieved to assist in the decision making process. Many bridge owners and consulting engineers employed by owners have developed bridgeworks databases; unfortunately there is at present little compatibility between different systems.

4 Maintenance Management Systems

The use of databases to store maintenance data is the first step in establishing maintenance systems, however the constraints of budget and the priority decision making required in bridgeworks maintenance means that management expertise is also needed. To handle and update large amounts of data quickly and easily necessitates the use of computer based systems.

In the US and increasingly in the UK, Management Maintenance Systems have been developed to consider the life cycle costing of structures, particularly for buildings and process plants. This concept is now beginning to be applied to a conventional database for the management of information in bridgeworks maintenance, contrasting the traditional 120 year design life approach with a life cycle costing philosophy.

4.1 DTp System
This system is still in the early stages of development, initially as a database but ultimately it has the potential for becoming a management maintenance system. It is known as the National Structures Database, [5], and stores data on DTp bridges which are the motorway and trunk road bridges and hence contains a higher proportion of large

span bridges than the national average.

It stores data including, the bridge location, as built details, history of defects, inspection records and budget information. There is a category for priority but this is a subjective assessment of the bridge into high, medium or low categories. The system is significant because it is the only national system to which other bridge owners can relate and is therefore likely to have a major impact on the future development of local systems.

4.2 The 'BRIDGIT' System
Surrey County Council and Howard Humphreys, Consulting Engineers have been collaborating on the development of a bridgeworks management maintenance system. This type of activity is being repeated by many different bridgeowners, DTp agencies and consultants in many parts of the UK. However, 'BRIDGIT' has progressed further than most of its competitors.

The system is based on three separate databases, storing details of location, condition and inspection, with an interlinked management system permit access to all data. The base data will be updated from a series of inspections and the system itself checks new data being entered for 'out of range' records. In this way the condition of the bridge is recorded and monitored and decisions can be based on a knowledge of the level of investment. The overall effectiveness of the system is assessed by calculating an asset value for the initial bridgeworks stock and checking that the investment decisions tend to increase the value of the stock.

The system includes a priority ranking facility to enable the user to determine which repairs to carry out when faced with the problems of resource or budget constraints. The ranking calculations are based on three factors; the location factor, the road factor and the condition factor. As the system is being developed commercially the details of these factors together with any weighting parameters are not readily available. However, it can be seen that the decision making process is based on a narrow set of criteria.

4.3 Review of Systems in use in the US
The US Federal Highway Administration has been encouraging the development of bridge maintenance and management systems for several years, [6]. At the national level a system of Sufficiency Ratings was proposed. To calculate priorities the system considers a number of factors; the cost, the existing loading limit, safety, the route, the remaining life, the class of highway, the traffic and HGV's

and the possible co-ordination with other construction works. Once established and validated the system would have a major role in predicting the future condition of bridges. Whilst this scheme has been well received many States have developed individual systems.

Pennsylvania State has produced a bridge management maintenance system, [7], which helps to justify expenditure on bridgeworks maintenance. This is perceived to be necessary as bridge defects are not always obvious to the public and their elected members, and the 'betterment' which results from the maintenance work cannot be readily recognised. The aim of the system is to plan, decide and effect bridgeworks maintenance works. This system also incorporates a priority system which is based upon considerations of safety, of the urgency of carrying out the remedial works, of the importance of the road and the effects of its closure and of the bridges rate of deterioration.

Many other states have commenced work on systems and some details of the respective key features of the priority ranking techniques are described, [6]. In Kansas two algorithms are used to examine the combined condition of roads and bridges and the traffic, safety and structural condition of the bridge. The resulting figures are then weighted to give a priority ranking. A Critical Bridge Rating is calculated in Michigan, based upon the safety and physical condition of the bridge, the importance of the route and the finances. This is the only system to include monetary values directly in the priority system. New York State assess the bridge condition and the traffic flows as the two key parameters in determining priority. A life cycle costing approach has been used by Wisconsin, amongst others, noting that the relationship between condition and performance is not linear.

4.4 Danish State Railway System

The aims of a management maintenance system are twofold; in the short term considering the safety of the structure and in the long term considering the rate of deterioration. The Danish System, [8], is based on a database with a management system incorporating priorities to permit continued performance and to ensure the optimum economic service life of the bridge.

The system receives input from the bridgestock inventory, inspections and special inspections which are stored in the bridge data base. A management system is then applied to produce two forms of output. Firstly, maintenance decisions based on a ranking scheme and secondly bridge element records indicating the type and frequency of defects. However, the comment is made that under strict budgetary constraints only collapses and serious defects can be tackled.

5 Further Developments

5.1 Guidelines
As a result of this general review of management maintenance systems there are certain factors which appear common to most systems and other factors which are seldom if ever include but which might appear to be important. All the systems used some form of database but the actual data recorded under similar headings varies widely. To facilitate interaction between systems it would seem desirable that some minimum standards and formats for bridge data need to be agreed.

The core of the management module is the priority calculation which is based on many different criteria and weighted in different ways. Some systems consider many parameters whilst others are constrained to two or three; nevertheless all systems include safety and bridge condition in some way in the assessment of priority.

There are some additional factors not directly considered in any of the systems outlined above, which may be worthy of further attention. The decision taken about maintenance on a particular bridge is affected by the close proximity of an alternative crossing. Although one system mentioned the length of detour, no formal assessment of alternatives has been included. There are also, in the UK, two more considerations which should be noted; firstly is the bridge short span or narrow deck or likely to be considered as part of the deck strengthening programme prior to 1999 and secondly is work to be expected on the public utilities contained within the bridge deck. The effectiveness of these factors and others will be considered in the current research work being undertaken at UMIST.

5.2 Current Research
A major study is being carried out by UMIST, funded by the Repair, Operations and Maintenance Programme of the Science and Engineering Research Council. The research work at UMIST is concerned with the management of the maintenance process for bridgeworks, and gives particular attention to information from a specific sub-division of this sector, the bridge maintenance sections of Sheffield City Council and Manchester City Council.

The principal objectives of the work involves the investigation of existing management procedures, the examination of the criteria used in the decision making process and consideration of the methods of assessing potential maintenance requirements in the appraisal of new works.

5.3 Implications for new bridges

Maintenance procedures have significant implications for the design of new works; design being considered in its widest sense, incorporating feasibility, technical assessment and detailing. There are two main areas of interest; firstly design which prolongs the life of the bridge and prevents deterioration and secondly design features which make routine inspections and routine maintenance, such as replacing bridge bearings, simple and low cost operations. The main conceptual design is usually well prepared and considers the 120 year design life, however it is in poor detailing, usually carried out by junior engineers that the source of future maintenance problems can be found. The cost of including features in a bridge design to facilitate maintenance are small if these requirements are identified early in the design process.

6 Acknowledgements

The authors gratefully acknowledge the assistance of the Science and Engineering Research Council, the Bridges Sections of the City of Sheffield and the City of Manchester in the current research work.

7 References

1. Jones, C.J.F.P., Bridgeworks Maintenance, Lecture to the Yorkshire Association of the Concrete Society, November 24th, 1988.

2. Leadbeater, R.D., A Local Authority Viewpoint, Concrete Bridges - Management, Maintenance and Renovation, Concrete Society, 23rd February, 1989.

3. Leadbeater, A.D., The Practical use of Bridge Assessments, National Workshop on Bridge Maintenance Initiatives, Leamington Spa, Institution of Highways and Transportation, 8th April, 1986.

4. Smith, N.J., Management of Bridgeworks Maintenance, Proceedings of an International Conference on Bridge Management, University of Surrey, March, 1990.

5.   DTp, National Structures Database, 1989.

6. Saito, M. and Sinha, K.C., Review of current practices of   bridge management at state level, Transportation Research Record 1113, USA, 1987.

7. Arner, R.C., Kruegler, J.M., McClure, R.M. and Patel, K.R.,   The Pennsylvania Bridge Maintenance Management System, Transportation Research Record, 1083, USA, 1986.

8. Rostam, S.,   The European Scene ; Assessment and Strategy for   Highways and Road Bridges, Concrete Bridges - Management,   Maintenance and Renovation, Concrete Society, 23rd February, 1989.

THE STRUCTURAL ASSESSMENT AND STRENGTHENING OF KALEMOUTH BRIDGE, SCOTLAND

P. JACKSON
E.D.S.-Unilink, Heriot–Watt University, Edinburgh, Scotland, UK

Abstract
This paper describes the structural assessment and subsequent repair of
the Kalemouth Suspension Bridge, now carrying traffic loads which are
totally different to those which existed when it was built around 1835.
Keywords: Structural Assessment, Refurbishment, Replacement
Components, Cost Effective Repair, Old Bridges.

1 Introduction

The Kalemouth Suspension Bridge crosses the River Teviot in the
Scottish borders alongside the A698 road and at a point approximately
midway between Jedburgh and Kelso. It was designed and built by
Captain Sir Samuel Brown around 1835 and has undergone little maintenance since then. This short paper outlines the recent structural
appraisal and subsequent refurbishment.

2 The Designer of the Kalemouth Bridge

Captain Sir Samuel Brown was born in 1776, joined the royal navy in
1795 and rose to Commander by 1811. In May 1812 he left the navy,
accepting the rank of Retired Captain. By 1805, Brown was already
actively engaged in experimenting with his ideas for wrought iron
chain. These led to his first award of Letters Patent of Invention on
February 4 1808 for 'Rigging of Ships or Vessels'. The superiority of
chain cable over hemp rope was recognised by the Navy and in 1811 chain
ground tackle was adopted into the Service.
  Brown's interest in suspension bridge techniques started in 1808.
He recorded '...about that period I made drawings and calculations of
the strength of bridges of suspension.....but it was not until 1813
when I constructed a bridge of straight bars for this purpose on my own
premises....the span of this bridge is one hundred and five feet, and
although the whole of the ironwork weighs only thirty- seven hundred
weight, it has supported loaded carts and carriages of various descriptions'. This early experimental work was made just two years before
James Finley's published accounts of experiments in 1810, and was most
likely initiated by Parliamentary Acts to improve roads and bridges,

rather than his probable knowledge of Finley's work.

In 1806, Brown formed a partnership with his cousin Samuel Lennox and the firm of Samuel Brown, Lennox & Co. prospered. In 1817 Brown obtained patent number 4137 for his 'Invention or Improvement in the Construction of a Bridge by the Formation and Uniting, of its Component Parts, in a Manner not Hitherto Practised'.

3 The Kalemouth Bridge

One of Brown's many bridge designs was that at Kalemouth (Photographs 1 and 2). The bridge spans approximately 57m (186ft) between suspension points, has a width between hangers of 4.6m (15 feet) and the suspension points are 4.9m (16 ft) above road level. Each side of the bridge has a suspension chain consisting of pairs of links, each 51mm (2 inch) diameter and approximately 3.05m (10ft) long, connected together by three short coupling links and two nominally 54mm (2.125 inch) diameter common cross pins (Photograph 3). The deck hangers, 25mm (1 inch) diameter, are hung from each pinned connection and at the mid length of each main bar link. This intermediate hanger introduces bending but has worked well for 153 years. A more worrying point is Brown's known policy to use abnormally low factors of safety in his designs. Furthermore, he applied these to the ultimate strength of the wrought iron and not to the yield strength.

The timber deck has an upward camber which Brown saw as a means of resisting wind induced oscillations. More importantly, he included timber handrails in the form of 0.76m (2'6") high by 1.52m (5ft) diagonally braced panels. In effect, these provide stiffening trusses to both sides of the bridge.

4 The Structural Condition of Kalemouth Bridge in 1985

By 1985 the timber deck had deteriorated significantly and some deck planks had been replaced. Most of the joints in the timber handrails were loose and were providing little stiffness to the bridge. Many of the horizontal diagonal wrought iron braces beneath the deck were loose. A car crossing the bridge produced a 150mm (6") bow wave in the deck. A superficial inspection of the pinned joints of the main chains suggested that significant corrosion may exist and there had to be doubts concerning the condition of the wrought iron below ground level at the anchorages and encased in the head of the piers.

The bridge carries a weight restriction of 3 ton per vehicle. A check calculation indicated that the suspension chains on each side of the bridge were carrying around 400kN (40 tons). This is equivalent to a stress in the wrought iron of approximately 100 N/mm^2 (6.4 ton/sq.in), giving a factor of safety of around 3.5 on the typical ultimate tensile strength of 350 N/mm^2 (22.5 tons/sq.in). This is a low factor of safety compared to a normally acceptable value of 5.0, but is better than the 2.25 which results from Brown's normally adopted working stress of 155 N/mm^2 (10 ton/sq.in). The three small links at each joint have a total cross section equivalent to that of the two main bars and are

thus at the same nominal tensile stress. The shear stress in the pins
is 44 N/mm^2 (2.82 ton/sq.in). Of course, this just adequate performance
assumes that the wrought iron is of reasonable quality and that there
is no significant loss of cross section due to corrosion. Recently
these were checked by removing 'shavings' of metal for microscopic
examination, by visual inspection of the joints using an optical probe
and by x-ray photography of the joints using a portable radio active
cobalt source. The microscopic examination indicated that there was
minor surface corrosion of the wrought iron but that this had not propo-
gated and that the material was likely to have an ultimate tensile
strength of around 350 N/mm^2 (22.5 tons/sq.in). Unfortunately, the
condition of the joints was of concern. There had been insignificant
corrosion of the eyes at the ends of each long link, but the small
links had lost up to around one third of their cross section and the
pins had lost up to one quarter of their cross section at positions
where rain water and debris had collected over the years and where it
had been impracticable to protect by paint.

5 General replacement of joints in main chains

It was decided that all the small links and pins should be replaced.
   A destressing rig was designed, manufactured and tested in the
laboratory to support the load across each joint (Photograph 4). The
three small links and two pins were removed and replaced by new components. In the absence of modern wrought iron it was decided to manufacture the new components in speroidal graphite cast iron. Mechanical
properties of grey irons may be greatly improved if the graphite shape
is modified to eliminate planes of weakness caused by continuous flakes.
Such modification is possible if molten iron having a composition in
the range 3.2-4.5% carbon and 1.8-2.8% silicon is treated with magnesium
or cerium additions before casting. This produces iron with graphite in
spheroidal form instead of flakes and this is known as nodular, spheroidal graphite (SG) or ductile iron. Nodular irons are available with
matrices that consist of ferrite, pearlite, or mixtures of these, or
austenite.
   The compact shape of the carbon does not reduce ductility of the
matrix to the same extent as graphite flakes, so that useful ductility
is obtained. Nodular irons offer a range of ductilities and tensile
strengths considerably higher than those of grey irons.
   Nodular irons are made with lower sulphur and phosphorous content
than grey irons because these elements tend to restrict formation of
nodular graphite. The grade chosen for the replacement components has
an ultimate tensile strength similar to wrought iron but a slightly
smaller elongation. As the name of the material implies, all replacement components were manufactured by casting. All pins were then turned
down to diameters of 53, 54 and 55mm and sets of three small links were
machined at their internal ends to give as uniform as possible contact
with each pin diameter. Specimens representing 15% of the total number
of joints were selected at random and tested to failure to ensure that
there was a 95% statistical probability of a minimum factor of safety of
3.5. As a further precaution all sets of three links and two pins were
proof loaded in the laboratory to check that their yield load was above

600kN, i.e. 1.5 x likely maximum working load. It could be argued that these load tests were unnecessary, particularly bearing in mind that the guarantee of a minimum yield of 1.5 x working load is not a particularly demanding standard. The supplier of the castings was chosen because of the firm's known high standard of workmanship and quality control, and the testing of samples to destruction confirmed the chosen factor of safety. Nevertheless, the engineer responsible for the refurbishing wished for his own peace of mind that all components should undergo some proof loading. 'Engineering commonsense' led him to argue that defects could occur in some of the castings and if they did exist they could probably be identified at relatively low loads. Hence he implemented the proof loading of each joint assembly to 600 kN. In the event, none failed. Each joint in the bridge was fitted with the largest of the three diameters of pins which would pass through the eyes in the ends of the long chain links. The refurbished joint was coated with an anti-corrosion wax.

    Test specimens were cut from the removed wrought iron links and pins and when tested gave ultimate tensile stresses between 340 and 380 N/mm^2 and yield stresses from 230 to 270 N/mm^2. The elongations at failure were between 26% to 30%, with corresponding reductions in area between 33% and 46%. Ignoring loss of material due to corrosion, the minimum yield strength of a joint should have been around 900 kN (30 tons), or 2.25 times the maximum working load. Samples of the joints removed from the bridge, and including corroded components, were tested and were found to fail at loads as low as 700kN (70 tons). Thus, before the refurbishing of the bridge, some joints with corroded components had failing strengths only twice the maximum likely working load, and yield strengths which were around 1.5 times the working load. Vehicles heavier than the 3 ton limit would reduce these factors of safety further.

6 Strengthening of joints in main chains at suspension supports

Where the suspension chains pass over the supporting towers they consist of a number of short links, similar to the geometry of a cycle chain, which are supported on cast saddles. Despite a total lack of maintenance, their protected location had led to relatively little corrosion. Rather than replace these links it was decided to provide a system of strenghening. This consisted of stainless steel bars, clamped to the ends of the adjacent long links and bearing on curved stainless steel plates at each side of the cast saddle. The strengthening bars were merely stressed to a nominal load, so that in the short term the original wrought iron components continue to provide the strength. If major corrosion occurs in the future, the high elongation of the wrought iron would gradually transfer the load to the stainless steel components. These are of a sufficient size to carry the whole load, if that should ever be necessary.

7 Strengthening of ground anchorages

The ground anchorages present a problem because, clearly, they can not be exposed for inspection. At both sides of the bridge the wrought iron was exposed for a depth of approximately 600mm. At one side this merely exposed material around the mid lengths of the main links. The wrought iron was in good condition. At the other end of the bridge two joints were exposed, one at each side of the roadway. Both had significant loss of material from the small links due to corrosion and there was some corrosion around the end 'eyes' of the long links.

The exact form of the anchorage is not known and it may be that the chains are anchored into the rock which exists on both sides of the river. On the other hand Brown's Specification for a bridge at Montrose (1825) refers to '...one of the iron plates which is to be laid under ground, of sufficient depth to resist the drag of the back-stay chains, which pass through the plates, and are secured by means of a bolt. The plate will be 6 feet by 8, and about 4 inches thick in general section'.

To remove uncertainty, it was decided that the best solution would be to recommend new reinforced concrete compression and tension piles and to connect to the joints near ground level. In effect, the existing anchorages would become redundant.

8 Deck

Although the timber deck had suffered major deterioration, its basic design is adequate for current use. Furthermore, there was an aesthetic requirement to maintain the original appearance. Therefore the whole of the deck and handrails were rebuilt to the original dimensions. Canadian douglas fir was used instead of the original oak, and galvanised mild steel fixings and below-deck bracing were used instead of wrought iron.

9 Performance Check

For a major structural repair of the type carried out at the Kalemouth Bridge, it is desirable to carry out a performance check at around 18 to 24 months after completion of the repairs. The purpose of this check is not one of defect repair as would occur within a maintenance period but rather to confirm that the modified and repaired structure is behaving as intended. The cost of this check should be included in the original contract price.

This check was carried out on the Kalemouth Bridge during the summer of 1989.

10 Comments

Work such as described above can be amongst the most enjoyable that an engineer is asked to undertake, but must be approached in a somewhat different manner to new works. A useful analogy is that of the medical profession. When we take our deteriorating bodies to the medical professionals we hope for a complete cure but often are willing to settle for a repair which merely arrests further deterioration. We do not expect to be rejuvenated to the fitness of a young person and eventually the stage is reached when we and the medical profession must accept that we have come to the end of our lives. Old buildings or structures are very similar. We must not try to preserve all structures, and those which we do repair will have their performance limited by the original components which are allowed to remain. If the resulting performance is sufficient for current and anticipated requirements, then the repair is worthwhile. This has been the situation at the Kalemouth Bridge.

The assessment of old buildings or structures should not be limited to tests and calculations. The circumstances of the original design and construction should be investigated and these often will cast important light on the interpretation of the current performance. Poor construction and 'short cuts' will be found in all construction. At Kalemouth, timber packing had been used to accommodate poor fit of some of the components in the suspension chains. In this instance it was of no structural importance. In other situations it could be critical.

For more recent construction it is sometimes possible to trace people who worked on or were associated with the original design and/or construction. Their recollections can be invaluable.

To those who own, or are responsible for, such construction and who commission repairs, the writer would ask that they allow ample time for the work to be carried out. Unless there is an obvious risk of structure failure, it is unnecessary to have the speed which, quite correctly, is associated with new construction. The assessment and repair of old construction is generally cheaper and better if it can be carried out at a slower pace.

11 Bibliography and sources of information

Sources 2 and 3 below contain a large number of references.
1 Cowie F. M. Personal Files
2 Day T. (1985) 'Samuel Brown in North-East Scotland' **Industrial Archaeology Review, Vol. 7, No.2,** 171-89
3 Day T. (1983) Samuel Brown : his Influence on the Design of Suspension Bridges, **History of Technology,** 61-90
3 Miller G. Retired Archivist.

PHOTOGRAPH 1

PHOTOGRAPH 2

PHOTOGRAPH 3

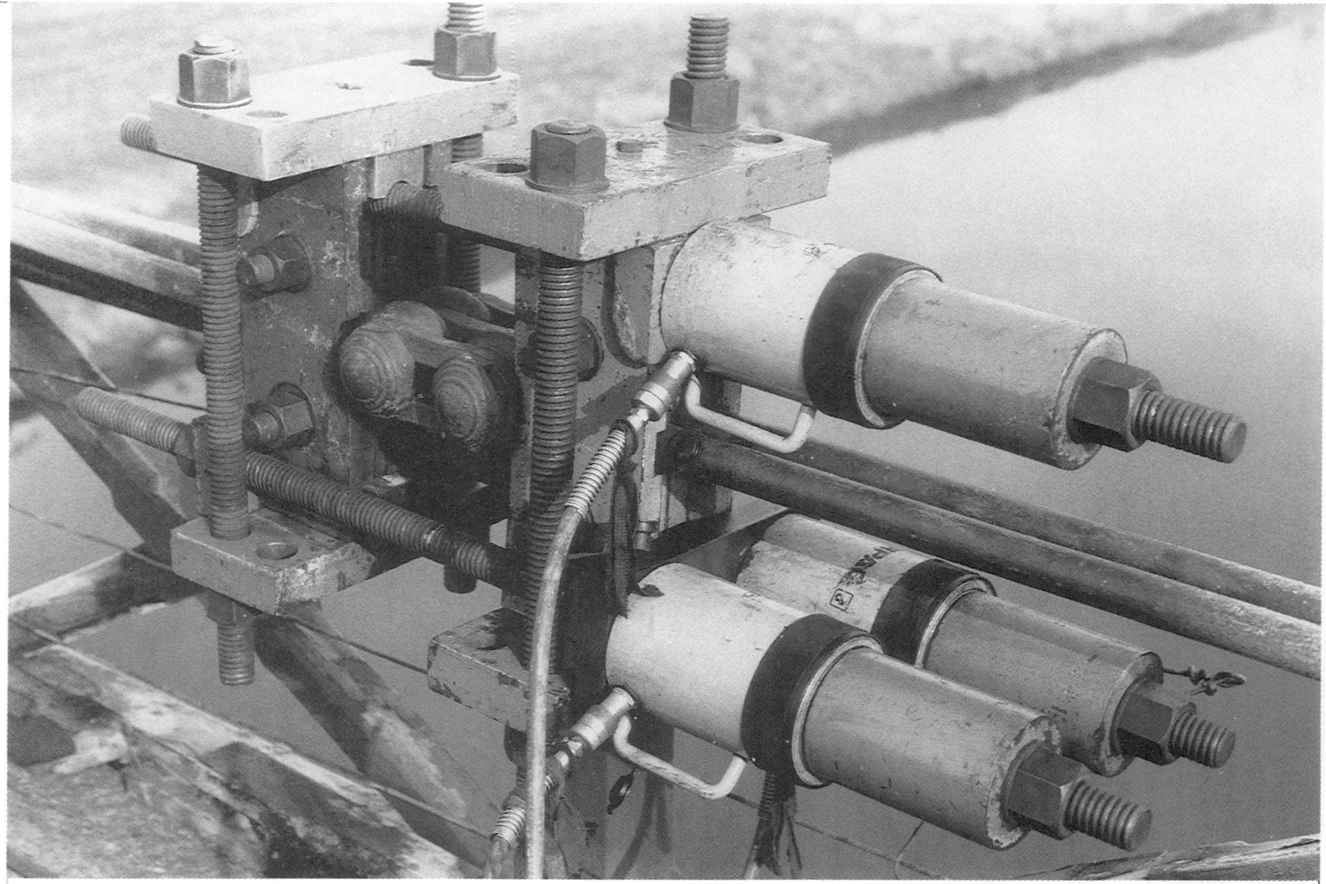

PHOTOGRAPH 4

STEEL TRUSS BRIDGE REHABILITATION TECHNIQUE USING SUPERIMPOSED STEEL ARCHES (THE APPLICATION TO A MONTANA RAILROAD DECK TRUSS BRIDGE)

R.H. KIM
BKLB Inc., Consulting Engineers, Pennsylvania, USA
J.B. KIM
Bucknell University, Pennsylvania, USA

Abstract
This paper describes a patented economical and reliable
way to rehabilitate old steel truss bridges. This
method has been applied to many highway bridges and,
most recently, a railroad truss bridge. The paper also
includes why this method of rehabilitation has
advantages over conventional truss bridge rehabilitation
techniques and bridge replacement.

Highway bridge problems

The United States government has neglected its
infrastructure for many years and is only now starting
to give it proper attention as structures get older and
deteriorate at increasingly faster rates. Among all
structures, bridges have been the most neglected. The
Federal Highway Administration (FHWA) has published
reports listing as many as 40%, or 240,000, of all
bridges as being deficient in the United States.

Steel truss bridges are prevalent throughout the
United States and many have aged to the point where
bridge owners must repair or replace them. Many years
ago engineers did not design bridges for heavier, modern
loads. The many deteriorated members and connections
make these bridges extremely difficult to analyze with
any accuracy. Factors contributing to the difficulties
of analysis are, but not limited to, the following:

1. Many members are loose under both dead load and
   live load conditions. This shows that other
   members are theoretically overstressed under
   design load conditions.

2. Corrosion freezes joints that are intended to act
   as frictionless pins. Among other consequences,
   this adds considerable bending stresses to truss
   members that were designed to resist only axial

stresses.

3. Inspectors can not positively determine the extent of corrosion and fatigue damage to the connections without dismantling them which is dangerous and expensive.

4. Bridge bearings are corroded and frozen.

5. Many states use salt for deicing roadways during winter. The salt accelerates corrosion along bottom chord members and connections to varying degrees of severity.

In actual comparisons between theoretical analyses and field testing of a highway truss bridge, results differed by as much as 800%. Out of all truss elements, the connections are by far the most critical. Bridge collapses are almost always due to failure at a connection, not in a member.

Typical truss bridge rehabilitations consist of strengthening individual connections and members that appear deficient. This method is very labor-intensive, expensive, and dangerous. The slightest mistake by the contractor can easily collapse the bridge. Repairs done in this manner are also of a dubious nature. Uncertainties remain with the connections and members that have not been rehabilitated. Bridge inspectors must constantly check on and maintain these remaining truss elements for years to come. Since bridge owners must assume the expenses and risks to public safety, rehabilitations of truss bridges in this manner are undesirable.

Arch reinforcement of highway truss bridges

To meet the requirements of safely, economically, and reliably rehabilitating old steel truss bridges, we have been applying a patented system to rehabilitate old steel trusses. Steel arches superimposed over the existing trusses can structurally upgrade these bridges to modern traffic standards. By themselves, the arches share a significant portion of live loads with the trusses and greatly reduce stresses in existing truss members and joints. By posttensioning the arches, we can negate dead load stresses in bottom chord members and connections, which are the most susceptible to corrosion and fatigue damage. After mounting the arches on the trusses, it is easy to make minor modifications to the floor systems if necessary. The addition of new intermediate floor beams supported by the arches can greatly reduce the span lengths of the stringers. In some cases where it would have been extremely difficult to replace existing floor beams, the addition of two new intermediate floor beams per truss panel allowed existing floor beams to remain in place. This reduced costs by omitting labor for difficult floor beam removals.

Among advantages realized by rehabilitating truss bridges with steel arches are:

1. The entire bridge is strengthened, not just individual members and connections.

2. It is possible to upgrade the load-carrying capacity of the bridge to any desired load capacity.

3. The bridge owner can reuse most, if not all, of the existing bridge with no modifications required for the approaches, resulting in costs much less than conventional rehabilitations or replacements. Typically, additional costs incurred by bridge replacement include environmental impact studies, acquisition of right-of-way properties, demolition and removal of the existing bridge, modification of bridge approaches, disruption of business due to bridge closings for construction, new substructures and superstructures, detours and traffic controls, temporary bridges for emergency vehicles, and long construction periods.

4. There is no additional encroachment on clearance below the bridge.

5. The contractor can maintain traffic on the bridge during the rehabilitation process.

6. The period of construction is short.

7. The arches introduce redundancies in the structural system. In the unlikely event that a member or connection should fail, the bridge will still be able to sustain traffic loads.

8. It is possible to preserve the original architecture of the bridge for historical purposes.

9. The arches significantly extend the bridge's life span to that of a new bridge. Since the arches are new steel, the rehabilitated truss bridge will last as long as a new bridge.

We have rehabilitated highway truss bridges in
Pennsylvania, Connecticut, Maryland, New York, Idaho,
Kentucky, and Oregon by superimposing steel arches on
the existing trusses. Figures 1 and 2 show a truss
bridge in New York before and after rehabilitation with
steel arches. Before rehabilitation, the maximum live
load allowed on the bridge was only 44.48 kN. After
rehabilitation, the allowable live load is 320.26 kN.

Figure 1
49 m Long Pratt Truss Bridge Built 1902 Posted for 44.48kN
Weight Limit, Cortland, New York

Figure 2
Arch-Reinforced Pratt Truss Bridge Upgraded in 1986 to
320.3 kN Weight Limit, Cortland, New York

Arch reinforcement of a railroad deck truss bridge

In the past year, we have applied arch rehabilitation to
a railroad deck truss bridge in St. Regis, Montana. The
bridge, formerly built by Northern Pacific Railroad in
1908, consists of four 18.19 m approach spans and two
46.25 m truss spans made out of open hearth steel.

The railroad owner required a bridge that would last
another 50 years under the American Railway Engineering
Association's Cooper E 80 live loading standard and also
wanted to replace the existing open timber deck system
with concrete ballast boxes. The ballast and concrete
boxes would add considerable dead load to the structure.

The approach spans could be reinforced easily to
accommodate the heavier loads for another half century
but the truss spans posed a major problem. Trains
running at track speed, 80.45 km/hr, had to slow down to
40.22 km/hr upon reaching the truss spans due to the
excessive deflections and other uncertainties with
regard to the trusses. Fatigue strains and corrosion
had accumulated in members and joints for more than 80
years. Since engineers did not originally design the
truss spans to carry heavier, modern loads, the trusses

underwent excessive swayings and deflections. As a
result of these deflections, many rivets failed at the
stringer/floor beam connections. Additionally, the
lateral bracings were cracking in the upper regions of
the trusses and bottom chord members were loose, even
under live load conditions. Finally, inspectors could
not economically and reliably determine the conditions
of the many connections while the bridge was still in
use.

The options available were to replace the truss spans
with plate girder spans or reinforce the truss spans
with steel arches. The plate girder option would have
required building intermediate piers more than 15 m tall
to meet track elevations and the detouring of train
traffic to other lines. The railroad company could not
afford to experience the projected loss of revenues and
customers to other railroad companies by closing this
main line. After also considering the additional
expenses associated with the new plate girder option
with new additional piers, they opted for us to design
the reinforcement of the trusses with steel arches and
maintain the continuity of train traffic during
construction.

The application of the Cooper E 80 live loading on
the truss spans meant the overall application of 6,574.1
kN of live loading alone on each truss span with a
specified impact of more than 42% applied to each
member. This gave the effect of an overall live load
per truss greater than 9,335.3 kN. The new concrete
ballast boxes and ballast added 1,729.8 kN to the dead
load of the original structure. After thorough
analyses, we decided to use four W24x62, A572 Grade 50
rolled steel sections (yield stress of 344.8 MPa) for
the arches with eight #18 Grade 75 rebars (yield stress
of 517.1 MPa and 5.733 cm in diameter) as posttensioning
rods. All new steel met American Institute of Steel
Construction standards. The truss spans are on slender
piers so posttensioning rods not only reduce the dead
load stresses in existing truss members, but also resist
the outward thrust of the arches and prevent overturning
of the piers. Figures 3 and 4 show the railroad deck
truss spans before and after arch reinforcement.

The railroad company hired a testing firm to study
the load distributions between the arches and trusses.
As of this writing, those results are not officially
available. However, preliminary results show that the
arches are performing as designed. We expect this deck
truss bridge to support trains reliably for at least
another 50 years.

Figure 3
165 m Long Railroad Bridge Built 1908, St. Regis, Montana

Figure 4
Arch-Reinforced Railroad Bridge Upgraded in 1989 to Carry
9,335.3 kN Live Load Plus 1,729.8 kN Additional Dead Load
Per Truss Span, St. Regis, Montana

PART ELEVEN
CABLE AND MEMBRANE STRUCTURES

STATE OF THE ART: NONLINEAR DYNAMIC RESPONSE OF MEMBRANES

C.H. JENKINS, J.W. LEONARD
Ocean Engineering Programme, Dept Civil Engineering, Oregon State
University, Corvallis, Oregon, USA

Abstract
Modern methods for the analysis of membrane structures are evolving.
Two problems of analysis are associated with membrane structures: (i)
shape (or form) finding; (ii) response (deformation and/or stress)
analysis. Shape finding is a non-trivial static problem that is not
considered in this paper. This paper is concerned with reviewing
methods of nonlinear dynamic analysis of membrane structures. Attention is focused on formulation of field equations, wrinkling analysis,
fluid/structure interactions, material nonlinearities, and computational methods.
Keywords: Membranes, Nonlinear, Dynamics

1 Introduction

In modern times, membranes have seen increasing use in building
structures such as radar domes, temporary storage, and aero-space
structures. Novel applications abound. For example, in the ocean
environment, membrane structures have been considered for use as
breakwaters at least since 1960 [NCEL (1971)]. Modi and Poon (1978)
investigated inflated viscoelastic tapered cantilevers for underwater
and offshore applications. Szyskowski and Glockner (1987a) have
considered static elastic models of membrane structures as floating
storage vessels. Leeuwrik (1987) and Bolzon, et al. (1988) considered
flexible membrane dams.

2 Nonlinear Membrane Field Equations

2.1 General
Structures that **cannot** support stress couples **anywhere** are by definition 'true membranes' and are inherently nonlinear, with the degree of
nonlinearity incorporated into the field equations dependent on the
formulation philosophy. Thus, theories may be developed ranging from
linear to small strain − finite rotation (geometric nonlinearity only:
hereafter called 'large deflection') to fully nonlinear both in
geometry and materials (geometric and physical nonlinearity: hereafter called 'large deformation'). The linear theory will not be
discussed here [see, e.g., Leonard (1988)].

The governing field equations are most easily analyzed in static loading configurations, and constitute the bulk of work reported. This is not necessarily as restrictive as it may seem since many dynamic loading regimes can be considered as quasi-static, i.e., when the inherent time scale of the loading is such that inertial effects can be ignored. The slow inflation of a plane membrane would be of this type. Furthermore, since membranes are inherently thin, light structures, their inertial effects may be ignored even for more rapid loading schemes. The more recent investigations into the fully dynamic problem will be reported separately.

2.2 Föppl–Hencky Theory

By equating the stiffness in the Föppl–von Kármán nonlinear (small strain, 'moderate' rotation) plate equations to zero, the equations for the Föppl–Hencky theory are obtained. The coupled nonlinear partial differential equations are difficult to solve in closed form. Berger (1955) used the Föppl theory to formulate the strain energy density of a deformed plate and then made the simplifying (but non-rational) assumption of ignoring the term containing the second invariant of strain. This approach has been applied to large deflections of membranes by Jones (1974) who used a Prandtl stress function solution and by Schmidt and DaDeppo (1974) who used a perturbation technique. For further discussion on Berger's hypothesis and equations, see Mazumdar and Jones (1974).

Numerical methods are usually employed to solve the governing equations. Shaw and Perrone (1954) used a finite difference–relaxation iteration method after recasting entirely in terms of displacement. They considered a rectangular membrane and calculated stress contours. See also Kao and Perrone (1972). Allen and Al–Qarra (1987) have used an extension of the theory, solved by the finite element method, and consider both square and circular membrane problems.

Caution must be exercised in the application of the Föppl–Henky theory to the large deflection problem. For the case of the axisymmetric deformation of an annular membrane, Weinitschke (1980) compared the Föppl–Henky and large rotation Reissner shell theories. In some cases, the difference between theories exceeds 10%. The appropriateness of the Reissner theory for large rotations is reiterated in Grabmüller and Weinitschke (1986), and Weinitschke (1987).

2.3 General Membrane Shell Theory

A considerable amount of work in nonlinear membrane response follows from the theory for the large deformation problem as derived from classical shell theory. [For a review of work preceding 1973 see Leonard (1974).]

A tractable application of the above is the case of axisymmetric deformations of an isotropic incompressible membrane of uniform undeformed thickness. Many investigators have studied the quasi-static inflation of a plane circular elastic membrane. Wineman (1976, 1978) and Roberts and Green (1980) have considered similar problems but with a nonlinear viscoelastic constitutive relation.

Finite element solutions of the quasi-static inflation of a plane circular membrane were given by Oden and Sato (1967), Leonard and Verma (1976), and Leonard and Lo (1987). Warby and Whiteman (1988) present a finite element solution for the quasi—static inflation of an initially plane circular linear viscoelastic membrane up to a rigid obstacle. Haug and Powell (1972) include wrinkling analysis in the finite element membrane solution. Douven, et al. (1989) consider a finite element formulation for a viscoelastic transversely isotropic membrane.

2.4 Additional Formulation Philosophies

Nonlinear membrane equations in terms of Lagrangian displacement components were formulated by Glockner and Vishwanath (1972) and applied to a spherical Hookean membrane. Closed form series solutions were generated by perturbation methods.

A new formulation was developed [Malcom and Glockner (1978)] to study the response of spherical inflatables to ponding collapse. See also Szyszkowski and Glockner (1984a,b). The above membranes are considered to be inextensible and wrinkling analysis is included. See the review article by Glockner (1987). Dacko and Glockner (1988) include extensibility of the membrane in order to adjust for difficulties in some of the previous work.

2.5 Nonlinear Dynamics

The quasi—static approximation to the dynamic membrane problem has already been previously discussed. Dynamic relaxation is a solution technique for the quasi—static problem that borrows from computational methods of solution for structural dynamic response. In this explicit, iterative method, fictitious values of mass and damping are chosen so that the static solution is achieved with the smallest number of steps [Barnes (1980a); Barnes and Wakefield (1988)]. The complete dynamic analysis takes into account the inertia of the membrane and its surrounding medium (added mass) as well as the damping of the membrane and its surrounding medium (radiation damping) [Davenport (1988)]. As in the quasi—static case, various approximations are made.

Nonlinear membrane equations following from the large deflection theory have been used by Chobotov and Binder (1969) and Yen and Lee (1975) to study the free and forced vibrations of a plane circular membrane using perturbation methods. Plaut and Leeuwrik (1988) derive a nonlinear equation of motion based on the inextensibility assumption and use Galerkin's method to analyze the nonlinear oscillations of a cylindrical membrane.

The equations of motion for the large deformation problem can be written using the principle of virtual work [Leonard (1988)]. The equations are usually solved by a spatial discretization using the finite element method and a temporal discretization using either an implicit or explicit difference method [see, e.g., Barnes (1980)]. Since the system stiffness is usually non—constant, iteration (e.g., Newton—Raphson) [Barnes (1980a)] during each time step will be required.

Implicit methods require equation solutions at each time step but are unconditionally stable. Explicit methods find the accelerations from the equations of motion and are then integrated to obtain the displacements without the need for solving equations (provided the mass matrix is diagonal). Explicit methods are conditionally stable and thus require smaller time steps.

Oden, et al. (1974) considered the problem of a plane square membrane subject to a central impulsive load using triangular elements and central difference method (explicit). Benzley and Key (1976) used cubic quadrilateral elements and the central difference method to examine the free vibration of a square membrane. Leonard and Lo (1987) considered the same problem using a variety of element types, and showed quadratic quadrilaterals to be advantageous. Forced vibration problems were considered using Newmark's method (implicit). Hsu (1987) describes use of a commercial nonlinear finite element code to study the dynamic response (explicit) of a tension fabric structure. Notable in his work are the use of a viscoelastic constitutive equation and stress relaxation analysis.

3 Constitutive Relations

Generalized linear elastic relations, while suitable for large deflection problems, will not be discussed here. The more general large deformation problem uses constitutive relations based on the strain energy function, i.e., the so-called 'hyperelastic' materials. The strain energy function is taken to be some function of the strain invarients, where the strain invariants are functions of the extension ratios. Many forms have been proposed [for a brief summary, see Libai and Simmonds (1988)].

For materials exhibiting elastic time dependency, a viscoelastic constitutive relation is in order. Certain membrane materials, like polymers at small strain, may be modeled for some purposes by a linear viscoelastic relation. Others at small strain, and most at large strain, require a nonlinear formulation. Linear viscoelastic relations may be of two types, differential or integral. Differential equations are written for mechanical models consisting of combinations of springs and dashpots. The alternative integral form is based on the so-called 'hereditary integral'. For uniaxial stress/strain, this may be shown to be expressible in terms of the creep compliance and the relaxation modulus. All of the above are generalizable to multi-axial stress/strain. With suitable selection of the creep and relaxation functions, correspondence between the differential and integral forms can be shown.

Warby and Whiteman (1989) have proposed a generalization of the Mooney-Rivlin relation to include viscoelastic effects. Nonlinear viscoelastic constitutive relations are fundamentally of the multiple-integral type. However, for tractability, certain assumptions (e.g., superposed small loading on a large constant loading) can lead to single integral representations.

Creep problems of membrane structures are documented and range in extremes from aesthetic considerations to loss of prestress [Liddell (1986); and Srivastava (1987)]. Viscoelastic effects are usually temperature dependent as well. Coated fabrics deserve consideration as composite structures and also include the problem of fracture [Racah (1980)].

4 Wrinkling

Thin membranes are inherently no-compression structures. Potential compressive stress and/or loss of prestress are handled via changes in membrane geometry, i.e., large out-of-plane deformations. These 'wrinkles' are a localized buckling phenomenon. Analysis of the wrinkling response is important to prediction of structural instabilities and fatigue life.

Wagner (1929) introduced the ideas of wrinkling and 'tension field theory': under the action of a specific loading, one of the principal stresses goes to zero, the other remains nonnegative and defines a 'tension field.' The crests and troughs of 'wrinkle waves' align with the direction of the nonzero principal stress. Reissner (1938) generalized Wagner's results by introducing an artificial orthotropy.

In most wrinkling analyses, results are in terms of average strains and displacements, while no detailed information is generated for each wrinkle. See Wu (1974, 1978), and Wu and Canfield (1981) wherein a 'pseudo deformed surface' and 'pseudo stress resultants' are defined. They also defined the 'wrinkle strain' which is the difference between stretch ratios in the pseudo deformed and actual deformed surfaces.

Croll (1985) described the wrinkling of a plane membrane with in-plane elastic constraints. Contri and Schrefler (1988) used geometrically nonlinear finite element analysis to study the wrinkling of an inflated airbag. A membrane need not be wrinkled over its entire surface. Stein and Hedgepeth (1961) study partly wrinkled membranes. Mikulas (1964) extended the work and provided experimental details; finite element implementations are presented in Miller and Hedgepeth (1982), and Miller, et al. (1985). Mansfield (1968, 1970) reformulated the theory using an energy approach. Other studies of interest include: Zak (1983); Ikemoto (1986); Pipkin (1986); Lukasiewicz and Glockner (1986); Glockner (1987); Szyszkowski and Glockner (1987b,c); Pipkin and Rogers (1987); Honma, et al. (1987); Roddeman, et al. (1987).

The creep/relaxation response leading to a loss of prestress in viscoelastic membrane structures should accelerate the formation of wrinkles. This phenomenon has been little studied, as has the relation between 'wrinkle wave length' and material properties, or the problem of dynamic wrinkling (panel flutter) [Srivastava (1987); Leonard (1988)].

5 Fluid-Structure Interaction

Since most membranes will be located in a fluid (either gas or liquid), the fluid-structure interaction is of interest. Assumptions for the fluid field include considering the associated pressure field to be unaffected by structure deformations, interactive analysis whereby the deformations do affect the fluid field, and real and inviscid fluid models.

Large deformations of both fluid and solid structure have only recently been investigated, and then more often for frequency rather than time domain [Huang, et al. (1985); and Minakawa (1986)]. Lee and Leonard (1988) report on nonlinear time domain models for floating body problems. Some work has been motivated by the areas of suspension rheology and biomechanics [Zahalak (1987)]. Wind—membrane structure interaction has recently been investigated by Kunieda, et al. (1981) and Han and Olson (1987).

There has been an interest in the interaction of ocean waves and highly deformable structures. Here the problem is fully nonlinear in the sense that the fluid, the structure, and the coupling (via boundary conditions) are all governed by nonlinear equations. This coupled problem has recently been reviewed by Broderick and Leonard (1990).

6 Experimental Methods

Experiments play many roles in the nonlinear membrane problem. Due to the low rigidity/low elastic moduli of typical membrane materials, non—invasive, non—contact methods of measurement are desirable, such as optical grid and moiré methods. Problems associated with the above methods are the difficulty of preparing an accurately ruled specimen, the distortion associated with representing a curved surface in a plane for viewing, and the need for a stable processing environment. Further details are outlined in: Dally and Riley (1978); and Laermann (1988).

Digital imaging techniques have been applied to the large deformation problem [Peters and Ranson (1982)]. Applications of the method are demonstrated by Chu, et al. (1985). Perry (1985) discusses the use of strain gages on low—modulus materials. Other methods used specifically on membranes are capacitance type displacement transducer on a conductive mylar membrane [Chobotov and Binder (1964)] and a combination of strain gages and high speed film camera [Konovalov (1987)].

Experimental determination of viscoelastic parameters of bulk materials is well known. Corresponding procedures specifically for thin membrane materials are not well documented, especially for differences in the static and dynamic regimes. Experiments may also reveal structural characteristics, e.g., vibration frequencies and mode shapes. Takeda, et al. (1986) performed vibration tests on a large—scale structural model. Numerous model tests were conducted by Yoshida, et al. (1986) and Ban, et al. (1986). Ishizu and Minami (1986) made in—situ tests of membrane tensile stress.

Experimental data concerning environmental loads on membrane structures are sketchy. In the case of wind loads, this is partly due to the fact that membranes are often large structures that disturb the flow fields, thus rendering the pressure distributions configuration specific. Wind tunnel testing of scaled aeroelastic models of membrane structures is needed. An early (ca. 1954) 'radome' wind tunnel test is discussed by Newman and Goland (1982) and a wind tunnel test on a cylindrical 'tent' (ca. 1963) is discussed by Ross (1969). Other reports of wind tunnel testing and wind loading may be found in Howell (1980); Turkkan, et al. (1983); Fukao, et al. (1986); Ikemoto, et al. (1986); Ikoma (1987); and Miyamura, et al. (1987).

7 Conclusions and Future Studies

The use of membrane structures has been outlined. Certainly, their use is not quiescent. Some aspects of the membrane problem are well investigated, especially quasi-static axisymmetric deformations and hyper-elastic and linear visco-elastic constitutive theory.

Other areas are not so well studied, including biaxial material properties of thin membranes; non-invasive experimental techniques for large out-of-plane membrane deformations especially for field testing and in real time; and documentation of environmental loads.

Current studies at Oregon State University include the fluid-membrane structure interaction problem using a coupled boundary element/finite element method for wave loads on membranes, analysis of hydrodynamic response of visco-hyper-elastic membranes, and advances in the wrinkling theory of membranes.

8 Acknowledgement

This material is based upon work supported by the USN Office of Naval Research under the University Research Initiative N00014-86-K-0687.

9 References

Allen, H.G. and Al-Qarra, H.H. (1987) Geometrically nonlinear analysis of structural membranes. **Comput. Struct.**, 25, 871–876.

Ban, S., Tsubota, H., Kurihara, K., and Yoshida, A. (1986) Experimental investigation of a tensile fabric structure, in **Proc. of the IASS Symp. on Membrane Structures and Space Frames**, Osaka.

Barnes, M. (1980a) Non-linear numerical solution methods for static and dynamic analysis of tension structures, in **Proc. of the Symp. on Air-Supported Structures: The State of the Art**, London.

Barnes, M. (1980b) Explicit dynamic analysis of an impulsively loaded pneumatic dome, in **Proc. of the Symp. on Air-Supported Structures: The State of the Art**, London.

Barnes, M.R. and Wakefield, D.S. (1988) Form-finding, analysis and patterning of surface-stressed structures, in **Proc. of the First Oleg Kerensky Memorial Conf.**, London.

Benzley, S.E. and Key, S.W. (1976) Dynamic response of membranes with finite elements. **J. of Eng. Mechs.**, 102, 447–460.

Berger, H.M. (1955) A new approach to the analysis of large deflections of plates. **J. of Appl. Mech.**, 22, 465–472.

Bolzon, G., Schrefler, B.A., and Vitaliani, R. (1988) Finite element analysis of flexible dams made of rubber-textile composites, in **Computer Modeling in Ocean Engineering**, Proc. of an International Conference, Venice.

Broderick, L.L. and Leonard, J.W. (1990) Review of boundary element modeling for the interaction of deformable structures with water waves. To appear in **Eng. Struct.**

Chobotov, V.A. and Binder, R.C. (1964) Nonlinear response of a circular membrane to sinusoidal acoustic excitation. **J. of the Acoustical Society of America**, 36, 59–73.

Chu, T.C., Ranson, W.F., Sutton, M.A., and Peters, W.M. (1985) Applications of digital–image–correlation techniques to experimental mechanics. **Exp. Mech.**, 24, 232–244.

Contri, P. and Schrefler, B.A. (1988) A geometrically nonlinear finite element analysis of wrinkled membrane surfaces by a no–compression material model. **Commun. in Appl. Numer. Methods**, 4, 5–15.

Croll, J.G.A. (1985) A tension field solution for non–linear circular plates," in **Aspects of the Analysis of Plate Structures**, Oxford University Press, Oxford.

Dacko, A.K. and Glockner, P.G. (1988) Spherical inflatables under axisymmetric loads: another look. **Int. J. of Non–Linear Mech.**, 23, 393–407.

Dally, J.W. and Riley, W.F. (1978) **Experimental Stress Analysis (2ed)**. McGraw–Hill, New York.

Davenport, A.G. (1988) The response of tension structures to turbulent wind: the role of aerodynamic damping, in **Proc. of the First Oleg Kerensky Memorial Conf.**, London.

Douven, L.F.A., Schreurs, P.J.G., and Janssen, J.D. (1989) Analysis of viscoelastic behaviour of transversely isotropic materials. **Int. J. Num. Mthds. Eng.**, 28, 845–860.

Fukao, V., Iwasa, Y., Mataki, Y., and Okada, A. (1986) Experimental test and simulation analyses of the dynamic behavior of low–profile, cable–reinforced, air–supported structures, in **Proc. of the IASS Symp. on Membrane Structures and Space Frames**, Osaka.

Glockner, P.G. (1987) Recent developments on the large–deflection and stability behaviour of pneumatics, in **Developments in Eng. Mech.**, (ed. A.P.S. Selvadurai). Elsevier Sci. Pub., Amsterdam, 261–283.

Glockner, P.G. and Vishwanath, T. (1972) On the analysis of non–linear membranes. **Int. J. of Non–Linear Mech.**, 7, 361–394.

Grabmuller, H. and Weinitschke, H.J. (1986) Finite displacements of annular elastic membranes. **J. of Elast.**, 16, 135–147.

Han, P.S. and Olson, M.D. (1987) Interactive analysis of wind–loaded pneumatic membrane structures. **Comput. and Struct.**, 25, 699–712.

Haug, E. and Powell, G.H. (1972) Finite element analysis of nonlinear membrane structures, in **Proc. of the IASS Pacific Symp. Part II on Tension Structures and Space Frames**, Tokyo.

Honma, T., Nobuyoshi, T., and Nishimura, T. (1987) Static analysis of membrane structure with large deformation and wrinkling waves, in **Proc. of the Int. Colloquium on Space Structures for Sports Buildings**, London.

Howell, J.F. (1980) Definition of external forces, in **Proc. of the Symp. on Air–Supported Structures: the State of the Art**, London.

Hsu, M.B. (1987) Nonlinear analysis of tensioned fabric roofs, in **Proc. of the Int. Colloquium on Space Structures for Sports Buildings**, London.

Huang, M.C., Hudspeth, R.T., and Leonard, J.W. (1985) FEM solution of 3–D wave interference problems. **J. of Waterways, Ports, Coastal and Ocean Eng.**, 111, 661–677.

564

Ikemoto, N., Mizoguchi, Y., Fujikake,M., Kojima, O., and Hirota, M. (1986) Development of a design system for fabric tension structures, in **Proc. of the IASS Symp. on Membrane Structures and Space Frames**, Osaka.

Ikoma, T. (1987) Structural properties of air-supported structures. **Space Struct.**, 2, 195–203.

Ishizu, N. and Minami, H. (1986) A method of measurement of actual membrane tensile stress" in **Proc. of the IASS Symp. on Membrane Structure and Space Frames**, Osaka.

Jones, R. (1974) A simplified approach to the large deflection of membranes. **Int. J. of Non–Linear Mech.**, 9, pp. 141–145.

Kao, R. and Perrone, N. (1972) Large deflections of flat arbitrary membranes. **Comp. and Struct.**, 2, 535–546.

Konovalov, E.L. (1988) The deformed state of a membrane subject to cyclic loss of stability. **Soviet Eng. Research**, 7, 24–28.

Kunieda, H., Yokoyama, Y., and Arakawa, M. (1981) Cylindrical pneumatic membrane structures subject to wind. **J. Eng. Mech.**, 107, 851–867.

Laermann, K.H. (1988) Hybrid techniques for analyzing nonlinear problems in solid mechanics, in **Proc. of the Int. Conf. on Comput. Eng. Sci.**, Atlanta, 2, pp. 1–7.

Lee, J.F. and Leonard, J.W. (1988) A finite element model of wave-structure interactions in the time domain. **Eng. Struct.**, 10, 229–238.

Leeuwrik, M.J. (1987) Nonlinear vibration analysis of inflatable dams. M.S. Thesis, Virginia Polytechnic Institute and State University, Blacksburg, VA.

Leonard, J.W. (1974) State-of-the-art in inflatable shell research. **J. of Eng. Mechs.**, 100, 17–25.

Leonard, J.W. (1988) **Tension Structures, Behavior & Analysis**. McGraw-Hill, New York.

Leonard, J.W. and Lo, A. (1987) Nonlinear dynamics of cable-reinforced membranes, in **Proc. of the Structures Congress 1987 Related to Buildings**, Orlando, FL.

Leonard, J.W. and Verma, V.K. (1976) Double-curved element for Mooney-Rivlin membranes. **J. of Eng. Mech.**, 102, 625–641.

Libai, A. and Simmonds, J.G. (1988) **The Nonlinear Theory of Elastic Shells**. Academic Press, Inc., Boston.

Liddell, W.I. (1986) Current developments in materials for fabric structures, in **Proc. of the First Int. Conf. on Lightweight Structures in Architecture**, Sydney.

Lukasiewicz, L.B. and Glockner, P.G. (1986) Stability and large deformation behavior of nonsymmetrically loaded cylindrical membranes. **J. Struct. Mech.**, 14, 229–255.

Malcolm, D.J. and Glockner, P.G. (1978) Collapse by ponding of air-supported membranes. **J. of Struct. Eng.**, 104, 1525–1532.

Mansfield, E.H. (1970) Load transfer via a wrinkled membrane, in **Proc. Roy. Soc. Lond. Ser. A.** 316, 269–289.

Mansfield, E.H. (1968) Tension field theory, in **Proc. of the Twelfth Int. Cong. of Appl. Mech.**, Stanford.

Mazumdar, J. and Jones, R. (1974) A simplified approach to the analysis of large deflections of plates. **J. of Appl. Mech.**, 41, 523–524.

Mikulas, M.M. (1964) Behavior of a flat stretched membrane wrinkled by the rotation of an attached hub. **NASA Technical Note TN D–2456**, NASA, Washington, DC.

Miller, R.K. and J.M. Hedgepeth (1982) An algorithm for finite element analysis of partly wrinkled membranes. **AIAA**, 20, 761–763.

Miller, R.K., Hedgepeth, J.M., Weingarten, V.I., Das, P., and Shahrzad, K. (1985) Finite element analysis of partly wrinkled membranes. **Comp. and Struct.**, 20, 631–639.

Minakaw, Y. (1986) Lagrangian functions of the interactive behavior between potential fluid and elastic containers in fields of finite deformations, in **Proc. of the IASS Symp. on Membrane Structures and Space Frames**, Osaka.

Miyamura, A., Tagawa, K., Mizoguchi, Y., Osame, K., Masahisa, F., and Murata, J. (1987) A case study of the design and construction of a tension fabric structure, in **Proc. of the Int. Colloquium on Space Structures for Sports Buildings**, London.

Modi, V.J. and Poon, D.T. (1978) Dynamics of neutrally buoyant inflated viscoelastic tapered cantilevers used in underwater applications. **J. of Mech. Design**, 100, 337–346.

NCEL (1971) Transportable breakwaters—a survey of concepts. **Technical Report No. R 727**, Naval Civil Engineering Lab., Port Hueneme, CA.

Newman, B.G. and Goland, D. (1982) Two–dimensional inflated buildings in a cross wind. **J. Fluid Mech.**, 117, 507–530.

Oden, J.T. and Sato, T. (1967) Finite strains and displacements of elastic membranes by the finite element method. **Int. J. Solids Struct.**, 3, 471–488.

Oden, J.T., Key, J.E., and Fost, R.B. (1974) A note on the analysis of nonlinear dynamics of elastic membranes by the finite element method. **Comp. and Struct.**, 4, 445–452.

Perry, C.C. (1985) Strain–gage reinforcement effects on low–modulus materials. **Exp. Tech.**, 25–27.

Peters, W. and Ranson, W.F. (1982) Digital imaging techniques in experimental stress analysis. **Opt. Eng.**, 21, 427–431.

Pipkin, A.C. (1986) Continuously distributed wrinkles in fabrics. **Arch. Rational Mech. Anal.**, 95, 93–115.

Pipkin, A.C. and Rogers, T.G. (1987) Infinitesimal plane wrinkling of inextensible networks. **J. of Elast.**, 17, 35–52.

Plaut, R.H. and Leeuwrik, M.J. (1988) Non–linear oscillations of an inextensible, air–inflated, cylindrical membrane. **Int. J. of Non–Linear Mech.**, 23, 347–353.

Racah, E. (1980) Fracture of coated fabrics used in airhouses, in **Proc. of the Symp. on Air–Supported Structures: the State of the Art**, London.

Reissner, E. (1938) On tension field theory, in **Proc. of the Fifth Int. Cong. For Appl. Mech.**, Cambridge, MA.

Roberts, D.H. and Green, W.A. (1980) Large axisymmetric deformation of a non–linear viscoelastic circular membrane. **Acta Mech.**, 36, 31–42.

Roddeman, D.G., Drukker, J., Oomens, C.W.J., and Janssen, J.D. (1987) The wrinkling of thin membranes: part I– theory; part II– numerical analysis. **J. of Appl. Mech.**, 54, 884–892.

Ross, E.W. (1969) Large deflections of an inflated cylindrical tent. **J. of Appl. Mech.**, 36, 845–851.

Shaw, F.S. and Perrone, N. (1954) A numerical solution for the non-linear deflection of membranes. J. of Appl. Mech., 21, 117-128.

Shimamura, S. and Takeuchi, O. (1972) Mechanical behavior of selected coated fabrics used in membrane structures in Japan, in Proc. of the 1971 IASS Pacific Symp. Part II on Tension Structures and Space Frames, Tokyo.

Srivastava, N.K. (1987) Mechanics of air-supported single membrane structures, in Developments in Eng. Mech., (ed. A.P.S. Selvadurai). Elsevier Sci. Pub., Amsterdam.

Stein, M. and Hedgepeth, J.M. (1961) Analysis of partly wrinkled membranes. NASA Technical Note D-813, 2-32.

Szyszkowski, W. and Glockner, P.G. (1984a) Finite deformation and stability behaviour of spherical inflatables under axisymmetric concentrated loads. J. of Non-Linear Mech., 19, 489-496.

Szyzkwoski, W. and Glockner, P.G. (1984b) Finite Deformation and stability behaviour of spherical inflatables subjected to axisymmetric hydrostatic loading. Int. J. Solids Struct., 20, 489-496.

Szyskowski, W. and Glockner, P.G. (1987a) On the statics of large-scale cylindrical floating membrane containers. Int. J. Non-Linear Mech., 22, 275-282.

Szyszkowski, W. and Glockner, P.G. (1987b) Spherical membranes subjected to concentrated loads. Eng. Struct., 9, 45-52.

Szyszkowski, W. and Glockner, P.G. (1987c) Spherical membranes subjected to vertical concentrated loads: an experimental study. Eng. Struct., 9, 183-192.

Takeda, T., Kageyama, M., and Homma, Y. (1986) Experimental studies on structural characteristics of a cable-reinforced air-supported structure, in Proc. of the IASS Symp. on Membrane Structures and Space Frames, Osaka.

Turkkan, N., Srivastava, N.K., Barakat, D., and Dickey, R. (1983) Study of air supported spherical structures subjected to an experimentally obtained wind pressure distribution, in Proc. of the Int. Symp. on Shell and Spatial Structures, Brazil.

Wagner, H. (1929) Flat sheet girder with very thin metal web. Z. Flugtech. Motorluft-Schiffahrt, 20, 200-207, 227-231, 281-84, 306-314. ((1931) Reprinted as NACA Technical Memorandum 604-606, NACA, Washington, DC.)

Warby, M.K. and Whiteman, J.R. (1988) Finite element model of visco-elastic membrane deformation. Comput. Methods in Appl. Mech. Eng., 68, 33-54.

Weinitschke, H.J. (1980) On axisymmetric deformations of nonlinear elastic membranes, in Mechanics Today, Vol. 5. Pergamon Press, Oxford, 523-542.

Weinitschke, H.J. (1987) On finite displacements of circular elastic membranes. Math. Methods in the Appl. Sci., 9, 76-98.

Wineman, A.S. (1976) Large axisymmetric inflation of a nonlinear Viscoelastic membrane by lateral pressure. Trans. of the Soc. of Rheology, 20, 203-225.

Wineman, A.S. (1978) On axisymmetric deformations of nonlinear vis-coelastic membranes. J. of Non-Newtonian Fluid Mech., 4, 249-260.

Wu, C.H. (1974) The wrinkled axisymmetric air bags made of inextensible membranes. J. of Appl. Mech., 41, 963-968.

Wu, C.H. (1978) Nonlinear wrinkling of nonlinear membranes of revolution. **J. of Appl. Mech.**, 45, 533–538.

Wu, C.H. and Canfield, T.R. (1981) Wrinkling in finite plane–stress theory. **Quart. of Appl. Math.**, 39, 179–199.

Yen, D.H. and Lee, T.W. (1975) On the nonlinear vibrations of a circular membrane. **Int. J. NonLinear Mech.**, 10, 47–62.

Yoshida, A., Ban, S., Tsubota, H., and Kurihara, K. (1986) Nonlinear analysis of tension fabric structures, in **Proc. of the Int. Symp. on Shell and Spatial Structures: Computational Aspects**, Leuven, Belgium.

Zahalak, G.I., Rao, P.R., and Sutera, S.P. (1987) Large deformations of a cylindrical liquid–filled membrane by a viscous shear flow. **J. of Fluid Mech.**, 179, 283–305.

Zak, M. (1983) Wrinkling phenomenon in structures – part 1: model formulation; part 2: wrinkling criteria. **Solid Mech. Arch.**, 8, 181–216 (part 1), 279–311 (part 2).

DYNAMIC ANALYSIS OF TRANSVERSELY STIFFENED SINGLE CURVATURE CABLE SUSPENDED ROOF AND APPLICATION

T. JI
School of Civil Engineering, University of Birmingham,
Birmingham, UK
J. ZHAO
Institute of Building Structures, China Academy of Building
Research, Beijing, Peoples Republic of China

Abstract
The dynamic characteristics of transversely stiffened single curvature cable-suspended roof are studied and the analytical expressions of natural frequencies and some related conclusions are presented in this paper. The analysis of this type of structure is different from that of common structures in two aspects that the change of cable forces during the small-amplitude vibration must be considered and the dynamic analysis should be based on the static equilibrium position rather than the original position before loading. A comparison between theoretical prediction and the measurements of a test model gives very satisfactory agreement. Finally the first ten periods of this type of roof of a sport hall built in Anhui Province, China, are predicted by the proposed formulas.

Keywords: Single Curvature, Cable-Suspended Roofs, Small amplitude vibration, Dynamic Analysis, Dynamic Characteristics, Frequencies, Modes.

1 Introduction

A schematic diagram of transversely stiffened single curvature cable suspended roof structure is shown in Fig.1. Either beams or trusses may be employed as the transversely stiffening elements which are established orthogonally on the cables. By means of the forced displacements to the end supports of transversely stiffened elements, pre-tension is resulted in the cables which ensures the stiffness of the roof effectively and the pre-stress is also built in the transversely stiffened elements which will balance part of external loads and reduce structural weight.

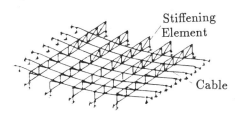

Fig.1 *Schematic Diagram of the Structure*

Cable-suspended roofs seem to have different free-vibration characteristics compared with common structures by reviewing previous theoretical analysis and experimental study on dynamic aspects. Geschwindner and West[1] numerically studied the natural frequencies and modes for Aden Airways Network with hyperbolic paraboloid cable surface. For the computational purpose, the network was discreted with 25 inner nodes. The results indicated that the first 25 mode shapes exhibit pre-dominately vertical displacements and the fundamental mode of the net shows an anti-symmetric shape. Krishina[2] also analysed a cable net in a rectangular plan and showed the first mode of the net is an anti-symmetric one. Shen et al[3] carried out a model test of a pre-stressed double layer cable system. Both measured and analysed results revealed that the fundamental mode presents an anti-symmetric shape. Lan and the authors[4] also accomplished a model study of transversely stiffened single curvature cable-suspended roof. The experimental measurements showed the first mode is an anti-symmetric one in the cable direction.

Ji and Lan[5] proposed a rational design and simplified analysis method for the transversely stiffened single curvature cable-suspended roof on static state which included determination of responses at both pre-stressing state and loading state. The static and dynamic structural behavior were also examined empirically in Ref[4]. On the basis of previous study, the dynamic characteristics of this kind of roof structure is studied theoretically in this paper. The study reveals that (1) the change of the cable forces, even during small amplitude vibration, must be taken into account; (2) because of the geometric non-linearity of cable type roof and the effect on dynamic characteristics by cable forces, it is required that the relative zero position for the dynamic study is the equilibrium position after loading rather than the initial position before loading.

For the simplicity, the *Transversely Stiffened Single Curvature Cable-Suspended Roof* is called TSSCCSR in following sections.

2 Dynamic Equilibrium Equation of TSS-CCSR

Except the assumptions made for the static analysis of TSSCCSR[5], the small amplitude vibration is assumed in the dynamic analysis. Because the trusses in the analysis can be equivalently expressed as beams, we simply mention beams instead of beams or trusses.

Taking any node (x_m, y_n) in the structure as the centre(Fig.2), an element ABCD is symmetrically cut and separated and the action between cut surfaces is represented by forces. The element ABCD is equilibrium at position $z + w$ where z and w are the initial position and the vertical displacement under loads respectively. The equilibrium equations at x and z directions are

$$
\begin{aligned}
H(x_C, y_C) \; - \; & H(x_A, y_A) = 0 \\
\left[H \frac{\partial(z+w)}{\partial x} \right]_C - \left[H \frac{\partial(z+w)}{\partial x} \right]_A \; + \; & Q_D - Q_B + \bar{q}(x_C - x_A)(y_D - y_B) = 0
\end{aligned}
\tag{1}
$$

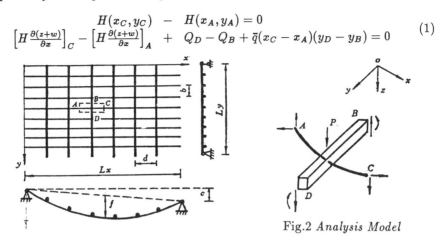

Fig.2 *Analysis Model*

Noticing $x_C - x_A = d$, $y_D - y_B = b$, Eq.1 can be alternatively expressed

$$
\frac{\frac{H(x_C, y_C)}{b} - \frac{H(x_A, y_A)}{b}}{x_C - x_A} = 0
$$

$$
\frac{\left[\frac{H}{b}\frac{\partial(z+w)}{\partial x}\right]_C - \left[\frac{H}{b}\frac{\partial(z+w)}{\partial x}\right]_A}{x_C - x_A} + \frac{\frac{Q_D}{d} - \frac{Q_B}{d}}{y_D - y_C} + \bar{q} = 0
\tag{2}
$$

Let

$$
\bar{H} = \frac{H}{b} \qquad \bar{Q} = \frac{Q}{d}
$$

It can be found that Eq.2 are the difference equations of following differential equations

$$
\begin{aligned}
\frac{\partial \bar{H}}{\partial x} &= 0 \\
\frac{\partial}{\partial x}\left[\bar{H}\frac{\partial(z+w)}{\partial x}\right] + \frac{\partial \bar{Q}}{\partial y} + \bar{q} &= 0
\end{aligned}
\tag{3}
$$

because

$$
\bar{D}\frac{\partial^3 w}{\partial y^3} = -\bar{Q}
$$

Eq.3 can be summarized as following one basic equation

$$
-\bar{D}\frac{\partial^4 w}{\partial y^4} + \bar{H}(y)\frac{\partial^2(z+w)}{\partial x^2} + \bar{q} = 0
\tag{4}
$$

This is the basic static equilibrium differential equation of TSSCCSR consisting of single direction membranes and single direction plate strips in which the cable material and the bending stiffness of beams are uniformly distributed. $\bar{H}(y)$ is only the function of y co-ordinate from the first formula of Eq.3.

When the roof is vibrated with a small amplitude $\bar{w}$ on the basis of the static equilibrium position $z + w$. the corresponding dynamic equation, considering the change of cable forces during vibration, is

$$- \bar{D}\frac{\partial^4 (w + \bar{w})}{\partial y^4} + [\bar{H}(y) + \Delta\bar{H}(y)]\frac{\partial^2 (z + w + \bar{w})}{\partial x^2} + \bar{q} - \bar{m}\frac{\partial^2 \bar{w}}{\partial t^2} = 0 \qquad (5)$$

Substituting Eq.4 into Eq.5, we have the free vibration equation of TSSCCSR

$$- \bar{D}\frac{\partial^4 \bar{w}}{\partial y^4} + \bar{H}(y)\frac{\partial^2 \bar{w}}{\partial x^2} + \Delta\bar{H}(y)(\frac{\partial^2 z'}{\partial x^2} + \frac{\partial^2 \bar{w}}{\partial x^2}) - \bar{m}\frac{\partial^2 \bar{w}}{\partial t^2} = 0 \qquad (6)$$

where　　$\bar{D}$　-　the distributed bending stiffness in a unit width

$\bar{H}$　-　the tension in a unit width

$\Delta\bar{H}$　-　the tension variation during vibration in a unit width

z'　-　the co-ordinate of the static equilibrium position, i.e. $z + w$

m　-　the distributed mass of the structure

There are two considerations in Eq.(6) different from common structures that

1. The change of cable forces $\Delta\bar{H}(y)$ during small amplitude vibration is taken into account. This is essential to reveal correctly the dynamic characteristics of this kind of roof structures.

2. The study of the dynamic characteristics is based on the static equilibrium position z' rather than original (before loading) position z. The corresponding terms in Eq.6 are $\bar{H}(y)$ and z'.

3　A Few Results from Static Analysis of TSS-CCSR

Ref.[5] had studied the static characteristics and rational design of TSSCCSR. Some results from Ref.[5] can be directly adopted for the purpose of dynamic analysis.

3.1　The surface equation of TSSCCSR after loading

After pre-stressing and loading stages, the surface of the roof can be approximately represented as

$$z' = \frac{4(f + \Delta_m)(L_x - x)x}{L_x^2}(1 + \frac{\Delta f}{f + \Delta_m} \sin \frac{\pi}{L_y}y) \qquad (7)$$

where f is the initial sag of cables before loading; Δ_m is the forced displacement of the end of the middle beam; $\Delta f = \delta_m - \Delta_m$; δ_m, obtained by Eq.11, is the displacement of the centre point of the roof after pre-stressing and loading.

3.2 Cable force distribution after loading

Under uniformly distributed loads, the cable force distribution can be described

$$\bar{H}(y) = \bar{H}_0 + (\bar{H}_m - \bar{H}_0)\sin\frac{\pi}{L_y}y = \bar{H}_0 + \bar{H}\sin\frac{\pi}{L_y}y \tag{8}$$

$\bar{H}_0$ and $\bar{H}_m$ in Eq.8 can be determined by following formulas

$$\bar{H}_j = \frac{(\bar{q}^0 + \Delta\bar{q}_j)L_x^2}{8(f + \delta_j)} \tag{9}$$

$$\Delta\bar{q}_j = \frac{64EA}{3L_x^4 b\alpha}\left[\delta_j^3 + 3f\delta_j^2 + 2f^2\mu\delta_j\right] \tag{10}$$

$$\delta_j = \Delta_m + \frac{4}{\pi}\left(\frac{q_y L_y^4}{D\pi^4} - \lambda_m\Delta_m\right)\sum_{n=1,3,\cdots}^{\infty}\frac{1 + n^2\gamma_m}{n(n^4 + \lambda_m\gamma_m n^2 + \lambda_m)}\sin\frac{n\pi}{L_y}y_j \tag{11}$$

$$\alpha = 1 + \frac{16f^2}{L_x^2} + \frac{c^2}{L_x^2} \qquad \mu = 1 + \frac{3H^0 L_x^2\alpha}{16EAf^2} \tag{12}$$

$$\lambda_m = \frac{128df^2 L_y^4 EA\mu}{3b\pi^4 L_x^4 D} \qquad \gamma_m = \frac{D\pi^2}{GAL_y^2} \tag{13}$$

$$E\bar{A} = EA/b \qquad \bar{D} = D/d \tag{14}$$

Where $\bar{q}^0$ and H_j^0 are the uniformly distributed load and the j th cable force before pre-stressing stage respectively; q_y is the product of the external uniformly distributed load $\bar{q}$ and interval of beams d; b is the interval of cables. c is the difference between the two fixed ends of cables. EA is the tension stiffness; D and GA is the bending stiffness and shearing stiffness of beams respectively. Let $y_j = 0$ and $y_j = L_y/2$ respectively, $\bar{H}_0$ and $\bar{H}_m$ can be calculated by Eq.(9-14).

3.3 The relationship between increment of cable force and displacement

The displacement $\bar{w}$ is produced based on the equilibrium position z', the cable force change is by the displacement

$$\Delta\bar{H}(y) = \frac{E\bar{A}}{L_x}\int_0^{L_x}\left[\frac{\partial z'}{\partial x}\frac{\partial \bar{w}}{\partial x} + \frac{1}{2}\left(\frac{\partial \bar{w}}{\partial x}\right)^2\right]dx$$

According to the assumption of small amplitude vibration, above equation is reduced to

$$\Delta\bar{H}(y) = \frac{E\bar{A}}{L_x}\int_0^{L_x}\frac{\partial z'}{\partial x}\frac{\partial \bar{w}}{\partial x}dx \tag{15}$$

Substituting Eq.7 into Eq.15, it yields

$$\Delta\bar{H}(y) = \frac{4E\bar{A}(f + \Delta_m)}{L_x^3}\left(1 + \frac{\Delta f}{f + \Delta_m}\sin\frac{\pi}{L_y}y\right)\int_0^{L_x}(L_x - 2x)\frac{\partial \bar{w}}{\partial x}dx \tag{16}$$

4 Dynamic Characteristics of TSSCCSR

When the first three terms in Eq.6 are the linear functions of $\bar{w}$, the total strain energy is

$$U = \frac{1}{2} \int_0^{L_x} \int_0^{L_y} \left[\bar{D} \frac{\partial^4 \bar{w}}{\partial y^4} - \bar{H}(y) \frac{\partial^2 \bar{w}}{\partial x^2} - \Delta\bar{H}(y)(\frac{\partial^2 z'}{\partial x^2} + \frac{\partial^2 \bar{w}}{\partial x^2}) \right] \bar{w} \, dx \, dy \qquad (17)$$

The kinetic energy of the system is

$$V = \frac{1}{2} \bar{m} \int_0^{L_x} \int_0^{L_y} \left(\frac{\partial \bar{w}}{\partial t} \right)^2 dx \, dy \qquad (18)$$

The vibration equation satisfying boundary conditions is assumed

$$\bar{w} = \sum_{i=1}^{\infty} \sum_{j=1}^{\infty} \bar{w}_{ij} = \sum_{i=1}^{\infty} \sum_{j=1}^{\infty} C_{ij} \sin \frac{i\pi x}{L_x} \sin \frac{j\pi y}{L_y} \sin(\rho_{ij} + \phi_{ij}) \qquad (19)$$

By using Lagrange Equation of a conservative system

$$\frac{d}{dt}\left(\frac{\partial V}{\partial \dot{\bar{w}}_{ij}} \right) - \frac{\partial V}{\partial \bar{w}_{ij}} + \frac{\partial U}{\partial \bar{w}_{ij}} = 0 \qquad (20)$$

and substituting concerned equations into Eq.20, we have the expression of the ij th frequency as following

$$\begin{aligned} p_{i,j}^2 &= \frac{1}{\bar{m}} \left\{ \bar{D} \left(\frac{j\pi}{L_y} \right)^4 + \left[\bar{H}_0 + \frac{2}{\pi} \left(\frac{4j^2}{4j^2-1} \right) \bar{H} \right] \left(\frac{i\pi}{L_x} \right)^2 + \right. \\ &\quad + \left. \frac{512 E \bar{A}(f+\Delta_m)^2}{(i\pi)^2 l_x^4} \left[1 + \frac{4}{\pi} \left(\frac{4j^2}{4j^2-1} \right) \left(\frac{\Delta f}{f+\Delta_m} \right) + \frac{1}{2} \left(\frac{\Delta f}{f+\Delta_m} \right)^2 \right] \right\} \end{aligned} \qquad (21)$$

The term containing $E\bar{A}$ only exists when the subscript i is odd number. The corresponding vibration mode is

$$\Phi_{i,j} = C_{ij} \sin \frac{i\pi x}{L_x} \sin \frac{j\pi y}{L_y} \qquad (22)$$

Table 1. Comparison between theoretical predication and measurements

mode order	vibration mode	theoretical values 1(HZ)	theoretical values 2(HZ)	measured values(HZ)
1	i=2, j=1	4.05	4.02	4.32
2	i=3, j=1	5.11	5.05	4.63
3	i=4, j=1	5.63	5.53	5.35
4	i=5, j=1	6.55	6.53	5.55(6.35)
5	i=1, j=1	6.51	6.51	5.92

Ref[4] completed a static and dynamic model test of TSCCSR. The model, $3.6 \times 4.8m$ in plane, consisted of 7 $\phi5$ high tensile wires and 5 $\phi53 \times 3$ steel tubes. A comparison between the theoretical prediction and the measurements of a real model test gives very satisfactory agreement. Table 1 shows the comparison. The *theoretical value 1* is predicated by eqs(8-14) and (21) and *theoretical value 2* is calculated by Eq.(21) based on the measured static results. The uniform load acting on the model is $4.17kg/m^2$

It can be found from Eq.21 and Table 1 that

1. Different from ordinary structures, it is interested to point out that the fundamental frequency of a transversely stiffened single curvature cable suspended structure corresponds an anti-symmetrical double half-wave mode, the symmetric single half-wave mode will occur at much higher frequency.

2. The variation of cable forces during vibration, even in smaller amplitude vibration, must be considered. Otherwise the fifth frequency in Table 1 will become the first one with the value $3.54/sec.$ and it is a obvious error both numerically and physically.

3. The reason that the symmetric structure presents the fundamentally anti-symmetric mode is that the cable forces will be changed when cables vibrate in a symmetric mode. The increased cable forces will contribute extra strain energy to the whole system and consequently the corresponding frequency will be higher. When cables vibrate in an anti-symmetric mode, the cable forces will keep constant, *i.e.*, $\Delta \bar{H}(y) = 0$ and no extra contribution to the strain energy. As a result, the frequency will be relatively smaller and easier to be excited.

4. The differences between the predicated and measurement are smaller in the first and the third frequencies than the rest frequencies. When the first or the third mode is produced, $\Delta \bar{H}(y) = 0$ and the assumed mode(Eq.22) is more close to the real one.

5. It should be pointed out that the fundamental mode also can appear a symmetric single wave in cable-suspended roof. The condition that the first mode presents a symmetric single wave in the cable direction of TSSCCSR is

$$3\pi^4 \left[\bar{H}_0 + \frac{8}{3\pi}\bar{H} \right] > 512 E\bar{A}\frac{(f + \Delta_m)^2}{L_x^2} \left[1 + \frac{16}{3\pi}\left(\frac{\Delta f}{f + \Delta_m} \right) + \frac{1}{2}\left(\frac{\Delta f}{f + \Delta_m} \right)^2 \right]$$
$$(23)$$

If the sag f of the cables goes smaller and consequently the cable forces will be increased, or alternatively, the smaller tension stiffness $E\bar{A}$ is adopted, Eq.(23)can be realized. The corresponding physical explanation is

If the difference of strain energy contributed by cable forces in the first two modes(i = 1, 2 j = 1) is greater than the extra strain energy produced by the change of cable forces in the mode (i = 1 j = 1) during vibration, the fundamental mode will presents the symmetric single wave in the cable direction

This conclusion is also valid for other types of cable-suspended roof though the formula may be varied according to different type structures. As a deduction, the condition that the first mode presents a symmetric single wave for a single cable under uniformly distributed line load q_x is

$$\frac{3\pi^4 q_x L_x^2}{f + \Delta f} > 64EA \left(\frac{f}{L_x}\right)^2 \left[1 + \frac{16}{3\pi}\left(\frac{\Delta f}{f}\right) + \frac{1}{2}\left(\frac{\Delta f}{f}\right)^2\right] \qquad (24)$$

5   Prediction of Periods and Modes on the TSS-CCSR of Anhui Sport Hall

The TSSCCSR has been employed in the design and construction of Anhui Gymnasium in China. The roof consists of 29 cables, spanning $72m$ and with an interval of $1.5m$, and 11 steel latticed trusses, at a distance of $6m$ and orthogonally located on the cables. The length of the trusses is varied according to the geometry of the plane(Fig.3) and the depth is changed linearly from $3.2m$ in the centre to $1.6m$ at the end[6]

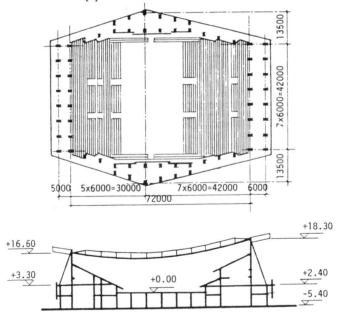

Fig.3 *The Anhui Sport Hall*

The design data for the prediction of natural periods of the real roof are listed as following

$$L_x = 72m \qquad L_y = 53.4m \qquad f = 4.5m \qquad c = 4.6m$$
$$b = 1.5m \qquad d = 6m \qquad \bar{q}^0 = 0.02t/m^2 \qquad \bar{q} = 0.133t/m^2$$
$$\Delta_m = 0.29m \qquad EA = 9918t \qquad D = 186186t - m^2$$

The first ten periods and the corresponding mode shapes of the roof of Auhui Sport Hall are listed in Table 2.

Table 2. The Prediction of Natural Periods and Mode Shapes

mode order	vibration mode	predicted values(s)	mode order	vibration mode	predicted values(s)
1	i=2, j=1	1.248	6	i=1, j=1	0.957
2	i=4, j=1	1.149	7	i=8, j=1	0.905
3	i=3, j=1	1.139	8	i=7, j=1	0.899
4	i=5, j=1	1.031	9	i=10,j=1	0.799
5	i=6, j=1	1.026	10	i=9, j=1	0.781

The actual roof would be a little more stiff than the calculated one since all the formulas for both static and dynamic analysis are derived on the basis of a regular plane. Consequently, the actual periods of the roof will be a little smaller than the predicted ones.

6 Conclusions

The analytical study of the dynamic aspect of TSSCCSR reveals the dynamic characteristics of this structure, explains the reason that fundamental mode of this kind of cable-suspended roof presents an anti-symmetrical shape in cable direction and provides the condition that the first vibration mode appears a symmetrical single wave.

The feasibility of the proposed analytical formulas for predicating the natural frequencies of TSSCCSR is verified by a real model dynamic test. Furthermore, these formulas are applied to predict the periods and mode shapes of the TSS-CCSR of Auhui Sport Hall.

7 References

1. L. F. Geschwinder and H. H. West (1979) Parametric investigations of vibration cable networks, **J. Strct. Div.**, 105, pp.465-479.

2. P. Krishina (1978) **Cable-Suspended Roofs**, Mc. Graw-Hill Book Company.

3. S. Shen, C. Xu and H. Guo (1984) Static and dynamic characteristics of pre-stressed double layer cable system, **Proceedings of 2nd National Conference on Space Structures** (ed. Tien T. Lan) (in Chinese).

4. Tien T. Lan, Jida Zhao and Tianjian Ji (1988) A study on the structural behavior of transversely stiffened cable-suspended roof and its application to sports buildings, **Space Structures for Sports Buildings**(eds. Tien T. Lan and Zhilian Yuan), Elsevier Applied Science, London, pp.524-533.

5. Tianjian Ji and Tien T. Lan (1987) Rational design and simplified analysis method for transversely stiffened single curvature cable suspended roof, **Proceedings of the International Conference on the Design and Construction of Non-Conventional Structures** (ed. B. Topping), Civil–Comp, **2**, pp.7-16.

6. Gymnasium Design Group, Anhui Design Institute of Architecture (1984) Anhui Gymnasium, **J. Building Structures**, 5, (in Chinese).

DESIGN, ANALYSIS AND RESPONSE OF DOUBLE-LAYER TENSEGRITY GRIDS

A. HANAOR
Dept Civil and Environmental Engineering, Rutgers University,
Piscataway, New Jersey, USA

Abstract
Tensegrity structures are free-standing, internally prestressed cable
networks. The cables (tendons) are prestressed against a
discontinuous system of bars (or struts). In double-layer tensegrity
grids (DLTGs), the bars (struts) are confined between two parallel
surfaces of cables (tendons). The analysis of cable networks,
including tensegrity structures, typically involves a shape finding
phase and a load analysis phase, both associated with geometric
nonlinearities. Two types of cable networks are distinguished —
geometrically flexible and geometrically rigid, with relatively large
and small deflections respectively.
  The paper presents two species of DLTGs constructed from prismatic
and pyramidal units — flat grids and domes. While flat grids do not
involve shape finding, domes do, and the shape depends on the
construction constraints. The stiffness of DLTGs varies largely,
depending primarily on topology, shape (flat or curved) and boundary
conditions. Domes are stiffer than flat grids when lateral movement
of supports is restrained, and geometrically rigid topologies are
considerably stiffer than geometrically flexible ones.
Keywords: Tensegrity, Tensegrity Structures, Cables, Tendons, Cable
Networks, Prestress, Nonlinear Analysis, Geometric Nonlinearity.

1 Introduction

Tensegrity structures are prestressed cable networks, in which the
cables (or tendons, in general) are prestressed against a
discontinuous system of bars (struts). Unlike conventional
prestressed cable networks, they do not require an external anchoring
system in the form of ring beams or masts. They are prestressed
internally and hence are free-standing. This feature makes these
structures inherently deployable, since in the unprestressed state
they possess no stiffness and are thus collapsible. Furthermore, the
discontinuity of the bar system, makes for very simple and reliable
connections between bars and cables. Tensegrity structures are thus
eminently suitable for deployable or temporary applications, but
permanent space coverings are also feasible.
  Although a number of tensegrity structures were constructed, most
notably by the artist Kenneth Snelson, these have been largely

decorative features. Early engineering structural concepts like R.B. Fuller's dome and O. Vilnay's tensegrity networks (Fig. 1) suffered from some technical problems, which may have contributed to the lack of interest in their application. These problems are associated mainly with bar congestion (interference of bars with one another) on the one hand (Fuller), and excessive bar lengths, causing buckling problems, on the other hand (Vilnay). Details of Fuller's and Vilnay's concepts can be found in Fuller (1963) and Vilnay (1977,1981).

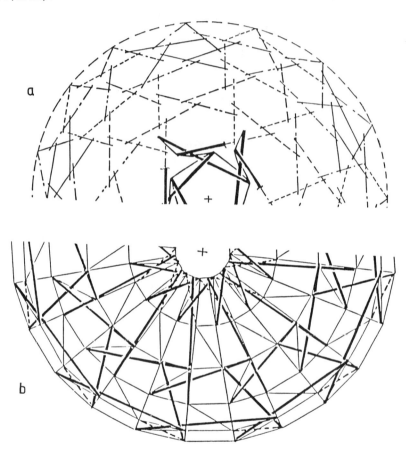

Fig. 1 a) Fuller's tensegrity dome (half plan)
b) Vilnay's tensegrity Dome (half plan)

These difficulties have led to the conceptualization of double-layer tensegrity grids (Hanaor 1987, 1990a, Motro 1987). In this concept, bars are confined between two parallel layers of tendons, thus providing a limit on bar lengths and adequate spacing of bars. Additional advantage is that flat as well as curved shapes are feasible, whereas the single-layer concepts require curvature to

provide structural depth. All *double-layer tensegrity grid* (DLTG)
configurations published to date are based on *tensegrity prisms* (T
prisms) as the building blocks. A triangular prism is shown in Fig.
2. The tendons lie along the edges of a regular prism, and the bars
along the diagonals of the side faces, in a consistent sense
(clockwise or anticlockwise). Elongation of the bars (struts) causes
the two bases to rotate until prestress is achieved at a
characteristic angle of rotation equal to half the angle between two
adjacent sides of the base polygon (30o for triangular prism, 45o for
square etc.).

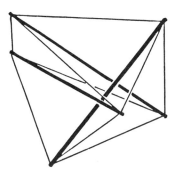

Fig. 2   Tensegrity prism

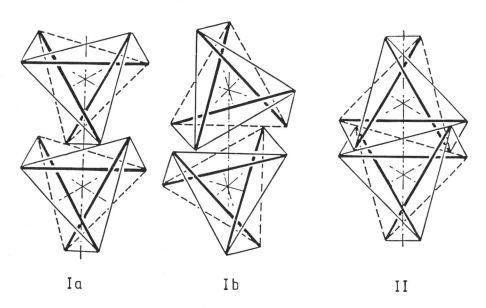

Ia   Ib   II

Fig. 3   Methods for forming DLTGs

Motro's and Hanaor's concepts differ in that Motro (1987) produces
regular patterns of the chords, but the bars are joined together at
the common nodes. Hanaor's concept is geometrically more complex,
with chords forming less regular patterns, but it preserves the
principle that bars are joined only to cables. Three methods of
joining T prisms are illustrated in Fig. 3 using triangular prisms.
Sample grids can be found in Hanaor (1987, 1990a). Curved surfaces
are obtained by using truncated pyramids (T pyramids) instead of
prisms, with the two base polygons having different radii.

2 Analysis

The analysis of tensegrity structures is similar to the analysis of
conventional prestressed cable networks, with the exception that the
network contains bar elements, capable of sustaining compression, as
well as cable elements. Two types of prestressable networks can be
distinguished (Vilnay 1987) — (a) geometrically rigid and statically
indeterminate, and (b) geometrically flexible and kinematically
indeterminate (Pellegrino and Calladine, 1986). Most conventional
cable networks belong to the second category, but in many cases, a
network can be stiffened into the first category, by the addition of
cables. Fig. 4 shows a *reinforced tensegrity prism* (RT prism) in
which additional cables have been added along the diagonal of the
prism faces not occupied by bars. In this case the relative rotation
angle of the bases is not unique and can be varied between the
original angle, at which point the additional cables are not
prestressed, and double that value, at which point the bars intersect
(for the triangle, 30o to 60o). The prism shown in Fig. 4 has a
relative rotation of 45o.

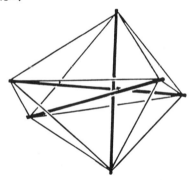

Fig. 4   Reinforced tensegrity prism

Geometrically flexible networks involve large deflections and are
therefore highly geometrically nonlinear. The analysis is typically
done in two phases. Phase I is a shape finding procedure, which
finds the deflections from an assumed initial geometry to a
prestressable geòmetry. Phase II is analysis under the applied load
of the prestressed geometry. An intermediate phase may be employed,

particularly for investigative purposes, involving analysis of the states of prestress and internal mechanisms of the prestressed geometry (Pellegrino and Calladine, 1986). Geometrically rigid networks involve only elastic deformations and a linear analysis is often adequate.

Several algorithms are available for the nonlinear analysis of cable networks. The most widely used is, arguably, a stiffness based procedure employing Newton iteration (Argyris and Scharpf 1972). Dynamic relaxation (Barnes 1984, 1987) is a very efficient method for shape finding, particularly where very large deflections are involved, which may cause convergence problems in the Newton method. Hanaor (1988) has developed a flexibility based method for the investigation of states of prestress and mechanisms and for first order analysis, when the prestressed geometry is known.

Flat DLTGs have a known prestressed geometry which is derived directly from the geometry of the constituent prisms (Hanaor, 1990a). Domes, on the other hand, require a shape finding phase and their prestressed geometry depends on construction constraints and boundary conditions (Hanaor 1990b). In the present study two method were employed: a first order flexibility method for investigating flat grids and a Newton iterative stiffness based methods for investigating domes, as well as for nonlinear analysis of flat grids. Detailed descriptions of the procedures are given in Hanaor (1988) and in Argyris and Scharpf (1972) respectively.

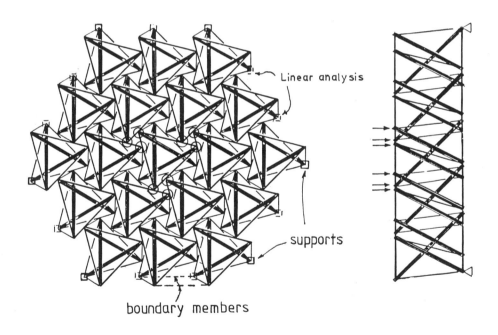

Fig. 6 Typical plane grid

3   Plane Grids

3.1   Analytical investigation
A first order parametric analytical study of the response of flat triangular grids (Fig. 6) was carried out on grids loaded with uniformly distributed load and simply supported at all boundary nodes along the edges (Hanaor and Liao 1990). The main results are summarised below:

a)   Member forces (particularly bars) vary approximately linearly with span, for constant depth/span ratio.
b)   The geometric deflections dominate. Their magnitude depends on the level of prestress. The analysis suggests that for a prestress evel which is a constant fraction of the total allowable bar force, the geometric displacement/span ratio is approximately constant (independent of span).
c)   The first order analysis overestimates the deflections and underestimates the member forces.
d)   The most highly stressed members − both bars and tendons − are located near the boundaries. This implies that shear forces rather than bending moments dominate the response.

Nonlinear analyses were carried out on a grid similar to that of Fig. 6 but supported at six bottom nodes only and with boundary members added along all boundaries (see Fig. 6). This grid was loaded at nine central nodes only (see also dome of similar layout, Fig. 10). A geometrically rigid grid of a similar layout was also analysed. To achieve geometric rigidity, it is not sufficient to use reinforced prisms but the connecting tendons have to attach to bars of different units (Fig. 7).

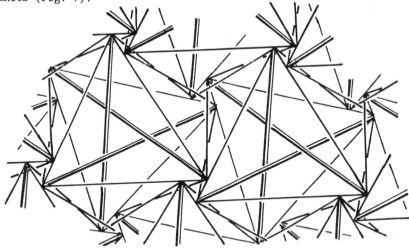

Fig. 7   Interior tendons in geometrically rigid grid.

The response of the two grids is compared in Fig. 8 in the form of

load - deflection curves, using arbitrary units. The thin dashed
lines indicate the practical load range, assuming a prestress force
level in bars of 1/3 the allowable, yielding a "practical"
displacement/span ratio of 1/25 for the flexible geometry and 1/20
for the geometrically rigid configuration, assuming the same
prestress level. The "practical" load of the "rigid" configuration
is, however, 1.75 times the "flexible" one. Eliminating the
prestress in the rigid geometry increases somewhat the deflections
but the load carrying capacity for the same design is also increased,
as more of the member capacity is available for the applied load.

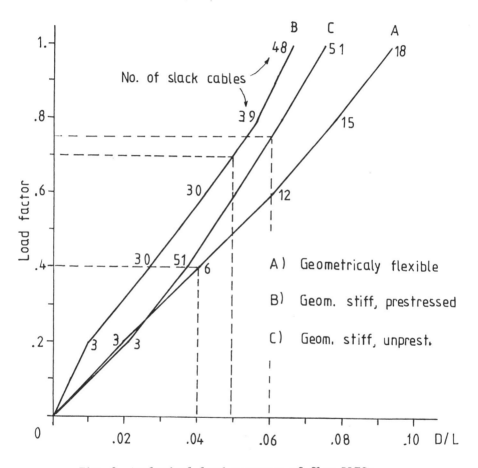

Fig. 8 Analytical load response of flat DLTGs

3.2 Experimental study

The small model shown in Fig. 9a was tested under three different
levels of prestress, with bar prestress forces of 1/12, 1/6 and
1/3 of the bar buckling load. A fourth test was carried out from
the unloaded state of the third test to failure. Results are

presented in Fig. 9b as load- displacement curves, together with
linear and nonlinear analytical results. Analytical results are
presented also for a grid of the same layout employing reinforced
prisms and for a geometrically rigid topology, with the third
prestress level.

The test results are in very good agreement with nonlinear
analysis. The failure mode of the grid in the fourth test was
rupture of cables. The test indicates that this type of structure
has inbuilt structural redundancy due to its composition of
individual units. Following first rupture, representing loss of one
unit out of seven, the structure recovered almost its full load
bearing capacity.

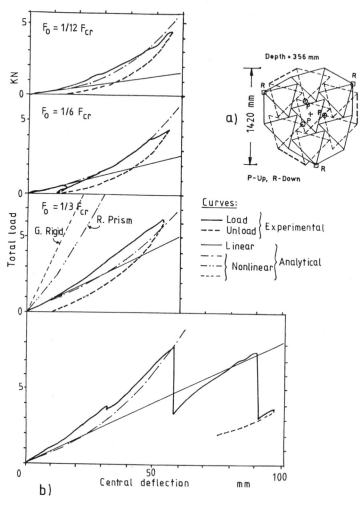

Fig. 9 a) Test model layout; b) Load response

4 Domes

4.1 Shape finding

Unlike flat DLTGs, the design and analysis of tensegrity domes
involves shape finding. The prestressed shape depends primarily on
the construction constraints and on the boundary (support)
conditions. Some typical construction constraints may include:

a) The dome is constructed from identical pyramidal units. This
is possible due to the geometric flexibility of the network.
b) Prestress is obtained by equal elongations of all bars.
c) Prestress forces are equal in all bars. This may be the case
in deployable domes with telescoping bars, deployed by hydraulic
or pneumatic pressure.
d) Minimum deviation of the prestressed surface from a desired
theoretical surface is required.
e) The prestress is done with the final boundary conditions (see
below) in place or, alternatively, with different boundary
conditions (e.g. without boundary members). In the latter case,
final support conditions are applied to the prestressed geometry.

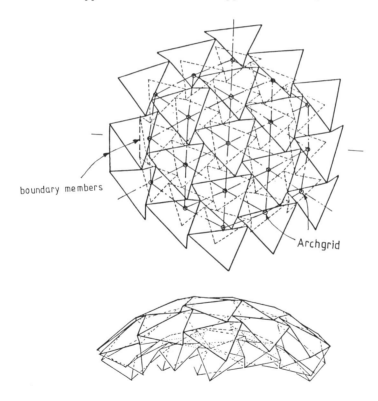

boundary members

Archgrid

Fig. 10 Initial dome geometry

Boundary conditions may include:

a) Free horizontal displacements of supports (other than restraints required to prevent rigid body motion).
b) Horizontal displacements are prevented (e.g. by tension ring).
c) Various combinations of boundary members (which do not form part of the grid pattern) and supports.

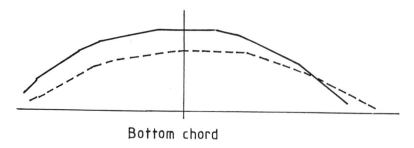

Bottom chord

Fig. 11 Deflections of prestressed geometry

A spherical dome is used as an example. The initial geometry is computed by assuming that the *archgrid* — the grid connecting the centroids of the truncated pyramidal tensegrity units — forms a geodesic subdivision of the regular hexagonal pyramid enclosed by the dome. Details of the algorithm are given in Hanaor (1990b). Fig. 10 shows the initial geometry of a dome with *subdivision frequency* of 2. Only the top and bottom tendon layers are shown, for clarity.

The Stiffness method was used to obtain the prestressed geometry under various constraints and boundary conditions. Fig 11 shows the vertical deflections across a centreline of the dome of Fig. 10, when it is prestressed with constant bar force, without boundary members and with only rigid body motion restrained. When prestress is with boundary members, as indicated in Fig. 10, the deflections of the prestressed geometry from the initial geometry are small (less than 1/100 of the span). Units and dimensions are the same as in the flat grid of section 3 (Fig. 6). Although these dimensions are arbitrary, the results are roughly representative of a realistic structure of this type.

4.2 Load response

The load response of domes strongly depends on the boundary conditions. Fig. 12 shows load -deflection curves for domes of the layout of Fig. 10, loaded at the nine central top chord nodes, with different boundary conditions. A geometrically rigid response is also shown. It can be seen that when true dome action is facilitated by restraining horizontal movement of the supports, the stiffness is substantially enhanced. Compression forces in the bars (not shown here) are also reduced under these boundary conditions compared with flat grids.

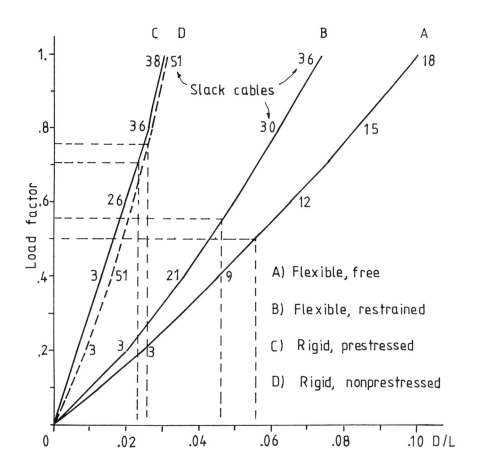

Fig. 12   Double-layer tensegrity dome load response

5 Acknowledgement

The work reported in this paper was partially supported by a National Science Foundation grant.

6 References

Argyris, J.H. and Scharpf, D.W. (1972) Large deflection analysis of prestressed networks. **J. Struct. Div.**, ASCE, 106 (ST3),633-654.

Barnes, M.R. (1984) Form-finding, analysis and patterning of tension structures, in **Proc. of the 3rd. Intnl. Conf on Space Structures** (ed. H. Nooshin), Elsevier, London, pp. 730-736.

Barnes, M.R. (1987) Form-finding and analysis of prestressed nets and membranes, in **Intnl. Conf. on the Analysis and Construction of Non-conventional Structures (Vol. 1)** (ed. H.V. Topping), Civil Comp Press, London, pp. 327-337.

Fuller R.B. (1962) Tensile-integrity structures, **U.S. Patent** 3,063,521.

Hanaor A. (1987) Preliminary investigation of double-layer tensegrities, in **Intnl. Conf. on the Analysis and Construction of Non-conventional Structures (Vol. 2)** (ed. H.V. Topping), Civil Comp Press, London, pp. 35-42.

Hanaor A. (1988) Prestressed pin-jointed structures – Flexibility analysis and prestress design, **Computers and Structures** 28 (6), 757-769.

Hanaor A. (1990a) Double-layer tensegrity grids: geometric configuration and behaviour, in **Space Structures: Theory and Practice** (ed. H. Nooshin), Multiscience, in press.

Hanaor A. (1990b) Aspects of design of double-layer tensegrity domes, **Int. J. Space Structures**, forthcoming.

Hanaor A. and Liao M.K. (1990) Double-layer tensegrity grids: static load response, analytical study., **J. Structural Engineering**, ASCE forthcoming.

Motro R. (1987) Tensegrity systems for double-layer space structures **Intnl. Conf. on the Design and Construction of Non-conventional Structures (Vol. 2)** (ed. H.V. Topping), Civil Comp Press, London, pp. 43-52.

Pellegrino, S. and Calladine, C.R. (1986) Matrix analysis of statically and kinematically indeterminate frameworks, **Int. J. Solids Structs.**, 22, 409-428.

Vilnay, O. (1977) Structures made of infinite regular tensegric nets, **IASS bulletin**, 18 (63), 51-57.

Vilnay O. (1981) Determinate tensegric shells, **J. of the Struct. Div.**, ASCE, 107 (ST10), 2029-2033.

Vilnay O. (1987) Characteristics of cable nets, **J. Structural Engineering**, ASCE, 113 (7), 1586-1699.

PRESTRESSED TRUSSES: ANALYSIS BEHAVIOUR AND DESIGN

G.G. SCHIERLE
Director, Graduate Programme of Building Science, School of
Architecture, University of Southern California, Los Angeles,
California, USA

Abstract
Prestressed truss concepts are presented, their structural behavior evaluated, and integration with functional design objectives described. The Evaluation is based on computer analysis, static simulation models, and prototype structures. Integration of prestressed trusses for synergy with functional objectives is described on hand of designs for two university sports centers. Cable trusses are light-weight, which is important for long-span structures, especially in seismic zones to minimize seismic forces. The collapse of a heavy double-deck concrete freeway in the recent San Francisco earthquake dramatized this point.
<u>Keywords:</u> Cable Truss, Computer Analysis, Prestress, Space Truss, Test Model.

1 Introduction

Prestressed cable trusses of curved configuration have been proposed or built for various longspan structures; some based on the Jawerth System (Bach, 1975, Hottinger, 1963). The rational for such trusses is based on using members in tension to avoid buckling and bending and on using high-strength cables or strands, all to reduce dead weight and to allow for potentially longer spans. Yet these benefits are partly offset by the need to resist horizontal reactions, and by possible flutter (Zetlin, 1963). This paper includes a new truss concept of straight chord members with diagonal bars, to provide a triangular configuration for better stiffness and stability; combining straight system lines of conventional trusses with flexible tensile members of cable trusses.

1.1 Basic Principles
Compared to conventional trusses, prestressed trusses have some or all rigid bars substituted by flexible tension members (cables, strands) which absorb compressive forces through reduction of prestress. The effect of prestress is shown by Fig 1, a flexible wire supported at both ends and with a load P that is applied at its center. The wire on the left, without prestress, strains an amount e under the load P. The wire on the right is prestressed and strains only e/2 under the same load P, because half of its load is carried by the lower link through a reduction of prestress. For prestressed cable systems of similar symmetry conditions these observations can be made:

Prestress reduces strain deformation to half.
Compressive forces are absorbed through reduction of prestress.
Prestress must be 50% of design stress to prevent slack members.
Prestress above 50% compensates for creep and temperature changes.

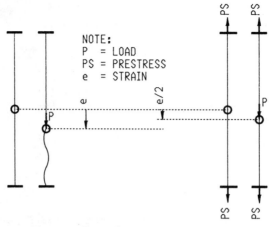

Fig. 1. Effect of prestress on strain

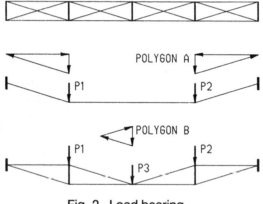

Fig. 2. Load bearing

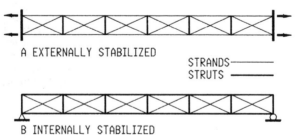

Fig. 3. Support conditions

Replacing rigid struts by flexible tension members eliminates buckling, for better efficiency. This, and substituting high-strength cables or strands for mild steel, substantially reduces the dead weight of prestressed trusses; thus reducing seismic forces, an important consideration in seismic zones. Yet the benefits are partly offset by the need in some prestressed truss systems to resist horizontal reactions, which typically are about twice the amount of the vertical reactions. To resist the horizontal reactions by economical means is important for prestressed trusses to be economically viable. Design options include circular compression rings, available infrastructures, such as structural slabs, shear walls; and soil or rock anchors, given a firm soil.

1.2 Load Bearing
Parabolic trusses (Fig 4) absorb loads in their bottom and top cable polygons for gravity and uplift loads respectively. The load bearing mode of parallel chord trusses (Fig 5) is more complex, as illustrated in the abstract truss of Fig 2. Loads P1 and P2 are carried by polygon A. Load P3 is carried by a second polygon B. Inward sloping diagonals are active, as are the top chords near the supports and the bottom chords at mid-span. This force flow reflects shear and moment distribution of an equivalent beam with fixed end supports. The truss members not shown carry uplift load and part of gravity load through reduction of prestress.

1.3 Classification
Prestressed trusses can be classified by support condition, configuration, and dimensionality, with variations of each. Regarding support, they may be either externally stabilized (horizontal reactions resisted externally) or internally stabilized (horizontal reactions balanced by internal compression members) as shown in Figs 3 A and B respectively. Configuration may either be based on curved or straight system lines (Figs 4 and 5). The latter may also be grouped by systems with diagonal members (Jawerth) or with vertical struts/strands only. Dimensionality may either be one-way (Figs 12-18) or two-way (Figs 4-5). The latter have inherent stability against buckling but are more complex to built and are only efficient if the spans are about equal both ways.

2 Static Simulation Models

Static simulation models are means to explore new structures (Schierle, 1986). They visualize spatial qualities and provide information on deformation and stress distribution, for intuitive understanding of force flows, reinforced by touching models under various loads. Seven simulation models were built and tested in a seminar conducted by the author to develop, test, and compare various prestress truss concepts; two of which are shown in Fig 4 and 5. The models were related to original structures by three scales: geometric scale S_g, force scale S_f, and strain scale S_s. The latter was chosen 1:1 to prevent errors due to geometrically non-linear behavior. Since all members were subject to axial load only, without bending, the force scale S_f was defined as

$$S_f = P_m/P_o = A_m E_m/A_o E_o = \text{model force/original force, where} \qquad (1)$$

A = cross section area of members,
E = modulus of elasticity,
m = subscript for model,
o = subscript for original structure.

Fig. 4. Parabolic truss

Fig. 5. Parallel chord truss

The models were built with piano wire of .01 inch (.25 mm) diameter and tubing, to represent cables and struts respectively. Prestress was applied by turnbuckles and measured by comparing the sound of wires with a tension gauge of equal wire length and cross-section area. Tests were conducted to measure deformations and forces in truss members under load. Combined dead and live loads of 30 psf (1.4 kPa) on the original structure were simulated on models by sand cups. The encouraging test results prompted further investigation (by computer simulation and prototype structures) of concepts with parallel chords (Fig 5 and a similar one-way truss). While more complex, the parallel chord truss with diagonals had greater stiffness than the parabolic one without diagonals; and had better dynamic stability (no flutter). Also, trusses with parallel chords may be architecturally advantageous in certain situations.

3 Computer Simulation

The systems of Fig 6 (B: 4-bay and C: 8-bay) were studied to determine how forces and deflections are effected by the following design variables:

Variable depth/span ratio (Fig 6B).
Variable prestress (Fig 6B).
Variable safety factor (Fig 6C).

The computer graphs visualize results in Fig 7 and show typical member forces and maximum deflections under applied gravity load. Horizontal reactions were assumed as the sum of top and bottom chords and diagonal members at supports. The iterative analysis program TRITRS (Haug, 1975) was used for the simulation, based on the following assumptions (except where the values are design variable):

Strand cross section: 1.86 sq.in. (12 cm^2).
Strut cross section: 7.8 sq.in. (50 cm^2).
Depth/span ratio: 1:12.5.
Prestress: 95 k (423 kN) = 44% of max. force under design load.
Safety factor: 2 for maximum stress and 3.9 for mean stress.
Uniform gravity load: 30 psf (1.4 kPa) combined dead and live load.

Evaluating the computer graphs of Fig 7, these observations are of interest:

3.1 Variable Depth/Span Ratio
The median force increase is only 11 %, for a 50 % depth/span ratio decrease from .1 (1:10) to .05 (1:20). Deflections caused by live load are tabulated below for three depth/span ratios and four safety factors (four member sizes).

Safety factor	depth/span ratio:	1:10	1:15	1:20
2	defl./span ratio:	1:480	1:240	1:160
3	defl./span ratio:	1:727	1:362	1:242
4	defl./span ratio:	1:960	1:480	1:320
5	defl./span ratio:	1:1100	1:550	1:366

This deflection table illustrates an impressive performance, and suggests that depth/span ratios of 1:15, or 1:20, are feasible. Hence prestressed trusses could be used for very long spans, for a small depth implies short struts with reduced buckling. Depth/span ratios under 1:20 are too shallow for most applications.

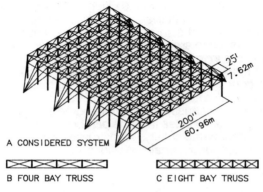

A CONSIDERED SYSTEM

B FOUR BAY TRUSS   C EIGHT BAY TRUSS

Fig. 6. Systems used in computer simulation

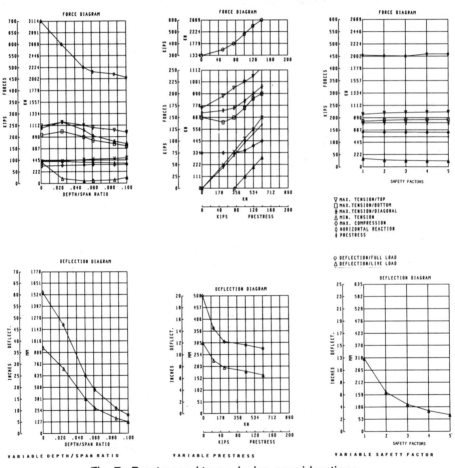

Fig. 7. Prestressed truss design considerations

3.2 Variable Prestress

To prevent slack members, prestress must be at least 34% of maximum member stress and 50% of the mean stress of all members under design load. Deflection increases sharply when prestress reaches 0 under applied load (slack members). Those observations correspond with observations on Fig 1. Also, when slack members occurred, stability was impaired by a strong flutter tendency.

3.3 Variable Safety Factor

Increasing safety factors from 2 to 4 reduces deflections by 50 %, yet causes only minor force increases. A safety factor of 5 reduces deflections by 66%.

4 Prototype Structures

Two prototypes, a one-way and a two-way truss (Figs 8 to 11), were built with the objective to explore details, fabrication and erection procedures, test buckling, and verify model and computer simulations. The one-way system was stabilized against rotational buckling by lateral bracing at alternate joints. For the two-way truss a joint was developed and tested to connect continuous cable members to vertical struts. The diagonal cables were connected in pairs (Fig 11) to provide concentric connections and a smaller joint radius. Lead inlets, inserted between cables and strut brackets, provided slip resistance needed at those connections. Based on the test model (Fig 8) and computer analysis the truss was partly pre-assembled in the shop with loosely connected joints. The erection was started by first installing helix ground anchors and erecting four support trusses, held by temporary guy ropes. Next, the cable trusses were spread out on the ground, then gradually lifted by a boom and pulley system and connected to the support trusses. After some prestress had been applied, by means of threaded stud connections, the vertical struts were aligned to their final position. Prestress was then fine-tuned and checked with a tension gauge. Finally the strut joints were firmly tightened for slip resistance between cables and struts to achieve stiffness and stability for the system. The erection took about six hours. Due to experience, a second erection took only about three hours. Load tests confirmed the model and computer simulations with remarkably good accuracy. But most impressive, the system proved incredibly stiff when several students bounced jointly from a single joint, without visually noticing deflection or dynamic flutter.

5 Projects

The following projects illustrate the integration of prestressed truss systems with architectural and other design objectives. Given the location of both projects in very active seismic regions, reducing dead weight and thus lateral seismic forces was an important factor in system selection and design.

5.1 Sports Center University of California Berkeley

A prestressed roof truss was proposed for the master plan of this addition to an existing gym (Figs 12 and 13). One-way trusses allow modular expansion of the facility. Horizontal reactions are transferred to a structural slab by guy cables which are slanted two ways to provide stability against seismic forces as well. The guy cables are expressed in the entry lobby for spatial definition and externally to articulate the facade, similar to buttresses in Gothic cathedrals. Exterior walls follow the inward slope of guys to visually open the adjacent street space, while

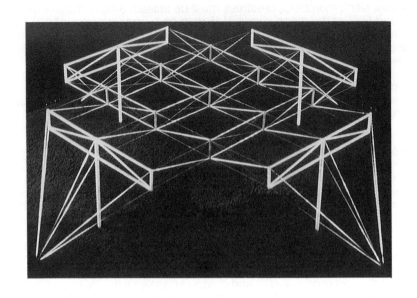

Fig. 8. Prototype model

Fig. 9. Prototype structure

Fig. 10. Prototype structure

Fig. 11. Protoype - strut joint

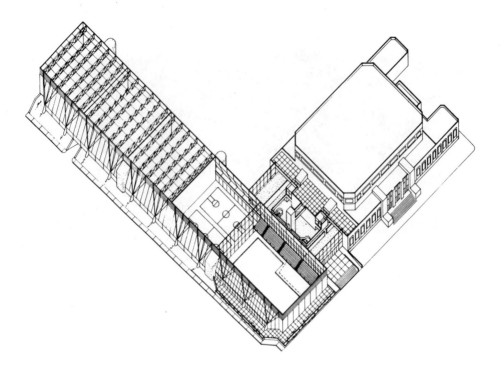

Fig. 12. UC Berkeley sports center - axon

Fig. 13. UC Berkeley sports center - elevation

reducing the interior volume for economy in air handling. A band of louvers at the truss level serves for natural ventilation and further articulates the exterior.

5.2 Loyola Marymount University Pavilion

This sports center (Figs 14 to 18) features parabolic cable trusses that span the long way, to provide large side doors as visual link with outdoor seating, for occasional large events. Exposed guy cables transfer horizontal reactions on one side to shear walls which also serve as external handball courts and on the other side to wind scoop walls facing the Pacific for natural ventilation. The parabolic cables reinforce the interior space profile, formed by the gym floor and bleachers. Guys for the main cables and radiating truss cables are connected to common anchor points to balance forces that alternate between bottom and top cables for gravity and uplift loads respectively. Computer plots of some load response cases are illustrated in Fig 17. The CAD studies (Figs 16 and 18), based on the analysis data, visualize the structure.

6 Conclusions

The prestressed trusses presented include a new concept with parallel chords, along with design data and evaluation. This truss has better stiffness and appears less susceptible to dynamic flutter than other cable systems. For some applications straight system lines may be desirable. While the system's complexity may result in higher costs, it has good potential from a structural point of view. As for most cable trusses, efficient resolution of horizontal reactions is critical. The design projects describe the creative integration of structures with architectural objectives; based on the author's belief, reinforced by historic examples, that an integrative design philosophy can enrich architecture, by techtonic integrity and synergy of form and structure.

7 Credits

Student teams of models, Fig 4: Miller, Suekama; Fig 5: Fung, Gee, Jordan.
Grant for prototype Figs 8-11: American Iron and Steel Institute.
Design team for project Figs 12-13: Schierle Associates, Architects;
    T.Y. Lin International, Structural Engineers.
Design team, project Figs 14-18: Kahn, Kappe, Lotery, Boccato, Architects;
    Reiss and Brown Structural Engineers;
    Dr. Schierle, consultant, CAD/analysis.

8 References

Bach, K. et al (1975) **Nets in Nature and Technics**, Institute for Lightweight Structures, IL 8, University of Stuttgart.
Haug, E. and Powell, G.H. (1975) Analytical Shape Finding of Cable Nets, **Proceedings, IASS Pacific Symposium**, Tokyo.
Hottinger, H. (1963) Concepts For The Design Of Structures Using Tensional Cable Systems, **Hanging Roofs** (eds Esquillan, Saillard) John Wiley, New York
Schierle, G.G. (1986) Design Considerations for Lightweight Structures, **LSA 86 Proceedings**, University of New South Wales, Sydney.
Zetlin, L. (1963) Elimination Of Flutter In Suspension Roofs, **Hanging Roofs**, (eds Esquillan, Saillard) John Wiley, New York.

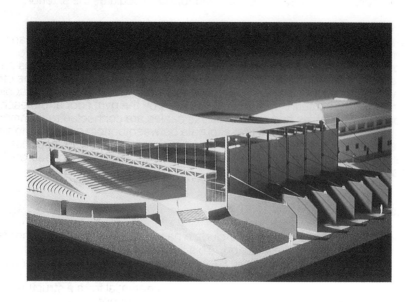

Fig. 14. Loyola sports center

Fig. 15. Loyola sports center

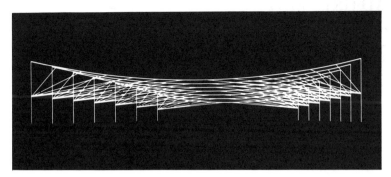

Fig. 16. Loyola sports center - truss system interior

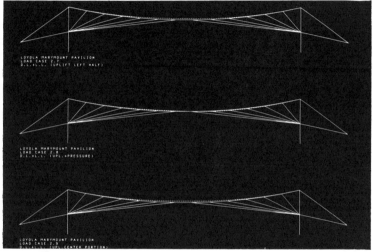

Fig. 17. Loyola sports center - load response plots

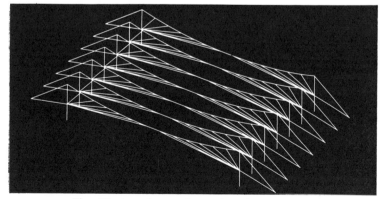

Fig. 18. Loyola sports center - truss axon

A CAD GRID METHOD FOR TENSILE PRESTRESSED MEMBRANE STRUCTURE CONCEPTS

D.A. CHAMBERLAIN
Lightweight Structures Unit, Dept Civil Engineering, City University, London, UK

Abstract

This paper describes a grid scheme by which membrane fields can be defined according to predetermined constant prestress ratios. Both integer and non-integer ratio cases are covered, these respectively determined from explicit and numerical based solutions of the derived, governing, grid generator curve. The proposed device is implemented within a PC based CAD system, which provides the means of addressing and scribing on the grid surface. Worked examples are illustrated for single and multiple field systems, including consideration of the generation of geodesics, which are associated with surface pattern production.

1   Introduction

The author is a member of the lightweight structures unit in the Department of Civil Engineering, City University, London, which has ongoing interests in the development of expert systems for tension structure technology. The content of this paper, together with previously reported work,[1,2] represents some enabling devices for the constitution of an advisory system for the conceptual design of prestressed membrane forms. This is aimed at those wishing to experiment in membrane forms principally by geometrical means.

Whilst an investigator might prefer to be free to define any surface which possess the essential anteclastic character, workable designs are related to tenable prestress regimes. Considering the pure prestress state, the uniform prestress condition is often preferred on both esthetic and economic grounds. Such surfaces are thought to be natural in appearance, and have the advantage of minimum prestress requirement. However, factors such as the general plan shape and the requirement for more contrast in the form, may lead to non-uniform prestressing.

Consideration of the membrane fabric supply, the weave and its stretch behaviour, leads to the observation that prestress variation within a single field should be restricted where possible. Practical states would include uniform stress, constant prestress ratio and non-linear of the type arising in conical forms. The latter is the most general case, and is beyond the scope of the work currently reported. In this work, a general approach is presented for the constant prestress ratio state, be this unity or any other desired value.

The solutions presented have been prepared using Fortran and Lisp code run on an IBM PC AT microcomputer supporting VGA level graphics. Grid generation and surface scribing are supported within the framework of AutoCAD version 10, a well known CAD drawing package, with hardcopy output to a Roland DXY1300 pen plotter.

2   Prestress regimes

Figure 1 illustrates a portion of an anticlastic surface grid formed by revolving the grid generator profile curve abc, which lies in the XY plane, about the X axis.

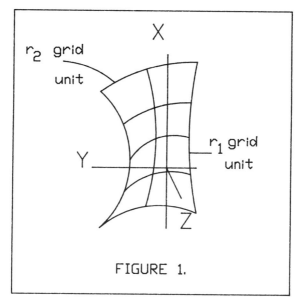

FIGURE 1.

Referring now to figure 2, description of the generation is apparent in the local profile and offset radii r_1 and r_2 respectively.

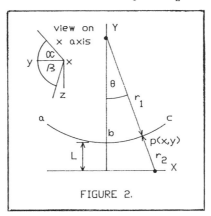

FIGURE 2.

Since, under pure prestress, the ratio of principal prestress values must be equal to the inverse principal curvature ratio, we have:

$$\frac{r_1}{r_2} = \frac{\sigma_1}{\sigma_2} = N \qquad (1)$$

Where N is the pre-determined constant prestress ratio. Considering a general point $p(x,y)$ on the profile, we can express r_1 and r_2 as follows:

$$r_1 = \left[1 + \left(\frac{dy}{dx}\right)^2 \right]^{3/2} \bigg/ \left(\frac{d^2y}{dx^2}\right)$$

$$\text{and } r_2 = y/\cos \theta \qquad (2)$$

where θ is defined in figure 2.

Using (2) in (1) and simplifying, we arrive at a second order differential equation for the profile abc:

$$N.y.y\,'' - (y')^2 - 1 = 0 \qquad (3)$$

By substitution $q = y'$ and use of the boundary conditions $q = 0$ at $y = L$ we have

$$\int_{x=0}^{x=x} dx' = \pm \int_{y'=L}^{y'=y} \frac{dy'}{\left(\left(\frac{y'}{L}\right)^{2/N} - 1 \right)^{1/2}} \qquad (4)$$

Analytical solutions are readily available for (4), where N is a positive integer, examples being

$$y = L.\cosh (x/L) \quad \text{for } N = 1,$$

and

$$y = \left[\left(\frac{x}{2L}\right)^2 + 1 \right].L \quad \text{for } N = 2 \qquad (5)$$

However, for non-integer positive values for N, solutions are computed numerically from the integration formula (4).

3 Grid generation

Having defined the grid generator profile at a series of points, these are then used to create the grid surface by revolution about the x axis. With respect to the positive sense of the x axis, revolution requirements are specified as $-\alpha$ to $+\beta$. If required, the r_1 grid

606

unit can be held constant, by determining generator points at equal curve intervals. The r_2 grid is obviously not constant, and depends on the integer division of $\alpha + \beta$. Complementary to the setting of the uniform r_1 grid unit, it is an easy matter to determine the curved distances along any particular r_2 grid curve. A knowledge of grid unit spacing is fundamental to the generation of geodesic curves, these by definition processing the property of maximum or minimum path lengths between two points on the surface.

Figure 3. illustrates provisions for dual field generation together with a portion of the resulting surface grid. The profile A is based on N = 1 and that of B on N = 2, and they are set to intersect at the common offset point 1. In this example, A and B fields are united by an arc ridge.

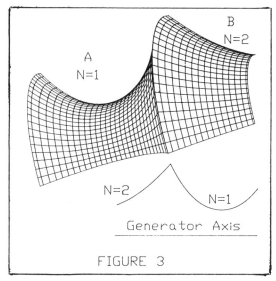

FIGURE 3

In the grid generation, the mesh is actual generated on the basis of polygons, which are subsequently smoothed by fitting a cubic B - spline surface[3]. Though computationally slower, the true surface generated from (4) could be used for greater accuracy.

4   Field production

The user grids previously described are constructed using Lisp coding run within the framework of AutoCAD. Once the prestress ratio N, has been selected, and the appropriate grid displayed, the user is able to window and zoom on the portion of the surface upon which the design field is to be enscribed. By rotation of the grid surface, it is possible to access obscured regions, and also overcome the problem of multiple layers, though with some computational delay, this can be avoided by processing a hidden line facility. When the design is complete, and the grid surface is removed, the design can be processed for geodesic production, surface shading or other requirements.

Options in the CAD system permit the user to snap on to grid intersection points and intermediate points on grid lines, on a nearest point basis. Each snap selection provides 3D coordinate feedback, thus the surface is addressable by means of its definition grid system. Further refinement is possible by setting a finer mesh over a particular region of interest. Cable type scallop features can be represented by spatial B - spline curve smoothing for data points selected between scallop ends. Although resulting in difficult prestressing requirements, the system also allows the user to pull out a grid intersection point and resmooth the surface.

Whilst grids are readily established and conveniently addressed, it is difficult to satisfy fixed control point requirements with the current

system. In practical designs, such points arise in the form envelope on account of plan and elevation restrictions, access and functional requirements. Ideally, these would be accepted as boundary conditions in the initial grid generation procedure. Unless boundary conditions are particularly numerous or restrictive, the user would still remain with considerable scope for experimentation. Using the current facility, three control points can be matched by rotations of the local coordinate system and overall scaling. In the conceptual design process, the user may be more interested in the surface form however, than meeting all control point requirements. Control point matching is the subject of ongoing study.

5   Patterning

Anteclastic surfaces are non-developable, and require careful patterning for their production. Production fabrics are typically manufactured in roll form, to about 2m width, thus patterning tends to be derived by subdivision into strips. Depending on the degree of fabric translucency and backlighting, the overlay of fabric at seams, results in pronounced lines over the surface. Apart from the most functional of structures, the architect may wish to use these lines to influence the character of the form.

Disregarding the mechanical properties of the actual woven fabric, there is little restriction on the setting of pattern arrangements, other than the advisability of following surface geodesics. Failure to observe the latter may result in panels needing to be sheared into place, a limited prospect with coated fabrics.

Turning to the matter of material behaviour, fabrics exhibit marked orthotropy in respect to the warp and weft directions of the weave. Under reasonable stress levels, the warp direction is noticeably stiffer than the weft direction, thus compensation for prestress stretch is differential even under uniform stress conditions. Compiling the complications arising in the Poison's ratio effect, the pattern seams are often set to correspond with either of the principal prestress directions. For forms derived with high prestress ratios, the orientation of patterning is thus likely to influence the form. Only with near uniform prestress is the form fairly independent of the patterning arrangements. Warp and weft stretching is typically of the order of ± 1% and 0-5% respectively, depending on the level of prestress and material type.

In the current facility, geodesic lines can be approximately determined between any two given points on a single field, using a series of conical helix[4] elements. Referring to figure 4, the procedure is as follows:

(i) Two points a and b are selected on the field, these   referenced by profile position u and profile inclination v.

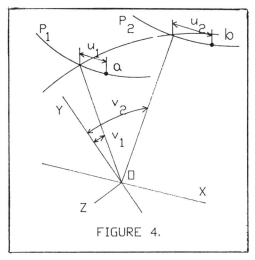

FIGURE 4.

(ii) Superimposing the profile curves P_1 and P_2 as shown in figure 5a, and creating n equal subdivisions (n = 4. illustrated), n + 1 number R circles are apparent.

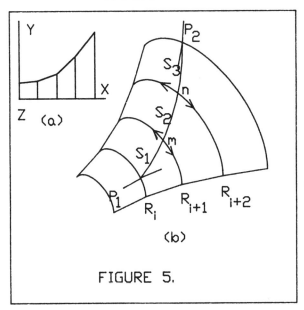

FIGURE 5.

By these means, a series of conic annuli provide an approximation to the curved surface as shown in Figure 5b.

(iii) The geodesic between points 1 and 2 is now fitted such that $\sum S_i$ is minimum, using the length notation shown in Figure 5b. Since each intermediate point m & n is required to remain on a particular R circle, the length is minimised by the v angular displacement of these points. This is achieved by an iterative algorithm.

By the above means, the non-developable surface is approximated by a series of developable conic annuli, and the accuracy is dependant on the number of annuli set in the representation. The patterning is thus set in the definition of the geodesics, the seams. Between the R circles corresponding to the chosen points 1 and 2, further geodesics could be prepared on an offset basis, though these would require extrapolation to field boundaries.

6   Case studies

Figure 6 shows a single field enscribed on an N = 1.5 grid, figure 7 a double field on N = 2.0 and figure 8 triple fields prepared on a combination of N = 1.0 (uniform stress) and N = 2.0 grids. In each case, the fields have been rotated to a position of interest.

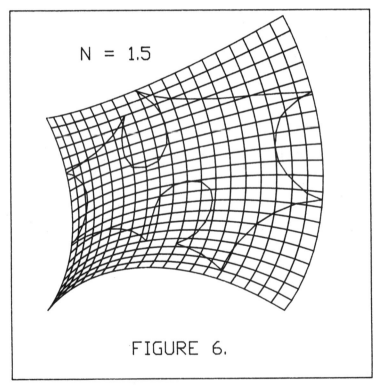

N = 1.5

FIGURE 6.

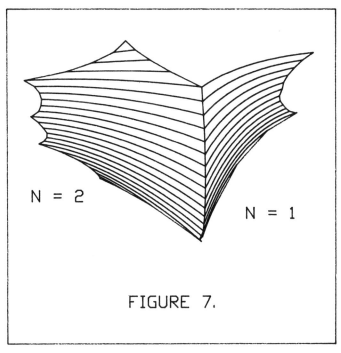

N = 2   N = 1

FIGURE 7.

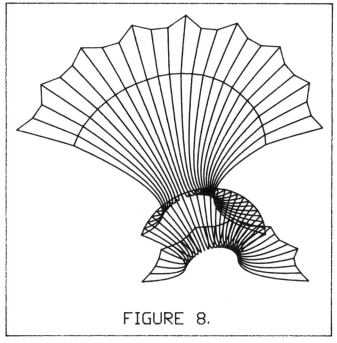

FIGURE 8.

611

7 Conclusions

A method has been described by which approximate prestress membrane forms can be constructed under constant prestress ratio conditions, the user setting the required ratio N. Principal curvature grid surfaces are automatically generated and can be addressed using CAD drawing software. An approximate means of generating geodesics between selected points has been described.

A matter which has not received attention is that of force equilibrium. Holding the weft prestress at some constant value throughout a field, the warp stress is similarly known from N. Using the geometry of the field and boundaries, it is possible to determine boundary reactions etc. In the case of the multiple fields of figure 8, the prestress levels could be adjusted to provide equilibrium for the arches at the field interfaces. Further study is necessary in order to properly exploit the combination of prestress and surface geometry.

A useful extension would be to provide grids for variable prestress ratio, as arising in the generation of smooth conic tents. The means of creating transition between different field settings is also of interest, though this is known to be a complex matter.

8 References

1. Chamberlain, D.A., Aids for the Conceptual Design of Pre-Stressed Membrane Structures, 4th Int. Conf. on Civil and Structural Engineering Computing, Civil Comp. Press, Edinburgh, 1989.

2. Chamberlain, D.A., A Geometrical Approach to Smooth Conic Tents, 30th Anniversay of IASS, Conf, Cedex, Madrid, 1989.

3. DeBoor, C., A Practical Guide to Splines, Springer, Berlin, FRG, 1978.

4. Nutbourne, A.W. and Martin, R.R., Differential Geometry Applied to Curve and Surface Design. Vol. 1 Pub. Ellis Harwood, Chichester 1988.

THE COMPUTER AIDED DESIGN OF MEMBRANE STRUCTURES

A.H. YOUSUFI
Faculty of Architecture, Delft University of Technology, TUD, The Netherlands

Abstract
The faculty of architecture of Delft University of Technology in the Netherlands has stimulated a specific research on the field of fabric structural design within our technological development in general, and the use of fabric as a building material towards many social and economical demands which require in physical form rapid response to a recognised social, economical or strategic need in specific.
The design of prestressed coated fabric structures are usually started from an architectural sketch te determine and develop the design concept. However, it is useful to employ computer simulation of the structural form which will display the accurate stress distribution throughout and a further investigation of the whole structure.

Introduction

The recent advances in design techniques and increasing utilisation of coated fabric and constructional materials have made possible to construct large span structures in a variety of geometrical forms. Most fabric structures, air supported, tension or frame supported fabric structures must be analyzed structurally through mathematical methods, which require computer programme to determine the accurate shape that result from the prestressed forces. The mathematical model with accuracy and exact shape at given stress patterns would permit the designer to find and explore a new shape and combination of shapes. Computer model incorporate the consideration of form and patterning, which influences the form and deformability, and allow the efficient change of element and a system of geometry.
The fabric and cables used to build structures are extremely light, they have no rigidity or stiffness, therefore they can carry loads only in tension and must be kept in tension to stabilize the structure and prestress shape.
Prestressed tension structures need guy ropes and anchors to lift up the horizontal components which are under prestress and external forces, and to transform these forces into the ground directly or indirectly.
Usually the self supported or prestressed structure which the forces are carried by the fabric itself and anchored to the ground, and cable stayed structure which the structure is stabilized by cables and ropes, may cause

some functional problems. Fig.(1) Therefore, some other forms of fabric structures have been developed to eliminate the need for external guys. Fig.(2).

As far as the anchors and guy ropes are the essential part of the structure, their elimination will change the form of the structure which has been expressed as skeleton or frame structure and braced fabric structure, that the fabric is planer or single curvature, and is used as cladding material. The variety of possible forms is infinite, any manipulation within the structural system, require the knowledge of the interrelations between the form and the mass of the structure and its ability to transmit forces. Fig.(3).

There is a potential for obtaining very economical structures using the fabric and plastic sheeting by the availability of powerful microcomputer systems which makes physical modelling very attractive for expediting and development of analytical procedures for modelling structures.

Computer aided design

When we talk about design in this field, it is not just the structural design but the total design which is Architectural, Structural and constructural.

The form of Tent and complex structures can be determined on the same principle of empirical methods used in conventional structure. Today the application of Software Computer Graphic has made possible to generate a coplex structure and form, that the output of the programme is a set of points co-ordinates which can describe the form, from this all other values can be determined, such as ground plan, elevations, sections, prespective and a three dimensional working model.

The application of computer to the Architectural Design and analysis of complex structures is one of the fields that attracted a great deal of research efforts, since the introduction of the computer as a tool of Architectural Design consideration, recently there are a number of computer programmes available for the Architectural Design, research and practice.

The ARIADNE programme is an integrated system for building structures which programmed to meet the needs of the computer Aided Architectural Design. The system which is still in its initial stage of development, includes the following sub-systems, each of which is programmed to have various graphical computation capabilities in 3D as :

- Manipulation and analysis of building components.
- Preprogrammed generation of building components.
- Combination of building components.
- Architectural drafting, presentation, and visualization in 3D.

The Ariadne programme has the capacity of three dimensional analyses of building system with more improved accuracy. The programme is written in FORTRAN language, operated with UNIX system, and possibly linked to the other software systems.

The geometric solid modelling of a structure consist of the out-line configuration and the geometric modelling of the building components. The designer has usually no need to search for the best outline configuration

614

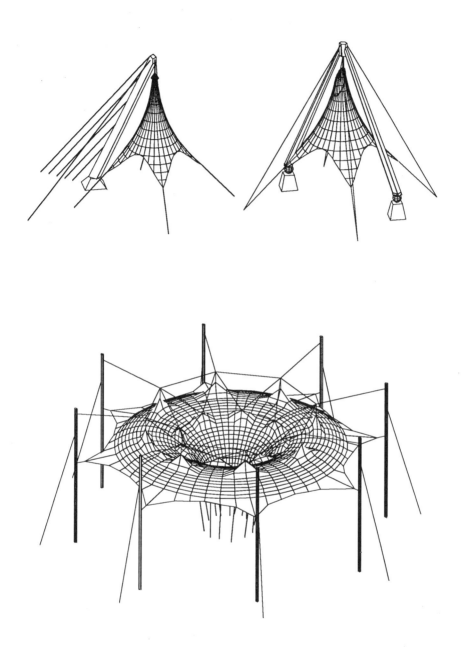

Fig.(1) The various possible type of cable stayed membrane structures generated by CAAD.

because it comes directly from the form of the structure, instead the designer has to select the size of the components and the number and situations of the building components, that the final result is structurally acceptable. The following page describe the design of tents under the effect of the load symmetrical in regard to columns, in square and polygonal plan which are reinforced by means of cable or braced stabilized by cables. Each unit of the structure is stabilized by itself, and can be erected as a single unit or as many as required. The structure can be erected by Prestressing the surface fabric upon upper and lower portion of the support. When the desired prestressing is achieved, the coated fabric and cables can be clamped to its support. Fig.(4).
Inorder to achieve with this system the following parameters should be taken into account.

a Various components used in a structure might have different sizes, section and jointing technique, which can change with different manufacturers.
b The geometrical solid components generated by computer should be handled like handling the real component in practice.
c The software computer should be suitable for moddelling of geometrical solid that combined together and made the final output as an object.
d As far as the solid modelling is concerned the combination of the component must be as accurate as possible.
e Finding the order, arrangement and pattern of structure, for ease of combination.
f Presentation of sequential drawings in 3D by means of bottom-up or top-down systems.
g Documentation of data for each different components and the structure as a whole.

In the geometrical modelling of structural components the effective use of computer requires a knowledge in the geometry, the geometry of form giving possibilities, and techniques which makes possible to deal with large combination at a time.
The designer can modify the model or component by changing the values of parameters in connection with size, cutting mode, existence of the separate components etc., changing the co-ordinates of the relative origin of the component, the joint can move along the absolute origin of the final object, that makes easier to change the outline configuration at any moment.

By using this system of computer the user can create structures very effectively. The process is like dealing with real structural components on site, beacause modelling progress is seen continuously in three dimension on the screen with its data value. The most common models are preprogrammed and can be used by pressing a corresponding special commands and inserting limited values. All the parameters of these common models are open to change, the user has an infinite number of possibilities to gain an optimum form and structure.

The use of solid modelling programme allows easy sectioning of models to reveal hidden detail and check interferences and examine relationships between components in assembllies particularly using colour rendering to

616

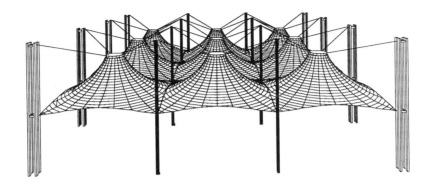

Fig.(2) Cable stayed membrane structure, a prototype of tent designed for Haj Terminal at the Jeddah airport in Saudi Arabia.

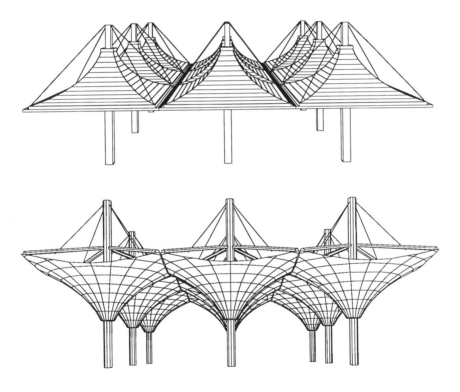

Fig.(3) Braced membrane structures, as alternative of the tent of Haj Terminal at the Jeddah airport in Saudi Arabia. Top; single curvatures, Bottom; double curvatures.

differentiate components. These techniques imply a further step in visualization of building structure.

There are many advantages achieved by CAAD besides the main benefits related to the cost reduction in the design and drafting process.

- Accuracy, quality and visualization .
- Reduction of time in the design process.
- Ability to produce alternative design.
- Avoiding the inconsistencies on complex structures and forms.
- The speed and control of design modifications.
- Estimating the precise components, materials and quantity.

References

Sedlak, V.F.J. (1981) Membrane structures, constructural and design aspects of some recent membrane structures in Australia. **P. Kneen** p.42.

Topping, B.H.V. (1987) Non-conventional structures. Vol.-1. About the prinzip Leichtbau. **J Hennike**, p. 277.

Wagner, W.F. (1980) Architectural Record, Tent structures: Are they Architecture,.p. 127.

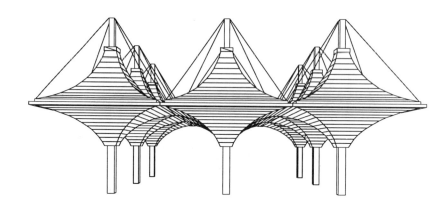

Fig.(4) Double layers, single curvature membrane structure.

PART TWELVE
SPACE STRUCTURES

A QUEST FOR GLASS SPACE STRUCTURES

M. EEKHOUT
Octatube Space Structures, Delft, The Netherlands

ABSTRACT

This contribution describes the development process of real structural glass as performed at Octatube of Delft NL. The described process includes several methods of structural use of glass panels in spatial structures, ranging from non-loadbearing glass panels in conventional skylights around space frames, via structurally sealed glass panels with an additional stabilising function to structurally load-bearing glass panels, where metal elements are only used as connection elements and additional stability.   The point of view of the author is that of a <u>product-architect</u>, who combines the profession of an architect, a structural engineer, an industrial designer on the one hand, and a specialist-producer on the other hand.

1 WHAT IS STRUCTURAL GLAZING?

The description 'Structural Glazing' has a number of different interpretations in the architectural profession, departing from the architectural or the structural point of view.

1.1 Architectural meaning

In the first place *'structural'* is used as the abbreviation of the building technical word 'structu-rally sealed' meaning a method of attachment of glass panels to the sub-structure in curtain walling. In the early days of curtain walls it indicated anything else than fixing by screw strips.   The growing tendency amongst architects to design in an abstract and non-material-bound way has strongly stimulated the use of glass as a cladding material for exterior walls and roofs. This tendency was reinforced shortly after the first oil crise in 1973 when in European glass industries a search began for glass panels with grossly improved building physical quality in order to beat the enhlighted energy costs of glass-clad buildings. The demand for slick building surfaces and low maintenance costs led to the application of glass panels that were structurally sealed to the aluminium sub-structure with silicone and were provided with a silicone watertight seal between the glass panels. These techniques were exploited on a large scale in the USA even before 1973, forced by the implications of high-rise building technology and imported into Europe only later. In fact, even after 20 years of experience in the USA, structurally sealed glass is still not permitted in some countries like West-Germany. In those cases the silicone sealing and glueing techniques still have to be combined with mechanical screwing techniques that have a conventional safety. These descriptions all regard curtain walling and skylights. Loadings are only external loadings acting on one glass panel. No further structural loadings are taken. 'Structural' refers only to the mode of connection to the subframe: with sealant. The confusion grows when people commonly refer to any type of cladding in which sealant is used for watertightness only as 'structural'.

1.2 Structural meaning

Quite different and scientifically more interesting, is the line of thought following the structural engineering meaning of 'structural glazing': glass structures that bear external loadings over more than one glass panel, and contain bending and normal stresses. The strength properties of heat-strengthened or tempered glass challenge some designers to see where the limits of suitability are for glass panels to be used as load bearing structural elements in structures that only contain an absolute minimum of metal components. This article explains the development process of glass panels from the starting point of a non-structural via a half-structural to a current state-of-the-art of completely structural use in the structural engineering sense. In particular those cases of an increasing degree of experimentation and difficulty are revealed as they were processed in the Company of Octatube Space Structures bv in Delft NL during the last few years. The state-of-the-art is March 1990. This process of Design + Development + Application is described from a broader point of view of the author being a product-architect combining the skills of an architect, a structural designer and an industrial designer, combined with the possibilities and responsabilities of the specialist-producer, rather than that of only a broad vision of an architect, a more deep and narrow interest of the structural researcher or the economical interest of the producer.

2 THE PRODUCT-ARCHITECT

In the Netherlands the term 'product-architect' has been known ever since november 18, 1988 when the author proposed it in the first Booosting congres in Rotterdam [ref 1]. It was there described that the function of the product-architect is to design, research and develop components of buildings, independant from the actual design of buildings by project-architects, and to apply these building products in the overall-design of buildings that are normally designed by project-architects.

   The product-architect tries to complete the potentialities of new materials (Material Science), production techniques (Material Processing) and of application systems (Structural Engineering and Architecture) with the analytical approach of the industrial designer, and the know-how on the architectural building site of the architect, and the creativity of both. So the field of action and also the

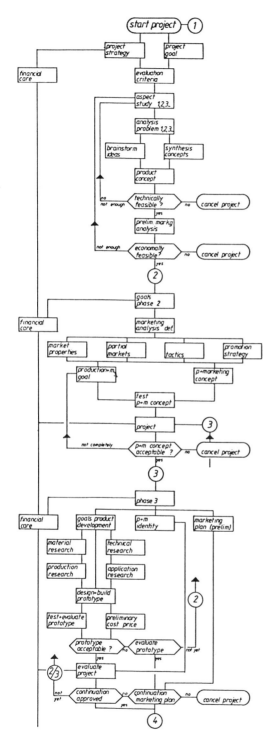
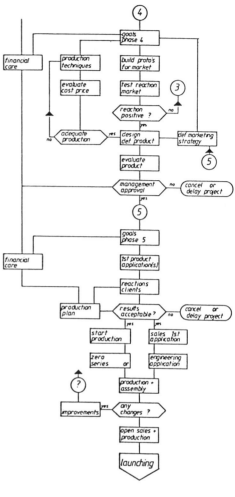

Organogram of a Design and Development Process for new products in the building industry.
Phase 1: 'Orientation and Product Concept' for a normal design + development process for new products in the building industry as a preliminary design phase with provisory market evaluation.
Phase 2: 'Testing Market on Product Concept' showing the market research on the first product concepts.
Phase 3: 'Techniques and Costs of Prototypes' showing the necessary in-house mainly technical developments to complete the prototype.
Phase 4: 'Prototype and Market' showing the confrontation in the market with the developed prototype and its evaluation.
Phase 5: 'Launching of Product' showing the process of production of the first application, with evaluation for duplication and further standard production.

abilities of the <u>product</u>-architect combine that of the architect, the industrial designer and the structural engineer.

3 DESIGNING + EXPERIMENTING + BUILDING

The position of the product-architect is producer-bound rather than consumer-bound. A large range of subsequent applications every specialist-producer normally is working on, enables him to design and develop a building component product into maturity within a reasonable short time without the danger that developments stop after the completion of only one prototype building, because one client stops a project or because a project has been completed.

The described process is a result of a slow step-by-step method. In detail and more generalised such a 'Design + Research + Development' process of building products has been worked out in fig 1. The Organogram in fig 1 gives a typical design and development process for building products and components (from 'Architecture in Space Structures' [ref 1]). This scheme is the result of analyses by the author of several product development processes. It has the advantage of visual communication for designers, needing only a brief comment:
. Phase 1: <u>Orientation and product Concept</u> showing a preliminary design phase with provisory market evaluation.
. Phase 2: <u>Testing Market on Design Concept</u> showing the market research on the first product concepts.
. Phase 3: <u>Techniques and costs of prototypes</u> showing the necessary mainly technical in-house developments to complete the prototype.
. Phase 4: <u>Prototype and Market</u> showing the confrontation in the market with the developed prototype and its development.
. Phase 5: <u>Launching of a product</u> showing the process of production of the first application, with evaluation for duplication and further standard production.
    This process scheme can be used for a wide range of new products. The 'smell' of materials and the physical presence of it, has always been the source of know-how of any specialist, and has proven to be very inspiring for designers. New opportunities begin and end with materials and production processes. Furthermore, the real building opportunities of a number of successive building products enables continuous feedback and product improvement.

4 MATERIALS + TECHNIQUES + SYSTEMS

4.1 Material Properties

For an architecturally trained designer like the author,the numerical side of the Science of Materials is not the most compelling aspect of the profession. Yet the relative large differences in properties of the different building materials provide a reasonable indication whether combinations of these materials are desirable, possible or whether they are not. By just comparing these properties, some basic and very logical conclusions can be drawn, that usually are never done because scientists hardly ever step over the borders of their specialist territory.

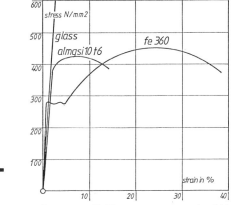

2. Strain-stress diagram of Glass, compared with Aluminium and Steel

4.2 Young's modulus of elasticity

Approximate E-values in N/mm2 are (see fig 2):
. Steel 210,000
. Aluminium 67,000 to 73,000
. Ordinary Glass 73,000 to 75,000
. Heat-strengthened glass 73,000 to 75,000
. PMMA 3,200
. PC 2,300
. Wood 14,000
Out of all transparant building materials glass is the best
suited for structural purposes. Both polycarbonate (PC) and acrylate
(PMMA) have a far less favourable modulus of elasticity (2,300 /
3,200 N/mm2 compared with 75,000 N/mm2 for normal / toughened glass),
meaning that the stiffness of these materials is far less; so
deflections under external loadings applied on PMMA and PC are much
larger.

4.3 Tensile strength

The maximal tensile strengths in N/mm2 of the different building
materials are:
. Mild Steel 360
. Alu alloy AlMgSi 0.5 215
. Ordinary glass 40
. Heat-strengthened Glass 200
. PMMA 70 to 110
. PC 60 to 100
. Wood 100
The maximal tensile strength of glass is 40 N/mm2, but that of
heat-strengthened glass goes up to even 200 N/mm2, while that of PC
is 60 to 100 N/mm2 and PMMA 70 to 110 N/mm2. More noticeable: the
maximal compressive strength of glass is in both cases 800 N/mm2
(without the influences of buckling).

4.4 Brittleness

The most deviant property of glass applied as an element of a primary structure, compared with other structural materials, is its *brittleness*, which turns out to be most dicisive for the use of glass as a structural material. The achilles heel, the weakness of glass (and to a far lesser extend also cast metals like some aluminium alloys), is caused by this brittleness. Very generally put, there are two fracture mechanisms competing to break a material:
. Plastic Flow
. Brittle Cracking.
   The material will succumb to whatever mechanism is weaker. If it yields before it cracks, it is ductile. If it cracks before it yields, it is brittle. The potentiality of both forms of failure is always present in most materials. Yielding is a safe and much desired property, spontaneous cracking is an undesirable property for a structural material.

4.5 Time and creep

Glass is a solidified liquid and not a crystallised solid. However, the tendency towards crystallisation is present, and given time, glass will crystallise. This is known as divitrification. It involves shrinkages, the glass is often weakened and sometimes falls into pieces during the process. It will always fracture in the same brittle way. Creep in glass and heat-strengthened glass is practically nil, but for PC and PMMA is fairly high (although no exact figures are known). And fArom here on only hearsay can be noted. Some glass industries tell us that glass in time will deform (liquid will flow): hence many old mirrors have become bobbly by now and old window panes are never flat. Only data about the time scale in which this happens are completely unknown. Some glazers state that old large and thick window panes always are difficult to remove because the lower end appears to be thicker than the upper end that sticks in the glass groove.
   Wood creeps, as we all know from heavily loades tiled roofs built from timber beams.

4.6 Thermal conductivity

The respective values for the respective materials in W/mK:
. Steel   40-50
. Aluminium 200
. Glass   0.80
. PMMA   0.07 to 0.21
. PC   0.12 to 0.19
. Timber 0.12 to 0.16

4.7 Thermal expansion

This thermal expansion is important for the mutual differences between structural and cladding materials and between co-operating materials in structure or cladding. Values are given in 10.-6 m/mK:

- Steel 12
- Aluminium 24
- Glass (all types) 9
- PMMA 70
- PC 65
- Timber 3 to 5

There will be trouble when values of co-opererating materials differ too much. PMMA and PC expand 7 times as much as glass. The combination of steel and glass causes less problems than aluminium and glass.

4.8 Specific gravety

Values in kN/m3:
- Steel 78
- Aluminium 27
- Glass 25
- PMMA 12
- PC 12
- Timber 7

4.9 Behaviour at high temperatures

The maximal workable temperature in degrees Celsius:
- Steel 550
- Aluminium 250
- Ordinary Glass 60-110
- Heat-strengthened Glass 270
- PMMA 80
- PC 115

These figures show why fire-resistant panels have to be made of steel and steel elements (wire mesh in glass). No ordinary transparant material will have a good fire rating. Only borosilicate glass has a better fire rating, but is much more expensive.

4.10 Production techniques

Since we are interested to see in how far glass could be used as a real structural material, the failure mechanism for glass can be seen as follows. Glass is cooled during fabrication so fast that the molecules do not have time to sort themselves out into crystals. So cooled glass is a solidified liquid, not a crystalised solid. However, the tendency into crystalisation is present, and given time, glass will crystallize. This is known as devitrification. It involves shrinkages, the glass is often weakened, and sometimes even falls into pieces during the process. It will always fracture on the same brittle way. In fact, if we want to prevent the glass from cracking, we have to put it under compression. This can be done by heating the glass panels again, and chilling the two outsides of the hot glass panels during fabrication very fastly, so that the two outsides form with the core a compression + tension mechanism. When the outer surfaces are cooled, they solidise, while the core is still hot (700°C). The shrinking during cooling of the core causes the outside skins to be compressed, while the central core will remain under tension. So the outside skin of heat-strengthened glass is under compression, the core is under tension. This is a mechanism to avoid surface cracks, but also hides internal cracks. The mechanism is the same for nodular iron and cast aluminium: the outside surface can be very smooth, while irregularities can be hidden inside the material.

After strengthening the outcome is a glass panel with higher tensile strengths, also a higher impact strength. Great care has to be taken that the glass surface is not scratched by a sharp tool, because then it will crack into thousands of small bits. Try to bring glass panels out the reach of vandals. Or reversedly put, sharp glass hammers are a safety tool to break out of an all glass cage. (For use in structural glass applications like the Glass Music Hall in Amsterdam, see par 12).

In heat-strengthened glass panels possible cracks are avoided by the compression mechanism. Using connections of the bolt-and-hole type in the glass can, consequently, be done best by a pretensioning type of bolt connection: in that case not the hole edges are loaded by the bolt on flush, (with the inherent danger of enlarging the micro cracks around the bolt hole by drilling), but the pretensioned bolt will compress the two outside metal rings or components on the glass where the mutual friction will bring over the connection force. The friction force can be enhightened by grinding or blasting the surface around the bolt hole. Alternatively, A flush-type connection will have to contain an intermediate material between bolt and glass hole to avoid local toptensions in the hole, from which micro cracks can lead to serious cracks. But another idea might be to fill in the irregular left-over spaces in the holes between the outsides of the holes and the bolts by liquid epoxy, in order to get a firm abrasion connection. This is a technique developed to renovate old nailed railway bridges with slotted bolt holes. In this case the method is a means to isolate metal from glass and to adjust glass panels exactly to the required sizes. The type of connection will decide on the vulnerability of the irregularity of the bolt shaft and the bolt hole.

On scale of the glass panels, a possible compression (normal) force introduces the danger of buckling the panel so that the largest commercially available panel thicknesses (12, 15 or 19 mm) will have to be used that are quite expensive per volume. It would be better then - if at all possible - to load the glass panels under tensile (normal) force in stead of compression: that is to suspend rather that to stack them.

The thermal-hardening process is one of the interesting production techniques of structural glass. It was invented decades ago, but is still the basis of the current structural glass types. Another remark has to be made on installation of glass panels: although glass panels are very strong they are likely to splinter up into thousands of small pieces for example when not full attention is paid during hoisting and installation.

4.11 Statical systems

The statical systems applicable for glass structures all will

depart from glass *panels*. Glass bars do exist but the nature of these
glassfiber bars filled with epoxy is not translucent. Moreover the
connections are still quite laboursome, so in this article glass is
only regarded in panelform, and not in glass bars. The most simple
solution is the classical vertical glass panel, able to make a
vertical span with or without metal or glass ribs. One of the first
architectural design priorities set by the author, and also by a
great number of his collegue-architects, is to develop structures
with minimal visual disturbances. As a consequence thereof the visual
minimal 2-dimensional structures (because of the flat plates) like
guyed structures are the most logical structures. Figure 3 with the
derivation of the guyed structure principle is self-explanatory. The
principle forming the base of guyed glass structures is that short
and slender metal compression bars are used, with long thin metal
tension bars and glass plates where invisible normal stresses in the
form of tension- or compression stresses are included. Fig 3 also
indicates the difference between *Open* and *closed* structural schemes,
that is important in regards to the connections with the substructure. Also an indication is given of the *single-sided* and the *double-
sided* schemes arising out of architectural reasons.

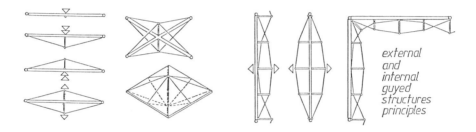

external
and
internal
guyed
structures
principles

3.
Derivation of guyed structures in cross section applicable
to space structures with glass panels as the main
structural elements and cross bars plus tensile rods as
auxiliaties

Although the structural principle is fixed now, it seems advisable to
develop new aspects and difficulties only at a modest speed: step-by-
step, where every step means only one or two new aspects compared
with the in-house technical state of the art. Thinking and developing
in this way has led to the rest of the description of this article.
The step-by-step-method is the only logical way of developing a new
technology above the level of the state-of-the-art, Only after
establishment of this technology the respective standards will
usually be developed. The involvement in an earlier phase of positive
product-research by the official Building research Institutes would
be most welcome.

5. SPACE FRAMES WITH SEPARATE GLAZING SYSTEMS

Fitting in the whole constellation of the building industry, the decision for a separate glazing system around or on a space frame is an accepted solution, ready for sub-contracting. Standard solutions are suited for normal applications, but often fail when geometrical complications arise. In fact very soon it appears that standard solutions and more experimental (or non-standard) solutions are a world apart and show large differences in approach. For example suspended glass surfaces, irregular geometical surfaces and facets usually require non-standard solutions. Apart from that, the resulting optical doubling of structural bars and glazing mullions can work very confusing. Along these considerations the author developed a new complete glazing system for the Raffles City Glass Atriums in Singapore (1983).

In case of the Raffles skylight system, there was no other way than to develop a separate skylight system on a separate space frame, as this space frame was very heavily loaded (high upward and downward windforces caused by the four large towers), and permitted only to rest on 3 or 4 points, to allow the towers to move freely without the danger of crushing the space frame. Figure 4 however, shows by the graphical play of lines that the doubling of space frame lining and skylight lining works a little confusing visually.

4.
Space frame and skylight of one of the Raffles City Centre Complex in Singapore

6. SPACE FRAMES WITH INTEGRATED GLAZING

When the glazing system is not seen as an independant or semi-dependant system on the space frame, the next step toward structural glass is to design and development an integrated system in which the glazing mullions do coincide with the space frame bars. A distinction has to be made here between space frames with square or rectangular modules and space frames with trapezoidal of triangular modules. Attention has to be paid to the fact that a definitive disadvantage is that the price of (trapezoidal but even worse:) triangular glass panels is more than double the price of square or rectangular panels. But by integrating the skylight mullion and the space frame tube, the visual aspect has been improved 100%. Following this route, in the Tuball space frame system a new line of profiles has been developed called the OT-profiles in which the functions of structure and cladding are clearly readable: the circular section carries normal forces, the T-flange on top carries the glass (or cladding) panel. The total system is called the Tuball-Plus system.

5 and 6.
Overall and inside view of the music pavillion geodesic dome in Haarlem, 9 m diameter, 7.5 m height. A 3-frequency icosahedron. Rib length 1.7 m, covered with laminated clear glass in the Tuball-Puls system.

7.
Detail of the Tuball-Plus system.
8 and 9.
Arcade in Tuball-Plus for the shopping centre 'de
Amsterdamse Poort' in Amsterdam Zuidoost.

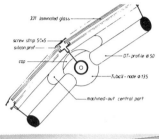

The music dome in Haarlem NL, designed by town-architect Prof
Wiek Röling in co-operation with the author as a product-designer has
been built in 1984 as the first application of a Tuball-Plus system
of integrated structure and cladding elements. See fig 5 and 6. This
dome is a geodesic dome with a 3-frequency icosahedronal subdivision.
All glass panels have the same triangular form as the structure. The
design of the Tuball-Plus node with machined-out ends of the bars
started as a brainwave behind the drafting table, strongly influenced
by the possibilities of a prototype laboratory. For all completeness
one should mention two typical technical contradictions, giving a
clue of the immanent battle between the designer and the structural
engineer behind the system. Firstly it is not a custom to introduce
bending stresses in the space frame bars apart from the normal
stresses. Secondly on the place where the largest shear forces are
acting because of the bending moments, the most material at the end
of the OT-bars has been machined away (see fig 7). These interven-
tions are calculated out in the stress analysis of the different
elements, but visually for the structural engineer they seem illogi-
cal. It is the product-architect/industrial designer who decided
here.
The arcade of the Amsterdam shopping centre 'De Amsterdamse Poort'
has been designed and built in the same Tuball-Plus system in square
panels. See fig 8 and 9. The arcade is composed of delta trusses
supported on each lower node, covered with laminated glass. Module of
space frame abnd glazing: 1.7 x 1.7 m. Span 11 m, length 34.7 m,
height 5.5 m. Designed by architect Ben Loerakker and the author.

7. SPACE FRAMES WITH STRUCTURALLY SEALED GLASS

A next step foreward is to change the mechanical screw connection between glass and stucture into structural sealant with silicone. The fixation is by glueing in stead of screwing, while the watertightness seam is again a separate silicone seal in stead of a rubber strip. So in this case there are 2 types of sealant: structural adhesive and weather seal, separated by a foamband so that structural movements in 2 directions are independant. The result is a flush outside surface without any screw strips, giving a dome even more a crystalline character. Irregularities that cause filthiness on the surface like screwstrips have been removed.

Example of this is the canopy of the Raffles City Hotel Complex in Singapore, where flat laminated clear glass panels 6.6.2 were sealed directly onto the aluminium Tuball-Plus profiles. See fig 10 and 11. The flat roof plane has only a camber of 1%. On top of the space frame square panels were used; on the sides triangular panels. The outside of the triangular glass panels form one flush glass surface 1.5 x 42 m long. Module length 1.9 m.

10 and 11.
Pictures of the Raffles City entrance canopy in Tuball-Plus with laminated glass sealed with silicone sealant on top of the OT-profiles

Glueing glass panels with structural sealant on the building site is not a real gain in assembly technology as it is very sensible for low temperature and humidity / wetness. Therefore, further development led toward glass elements provided with glued-on aluminium profiles in the factory (under ideal climatic conditions) with a structural sealant, that can be screwed with mechanical means on the structure on the building site, and subsequently weathersealed with silicone when the humidity allows it. Alas in this case the visual profile thickness is larger , and the detailling has not a minimal slenderness any more.

8. SPACE FRAMES AND STABILISING SEALED GLASS

The next step after the structural sealing of glass panels seems logical now. The idea is to give the glass panels also a stabilising function in the form of addition shear strength. At this moment is should be possible to design a rectangulated hinged space frame (actually a space 'truss'), where the horizontal stability can be taken by the glass panels sealed with structural sealant on the metal frame, preventing horizontal deformations. The consequence of this is that not only triangulated dome and saddle-shaped structures can be built, but also rectangulated or trapezoidal subdivisions, that are cheaper than the triangulated structures by the cladding. (Triangulated glass panels are 2.6 times as expensive as rectangular or square panels, and so are more decisive in price that the structure itself. Also in general one could state that the covering skylights are more expensive that the space frame underneath).
    Using the shear strength of glass panels is an idea that, unconsciously, was used in the last century in glass houses. These glass panels ensured the majority of the horizontal stability because they were fixed in rigid putty. Detail study of these glass houses like the Crystal Art Palace in the Botanical Garden in Glasgow show by the curvatures in the domes with their horizontally twisted lines, that the shear resistance came from the glasss panels and not from the metal structure below. See fig 12. There are no metal wind bracings in these structures.
    One example of a stabilised spatially curved roof is given in the saddle shaped roof of the Entertainment Hall in Zandvoort designed by architect Sjoerd Soeters and the author in 1987, that with a possible infull of structurally sealed glass panels could have been stabilised in any horizontal direction. See fig 13. The actual roof was designed as a double prestressed membrane, and as a single layered space frame with plywood infill panels, but structurally sealed glass panels would have been possible, too.

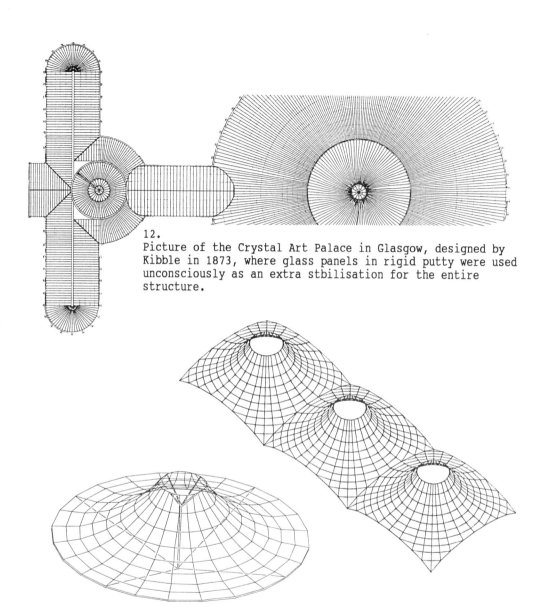

12.
Picture of the Crystal Art Palace in Glasgow, designed by
Kibble in 1873, where glass panels in rigid putty were used
unconsciously as an extra stbilisation for the entire
structure.

13.
Isometry of the anti-clastical roofs of the Entertainment
Hall in Zandvoort: 15 x 45 m
14.
Isometric view of a glass tent structure designed by the
author in 1989.
Another example is a tent-like polygonal atrium skylight that is
pretensioned by a central hanging mast and guying cables, as designed
by the author in 1989 (fig 14).

9. SPOKED GLASS PANELS: OSAKA PAVILIONS

It is only a very simple step forward from glass panels stabilising
space frames to the stabilisation of glass panels by metal components. But in doing so the importance of the two elements are
interchanged: glass plates get the primary function, while metal
components get the secondary function.

So this step is the most crucial one in the process described in
this booklet as it turns the normal way of thinking upside down!

The most simple form is an assembly of 4 glass panels in a metal
frame that are stabilised in the centre by two   cross-bars,
stabilised on their turn by 2 x 4 tensile bars to the 4 corners. The
whole assembly resembles a square bicycle wheel with 2x4 spokes. The
central nave really consists of 2 half elements compressing the glass
panels in between together. The seams between two adjacent panels are
formed by acrylate H-profiles that even do not give a shadow line and
are almost invisible. (Fig 15). This system has been designed as the
glass facade of a modular exposition pavilion for a Dutch pavilion
built in Osaka in March 1989 for the EVD (a service of the Dutch
ministry of Economic Affairs), by project-architect Frans Prins and
the author. A similar pavilion was later built for Heineken Japan,
(see fig 16) in a slightly modified form in Osaka in march 1990.

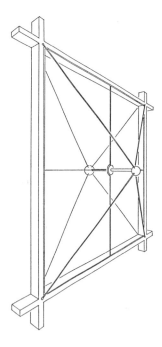

15.
Exhibition mock-up of one fascia unit of the Osaka-
pavilion, displaying the 2 cross bars and 8 tensile
spokes.

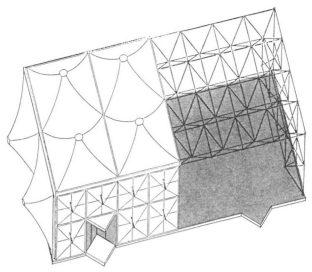

16.
The Osaka pavilion consisting of a tunnel shaped space
frame clad with prestressed membrane elements and stabilised glass facades as a tribute to traditional Japanese
Architecture; the fascia elements are composed of 4
stabilised glass panels (1.2 x 1.2 m) each with 2 cross
bars and 2 x 4 spokes. Designed by architect Frans Prins
and the author.

The original design consists of a modular Tuball space frame in
portal form 4.8 m high, 9.6 m deep and 19.2 m long. All three space
frame planes are covered with white prestressed PVC/PS membrane
elements, while the long fascia's are covered with the above
presented stabilised glass system. When the system should be built
without the surrounding metal frame, the glass panels would be
compressed in their plane towards each other, with the acrylate H-
profiles in between. The Osaka pavilion has been designed as an ode
to traditional Japanese modular architecture, in a modern form, with
modern materials and modern techniques: with a touch of western High-
Tech. The requirement of quick site assembly and demountability has
led to the specific choices of materials and components. The module
had to be 2.4 m all over, as the Japanese are very strict in the use
of regulations. As the shipping container size is outward 2.4 m but
inward 2.2 m, glass panels sized 2.4 x 2.4 m could not be
transported but had to be devided into 4 parts. So the size of the
glass panels is 1.2 x 1.2 m, in demountable window frames of 2.4 x
2.4 m. In consequence of the design of the Tuball space frame, the
nodal point of the top of the cross bars had the hollow sphere form
in which the 4 spokes run and can be prestressed internally without
visible nuts. This underlines the abstract character of the design.

10. PRESTRESSED GLASS PANELS

The next step in the development process is to give glass panels a
real primary structural function, by letting them function to pass
normal forces. For that aim a thorough theoretical study has been
made in 1988 in-house by Rik Grashoff, a student of Civil Engineering
at the TU Delft. The initiative was a result of writing the
dissertation 'Architecture in Space Structures' by the author,
started in jan 1988, and published in May 1989. [Ref 2]. The aim of
this structural feasibility study was purposely kept on technical
aspects, not financial or building-physical, as these aspects were
supposed only to restrict and endanger a possible technical step
foreward.   The material investigation showed indeed that glass is
still the only appropriate structural transparant material for
structures of the above described kind. (See par 4). For safety
reasons in roofs heat-strengthened glass panels could be laminated
and used as structural plates, although the lamination layer weakens
the structural capacity by 30 % in strength. Single heat-strengthened
glass is not safe (depending on the estimation of the danger of
vandalism or mechanical loadings and fall height). Duplex strengthened glass has problems with size accuracies of position of bolt holes
and panel sizes. Laminated normal glass is less expensive but not as
safe, but less expensive and cannot have bolt holes. The investigation has taken thick heat-strengthened glass panels as a base that
are laminated with thin normal glass panels for minimal security
reasons. The study resulted in a system of laminated heat-strengthened   panels in square sizes from 1.2 to 2.1 m, with thicknesses of
8, 10, 12, 15 and 19 mm.

17.
First Mock-up of the prestressed connection of 9 heat-
strengthened glass panels, 10 mm thick, as a first step
into the direction of load bearing glass structures.

The first prototype connection in the material-structural feasibility study was made with 4 (instead of glass) plywood panels and
double-sided 4 turnbuckles, to prestress the connection so that both
compression and tension forces could be transferred. In order to
avoid the danger of asymmetrical prestressing (with bending moments
on the glass panels), in the prototype at the end of the 3-months
study equilibrium saddles were built in, in order to obtain even
under strong asymmetrical pretensioning only normal forces in the
glass plates. The total size of this second prototype was 1.9 x 1.9 m
composed of 3 x 3 glass panels 630 x 630 mm (so in scale 1:3) to be
able to transport the prototype as a whole, and to use it for
exhibition purposes, while the mechanical connection nodes and
auxilates were made on real scale 1:1. See fig 17. The whole
assembly gave as a result of the mixture of scales a rather
mechanical outlook, asking for further refinement in design. Another
disadvantage of the 1:3 scale of the glass plated was their relative
stiffness, allowing loadings that normally could not have been taken
because of the high torsion rigidity of the scale model.
   The internal prestress-method had the aim to overcome the cutting
tolerances of the glass. In practice these glass tolerances seem
quite satisfactorally now (See par 12: Glass Music Box), so that the
tolerances in the metal components are most decisive now.

11. STRUCTURAL GLASS WALL: COOL CAT, GRONINGEN NL

The first contract for a prestressed structural curtain wall was
designed in 1988 by project-architect Paul Verhey for a fashion shop
of the Cool Cat concern in Groningen NL. See fig 18. The design
consists of 6 panels sized 2 x 2.25 m, in the total size of 4.5 m
high and 6 m long, as a suspended glass curtain on the first story of
the shop front, as shown on the perspective drawing nr. 18. The
detail of the joint is given on adjacent photograph 19. For this
first application the whole assembly has been built up in the
Octatube factory. The evaluation of this mock-up in the laboratory in
which both the glass panels as the joints are on real scale, proved
the visual correctness of the Minimal-Material hypothesis. The six
panels have been assembled into one consistent whole by a double-
sided guyed bar system on 2 x 3 cross bars. The size of the short
cross bars is 20 mm, the tensile guy bars 8 mm. The size of the glass
panels is so large compared with the joints that these joints have
been enlarged in the design phase out of visual considerations, in
order to obtain a visually credible structure from 40 mm to 50 mm
props. The structural assembly is a structural entity in the sense
that this closed system contends the neccesary tensile and compressive elements to form an independant whole. It could also work in
space. However, to function the structure has been suspended from a
steel portal frame. Windforces are taken to the 4 sides of the
structure; on the lower side a metal bridge takes over the horizontal
windforces of the both sides, but vertical forces are not taken over
due to the vertical sleeve holes. In this way also the extra loads
on the bridge can not cause extra vertical stresses in the curtain
wall. The detailling of the nodes, the cross bars and the guy bars is

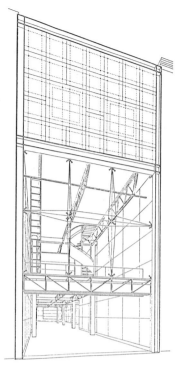

18.
Perspective view of the Coll Cat shop facade in Groningen
NL designed by Paul Verhey. The curtain wall of the first
floor (4.5 x 6 m) is suspended from the steel portal and
brings over horizontal wind forces to the 4 sides of the
glass wall.
19.
Detail of the Cool Cat node, designed by the author.

very functional, but in its design also abstract. See fig 19. The
glass plates are under compression by the prestress mechanism, while
the wind loads result in extra compression forces in the glass plates
as well as extra tension in one of the tensile elements. These are
two different mechanisms acting.
    The mock-up in the Octatube factory clearly showed the inspira-
tion derived from such a structure. As to material application: *"there
is not one ounce too much fat"*. The transparancy of the structure
underlines one of the evaluation criteria of its product development:
minimal visual disturbances. It also underlines the over-value aimed
at in this process: giving structures a certain inspiring form.
Professor Ludwig van Wilder of the TU Delft Architecture Department
gave another point of view when he commented on the structure as an
example of *"exchanging material for brains"*, and clearly expressed
that the development of glass structures stems from an intellectual
challenge. Due to retardement in the overall building scheme of the
shop, the curtain wall will be installed in februari 1990.

12. GLASS MUSIC HALL IN THE BERLAGE EXCHANGE, AMSTERDAM

The Cool Cat wall is the predecessor of the glass walls around the Glass Music Hall built in the former Option Exchange Hall in the famous Exchange built by Dr. Hendrik P. Berlage between 1896 and 1903, one of the first modernist buildings in the Netherlands. In this hall a smaller volume was built to function as the acoustical rehearsal room for the Dutch Philharmonical Orchestra. See fig 20. This hall has been made of glass because of the desired dominance of the interior of the old building and because the glass box acts only as an acoustical envellope, while around it other activities still can take place without mutual disturbances. The size of the Glass Music Hall is 9 m high, 9/13/10.8 m wide (belly form) and 21.6 m long (see fig 21). The 4 walls are composed of glass panels sized 1.8 x 1.8 m. The walls are suspended from a table-formed rigid double layered space frame structure supported by 6 slender steel columns.

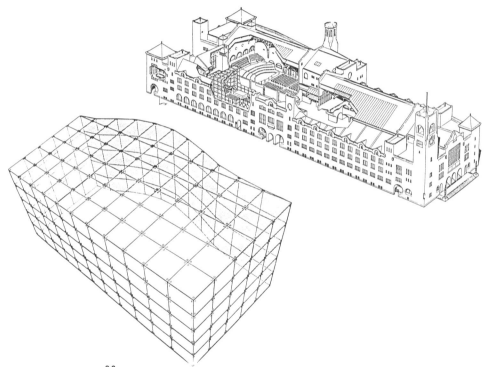

20.
Isometric view of Berlage's exchange with the Glass Music Hall inside, as originally designed by architect Pieter Zaanen.
21.
Bird eye's view of the Glass Music Hall sized 9/13/10.8 m wide, 9 m high and 21.6 m long, all covered with glass panels 1.8 x 1.8 m, as designed by the architect and the author.

22.
Cross section of the Glass Music Hall in Amsterdam, with
space frame roof structure and suspended glass panels in
both walls, stabilised by a guy truss system spanned
between space frame and reinforced concrete substructure

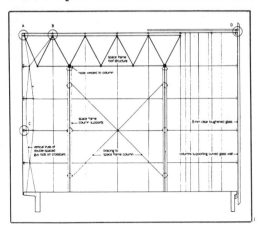

The glass walls are suspended from the edges of the space frame, and
stabilised by double (counter-)spanning guyrods on cross bars (fig
22). These stabilising systems are only present on the inside of the
hall, so that the outside has a slick glass surface. The roof plane
of the hall is covered with laminated glass 5.5.1, grey tinted. The 3
rectangular sides are 8 mm grey tinted heat-strenthened glass panels,
while the belly-formed (out of acoustical reasons) long wall has
clear glass 8 mm heat-strengthened glass panels. The original
suggestion by the author was to have an all-transparant clear glass
box, with only the curved wall as grey tinted, to emphasize the form-
deviation. However the architect decided to use the clear/grey glass
panels just reversed to enlarge the surprise effect when entering the
Glass Box.   The grey tinted glass has a remarkable cameleon effect.
Seen from the outside, with light outside the glass box is dark grey.
However, when sitting inside, and when the spotlights in the larger
Berlage space are lit on the walls, the glass all of a sudden seems
almost clear: It leaves a very good picture of the Berlage walls.
Darkening the room can hence be done by dimming the outside lights.
   The structure was completed in december 1989. The design of this
structure has been jointly made by project-architect Pieter Zaanen
and the author as a technical designer cum producer. The form of the
hall has been analysed and advised permanently during the design
phase by the acoustical advisor Peutz, and proved to give the desired
acousitcal values when measured after completion. Figure 23 to 26
give an overall view and some details.
   Building the Glass Music Hall has taught us one very important
lesson: the type of metal connection node does not permit large
deviations in size. Not in the overall size of the glass panel and
not in the seams in between. We have used glass panels very
accurately cut in the Swiss Securit factory with a water laser jet,

23 to 26.
Different details of the Glass Music Hall

giving bolt hole accuracies of +0,00 / -0,5 mm. The panel size
tolerance was +0,00 / -1,0 mm, orthogonally and diagonally. This of
course meant, together with the rigid connectors, that the max 1,0 mm
tolerance was to be met in the bolt hole. We also found that the
accuracy of these computer-cut glass panels was so high, that in the
total assembly the size of our steel structure became the point that
required the most attention, and appeared to be the most critical.
    During the installation of the 4 glass panel wall, we tested the
failure of one of the upper panels, which all carry the deadweight of
the lower panels. The structural design predicted a square chain,
action after collapse of an upper panel, so that the deadweight of
the 4 lower glass panels was carried over by panel nr. 4 to her
adjacent panels nr 4 and again up to panels nr 5. And this happened
indeed. There was no progressive collapse. But more of these
practical experiments will have to be made before these type of glass
structures will be accepted as being as safe as any other material.
It still will remain glass in all its properties, and its development
requires a lot of patience, trial and error and feeling for material
and structural behaviour from the side of the struggling designer.

13. FLOWER SHOP, HULST NL

The last step in this development report is the realisation of the
first outdoor glas roof in a guyed structure by the author. Not yet
in a form where the underspanning-and-glass cooperation is fully
actively structural, but in a way that the glass is passively
structural. The architect of the building is Walter Lockefeer, a
young Dutch architect working in the tradition of Dom van der Laan
(a school of Dutch architects where traditional materials and
proportions result in a very primary architecture). The design of
the glass roof has been made by the project-architect and the author
jointly, while the architect worked out the design very refined in
cooperation with the Octatube design team, making simultaneous
models, computer graphics, detail designs and statical analysis (fig
27).

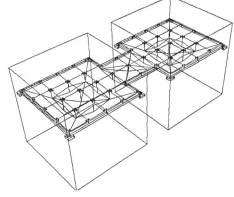

27.
Overall view of the glass roof between the brickwork cubes

645

28 and 29.
Two perspective views of the glass roof wioth the RHS
perimeter profile, inside the brickwork walls.

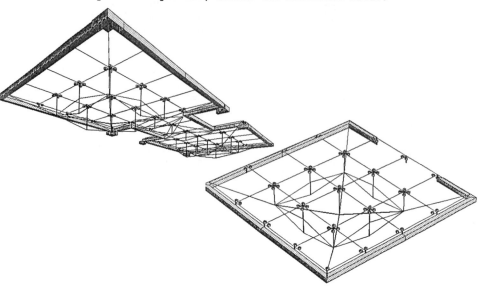

The Flower shop is a pavilion building, composed of 2 brickwork cubes
approximately 6 x 6 x 6 m, with an entrance bay of 2.23 m wide. The
two cubes are entirely closed in the outer walls. The glass roof is
at 5 m level and leaves a 1 m high parapet running around. The roof
is completely composed of flat double glass panels (composed as
reflective, heat-strenghtened upper panels 8 mm, 12 mm air and 3.3.1
laminated clear glass as the lower panels). The total size of the
roof is 6 x 15 m; the individual panels measure roughly 1.450 x
1.450 m. The panels are each supported at the 4 corners by support
brackets designed in the same mode as the 'frog fingers' or glass
brackets of the Glass Music Hall in Amsterdam. The glass brackets are
glued to the laminated lower glass planels, with a detail that still
enables turning of the support in vertical direction due to deflections. On the same spot the upper glass panel is also supported by a
solid prop in the 12 mm air cavity, sealed between upper and lower
panel. The steel support elements have been glued with a specially
tested type of glue to the glass inner pane to give a sufficient
adherence for a horizontal shear and vertical uplift. Each of the
described glass brackets has been elongated with a vertical 20 mm
thick pole that has been guyed by 8 mm steel rods in the node as
drafted in fig 25. The whole structure is surrounded by a RHS
profile, supporting a continuous edge gutter. Hence this system can
still be regarded as halfway between a closed roof system and an open
structure with a perimeter ringbeam that takes the compression
forces, omitted in the upper glass panels. The slope of the glass
panels is 1%. Completion of the stucture in March 1990. Fig 28 and 29
give some details.

REFERENCES

1. Booosting. *Booosting Experimenteert* Booosting, Den
Haag/IoN Rotterdam 1988

2. Eekhout, Mick. *Architecture in Space Structures* Uitgeve-
rij 010 Publishers Rotterdam 1989 ISBN 90-6450-080-0

ORTHOGONAL GRID SHELL: STATIC MODEL TESTS COMPARING DEFLECTIONS TO FORMS

P.H. KUO, W. WU, G.G. SCHIERLE
Graduate Programme of Building Science, School of Architecture,
University of Southern California, Los Angeles, California, USA

Abstract
Static model tests for an orthogonal grid shell, consisting of 25x25 timber bars (50x50 in the original structure), with spans of 20x20 meters (65.6 ft), are presented (Fig. 1, 2). The tests include 3 load conditions on 2 design cases of different cross-section heights (Fig.6). Test results are presented graphically to visualize a significant correlation of form and stiffness. This visualization was the prime objective of this research, to provide informed intuition regarding the interdependence of architectural and engineering design. Precise engineering analysis was not intended.
Keywords: Deflection, Form, Grid Shell, Model Test, Static Model.

1 Introduction

Grid shells have significant advantages over other shells. Being lightweight, easy to manufacture, and possibly translucent, if appropriate cladding is used, they can be employed for uses such as exhibition halls, auditoria, warehouses, gymnasia, aviaries and green houses; given the possible translucency or even transparency. The grid shell form is an important design consideration, not only regarding structural performance, but also aesthetic appearance, energy performance, functional and construction aspects.

2 Correlations of model and original structure

The test model is related to the original structure by three scales; namely

S_G = Geometric Scale = model dimension / original dimension = L_m / L_o
S_F = Force Scale   = model force / original force   = P_m / P_o
S_S = Strain Scale   = model strain / original strain   = $\varepsilon_m / \varepsilon_o$

The force scale S_F is derived as follows:

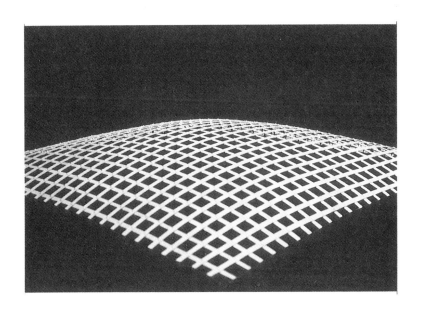

Fig. 1. Test model

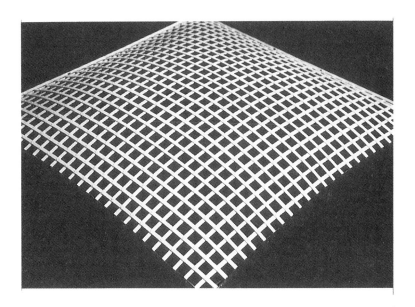

Fig. 2. Test model

unit strain $\mathcal{E}$ $= \dfrac{\Delta}{L}$, $\qquad$ strain $\Delta = K\dfrac{PL^3}{EI}$, $\qquad$ hence

Force P $= \dfrac{EI\Delta}{L^3 K} = \dfrac{EI}{KL^2} \times \dfrac{\Delta}{L}$, hence

Force Scale S_F $= \dfrac{P_m}{P_o} = \dfrac{E_m I_m}{E_o I_o} \times \dfrac{L_o^2}{L_m^2} \times \dfrac{(\Delta/L)_m}{(\Delta/L)_o} \times \dfrac{K}{K}$

Since $\dfrac{(\Delta/L)_m}{(\Delta/L)_o} = S_S$, and $\dfrac{L_o^2}{L_m^2} = \dfrac{1}{S_G^2}$

$S_F = \dfrac{E_m I_m}{E_o I_o} \times \dfrac{1}{S_G^2} \times S_S$, $\qquad$ or $\qquad$ (1)

$S_F = \dfrac{E_m I_m}{E_o I_o} \times \dfrac{1}{S_G^2}$, $\qquad$ if $S_S = 1$, or $\qquad$ (2)

$S_F = S_G^2$, $\quad$ if $S_S = 1$, $E_m = E_o$, and all member sizes are in geometric scale.

E = modules of elasticity
I = moment of inertia
K = constant of integration
m = subscript for model
o = subscript for original

For the tests presented here, the following scales, relating model to original structure, were selected and computed:

Geometric Scale $\qquad$ $S_G = 1:20$
Force Scale $\qquad\qquad$ $S_F = 1:1350$
Strain Scale $\qquad\quad$ $S_S = 1:1$

The strain scale was selected as unity to avoid errors that may be caused by geometrically non-linear behavior of the grid shell. The force scale was computed, based on the assumptions for materials and dimensions for systems and components tabulated as follows.

Table 3.   Comparing model to original structure

Item	Model	Original Structure
Length / Width	1000x1000 mm	20x20 m
Height Case 1 [1]	138 mm	2.76 m
Height Case 2 [2]	207 mm	4.14 m
Bar dimensions	1.78x8.89 mm	51x102 mm
Bar spacing [3]	41 mm	418 mm
Material	Ashwood	Ashwood

1. Case 1 has a depth / span ratio of 1:7.3
2. Case 2 has a depth / span ratio of 1:4.8
3. Bar spacing in developed (flat) position. (1 bar in model = 2 bars in original)

3 Form-Finding and Model Building

An orthogonal grid of square boundaries was built on a flat platform (Fig. 4). Bars were connected at each joint by a machine bolt. The bolts were loosely tightened to allow the square grids to deform into rhombuses and, thus, allow the flat grid to assume its synclastic curvature.

The form finding process began by placing the flat grid over a platform with a cut-out. 49 weights (cups with sand) attached to every fourth joint were temporarily supported by a platform, then gradually and simultaneously lowered to deform the grid (Fig.5). Additional weight was gradually added until the desired shape for the grid shell had been reached. The form-finding process was aided by gently vibrating the grid to reduce the friction between the grid and the platform cut-out. The exact form of the edge condition was measured and recorded for design and fabrication of the edge supports. The form-finding platform was then prepared as a permanent base for the grid shell. Continuous edge support was provided by 4 plywood segments. Those segments were attached to the base, slanted parallel to the grid tangent at the springing, to simulate an edge beam and earth berm in the original structure. Next, the grid shell was lifted in place, deformed, and attached to a groove along the edge segments by means of a continuous bar. The model then had assumed its permanent shape of test case 1 (Fig. 6). The bolts of all bar joints were tightened to improve stability.

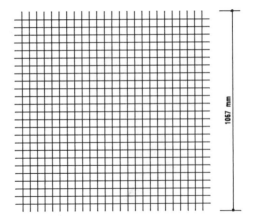

FIG. 4 FLAT GRID

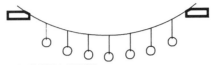

FIG. 5 FORM FINDING

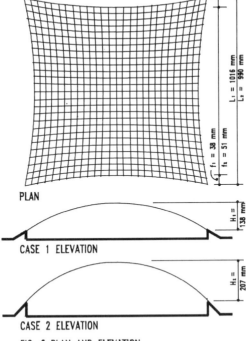

PLAN

CASE 1 ELEVATION

CASE 2 ELEVATION

FIG. 6 PLAN AND ELEVATION

Fig. 7. Model testing

4 Testing

Testing of case 1 began by suspending the same 49 weights used in the form-finding process from every forth joint and supporting them temporarily on a platform to avoid creep deformation during the test setup. The sand-cup weights were calibrated to 386g (0.85lb) each, to simulate a combined dead and live load of 0.77kPa (16 psf) in the original. The platform with the weights was then lowered (gradually to prevent impact load) to apply a uniform load over the entire grid shell (Fig.7). After vertical deformations were measured and recorded at 5 previously established measure points the weight-platform was raised for completion of test 1.A .

The above process was repeated for two additional load conditions, namely half load and diagonal half load (Fig. 8 , load conditions B and C) simulated with 28 point loads each. After completion of tests for design case 1 the previously described form-finding process was repeated for design case 2 , with a greater cross-section height. Load tests A, B, and C described above were then repeated for design case 2 to compare deformations as related to and affected by the form.

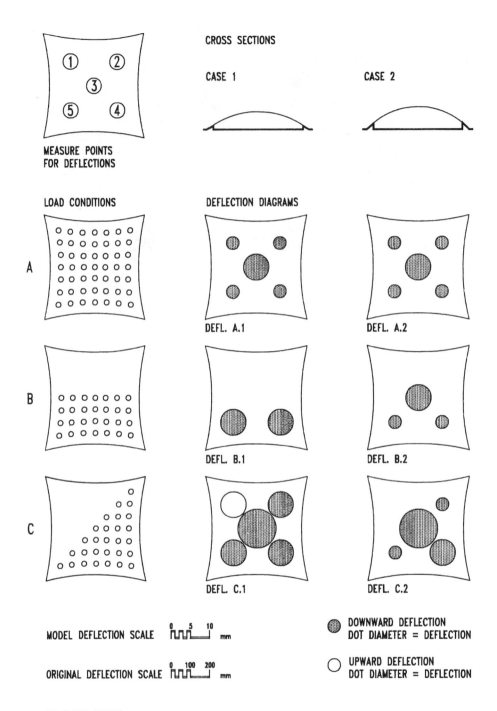

MEASURE POINTS
FOR DEFLECTIONS

CROSS SECTIONS

CASE 1

CASE 2

LOAD CONDITIONS

DEFLECTION DIAGRAMS

A

DEFL. A.1

DEFL. A.2

B

DEFL. B.1

DEFL. B.2

C

DEFL. C.1

DEFL. C.2

MODEL DEFLECTION SCALE   0 5 10 mm

ORIGINAL DEFLECTION SCALE   0 100 200 mm

DOWNWARD DEFLECTION
DOT DIAMETER = DEFLECTION

UPWARD DEFLECTION
DOT DIAMETER = DEFLECTION

FIG. 8 TEST RESULTS

5 Test Results

Test results for design cases 1 and 2 and for load conditions A, B, and C are illustrated in Fig. 8 for comparative visualization. The deflection diagrams show for each combination of design case and load condition the amount of vertical deformation on 5 measure points. The diameter of each dot reflects the amount of deformation on that point in the deflection scale. The deflection scale is 30 times exaggerated for better visualization and to facilitate scaling.

The results of Fig. 8 show no significant difference between design case 1 and 2 for load condition A (full uniform load). For load condition B the flat dome of case 1 has greater deformation in the corners and case 2 has greater deformation in the center, where case 1 has none. This observation seems to reveal a buckling tendency of case 1 under half load. A similar observation can be made for load condition C (diagonal half load), where case 1 also has a tendency to buckle at the corners; but also the center of case 1 deforms, while the corner opposite the loaded side deforms upward, another indication of a buckling tendency.

Both design cases appeared to have improved stability under load than without load applied. This could clearly be registered by touching the model. The greatest deflection under full load was 3mm (60mm for the original structure) at the center, which yields a deflection/span ratio of 1/330. The greatest deflection under half load was 4.8mm (96mm for original structure) at the center, which yields a deflection/span ratio of 1/206.

6 Conclusions

The test results reveal a significant correlation between form and deformation for asymmetrical but not for symmetrical load. Design case 1, with a depth/span ratio of 1/7.3 ,while still stable, appears to be too flat, and, under safety consideration, too vulnerable to buckling, given the assumed design conditions.

More tests should be conducted with larger test models and greater accuracy, as well as tests for lateral load and additional design cases.

Grid shells of square or similar forms have good potential for economical buildings with gracefully elegant appearance. Design information such as presented in this paper is important to make design decisions regarding architectural form based on informed intuition and to facilitate a design process toward synergy of form and structure.

7 References

Borrego,J. (1968) **Space Grid Structures**, Cambridge, Mass. The MIT Press.

Burkhardt, B.; Medlin, L.; Minke, G (1967) **Lattenkuppeln,** Montrealbericht of the Institute of Lightweight Structures, University of Stuttgart.

Hennicke, J. and Bubner, E. (1973) Grid Shells, Three-Dimensional Curved-Compression Stressed Lightweight Structures, **Proceeding International Symposium on Prefabricated Shells** Vol. III, IASS, Haifa, pp. 44-60.

Minke, G. (1968) Auditorium roof Montreal , **Architectural Design,** Vol. xxxvlll No. 4.

Otto, Frei et al (1974) **Grid Shells,** IL Report No.10, Institute for Lightweight Structures, University of Stuttgart.

SKYSCRAPERS OR SPACETOWNS

J. FRANÇOIS GABRIEL
Syracuse University School of Architecture, New York, USA

Abstract
The skyscraper is a vertical settlement which came about with the
development of the steel frame and the elevator. It is comparable to
a horizontal settlement in the sense that both exist along a linear
circulation system. The similarity ends there; for while roads can
branch out to permit lateral expansion, elevators must remain encased in the straitjacket of their vertical shafts.
   The spacetown is a large structural framework that supports all
the urban systems of a community. The form of the spacetown is not
fixed once and for all. Unlike the skyscraper, the spacetown can
expand laterally. Its form can also be modified during its life span
in a manner somewhat similar to urban communities built on the ground.
Keywords: Architecture, Structural Engineering, Urban Design, Space
Frames

1. Introduction

The archetypal skyscraper form reflects rigidly, and appropriately,
its circulation core: it is a solid looking, extruded form, seemingly proud of its finality. Even the most recent "post modern"
designs cling steadfastly to this imagery. Many architects currently practising feel drawn back to the idea of the wall which,
in skyscrapers, can only be an illusion. The cladding may be of
granite, but the structure is a steel skeleton.
   In a more forward looking trend led principally by architects and
engineers from the United Kingdom, there is a better understanding of
the relationship between structure and enclosure. Controversial
buildings such as the Bank of Hong Kong and the Lloyds of London
visually express the structural skeleton. They even celebrate it.
And the enclosures distinctly enclose rooms and do nothing else.
There is no ambiguity regarding structure and enclosure. In these
buildings and in others like them, architecture, which is essentially
an assemblage of rooms, is not defined by make-believe walls. There
are no walls, only lightweight enclosures.
   It is nearly seventy years since Le Corbusier formulated with great
insight the five points of modern architecture, two of which were the
free plan and the free façade made possible by skeletal structures.
But the first mature demonstration of these liberating points was made

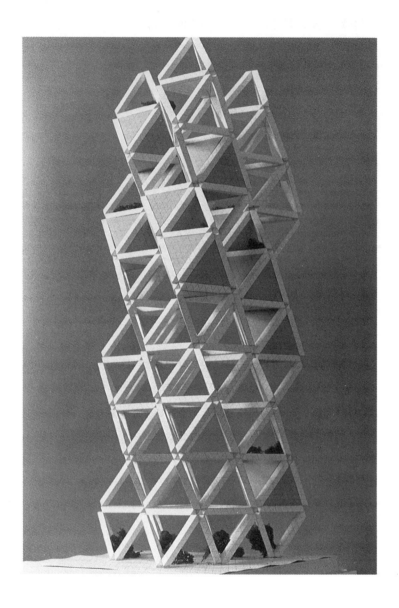

Fig. 1 A model of the 88-storey spacetown under discussion. The
structural framework consists of space trusses, 32 meters long,
forming octahedra and tetrahedra between them. Clusters of rooms
eight storeys high constitute the buildings proper. Clusters are
shown only in the upper part of the model.

Fig. 2 Top view of the three helicoidal towers forming our spacetown. The triangular space in the middle will receive elevators
and fire stairs. The free-standing condition of the octahedra where
clusters will be built can be better understood in this drawing. It
will also be seen that every octahedron is adjacent to a tetrahedron
in which the cluster can be expanded.

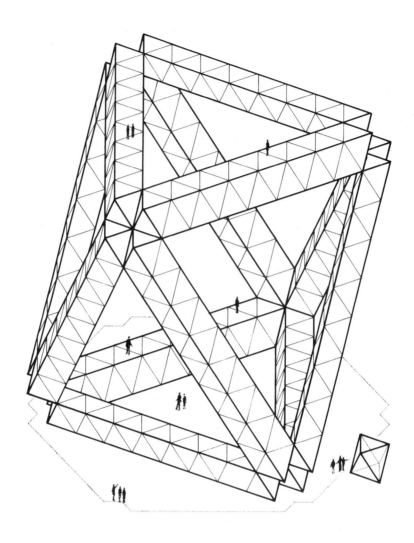

Fig. 3 Twelve space trusses form a ten-storey octahedron within
which an eight-storey cluster can be built. The detached octahedron
at bottom right is the module for both the space trusses and the
multilayer space frame that connects them.

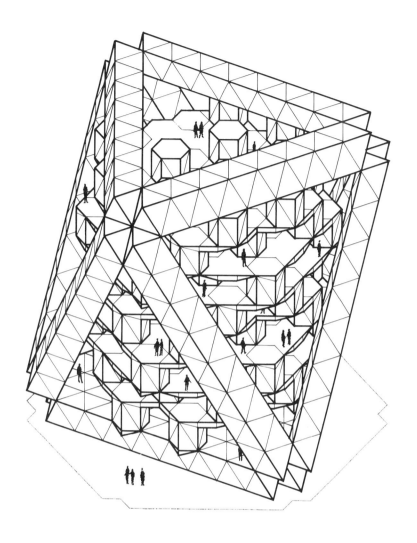

Fig. 4 An eight-storey cluster nestled within an octahedral framework of space trusses. The building system for the cluster is a
modified multi-layer space frame. The building blocks of the system
are represented as schematic solids, but in reality they are open
octahedral frames four meters wide.

in Centre Pompidou, where structure and enclosure are truly independent of one another.

What may appear to be the inordinately slow maturation of an idea is not unusual in architecture or in structural engineering. The first skyscraper to use the structural concept of the braced tube was the John Hancock Building, completed in Chicago in 1969. The braced tube was invented by Monsieur Eiffel, who used it eighty years earlier in the tower which bears his name.

Tour Eiffel claims our attention for other reasons as well. The greater part of the structure is poised over a very large void. Tour Eiffel is a lightweight structure that celebrates space, not mass. The rooms it contains are independent of the structural framework. Tour Eiffel is not a skyscraper, it is a precursor of the spacetown.

2. The Spacetown

The structural framework of the spacetown presented here is composed of octahedra and tetrahedra for several reasons. In a three-way space frame, vertical loads can be taken down to the ground in three directions. (In the post-and-beam system, which is still used in skyscrapers, vertical loads can be taken down in one direction only, that of the vertical columns.) The availability of choice for the direction in which loads can be taken down gives the designer a greater freedom.

In tall buildings, wind bracing is crucial. Tetrahedra and octahedra are inherently rigid. (Cubic frames are not rigid and must be triangulated as an afterthought, as it were.) A space frame is modular. Its components are standardized. Properly understood and applied, modularity renders possible the transformation of a built form by dismantling and re-building. Modularity is not an exclusivity of space frames, but the combination of structural freedom with rigidity of modules is, indeed, unmatched by any other structural systems.

The spacetown illustrated here is 88 storeys. It is organized in layers of nine storeys each. The standard structural member is 32 meters. It is called a space truss because it is composed of eight octahedra and 14 tetrahedra. Space trusses are hollow and fireproof, either made of reinforced concrete or of concrete encased steel (Ref. 3). Every ninth floor, where horizontal space trusses are found, the entire storey is devoted to mechanicals. Built as a double concrete shell, it forms a fire barrier.

Only the eight-storey octahedral spaces sandwiched between fire barriers are enclosed. It is in these spaces that the buildings proper exist, within the framework of the spacetown. We could call them clusters, or villages, to use Norman Foster's terminology in his description of the Bank of Hong Kong Building. Each cluster has its own self-contained system of stairs and elevators (Fig. 1).

Three identical, interconnected helicoidal towers form this particular spacetown. The towers surround a large open space, triangular in plan, where elevators and fire stairs are located. Elevators stop at every ninth floor, where a distribution platform provides access to three clusters, each one located in a different

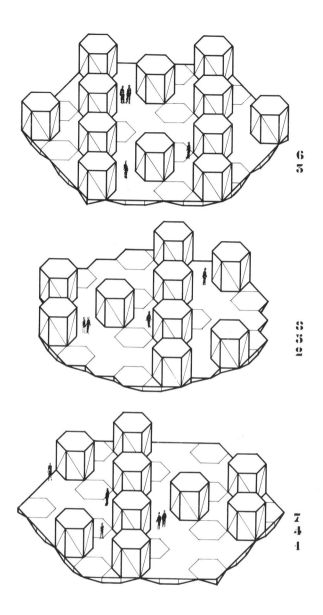

Fig. 5 Only these three patterns occur in the cluster shown in Fig.
4. They are shown here in actual sequence. The lower pattern is
repeated on the fourth and seventh levels, the second one on the fifth
and eighth and the third one on the sixth.

tower and therefore free-standing. The remarkable independence of the clusters from the structural framework is akin to the conditions of ordinary buildings built on the ground. This, of course, is very different from the impacted condition of the rooms in a typical skyscraper. The difference is so consequential as to justify the distinction we make between spacetowns and skyscrapers (Fig. 2).

As we have seen, each cluster is built within a large octahedron formed by 12 space trusses. Octahedra do not sit on top of one another; they are shifted in such a way that the tenth storey of one cluster, that is, the upper face of its octahedral frame, forms the lower face of an eight-storey tetrahedron. A cluster could, therefore, be expanded within the space of the adjacent tetrahedron or, alternatively, look out on a garden. The two faces of the tetrahedral frame that are open to the outside could be glazed to enclose a winter garden or, as it is sometimes called, an atrium. The decision would depend on the function of the building, the climate where the spacetown is to be built, the orientation of the particular octahedron under consideration and other variables.

How are the clusters built within the space truss framework? By using a multilayer space frame compatible with the space trusses. The octahedra and tetrahedra that compose the space trusses measure four meters. Therefore, the individual members of the space frame will also measure four meters. This standard dimension ensures a full storey height between chords.

In a three-way space frame, the space between chords is subdivided by the diagonals into a beehive patter of hexagonal prisms. We have found, however, that in multilayer space frames certain diagonals can be eliminated without affecting the rigidity of the structure (Ref. 2).

Another modification was made to the space frame configuration regarding the chords. These retain a triangular grid but it now matches the vertical projections of the diagonals (Ref. 1). One reason for the modification was to have floor beams directly under vertical enclosures. The diagonals that are retained form discrete octahedra separated by a distance varying between four and 12 meters (Fig. 3,4).

A building system results from our methodical elimination of diagonals and our modification of the chords. It is called the Hexmod System. The building block of the system is an octahedral frame four meters wide. Space flows freely between and around building blocks when assembled. Stairs and elevator shafts can be introduced anywhere by eliminating floor panels and joists from the storeys involved. Although building blocks cannot be removed, they are not obtrusive and almost any function can be accommodated in a cluster. Furthermore, the periphery of the cluster is not fixed and can be adapted to suit varied requirements (Fig. 5).

3. Conclusion

Put simply, architecture is the art of creating meaningful sequences of significant spaces. New materials and new building techniques have more than once in the past transformed architecture. In the last thirty years thin shells, reticulated domes, pneus, tents and

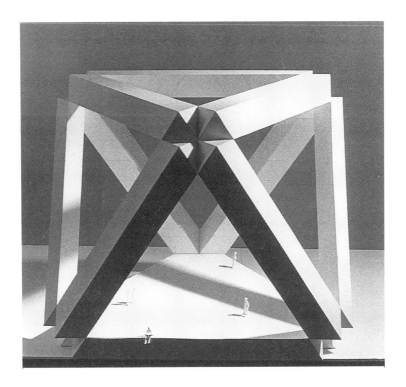

Fig. 6 The truly magnificent space created by eight space trusses on a triangular platform. A space like this will take its place among the great man-made spaces of the past when the advantages of space frame architecture are realized.

other space structures have opened new horizons for the making of architectural form and space. Among space structures, the space frame, or octet-truss as it is also called, occupies a special place. First used to span economically large column-free spaces, then later used as vertical enclosures as well, the space frame has in store an immense reserve of untapped structural and architectural possibilities. The space within, for one, with its own unique character, deserves more attention than it has received so far. Usually treated as a convenient bonus space for ducts and catwalks, it can now be raised to the status of a full-fledged architectural matrix, side by side with the best structural advances of the past (Fig. 6).

The space frame stands alone among the varied types of space structures when the problem at hand is the design of a multi-storey building. Its layered nature fits in perfectly with the requirements of a high-rise, as we have tried to show above. Structurally, a multi-layer space frame requires far fewer diagonals to maintain its rigidity than does a double or triple layer space frame.

The spacetown is made possible by the space frame. For sheer enrichment of the architectural vocabulary, few developments in structural engineering can compare with the coming of age of the space frame.

4. References

Gabriel, J.F. and Mandel, J.A. (1984) A space frame building system for housing. **Third international conference on space structures, Proceedings** (ed. H. Nooshin), Elsevier Applied Sciences Publishers, pp. 1052-1057.

Gabriel, J.F. (1985) Space frames: the space within--a guided tour. **J. Space Structures,** 1, 3-12.

Gabriel, J.F. (1986) Multi-layer space frames and architecture. **The first international conference on lightweight structures in architecture, Proceedings,** Sydney, Australia, 1, pp. 104-111.

5. Acknowledgements

The model illustrated in Fig. 1 was built with funds provided by the Office of the Vice-President for Research and Graduate Affairs at Syracuse University.

EXPERIMENTAL AND ANALYTICAL STUDY ON THE EFFECT OF JOINT PROPERTIES ON THE STABILITY OF SPACE STRUCTURES

F.A. FATHELBAB
Structural Engineering Dept, Faculty of Engineering, Alexandria
University, Alexandria, Egypt
R.E. McCONNEL
Engineering Dept, Cambridge University, Cambridge, England, UK

Abstract
The conventional procedure for the analysis and study of
space frames stability assumes that the joints of these
structures behave as either pure pins or as fully rigid.
This was adopted despite the fact that the joints of most
space structures are semi-rigid. The adoption of this
assumption was done due to the dificulty of considering
the effect of joint properties in a tangent stiffness
matrix for space frame member, and also due to the
dificulty of obtaining the actual properties of a space
frame joints without extensive experimental tests.

A general space frame tangent stiffness matrix which
includes in addition to the effects of stability and
bowing functions, the effects of joints size and
stiffness. An algorithm for nonlinear space frame
behaviour was developed.

The new tangent stiffness matrix and the developed
algorithm have been implemented in a computer program for
the geometrically nonlinear analysis of space frames.
Results obtained from this program were tested against the
available published results.

An experimental study was done by testing a shallow
single layer lattice dome models built using MERO jointing
system. In this models, various combination of member
size, joint type, and load pattern were used.

The experimental models were also analysed using the
developed computer program. The results obtained
demonstrate the great effect of the joint properties on
the behaviour of space frame structures specially on
single layer lattice domes.
Keywords: Stability, Space Domes, Joint Effect, Nonlinear, Tangent, Experimental.

1 Introduction

Space frame structures are used to cover large areas
without intermediate supports. These structures are built
from simple prefabricated units which can be easily and

rapidly put together on site. The small size of the basic units of these structures simplifies handling and transportation problems which makes these structures very useful in cases of rapid emergency installation. Another advantage of these structures, specially in the case of shallow single layer lattice domes, is that in the case of failure of one or more of the members or joints all the released stresses are often automatically redistributed in such way that the neighbouring members safely absorb the excess stresses imposed on them, thus preventing sudden collapse of the whole structure.

Because of the in-plane memberane action of shallow single layer lattice domes associated with their initial curvature, such structures are able to carry considerable out-of-plane loads compared with the self weight of these structures. However, for shallow single layer lattice domes with small rise to span ratio, the structure response is highly nonlinear even when the stresses in the members are small enough to insure that the material properties are well within the elastic range.

The two main components of any space structure are the members and the joints, and the advantages or disadvantages of any particular structure over the others depend largely on the type of joints used as this is the element that normally determines the flexibility and rapidity of the assembly of a space structure. Also, depending on how ais the joints connect the members together, the structure will behave.

Traditionally, structural steel research has been dominated by problems concerned with the behaviour of structural elements and their effect on the behaviour of the complete structure. Less attention has been paid to the behaviour of joints. While the members of space frame structures are, from a technical point of view, no more than a length of prismatic material with some structural cross section. On the other hand considerable improvements in the technology of space frame joints have been achieved. However, compared with the space frame members, this improvement in the technology and production of many new joints has not been associated with a similar improvement in investigation of the effects of these new joints on the behaviour and stability of space frames in general, and shallow single layer lattice domes in particular.

The design of joints of space structures has been constrained by lack of knowledge of the real behaviour of these joints in a structure, and of data relating to joint strength and stiffness. Almost all the joint design methods and their effects on the behaviour of space structures have been based on broad simplifications, and have often borne little relationship to real behaviour.

Attempts at semi-rigid analysis and design, for example, by Lightfoot and Messurier (1974), Lionberger and

Weaver (1969), Machaly (1986), and others have not been very successful mainly because all the methods proposed require a moment-rotation curve for each joint used. These curves are sensitive to a number of variables, and attempts to generalise moment-rotation characteristics have not been successful because of the large number of joint types and variables involved.

Many authers Birnstiel and Iffland (1980), Tong, and Prescott (1986), and Nethercot (1985) have recorded that joint behaviour makes an important contribution to the behaviour of space structures, and that uncertain joint performance has been one of the critical factors in many structral collapses.

Joints can be classified as pinned; semi-rigid, or rigid. The crucial joint characteristic that determines this classification is the moment-rotation relationship of the joint.

It is the purpose of this reseach reported here to provide some clarification and better understanding of the influence of joint characteristics on the behaviour of shallow single layer lattice domes.

2 Space frame element tangent stiffness matrix

A nonlinear space frame element tangent stiffness matrix which takes into consideration the effects of joint characteristics was deriven by Fathelbab (1987). This matrix allows for the standard causes of geometric nonlinearity, namely:

The effect of axial forces on the lateral stiffness (stability functions).
The effects of bowing on the member length (bowing functions).
The finite deflection of joints with small to moderate rotation.
The finite rotation of the joints and rigid body rotations of the members.

The adopted tangent stiffness matrix can be used to any of joint idealization shown in Fig. 1. Details of the adopted space frame element tangent stiffness matrix are given in Fathelbab and Mcconnel (1989).

Computer program

The adopted space frame tangent stiffness matrix has been implemented in a nonlinear incremental-iterative computer program. An algorithm based on the current stiffness parameter presented by Bergan (1980), and adopted by Fathelbab (1987) was used. Also a technique for the

669

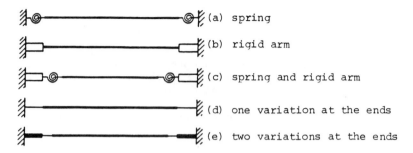

(a) spring

(b) rigid arm

(c) spring and rigid arm

(d) one variation at the ends

(e) two variations at the ends

Fig.1. Simulation of space frame joints

automation of load incrementation which enable the program
to follow post-buckling equilibrium paths, and for
detecting and classifying any instability may take place
at any stage of the loading process of the structure was
used.

4 Verification of the techniques used

The ability of the used technique which consider the
effect of joints on the behaviour of shallow single layer
lattice domes and the ability of the program used to trace
the complete equilibrium paths for these structures have
been tested against available results.

Fig. 2 shows the results obtained from the present
program compared with those obtained from a finite element
analysis, using this program also for a 2 member shallow
toggle. The toggle properties are those computed for MERO
members and joints. The finite element approach requires
the introduction of a node at each variation in cross
section along the member length. Simulation (e) of Fig. 1
was used in the program to represent MERO joints. From
Fig. 2, it is obvious that good agreement was achieved.
Table 1 shows a comparison between the number of free
nodes, the number of elements, and computer time required
for 50 load increaments using the present method and the
finite element technique.

Table 1. Comparison between the present and finite
element methods for 2 member toggle

Method	No. of Free nodes	No. of element	Computing time (sec)
Present	1	2	2.7
Finite element	9	10	10.95

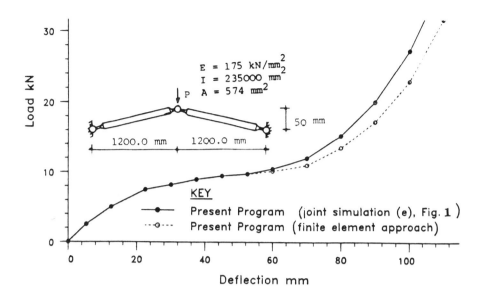

Fig.2. Two member shallow toggle using MERO joints

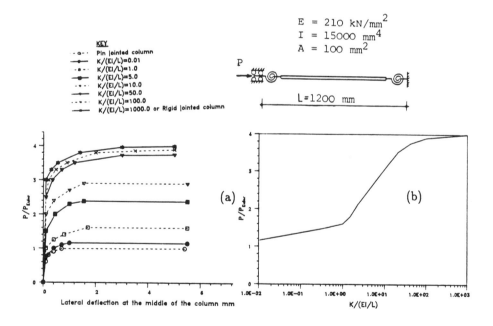

Fig.3. Elastically connected column

From Fig. 2 and Table 1, it is clear that the present approach is very economic in all aspects of computing, while at the same time it gives very good results compared with the finite element approach results.

Fig. 3 shows an elastically connected column and its buckling load for different values of joint bending stiffness. In the Figure, the backling load is ratioed by the Euler load against the lateral deflection at the middle of the column for different values of joint bending stiffness "K". An initial lateral imperfection at the middle of the column of 0.01 mm was introduced to initiate the mode of failure. From Fig. 3(a), it is clear that the buckling load increases with increasing joint stiffness with the buckling load of a fully rigid column as the upper limit and the buckling load of a pin jointed column as the lower limit. Also it is obvious that the buckling load changes rapidly near the lower limit for small changes in "K", and that no significant changes take place near the upper limit even for large changes in "K". The upper and lower values of the buckling loads for fully rigid and pure pinned columns are in agreement with appropriate hand calculations. From Fig. 3(b), which relates the joint bending stiffness to the buckling load, the effective length factor f of a column with flexible joints (varying from 1.0 for pure pinned column to 0.5 for fully rigid column) can be obtained from

$$P_{cr} = \frac{\pi^2 EI}{(fL)^2} \qquad (1)$$

where P_{cr} is the buckling load obtained from the computer program.

5 Joint continuity ratio (JCR)

It was found that for a certain value of joints bending stiffness, in case of a joint with a certain size, that the behaviour of the structure was almost exactly the same as that of a fully rigid jointed structure with normal members. For these joints, the bending stiffness of the joints is represented by the bending stiffness of a nondimensional spring (K_I), and their size is represented by a rigid length (λL), where λ is the ratio between joint size and member length. This particular value of joint bending stiffness is related to the flexural rigidity of the members EI and to the length of rigid parts λL at their ends, where

$$K_{IC} = \frac{EI}{\lambda L} \qquad (2)$$

where E is the modulus of elasticity, I is the member moment of inertia, and L is the member length. This special value of joint stiffness (K_{IC}) will be denoted the "joint continuity stiffness". The ratio between the bending stiffness of a joint (K_I) and the joint continuity stiffness (K_{IC}) can now be used as an indication for the degree of bending continuity of a joint. This ratio, which will be denoted the "the joint continuity ratio (JCR)", and is expressed as

$$ JCR = \frac{K_I}{K_{IC}} \tag{3} $$

Fig. 4 (a) shows a two member toggle. Joint simulation (c) of Fig. 1 was used to represent the toggle joints. A rigid part with length $\lambda L=57mm$ (so $\lambda=0.047$) plus a nondimentional spring with variable bending stiffness were used to represent the joints of the toggle. Curve (a) shows the relation between the joint continuity ratio (JCR) and the buckling load of the toggle for variable joint stiffness related to the buckling load of the fully rigid jointed toggle with no rigid lengths (the buckling load has been taken as the load corresponding to the minimum current stiffness parameter S_p). As can be seen, the load ratio of the toggle is greater than 1 for JCR greater than 1; also that the toggle buckles at a load level less than the buckling load of the pin jointed toggle for small values of JCR. It is interesting to note that a high percentage of the fully rigid jointed toggle strength is achieved for moderate value of JCR.

Curve (b) of Fig. 4 shows the results of a two member toggle as the toggle of curve (a) but with its joint size neglected. Joint simulation (a) of Fig. 1 was used to simulate the joints. It can be seen that the load capacity of the toggle increases with increasing joint bending stiffness. It is apparent that the behaviour change rapidly near the lower limit of a pin jointed structure and very slowly near the upper limit of a fully rigid jointed structure. Theoretically, the upper and lower limits of behaviour, in this case, can be obtained by setting the value of joint stiffness K_I to be ∞ and 0 respectively. Practically, a large value of $K_I/(EI/L)$ (200 in this case) was enough to get a behaviour very close to the upper limit, and a small value of $K_I/(EI/L)$ (0.0093 in this case) was enough to obtain the lower limit behaviour. In this joint simulation, the behaviour cannot be outside the upper and lower limits whatever is the value of joint bending stiffness not as the case of curve (a).

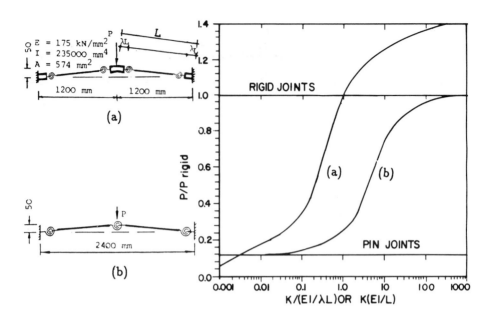

Fig.4. The effect of joint bending stiffness

6 Experimental results and their numerical predictions

Experimental dome models were tested. In these models various member size, joint type, and load pattern were used. The tested dome models varies from a three member toggles to a 90 member shallow domes with 7.2ms diameter in plan. Here only some of the experimental results and the corresponding computer results for some models are presented. The complete experimental and analytical results are given in Fathelbab (1987).

All the experimantal dome models were built using MERO joints. Two different size of joint sleeves (small size and large size) were used to connect the member to the ball joints. Also two different size of circular tube member were used to build the models, the small member size is 30x1.6 mm, and the large size is 60.3x3.2 mm. For each type of models a special test rig was built to support the models, also loading arrangement for each model type was done. Two different method of loading were used. The first one was to apply an incremental displacement for one node and record the corresponding level of load.The second loading technique was to control the value of the applied load instead of the displacement control.

The properties of the dome models components (the joints, and the members) were determined from experiments on individual joints and members. Fig. 5 shows the moment-rotation curves for some joints.

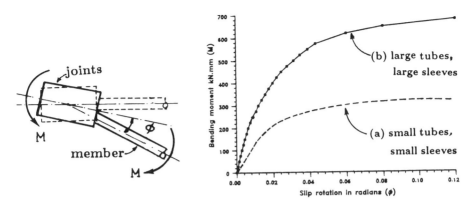

Fig.5. The moment-rotation curves of the joints

In all models the initial geometry before applying
loads was measured and used in the analysis. Transducers,
dial gauges and strain gauges were used to measure the
models response under the applied load at different load
levels.

6.1 Three member toggles

A number of 24 three-member toggle models were tested.
Fig. 7 shows the results of only one model. In this
model, large members and large sleeves were used. The
load applied to the middle node of the toggle was
controlled. The other three nodes were supported on the
rig in such way to be free to rotate and their trans-
lations were constrained. The measured relative heights of
nodes 1, 2, 3, and 4 of the toggle before applying any
loads were 2.6, 0.0, 31.3, and 0.6 mm respectively. From
Fig. 6, it can be seen that there is a good agreement
between the experimental and analytical results.
Fig. 6 shows also the computer results for two analyses
considering the extreme joint assumptions of either pure
pinned or fully rigid. It is obvious that the actual
behaviour is completely different from either of the two
extremes.

6.2   90-member dome model with small size members and
         joints

Fig. 7 shows the experimental and analytical load-
deflection curves of one node of a 90-member dome. This
dome was built using small size members and sleeves, and a
uniformly incremental load was applied at each of its 19
internal nodes. In the computer analysis, the actual
members, and joints properties from experiments were
used. The joint continuity ratio (JCR) of this model was
31%. From Fig. 7, it can be seen that there is good
agreement between the computer and experimental results.
Fig. 7 shows also the results of computer anlysis for the

675

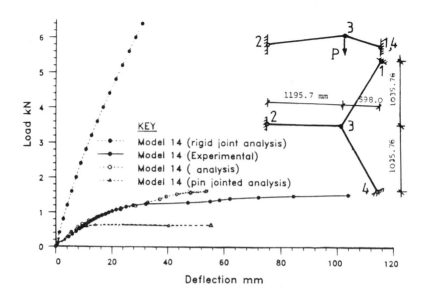

Fig.6. Load deflection curves of middle node of
three member toggle.

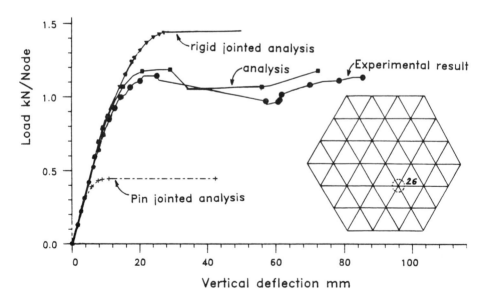

Fig.7. Load-deflection curves of node 26 in small
member shallow dome model.

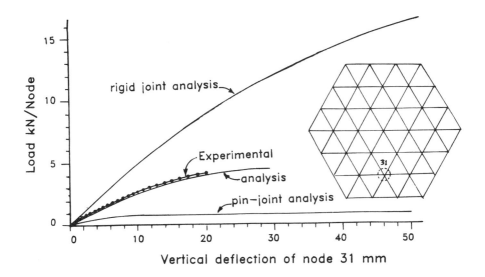

Fig.8. Load-deflection curves of large member dome model.

fully rigid and pure pinned jointed dome. It can be seen that a 31% joint continuity ratio provided about 76% of the difference between the buckling load of the two extreme analysis.

6.2 90-member dome model with large size members and joints

Fig. 8 shows the experimental and analytical load-deflection curves of one node of a 90-member dome model. In this dome large size members and sleeves were used to build the model. A uniform incremental loads were applied to all of the internal nodes of the model, and all the perimeter nodes were supported on the test rig in such way to rotate but not to translate.

Fig. 8 shows also the analytical results of the fully rigid and pure pin jointed analyses. The joint continuity ratio of this model was 4.5%. From Fig. 8, it can be seen that this small JCR provided 21% of the difference in buckling loads of the fully rigid and pin jointed dome.

7 Conclusion

From the results presented here of the behaviour of shallow single layer lattice domes, and the effect of joint properties on this behaviour, the following can be concluded:

The joints characteristics have a considerable effect

on the behaviour of this type of space structures.
For shallow single layer lattice domes with relatively flexible joints, a substantial increase in the load capacity can be obtained by providing small additional connection stiffness, in particular, a 4.5% JCR gave 21% improvement in performance between a pure pin and rigid joint behaviour.
For structures with large flexible joints, the load capacity of these structures can become less than that obtained by considering the joints of the structure to be pure pins.

8 Acknowledgement

The authers would like to aknowledge MERO-Raumstruckture of West Germany for providing the joints used in the experimental models.

9 References

Bergan, P.G. (1980) Solution algorithm for nonlinear structural problems, computer & structures, Vol. 12, pp. 497-509.

Birnstiel, C. and Iffland, J.S.B. (1980) Factors influencing frame stability, J. of Struct. Div., ASCE, Vol. 106, No.ST2, Feb., pp. 491-504.

Fathelbab, F.A. (1987) The effect of joints on the stability of shallow single layer lattice domes, Ph.D. thesis, Dept. of Engng., Univ. of Cambridge, U.K..

Fathelbab, F.A., and McConnel, R.E. (1989) Approximate tangent stiffness matrix includes the effects of joint properties for space frame member, proceeding, IASS Congress, 11-15 Sept., Madrid, Spain.

Jenkins, W.A., Tong, C.S., and Prescott, A.T. (1986) Moment-transmitting endplate connections in steel construction, and a proposed basis for flush endplate design, The Structural Engineer, Vol. 64A, No. 5, May, pp. 121-132.

Lightfoot, E. and Le Messurier, A.P. (1974) Elastic analysis of frame works with elastic connections, J. Struct. Div., ASCE, Vol. 100, No. ST6, June, pp. 1297-1309.

Lionberger, S.R. and Weaver, W. (1969) Dynamic response of frames with nonrigid connections, J. of Engng. Mech. Div., ASCE, Vol. 95, No. EM1, Feb., pp. 95-114.

Machaly, E.B. (1986) Buckling contribution to the analysis of steel trusses, computers & structures, Vol. 22, No. 3, pp. 445-458.

Nethercot, D.A. (1985) Joint action and the design of steel frames, The Structural Engineer, Vol. 63A, No.12, Dec., pp.371-379.